Techniques of circuit analysis

Techniques of circuit analysis

G. W. CARTER

Professor of Electrical Engineering,
University of Leeds

AND

A. RICHARDSON

Senior Lecturer in
Electrical and Electronic Engineering,
University of Leeds

CAMBRIDGE
At the University Press
1972

Published by the Syndics of the Cambridge University Press
Bentley House, 200 Euston Road, London NW1 2DB
American Branch: 32 East 57th Street, New York, N.Y. 10022

Library of Congress Catalogue Card Number: 79-183222

ISBN: 0 521 08435 0

Printed in Great Britain by
William Clowes & Sons Ltd., London, Colchester and Beccles

Contents

Preface

Give us the tools, and we will finish the job.

WINSTON S. CHURCHILL

Electric circuit theory is a foundation subject for all students of electrical and electronic engineering, in the first place on account of its direct engineering applications. But beyond this it has a quality which makes it one of the most fruitful of all subjects; the concepts which it embodies, and the analytical techniques which it employs, are valid far outside the boundaries of electrical engineering. From the teacher's point of view, circuits are linear systems *par excellence*; they are easy (within limits) to construct, convenient to experiment with, and frequently predictable in their performance. Through them the student gets the feel of how a linear system goes; of sets of quantities interlinked by groups of linear equations; of the response to an alternating stimulus long sustained, and to a sudden shock of the step or the impulse type; of the classification of systems; of resonance, stability, wave propagation, and many other general ideas.

As the title indicates, the emphasis of this book is upon inculcating the techniques of circuit analysis, rather than upon studying the general theory of circuits. For the student of electrical or electronic engineering there is no better way of learning about Fourier series or Laplace transforms, than through their circuit applications. We seek to leave with each reader a set of tools, useful in the first place for solving circuit problems, but useful for many other jobs as well. At this level we believe that there should be no difference between the tool-kits of electronic and power engineers; but few textbook writers are so catholic as to write without betraying the bias to one side or other which each man's experience has created. In many books on circuit theory this bias is all too evident; for example, writers eager to develop the light current aspects are sometimes content to ignore three-phase circuits completely. We have written this book in collaboration, primarily because we feel that our differing backgrounds enable us to do together what we could not have done separately – to develop a universal foundation course in this subject, from which it is expected that students will proceed to further studies more closely allied to their specialist interests. Thus we have tried to open as many doors as possible, even when we could not take our readers through them. When they seek to go on to explore wider fields, they should at least find the gates unlocked.

The practising engineer, accustomed to the daily manipulation of network analysis methods, easily falls into loose habits of thought which can mislead a student. For example, he may use the term 'voltage' indiscriminately to mean either the time-varying quantity $V_1 \cos(\omega t + \alpha)$, or the complex number $V_1 e^{j\alpha}$. Yet properly only the former quantity is a 'voltage', in the sense that it is susceptible of direct measurement by a voltage-measuring instrument; the other is a derived quantity, related to the voltage as the image in a mirror is related to the object which it reflects. We have made extensive use of this idea of two related worlds, and have used typographical devices to make the distinction clear. The concept is at least as old as K. W. Wagner,† but his *Oberbereich* and *Unterbereich* – 'upper world' and 'nether world' – hardly furnish suitable English terms; and after much consideration we have decided to follow van der Pol and Bremmer‡ in using the words 'original' and 'image'. Thus, in the instance given above, $V_1 e^{j\alpha}$ is described as an 'image voltage', at all events until it becomes clear that the entire analysis can take place in the image world. We do not think that this usage is likely to lead to confusion in the mind of a student who later encounters the 'image parameter' concept in relation to twoport networks. The two usages are widely different, van der Pol's being much more far-reaching, and more appropriate too; for an object and its mirror image are dissimilar in kind, the object coming first, whereas the so-called 'image impedances' of a twoport network form a reciprocal pair.

In conclusion we are glad to acknowledge the help we have received from the typists who have shared the typing of our work, and from Mr D. Dring for his valuable assistance in the preparation of diagrams. We also thank the Registrar of the University of Leeds for his permission to make use of University examination questions.

G. W. CARTER
A. RICHARDSON

April 1972

† K. W. Wagner, *Operatorenrechnung nebst Anwendungen in Physik und Technik*, J. A. Barth, Leipzig (1940).

‡ B. van der Pol and H. Bremmer, *Operational calculus based on the two-sided Laplace integral*, Cambridge University Press (1950).

1 Direct current circuits

1.1 Introduction

In principle, the analysis of direct current or d.c. circuits is simple. If the circuit elements consist of concentrated lumps of resistance interconnected by resistanceless wires, and if the lumps of resistance respond to a stimulus in a linear fashion, then only three simple laws are needed. Ohm's law is required to relate the current through an *element* of the circuit to the voltage across that element, and Kirchhoff's two laws are needed to determine what happens in the *circuit* as a whole; but unless these laws are applied in a systematic way, the solution may involve an undue amount of labour. The first object of this chapter is therefore to describe a systematic method (called the loop method) for the analysis of electric circuits. This is followed by some simple methods of reduction by which the labour of analysis may often be alleviated. Finally an alternative and equally important systematic method of analysis (called the node method) is introduced.

The second object of the chapter is to introduce some of the concepts of circuit theory. These concepts, presented in a chapter on d.c. circuits, are not hidden behind any mathematics more difficult than simple algebra and should therefore stand out clearly. They will be adopted in later chapters as an essential means of solving the more difficult problems in which the voltages and currents are time-varying.

1.2 Ohm's law

The work of the German physicist G. S. Ohm (1787–1854) on the flow of electric currents in metallic conductors is commemorated by the use of his name in what is now called Ohm's law. In 1827, Ohm postulated a linear relationship between the current flowing in a conductor and the strength of the device producing the current flow; indeed even the concept of current flow in a wire may be said to be due to Ohm. In modern terms Ohm's law is described by the equation

$$V = RI, \tag{1.2.1}$$

where R is a constant called the *resistance* of the element through which the current is flowing. The element itself is called a *resistor*. Although (1.2.1) is found to be valid for resistors made from a wide range of materials (principally metals such as copper, aluminium, iron, etc.) it is not universally valid. For this reason, (1.2.1) is nowadays used to define what is called a *linear* or *ohmic resistor*; a linear resistor is one which obeys (1.2.1). A resistor which does not obey (1.2.1) is called *non-linear* or *non-ohmic*. Since the vast majority of the components called resistors do in fact obey Ohm's law, (1.2.1) is used extensively in circuit theory, and the analysis of electric circuits which contain non-ohmic components is usually treated separately.

The application of Ohm's law to a circuit requires a convention for the direction of the current flow relative to the polarity of the applied voltage, but first of all it is useful to discuss the conventions for the directions of the currents and voltages themselves.

The arrowhead associated with the letter I in fig. 1.1(*a*) indicates the positive direction of current flow when I stands for a positive number, and

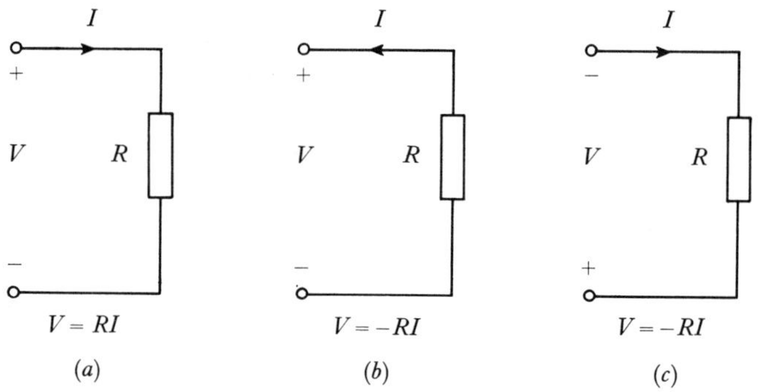

Fig. 1.1 Application of Ohm's law to a resistor.

the + and − signs associated with the letter V indicate that the terminal marked + is at a higher potential than the terminal marked − when V is a positive number. If the current is actually flowing in the direction opposite to that of the arrowhead shown in fig. 1.1(*a*) then I will have a numerical value that is negative. Similarly, if the terminal marked + is actually at a lower potential than that marked − in fig. 1.1(*a*) then V will be negative. In simple circuits the actual polarities of V and I may be known and the appropriate directions assigned to them, but in complicated circuits it may not be possible to tell in advance of the analysis what the polarity of V is and which way the current I flows. This does not matter. Polarities of V and I are assigned to the element in question and the actual polarities are determined by the sign of the numerical values that eventually result.

Once polarities have been assigned to V and I separately, Ohm's law is applied to a resistor in the manner shown in fig. 1.1. If I is assigned the direction *into* the resistor at the terminal marked +, then the law is given by $V = RI$, but if the assigned direction of I is *out* of the resistor at the terminal marked + then Ohm's law necessitates a minus sign in the equation as shown in fig. 1.1(*b*) and (*c*).

1.3 Terminology

The component parts of an electric network are given special names which it is convenient to know. The parts labelled a, b, c, d, etc. of the network shown in fig. 1.2(*a*) represent single *elements*, and the junction points of these elements, such as A, B, C, D, etc. are called *nodes*. Those elements that interconnect two nodes constitute a *branch*, provided that each of the two nodes has more than two elements terminated on it. Thus any one of the

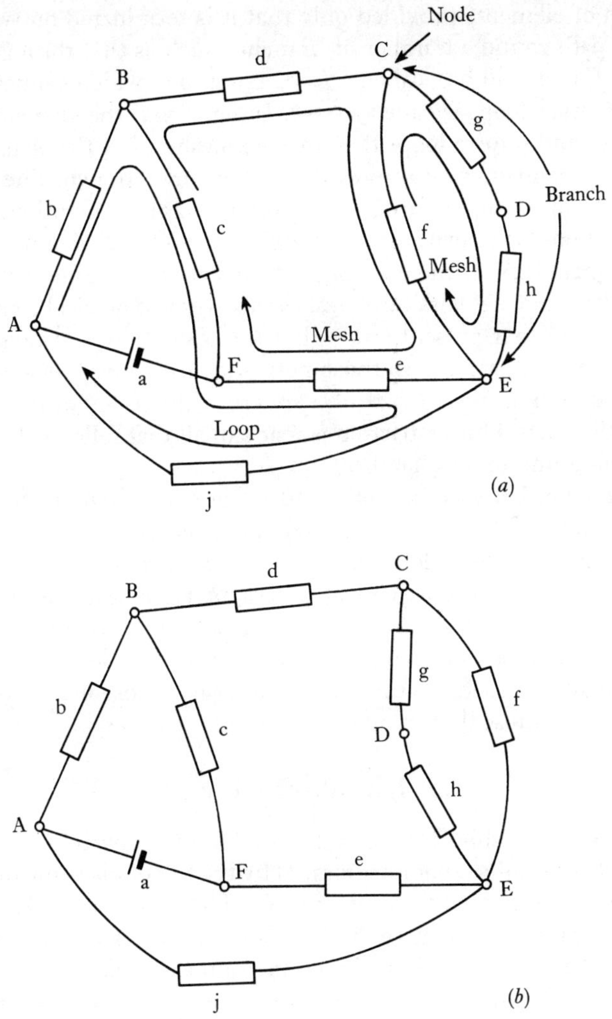

Fig. 1.2 (*a*) Network showing loops, meshes, nodes, and branches. (*b*) Identical network with element f repositioned.

single elements a, b, c, d, e, f, and j constitutes a branch, and the elements g and h in series constitute a branch. Unfortunately no definition of a branch appears to be universally accepted, because it is customary to change the definition to suit the context of the problem under discussion. Sometimes the word is found to be used synonymously with what is here defined as an element; on other occasions it may be found to describe any group of elements that are connected between two nodes. The latter usage will be replaced here by the word *arm*. In fig. 1.2(*a*) the elements f, g, and h between the nodes C and E might be called an arm of the network; so might the elements d, f, g, and h between the nodes B and E, or indeed any conglomeration of elements provided only that it is terminated on two nodes.

A closed path round a number of branches such as that through b, c, e, and j, or d, f, e, a, and b is called a *loop*, and a loop which cannot be subdivided into other loops is called a *mesh*. In fig. 1.2(*a*) the elements c, d, f, and e, or f, g, and h form loops that are also meshes, but the elements that form a mesh depend on the way in which the network is drawn. The elements c, d, f, and e, for example, form a mesh in fig. 1.2(*a*) as mentioned above, but if the element f is drawn as in fig. 1.2(*b*) on the other side of node D then the same elements c, d, f, and e no longer form a mesh. The general problem of the resolution of a network into meshes involves geometrical complexities which are beyond our present scope; but the resolution is always possible if consideration is restricted to those networks, called *mappable* networks, that can be drawn on the surface of a sphere without any of the branches crossing each other. This restriction is made in all that follows; fortunately it is not often found to be a handicap.

The distinction between *networks* and *circuits* has been left to the end. A *network* consists of two or more interconnected elements; a *circuit* is a network which contains at least one loop. Since all practical networks are also circuits the two words are almost always interchangeable in the context of this book. The reason for having the two words is that there are other contexts, such as those involving the geometrical complexities mentioned above, in which it is useful to be able to distinguish those networks that are also circuits from those that are not.

1.4 Kirchhoff's laws

Like Ohm, G. R. Kirchhoff (1824–87) was a German physicist whose great work on electric circuits is commemorated by the use of his name in association with two laws derived from his studies. The first of these laws, called Kirchhoff's current law or Kirchhoff's node law, states that the algebraic sum of all the currents entering a junction point (i.e. a node) in a circuit is equal to zero. This law is valid not only when applied to direct currents but also when applied to the instantaneous values of currents which vary with time and, since it is based on the principle of the conservation of electric charge, it is true of all circuits, not merely those in which the elements are ohmic. Equation (1.4.1) illustrates the law as applied to fig. 1.3(*a*)

$$I_1 + I_2 + I_3 - I_4 = 0. \tag{1.4.1}$$

This kind of equation is commonly written in the more compact form

$$\sum I = 0 \tag{1.4.2}$$

in which it is assumed that all of the currents I are designated as flowing into the node.

The second law is called Kirchhoff's voltage law or Kirchhoff's loop law; it states that the algebraic sum of all the voltages which exist in any closed

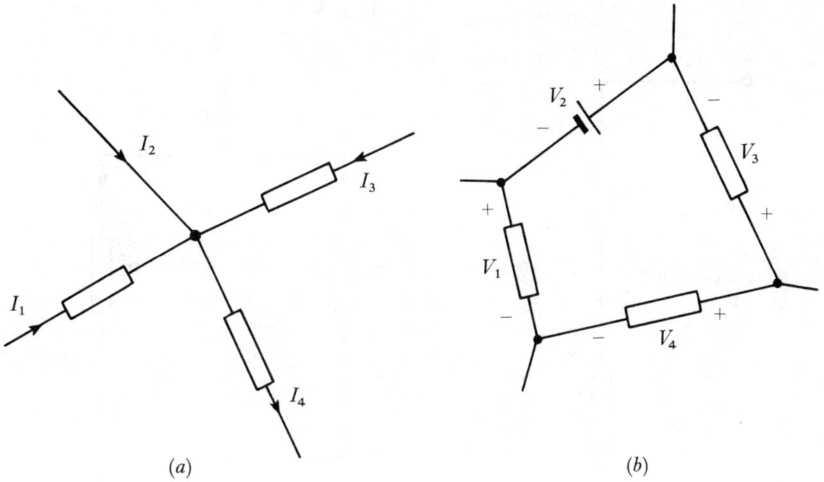

Fig. 1.3 Kirchhoff's laws: (*a*) node law ($\sum I = 0$). (*b*) loop law ($\sum V = 0$).

loop of a circuit is equal to zero. The word *voltage* is here used in the sense that includes both *electromotive force* and *potential difference*. A discussion on these terms is deferred to §1.6. Attention must be paid to the polarity of the voltages in the loop under examination. If voltages which are assigned in the clockwise direction round a loop are taken to be positive, (i.e. if the + sign on the element is further round the loop in the clockwise direction than the − sign on the other terminal of the same element), then those that are assigned in the opposite direction are negative. Application of this law to the circuit shown in fig. 1.3(*b*) gives the equation

$$V_1 + V_2 + V_3 - V_4 = 0, \tag{1.4.3}$$

which can also be written in the more compact form

$$\sum V = 0, \tag{1.4.4}$$

provided that all of the voltages V are assigned the same clockwise (or anticlockwise) direction.

The voltage law is based on the principle of the conservation of energy and is true of non-ohmic circuits as well as of ohmic circuits; it is also valid when applied to the instantaneous values of voltages which vary with time.

1.5 Applications of Kirchhoff's laws

Voltage and current dividing circuits

Some useful results follow the application of Kirchhoff's laws to the two specific circuits shown in fig. 1.4. Fig. 1.4(*a*) shows two resistors connected

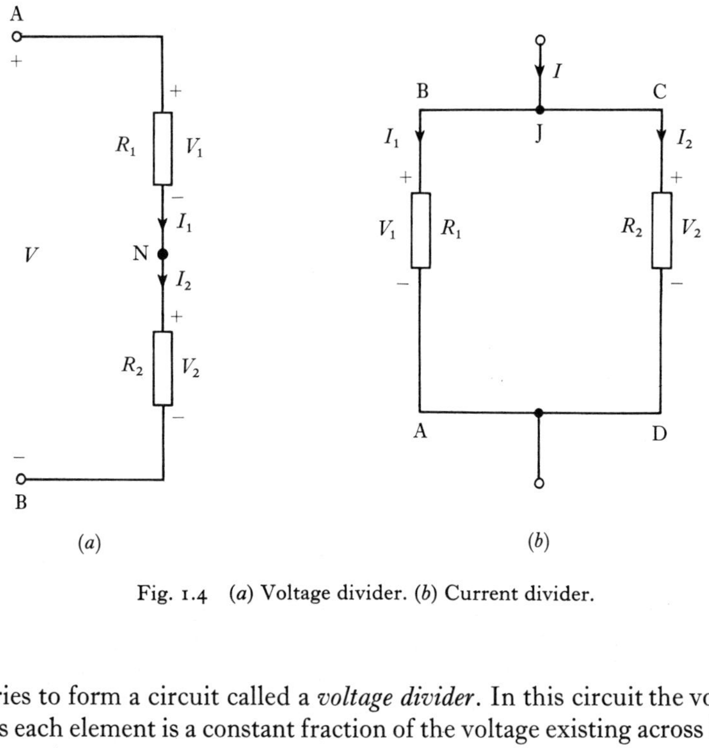

Fig. 1.4 (*a*) Voltage divider. (*b*) Current divider.

in series to form a circuit called a *voltage divider*. In this circuit the voltage across each element is a constant fraction of the voltage existing across both. This result emerges from the analysis given below. The application of Kirchhoff's current law to the node N gives

$$I_1 - I_2 = 0. \tag{1.5.1}$$

The application of the voltage law to the closed loop BANB gives

$$V - V_1 - V_2 = 0, \tag{1.5.2}$$

and Ohm's law applied to each element gives

$$\left.\begin{aligned} V_1 &= R_1 I_1, \\ V_2 &= R_2 I_2. \end{aligned}\right\} \tag{1.5.3}$$

The result of putting (1.5.1) into (1.5.3) and substituting in (1.5.2) is

$$V_1 = \frac{R_1}{R_1 + R_2} . V \tag{1.5.4}$$

and

$$V_2 = \frac{R_2}{R_1 + R_2} . V. \tag{1.5.5}$$

Since Ohm's law has been used in the analysis, these well-known equations are valid only when the two elements of the circuit are ohmic.

Figure 1.4(*b*) shows two resistors connected in parallel. This circuit may be visualized as a fragment of a much larger circuit (as also may that of fig. 1.4(*a*)). A current I from the larger circuit divides between the two elements R_1 and R_2 in such a fashion that the current in each element is a constant fraction of the total current I. The circuit is called a *current divider*, and the division of current between the two elements is dependent on their relative resistance values as shown below. From the current law, at node J,

$$I - I_1 - I_2 = 0. \tag{1.5.6}$$

From the voltage law, around loop ABCD,

$$V_1 - V_2 = 0, \tag{1.5.7}$$

and from Ohm's law,

$$\left.\begin{aligned} V_1 &= R_1 I_1, \\ V_2 &= R_2 I_2. \end{aligned}\right\} \tag{1.5.8}$$

Elimination of the voltages V_1 and V_2 from these equations gives the desired results:

$$I_1 = \frac{R_2}{R_1 + R_2} . I \tag{1.5.9}$$

and

$$I_2 = \frac{R_1}{R_1 + R_2} . I. \tag{1.5.10}$$

Similar methods may be used on circuits which have more than two elements. It can be shown, for example, that the voltage across the resistor

R_1 in fig. 1.5(*a*) and the current through the resistor R_1 in fig. 1.6(*a*) are respectively

$$V_1 = \frac{R_1}{R_1 + R_2 + R_3} . V, \qquad (1.5.11)$$

$$I_1 = \frac{R_2 R_3}{R_1 R_2 + R_2 R_3 + R_3 R_1} . I. \qquad (1.5.12)$$

Resistors in series and in parallel

There is a well-known equivalence between several resistors connected in series and one resistor whose resistance equals the sum of the resistances of the series-connected resistors. This result, which derives from Kirchhoff's

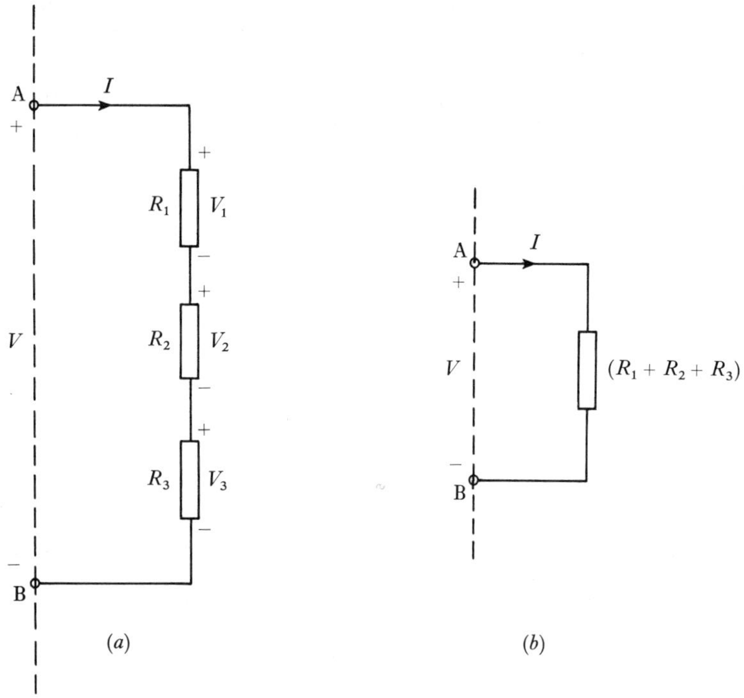

Fig. 1.5 (*a*) Resistors in series. (*b*) Circuit model.

laws, is an example of what is known as an *equivalent circuit* or *circuit model*. Fig. 1.5(*a*) shows three resistors in series; the loop equation for the circuit is

$$V - V_1 - V_2 - V_3 = 0. \qquad (1.5.13)$$

If the voltages V_1, V_2, and V_3 are now replaced by their ohmic equivalents R_1I, R_2I, and R_3I, the result is:

$$V = (R_1 + R_2 + R_3)I. \tag{1.5.14}$$

Since this result is the same as that which is obtained by applying Ohm's law directly to the circuit of fig. 1.5(*b*), the equivalence of the two circuits is established *from the point of view of the current and voltage at the terminals AB*. Provided that one is not interested in the individual voltages V_1, V_2, etc., the circuit of fig. 1.5(*a*) may be replaced by that of fig. 1.5(*b*) without affecting anything else in the larger circuit of which fig. 1.5(*a*) may form a part. Fig. 1.5(*b*) is known as an equivalent circuit or a model of fig. 1.5(*a*).

The single resistor R which is equivalent to n resistors R_1, R_2, R_3, etc. in series can be found by extending the process described above. The result is

$$R = R_1 + R_2 + R_3 + \ldots + R_n. \tag{1.5.15}$$

The concept of equivalence may be usefully applied to any network in which a path of interconnected resistors exists between two terminals A and B. Equation (1.5.15) expresses the equivalence between a single resistor and a series connexion of resistors. The equivalence between a single

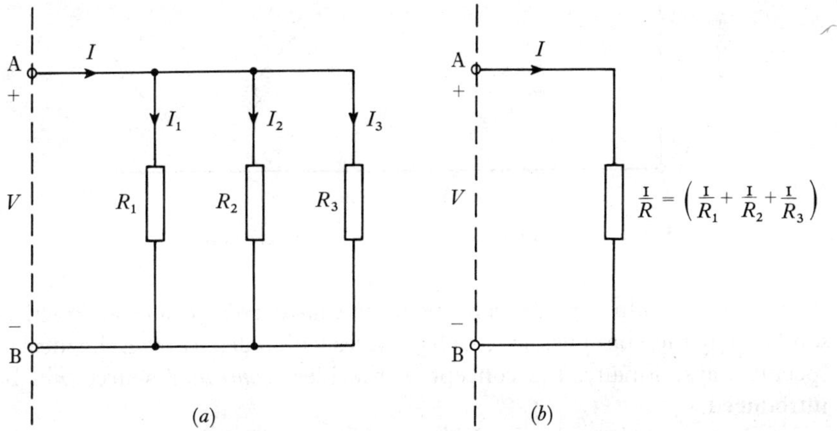

Fig. 1.6 (*a*) Resistors in parallel. (*b*) Circuit model.

resistor and a parallel connexion of resistors is derived in the same kind of way. If the two circuits shown in fig. 1.6 are to be equivalent, they must pass the same current I for the same applied voltage V. By equating the total current I in each circuit, the equation for equivalence is found; the equation for n resistors in parallel is

$$\frac{1}{R} = \frac{1}{R_1} + \frac{1}{R_2} + \frac{1}{R_3} + \ldots + \frac{1}{R_n}. \tag{1.5.16}$$

When only two elements are connected in parallel it is useful to remember the result in the form

$$R = R_1 R_2/(R_1 + R_2), \tag{1.5.17}$$

but for three elements or more it is usually best to use the form given by (1.5.16).

1.6 Voltage sources

The driving force or source of electric energy, such as a battery or a d.c. generator, has so far been neglected in this treatment of d.c. circuits, and it is the object of this section to consider the terminal characteristics of

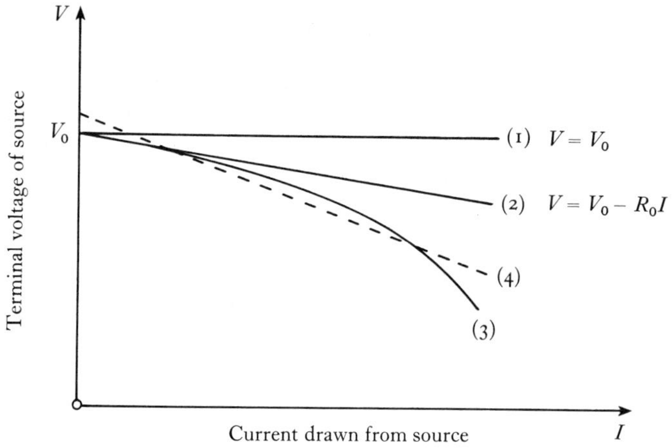

Fig. 1.7 Terminal characteristics of voltage sources.

sources by introducing the concept of the *ideal voltage source*. Practical sources may then be considered to be devices which depart from the ideal in specific ways. Finally, the concept of the ideal *controlled* source will be introduced.

An ideal device has whatever characteristics we choose to give it, and the principal feature of our choice for an ideal voltage source is a terminal voltage that is not influenced by the current drawn from it, no matter how large this may be; an ideal voltage source has the characteristic labelled (1) in fig. 1.7. The graphical symbol for the device is shown in fig. 1.8(*a*). This symbol is used to represent not only those ideal voltage sources that produce constant output voltages (i.e. d.c. sources) but also those that produce time-varying voltages. The symbol for a battery, shown in fig. 1.8(*b*), is sometimes used to represent an ideal d.c. voltage source but this symbol should strictly be used for the practical device, not the ideal version of it.

The + and − signs associated with the letter V in the diagrams follow the same convention as in fig. 1.1; the terminal marked + is at a higher potential than the terminal marked − when V has a positive value. V is sometimes called the *electromotive force* or e.m.f. of the source to distinguish it from the *potential difference* or p.d. across a resistor through which current is flowing and which is consuming energy. In this terminology, e.m.f. refers to a source of energy and p.d. refers to a sink of energy, but since both quantities are measured in volts it is nowadays common to refer to them both as voltages. Sometimes, however, a distinction is made by calling the e.m.f. or driving function the *voltage rise* and by calling the p.d. across a resistor the *voltage drop*.

The assigned direction of current flow in a source is usually outwards from the source at the assigned positive terminal. This is because the source

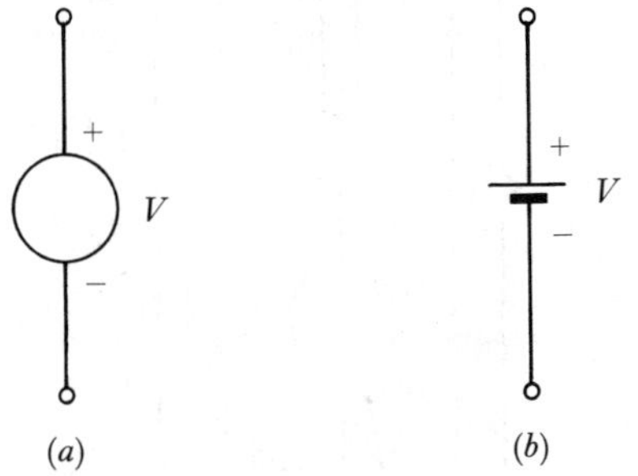

Fig. 1.8 Graphical symbols for voltage sources: (*a*) general symbol; (*b*) battery.

normally supplies energy to the external circuit and this will occur when V and I are both positive if I is assigned the direction described. If I is assigned the opposite direction, that is, *into* the source at the positive terminal, then when V and I are both positive, energy is being supplied to the source from the external circuit. This is quite possible if the external circuit itself contains an energy source of sufficient voltage to drive the current I into the first source as occurs, for example, in the charging of a battery.

A practical voltage source, such as a battery, does not behave like an ideal voltage source; its terminal characteristics may be similar to that labelled (2) in fig. 1.7. When this battery is on open-circuit (that is, when no current is being drawn from it), its terminal voltage is V_0, and when it supplies a current I to an external circuit its terminal voltage is given by the equation

$$V = V_0 - R_0 I. \qquad (1.6.1)$$

The two quantities V_0 and R_0 define this (linear) source, and it is possible to analyse any linear source into these two components as shown in fig. 1.9. The resistance R_0 is called the *internal resistance* or *output resistance* of the source. If R_0 is zero the terminal voltage is given by V_0, uninfluenced by the current I, and the source is an ideal one. Thus an ideal voltage source is one that has zero output resistance.

Some sources do not have the linear characteristic shown by curve (2) in fig. 1.7 but may, for example, be like curve (3). A non-linear characteristic of this kind cannot be represented by linear circuit elements, and the representation of such a source falls outside the scope of this book. It is a simple matter, however, to approximate a non-linear characteristic by the

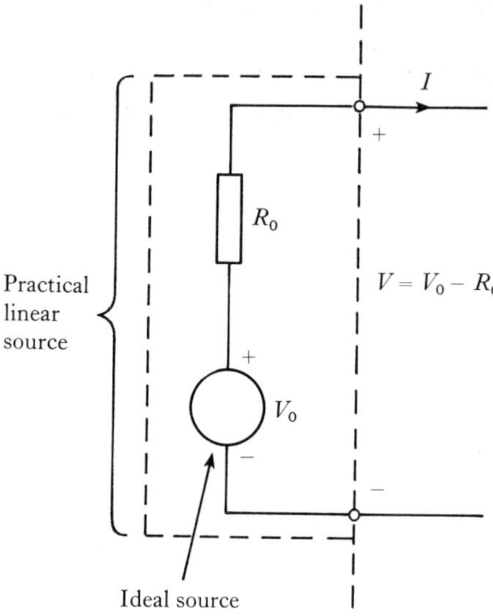

Fig. 1.9 Circuit model of a linear voltage source.

nearest straight line and hence to approximate the non-linear source by the nearest linear one. Such an approximation is shown by the dotted characteristic of fig. 1.7 labelled (4). This would be a suitable approximation to the non-linear characteristic (3) for currents ranging from zero to the full amount shown on the diagram. If only very small currents are to be drawn from the source, however, a better approximation would be (2) which is tangential to (3) at the origin.

The sources so far described are called *independent* sources because their outputs are independent of the circuits to which they are connected. Not all sources are of this kind. *Controlled* sources have two pairs of terminals, an output pair and a second pair, called the *input terminals*, which is used to control the magnitude of the output voltage. The controlling quantity, applied to the input terminals, may be either a voltage or a current and it follows that there are two forms of controlled voltage sources. An ideal *voltage-controlled* voltage source is shown in fig. 1.10(*a*). A voltage V_c applied to the input terminals AB controls the output voltage V_{ov} according to the equation $V_{ov} = \mu V_c$, where μ is a dimensionless constant. An ideal *current-controlled* voltage source is shown in fig. 1.10(*b*). Here, an input current I_c controls the output voltage V_{oc} in accordance with the equation $V_{oc} = rI_c$,

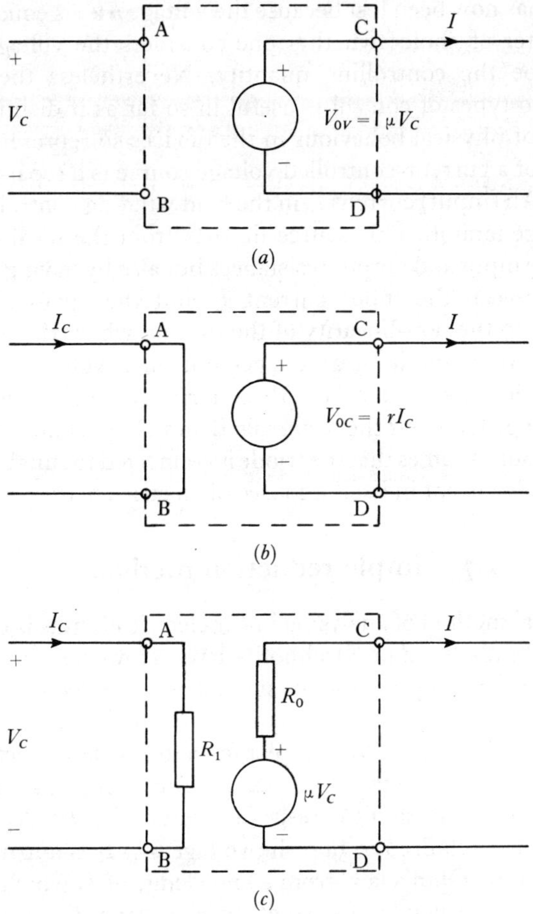

Fig. 1.10 Controlled voltage sources: (*a*) ideal voltage-controlled voltage source; (*b*) ideal current-controlled voltage source; (*c*) controlled voltage source with finite input and output resistances.

where r is a constant with the dimension of resistance. Both sources are ideal. In this context, the word *ideal* has two implications. First it refers to the fact that the source has zero output resistance, and second to the fact that the input resistance of the voltage-controlled source is infinite, and that the input resistance of the current-controlled source is zero. Practical controlled voltage sources depart from the ideal by having not only a finite, non-zero, output resistance but also by having a finite, non-zero, input resistance. Normally, the input resistance of a voltage-controlled source is large and that of a current-controlled source is small. The imperfections of a controlled voltage source are shown in fig. 1.10(*c*). It will be observed that the distinction between a voltage-controlled voltage source and a current-controlled voltage source has now been lost because the voltage μV_c is equal to $(\mu R_1)I_c$ and it is a matter of choice whether one considers the voltage V_c or the current I_c to be the controlling quantity. Nevertheless the distinction between the two types of control is useful in so far as it distinguishes two different types of physical behaviour in the devices so represented.

An example of a current-controlled voltage source is a separately excited d.c. generator. An (input) current I_c in the field winding controls the output voltage of the generator. This source departs from the ideal not only by possessing finite input and output resistances but also by having a non-linear relationship between the input current I_c and the open-circuit output voltage V_{oc} due to the non-linearity of the iron on which the field winding is constructed. An example of a voltage-controlled voltage source is the thermionic triode valve. This is a three-terminal device which can be represented by fig. 1.10(*a*) if the terminals B and D are connected together. The representation assumes that the triode is connected to suitable polarizing batteries; the valve is not in itself a source of electric energy.

1.7 Simple reduction methods

The fundamental method of analysing linear electric circuits is by the direct application of Ohm's law and Kirchhoff's laws. However, in many simple circuits which contain only a few loops and nodes the concept of equivalence described in §1.5.2 enables the analysis to be done more rapidly; the circuit is successively reduced to a simpler form to enable a direct calculation of the wanted current or voltage to be made. For example, suppose that in fig. 1.11(*a*) the voltage V_0 and the resistance values R_0, R_1, R_2, and R_3 are given, and that it is required to find the voltage V_3. It is apparent that V_3 can be calculated, by Ohm's law, from a knowledge of I_3 which in turn can be obtained, by the principle of current division, from the current I. The currents I and I_3 are calculated by means of the equivalences illustrated in fig. 1.11(*b*) and (*c*). Thus:

$$\begin{aligned} V_3 &= R_3 I_3, \\ &= R_3 . \frac{R_1}{R_1+R_2+R_3} . I, \\ &= \frac{R_1 R_3}{R_1+R_2+R_3} . V_0 \bigg/ \left(R_0 + \frac{R_1(R_2+R_3)}{R_1+R_2+R_3}\right), \end{aligned} \qquad (1.7.1)$$

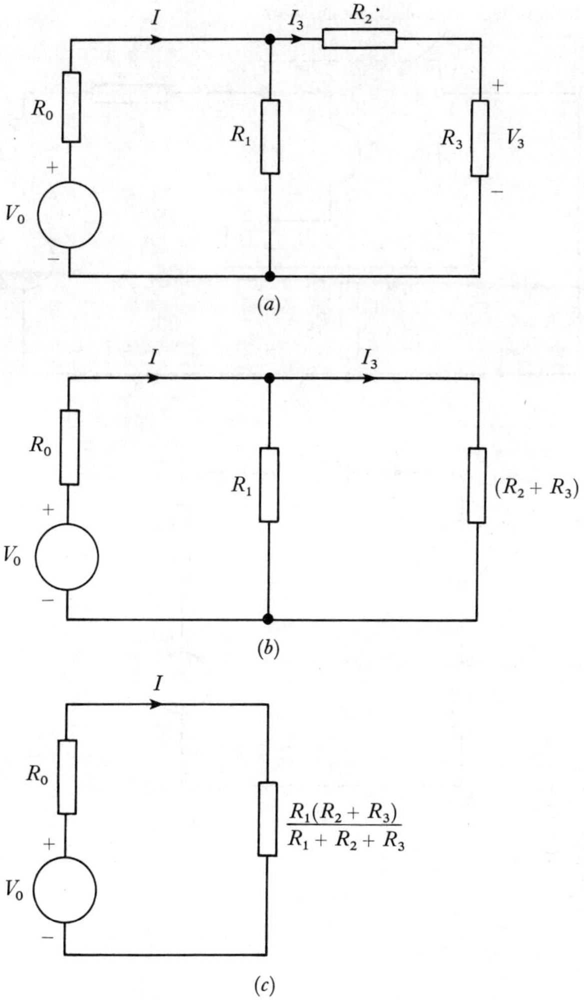

Fig. 1.11 Successive circuit reduction for the calculation of I.

$$= \frac{R_1 R_3}{(R_0 + R_1)(R_2 + R_3) + R_0 R_1} \cdot V_0. \qquad (1.7.2)$$

Reduction techniques can also be applied to circuits that contain controlled sources. The circuit shown in fig. 1.12(*a*) which is similar to that of a thermionic triode amplifier, is such a one. It is required to find the current through and the voltage across the resistor R_K. The significant feature of an ideal controlled source is its identification by two quantities, the output quantity (μV_c) and the controlling quantity (V_c). There is no need to draw

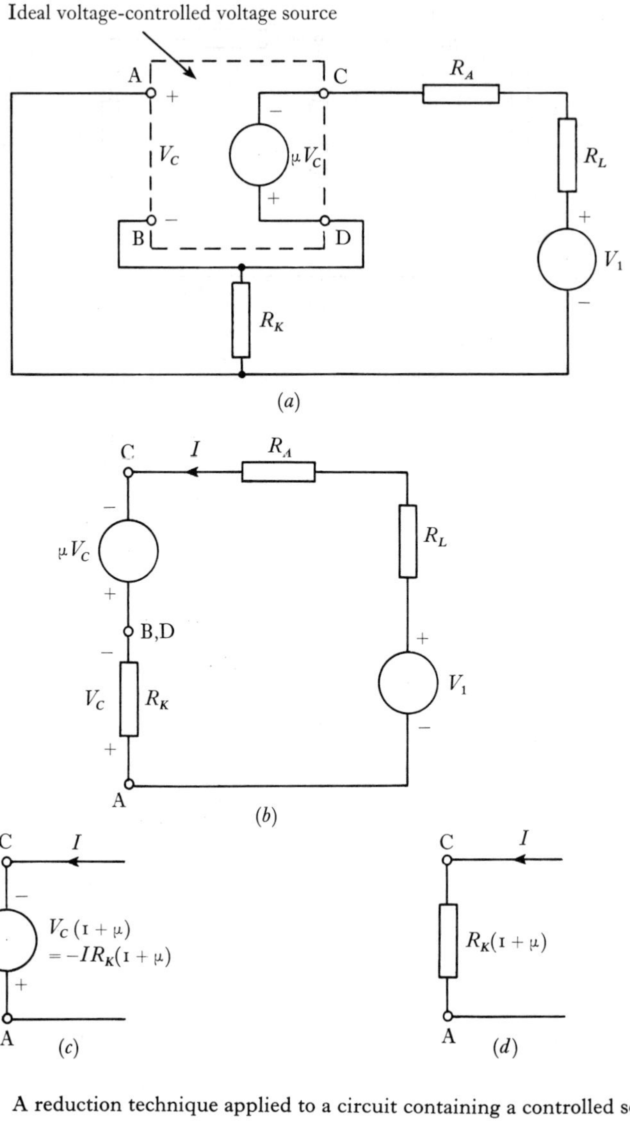

Fig. 1.12 A reduction technique applied to a circuit containing a controlled source.

it as a box with four terminals; it is better drawn as in fig. 1.12(*b*) where the terminals are rearranged in a more convenient manner. The absence of the graphical symbol for a voltage source corresponding to the quantity V_c in fig. 1.12(*a*), and hence in fig. 1.12(*b*), is due to the fact that V_c is not a signal impressed on the circuit from outside, but is just a voltage picked off a particular element (R_K) or arm of the circuit. It is now possible to reduce those elements of fig. 1.12(*b*) which lie between the terminals C and A (viz. the source and resistor R_K) successively to the forms shown in fig. 1.12(*c*) and (*d*) by means of the following arguments.

The voltage of the terminal C over that of A, V_{CA}, is equal to $-V_c(1+\mu)$. The current I will be undisturbed if an ideal voltage source of this magnitude replaces the elements between C and A as shown in fig. 1.12(*c*). But since $V_c = -R_K I$, the voltage V_{CA} is $+R_K(1+\mu)I$, a voltage proportional to the current flowing through the branch CA and having the same polarity and magnitude as that which would occur if a resistance of value $R_K(1+\mu)$ were present between the terminals C and A. Thus the controlled source together with the associated resistor R_K may be replaced by a resistor of magnitude $R_K(1+\mu)$ as shown in fig. 1.12(*d*). Once this reduction has been made, the current is seen to be

$$I = \frac{V_1}{R_L + R_A + R_K(1+\mu)}, \tag{1.7.3}$$

and the voltage across R_K, for example, is

$$V_c = \frac{R_K}{R_L + R_A + R_K(1+\mu)} \cdot V_1. \tag{1.7.4}$$

More sophisticated techniques of reduction are treated in chapter 6.

1.8 The Wheatstone bridge

Not all circuits are susceptible to the reduction methods used in §1.7. In any case, a more fundamental approach is sometimes advisable in complicated circuits. The Wheatstone bridge, named after the English physicist Sir Charles Wheatstone (1802–75) but invented by S. H. Christie in 1833, is an important example of such a circuit both because the reduction methods so far described are of no help in its analysis, and also because of the intrinsic importance of the circuit itself. Most applications of the bridge, which is shown in fig. 1.13, require a knowledge of the current in the element R_M which may be visualized as the resistance of an electric meter connected between the points B and D. If the currents in the elements R_1, R_2 and R_M are denoted by I_P, I_Q and I_M respectively, then the currents in the other elements can be obtained, in terms of these, by the application of Kirchhoff's current law to each node in turn. Since there are only three unknown currents, I_P, I_Q and I_M, only three independent equations are necessary for their solution, and these three independent equations can be found by applying Kirchhoff's voltage law in turn to each of three independent loops of the circuit. An independent loop is one that traverses at least one element not traversed by other loops. For example, loops FEABC, ADB and DCB give (1.8.1), (1.8.2) and (1.8.3) respectively which, by simultaneous solution

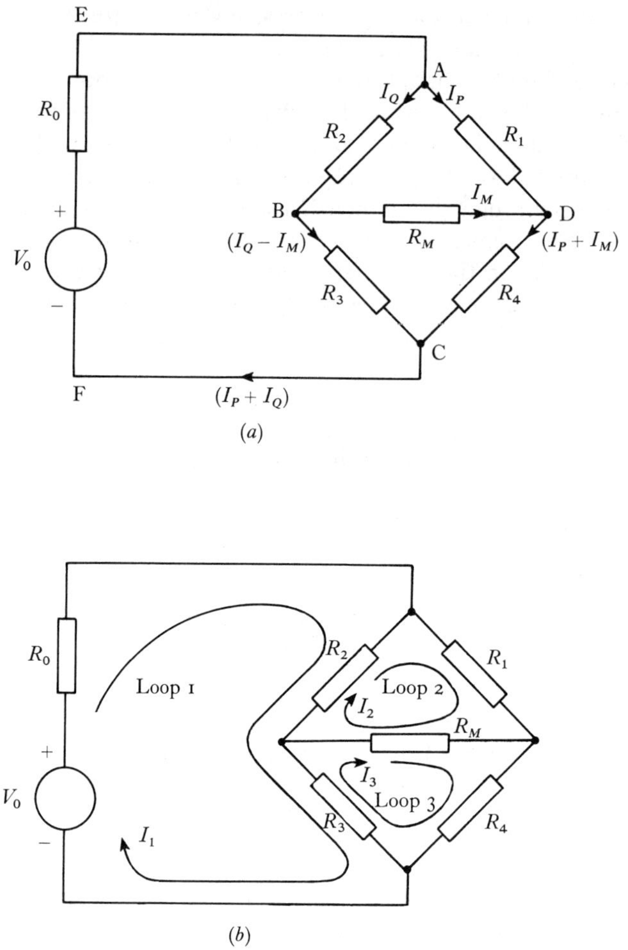

Fig. 1.13 Wheatstone bridge in which are postulated (*a*) branch currents; (*b*) loop currents.

produce answers for the three unknown currents in terms of the elements R_1, R_2, R_3, R_4, R_M, and the source V_0, R_0. This solution will not be pursued further because a more systematic approach to the problem is desirable.

$$V_0 = R_0(I_P + I_Q) + R_2 I_Q + R_3(I_Q - I_M), \quad (1.8.1)$$

$$0 = R_1 I_P - R_M I_M - R_2 I_Q, \quad (1.8.2)$$

$$0 = R_4(I_P + I_M) - R_3(I_Q - I_M) + R_M I_M. \quad (1.8.3)$$

1.9 The loop method of analysis

The systematic approach first to be described utilizes the concept of *loop currents*. Currents are postulated in independent loops of the circuit rather than in the branches, and equations are set up by applying Kirchhoff's voltage law to the loops containing the postulated currents. It is usually convenient, though not necessary, to postulate the currents in *meshes* (i.e. smallest possible loops) as shown in fig. 1.13(*b*) and also to use the same sense of direction (e.g. clockwise) in each mesh. This causes the resultant equations to appear in a form that becomes easily recognized and gives to the coefficients in the equations a useful physical significance. A loop equation is written down for each postulated loop current and the equations are solved for the loop currents. Branch currents are calculated from loop currents by taking their algebraic sum in the given branch. The loop method is illustrated below with reference to fig. 1.13(*b*). The three loop equations are

$$V_0 = (R_0 + R_2 + R_3)I_1 \qquad - R_2 I_2 \qquad - R_3 I_3, \tag{1.9.1}$$

$$0 = \quad - R_2 I_1 \qquad + (R_1 + R_2 + R_M)I \qquad - R_M I_3, \tag{1.9.2}$$

$$0 = \quad - R_3 I_1 \qquad - R_M I_2 \qquad + (R_3 + R_4 + R_M)I_3. \tag{1.9.3}$$

In each of them, the left hand side represents the algebraic sum of the voltage sources acting round the loop in the direction of the postulated loop current;

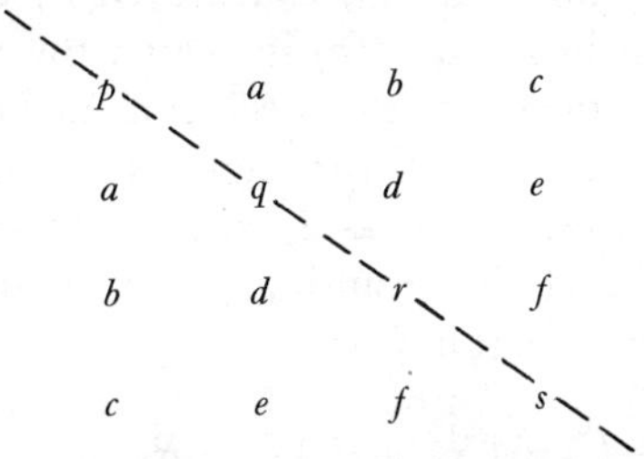

Fig. 1.14 Symmetry of resistance coefficients about the principal diagonal.

the right hand side represents the sum of the voltage drops in resistive elements due to currents flowing in these elements. The first term on the right hand side of (1.9.1), for instance, represents the voltage produced in loop 1 by the current I_1 flowing through all the elements of the loop; the coefficient $(R_0 + R_2 + R_3)$ is called the *self resistance* of the loop. The second term represents the voltage produced in loop 1 by the current I_2 flowing through elements common to loops 1 and 2, viz. R_2, and the third term represents the voltage produced in loop 1 by the current I_3 flowing through elements common to loops 1 and 3, viz. R_3. The coefficients (R_2) and (R_3) of these terms are called the *mutual* resistances of loops 1 and 2 and of loops 1 and 3 respectively, and the negative signs attached to the mutual terms in (1.9.1) are due to the consistent (clockwise) direction postulated for the loop currents. Equations (1.9.2) and (1.9.3) have a structure similar to (1.9.1), and it is worth noting that the array of resistance coefficients in these equations has a symmetry about the principal diagonal, illustrated in fig. 1.14 for a four-mesh circuit, which is a consequence of formulating the equations in the manner described. The symmetry only occurs if mesh currents with a consistent sense of direction are postulated and if the circuit contains no controlled sources.

1.10 Some examples of loop analysis

The loop method of analysis will now be demonstrated on some specific circuits. In the first two examples the circuits contain no controlled sources; in the final two, controlled sources are present.

1. The circuit first to be analysed is given in fig. 1.15(*a*) and it is required to find the current I_P in the resistor R_3. The initial task is to choose the loop currents, and a first choice might be the currents I_A and I_B in the two meshes. The disadvantage of this choice is that the wanted branch current I_P is the difference between the designated mesh currents I_A and I_B, and the calculation is thereby made longer than need be. An alternative would be to postulate the loop currents I_A and I_C; the wanted current is then directly equal to I_A. The same end is achieved by rearranging the circuit in the manner shown in fig. 1.15(*b*) in which the wanted current in the resistor R_3 is now identical with the mesh current I_2. It is a matter of taste whether one goes to the trouble of rearranging a circuit in this way or whether one employs loop currents such as I_A and I_C directly. The virtue of the rearranged circuit is that the resulting (mesh) equations have a familiar and consistent appearance. They are

$$\left.\begin{aligned} V_1 - V_2 &= (R_1 + R_2)I_1 - R_2 I_2, \\ V_2 &= -R_2 I_1 \qquad + (R_2 + R_3) I_2, \end{aligned}\right\} \qquad (1.10.1)$$

and the solution for I_2 is

$$I_2 = \frac{R_1 V_2 + R_2 V_1}{R_1 R_2 + R_2 R_3 + R_3 R_1}. \qquad (1.10.2)$$

2. The Wheatstone bridge of fig. 1.13 has the values $V_0 = 12$ V, $R_0 = 1\ \Omega$, $R_1 = 4\ \Omega$, $R_2 = 2\ \Omega$, $R_3 = 3\ \Omega$, $R_4 = 5\ \Omega$, and $R_M = 2\ \Omega$. What is the current I_M? For simplicity in formulating the equations, the circuit is

redrawn as shown in fig. 1.16 in which the wanted current I_M is equal to one of the postulated mesh currents I_1, I_2, and I_3. The mesh equations are

$$12 = 6I_1 - I_2 - 3I_3, \tag{1.10.3}$$

$$-12 = -I_1 + 10I_2 - 5I_3, \tag{1.10.4}$$

$$0 = -3I_1 - 5I_2 + 10I_3, \tag{1.10.5}$$

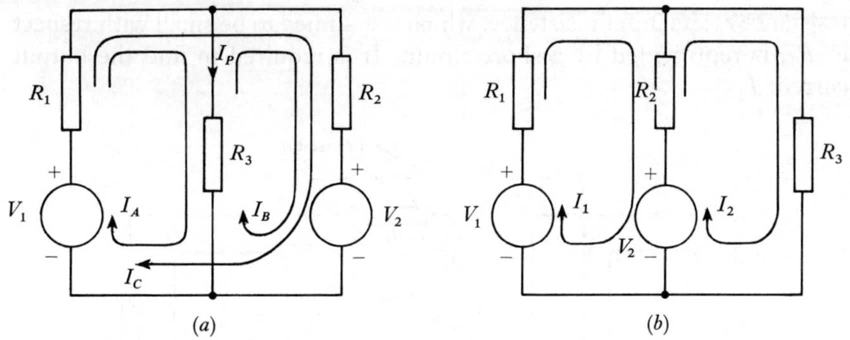

Fig. 1.15 (*a*) A two-mesh circuit. (*b*) Rearranged version.

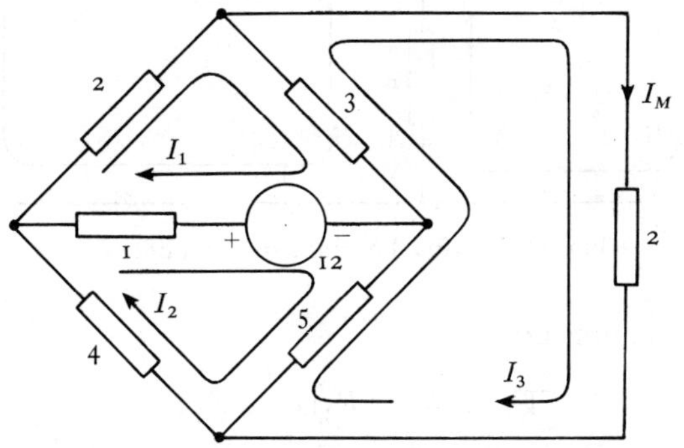

Fig. 1.16 Wheatstone bridge.

and the current $I_M(=I_3)$ can be found either directly by the method of determinants or by successive eliminations of I_2 and I_1 as follows. By subtracting (1.10.5) from 5 times (1.10.3), and adding (1.10.4) to 2 times (1.10.5) we eliminate I_2:

$$60 = 33I_1 - 25I_3, \tag{1.10.6}$$

$$-12 = -7I_1 + 15I_3. \tag{1.10.7}$$

Finally, by adding 7 times (1.10.6) to 33 times (1.10.7) the solution is obtained:

$$(60 \times 7) - (12 \times 33) = [-(7 \times 25) + (15 \times 33)]\,I_3,$$

and $$I_3 = 75 \text{ mA.}^\dagger$$

3. A transistor amplifier, under d.c. conditions, may be approximated by the model shown in fig. 1.17. The transistor, a three-terminal device, is represented by the current-controlled voltage source which has an output resistance r_0; its input resistance, which is assumed to be small with respect to R_1, is represented by a short-circuit. It is required to find the output current I_2.

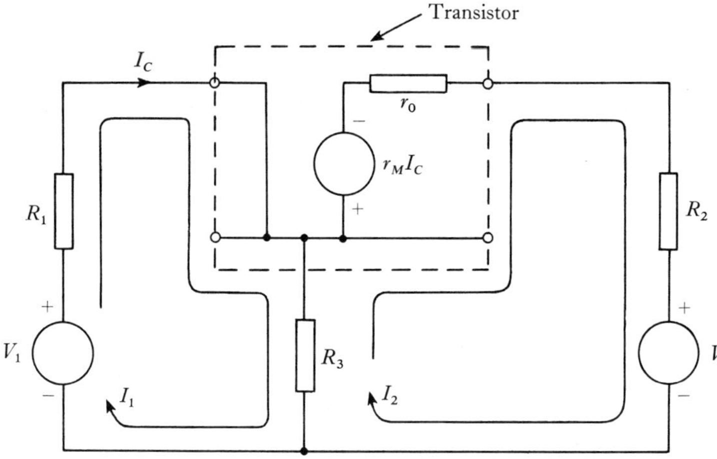

Fig. 1.17 Model of transistor amplifier circuit.

The mesh equations are

$$V_1 = (R_1 + R_3)\,I_1 - R_3\,I_2, \tag{1.10.8}$$

$$-(V_2 + r_M I_C) = -R_3\,I_1 \qquad + (R_2 + R_3 + r_0)\,I_2. \tag{1.10.9}$$

The terms on the right hand side correspond to the *passive* parts of the network, i.e. the non-source parts, those on the left hand side to the *active* or energy-producing parts; and symmetry about the principal diagonal exists. But in this example, as in all examples containing controlled sources, the overall symmetry is destroyed by terms arising from the controlled source. Here, the left hand side of (1.10.9) contains the term $r_M I_C$ $(= r_M I_1)$ which, on being transferred to the right hand side, destroys the previously existing symmetry. This is due to the unsymmetrical nature of a controlled

† The names of multiples and submultiples of units are formed by means of prefixes; the important ones are shown in appendix 2.

source which will transfer a stimulus from one loop into a second loop but not vice versa:

$$V_1 = (R_1 + R_3) I_1 - R_3 I_2, \qquad (1.10.10)$$

$$-V_2 = (r_M - R_3) I_1 + (R_2 + R_3 + r_0) I_2. \qquad (1.10.11)$$

The current I_2 is

$$I_2 = -\left[\frac{(R_1 + R_3) V_2 + (r_M - R_3) V_1}{(R_1 + R_3)(R_2 + R_3 + r_0) + R_3(r_M - R_3)}\right]. \qquad (1.10.12)$$

4. A thermionic triode valve, which acts like a voltage-controlled voltage source, is shown in fig. 1.18 in a d.c. circuit. The current I_2 is required.

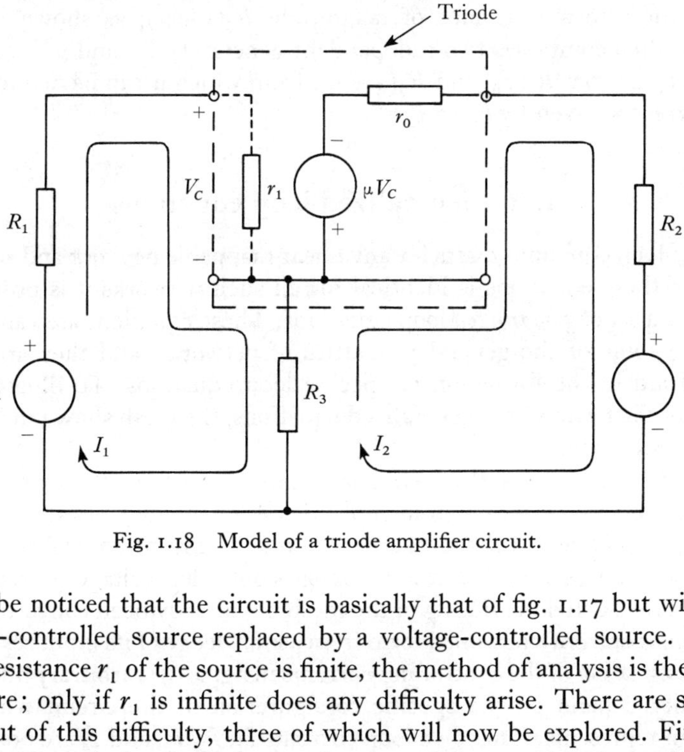

Fig. 1.18 Model of a triode amplifier circuit.

It will be noticed that the circuit is basically that of fig. 1.17 but with the current-controlled source replaced by a voltage-controlled source. If the input resistance r_1 of the source is finite, the method of analysis is the same as before; only if r_1 is infinite does any difficulty arise. There are several ways out of this difficulty, three of which will now be explored. First we can write the solution for I_2 by treating r_1 as finite, putting down the loop equations, and then letting r_1 tend to infinity. It is apparent from a comparison of the two circuits (figs. 1.17 and 1.18) that the solution for I_2 in fig. 1.18 is given by (1.10.12) if R_1 is replaced by $(R_1 + r_1)$ and if r_M is replaced by μr_1. Doing this and letting r_1 tend to infinity gives

$$I_2 = -\frac{V_2 + \mu V_1}{r_0 + R_2 + R_3(1 + \mu)}. \qquad (1.10.13)$$

A better way is to observe that since I_1 is zero (assuming r_1 to be infinite) the controlling voltage of the source is given by

$$V_C = V_1 + R_3 I_2. \qquad (1.10.14)$$

Substitution of (1.10.14) into the voltage equation for loop 2 gives

$$-V_2 - \mu(V_1 + R_3 I_2) = (r_0 + R_2 + R_3)\, I_2, \qquad (1.10.15)$$

which immediately leads to the solution for I_2 given by (1.10.13).

The final method is by use of a reduction technique already described. Since V_C is given by (1.10.14) the generator μV_C in loop 2 of fig. 1.18 may be split into two generators in series, one of value μV_1 and the other of value $\mu R_3 I_2$. This latter generator, taken in conjunction with series element R_3, reduces to a resistance of magnitude $R_3(1+\mu)$, as shown in §1.7. Loop 2 thus comprises two independent generators V_2 and μV_1 together with the resistors R_2, r_0, and $R_3(1+\mu)$, from which it can be deduced that the current is given by (1.10.13).

1.11 Generalized loop equations

A set of loop equations exists for any linear mappable network and since the form of these equations is identical for all such networks it is possible to produce a set of *generalized* loop equations. These equations are valuable in proving some of the general properties of networks, and they are also a mental aid in the formation of specific loop equations. To illustrate the origin of the terms in the generalized equations, the mesh shown in fig. 1.19 will be examined.

This is a typical mesh which is supposed to be a small part of a more extensive circuit, and each branch of the mesh contains a resistive element R, an independent voltage source V, and a current-controlled voltage source rI. It is assumed that any voltage-controlled voltage source which might have existed previously in the mesh has already been converted to its equivalent current-controlled form from a knowledge of its input resistance. The designation of the controlling currents in fig. 1.19 is arbitrary but would in reality depend on the specific configuration of the circuit which the diagram represents. Unknown loop currents I_1, I_2, I_3, and I_4 are postulated and Kirchhoff's voltage law is applied to the mesh. The resulting equation is

$$\begin{aligned} V_A + V_B + V_C + r_A I_1 + r_B I_3 + r_C I_5 \\ = -R_C I_1 + (R_A + R_B + R_C) I_2 - R_A I_3 - R_B I_4. \end{aligned} \qquad (1.11.1)$$

Now if the total circuit has n meshes, there will be n similar equations, and this set of equations is shown in (1.11.2) in which the following systematic terminology is used:

V_j is the sum of the independent voltage sources in loop j acting in the same direction as I_j,

$r_{jk}I_k$ is the voltage produced in loop j in the direction of the current I_j by a current-controlled voltage source whose controlling current is I_k,

R_{jj} is the self resistance of loop j,

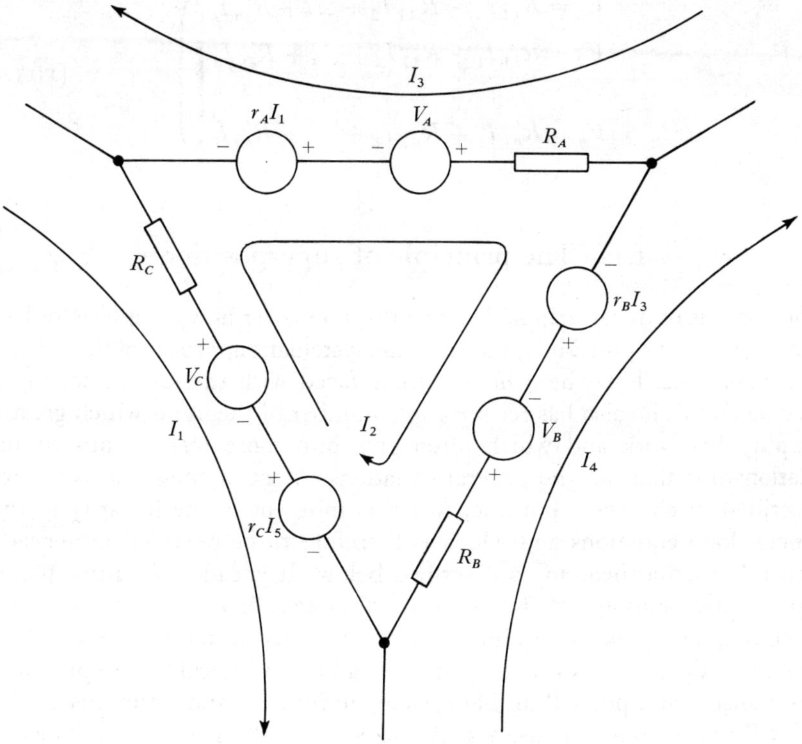

Fig. 1.19 A typical mesh.

R_{jk} is the resistance common to loops j and k, taken as positive if I_j and I_k traverse it in the same direction and taken as negative if I_j and I_k traverse it in opposite directions.

$$\left.\begin{aligned} V_1 + r_{11}I_1 + r_{12}I_2 + \ldots + r_{1n}I_n &= R_{11}I_1 + R_{12}I_2 + \ldots + R_{1n}I_n, \\ V_2 + r_{21}I_1 + r_{22}I_2 + \ldots + r_{2n}I_n &= R_{21}I_1 + R_{22}I_2 + \ldots + R_{2n}I_n, \\ \ldots \quad \ldots \quad \ldots \quad \ldots & \quad \ldots \quad \ldots \quad \ldots \quad \ldots \\ V_n + r_{n1}I_1 + r_{n2}I_2 + \ldots + r_{nn}I_n &= R_{n1}I_1 + R_{n2}I_2 + \ldots + R_{nn}I_n. \end{aligned}\right\} \quad (1.11.2)$$

It can be seen from the way in which (1.11.2) is formed that $R_{jk} = R_{kj}$ and that the right hand side of these equations is symmetrical about the principal diagonal. This symmetry is destroyed when the terms $r_{jk}I_k$, arising from the controlled sources, are transferred to the right hand side and are incorporated in new coefficients R'_{jk} where $R'_{jk} = R_{jk} - r_{jk}$ as shown in (1.11.3). This is because r_{jk} is not in general equal to r_{kj}.

$$\left.\begin{aligned} V_1 &= R'_{11}I_1 + R'_{12}I_2 + \ldots + R'_{1n}I_n, \\ V_2 &= R'_{21}I_1 + R'_{22}I_2 + \ldots + R'_{2n}I_n, \\ &\ldots \quad \ldots \quad \ldots \quad \ldots \\ V_n &= R'_{n1}I_1 + R'_{n2}I_2 + \ldots + R'_{nn}I_n. \end{aligned}\right\} \qquad (1.11.3)$$

1.12 The principle of superposition

The analysis of linear mappable networks, no matter how complicated they may be, is always possible by means of the systematic approach of the general loop equations, but when the analyst is faced with the task of solving a specific circuit he also has recourse to a number of theorems which greatly simplify his work and which often give him more insight into circuit relationships than do the general equations. Most of these theorems are described in chapter 6 but one, which is inherent in the linearity of the general loop equations and which is therefore to be regarded rather as a principle than a theorem, is described below. It is called the principle of superposition and applies to any physical system in which a stimulus and its consequent response are linearly related. In its simplest form it is this: if an individual stimulus X at a point A in a linear physical system produces a response x at a point P in that system, and if a separate stimulus Y at a point B produces a response y at the same point P, then the simultaneous application of the stimuli X and Y at their respective points A and B will produce a response $(x+y)$ at the point P.

In electrical terms, if a circuit is excited by several voltage sources $V_1, V_2, \ldots, V_n$, then the current at any point is the sum of the currents at that point due to each voltage in turn acting alone with all the others reduced to zero. This principle is expressed in the general loop equations, (1.11.3). If voltages $V'_1, V'_2, \ldots, V'_n$ produce currents $I'_1, I'_2, \ldots, I'_n$, and if voltages $V''_1, V''_2, \ldots, V''_n$ at the same points produce currents $I''_1, I''_2, \ldots, I''_n$, and if voltages $V'''_1, V'''_2, \ldots, V'''_n$ produce currents $I'''_1, I'''_2, \ldots, I'''_n$, etc., then by direct substitution in (1.11.3) it follows that voltages $(V'_1 + V''_1 + V'''_1$ etc.), $(V'_2 + V''_2 + V'''_2$ etc.), ..., $(V'_n + V''_n + V'''_n$ etc.) will produce currents $(I'_1 + I''_1 + I'''_1$ etc.), $(I'_2 + I''_2 + I'''_2$ etc.), ..., $(I'_n + I''_n + I'''_n$ etc.). The initial statement of the superposition principle is a special case of this result; the current I_j at any point is the sum of the currents I'_j, I''_j, I'''_j, etc. due to the individual voltages V'_1, V''_2, V'''_3, etc. acting singly with all the others made equal to zero. The principle is also valid when the stimuli are currents instead of voltages and also when the responses are voltages instead of currents.

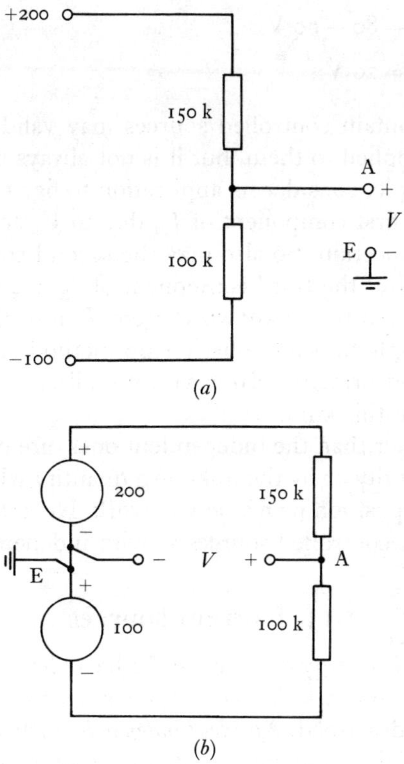

Fig. 1.20 Application of the superposition principle.

A simple application is given in fig. 1.20(*a*) in which the potential of each terminal is specified with respect to the earth terminal. What is the open-circuit voltage at the terminals AE? Notice first that when a terminal is held at a constant potential (e.g. 200 V) with respect to another it is as though an ideal voltage source equal to the potential difference between the terminals (200 V) were connected between them. Fig. 1.20(*a*) may therefore be redrawn as fig. 1.20(*b*), and since ideal voltage sources have zero internal resistance each source in fig. 1.20(*b*) acts like a short-circuit when, for the application of the superposition principle, its voltage is turned down to zero. Let V' be the open-circuit voltage at AE due to the 200 V supply acting alone and let V'' be the open-circuit voltage at AE due to the 100 V supply acting alone. Then

$$V = V' + V'',$$
$$= \frac{100}{100+150} . 200 - \frac{150}{100+150} . 100 \text{ V},$$
$$= 80 - 60 \text{ V},$$
$$= 20 \text{ V}.$$

Circuits which contain controlled sources may validly have the superposition principle applied to them, but it is not always useful in circuits of this kind. For example, consider its application to fig. 1.17 in order to find the current I_2. The first component of I_2, due to V_1 acting alone, may be written down by inspection. So also may the second component of I_2 due to V_2 acting alone. But the third component of I_2 due to the source $r_M I_C$ acting alone introduces the unknown current I_1 into the answer. In this situation it is as simple to use the basic loop method of analysis as it is to use the superposition principle. However, not all circuits containing controlled sources are of this kind.

If the sources, other than the independent ones, are controlled either by an independent quantity or by the unknown quantity which is being evaluated, then the superposition principle will work. Nevertheless, the method is not so useful with controlled sources as with independent sources.

1.13 Current sources

The sources of electric energy that have so far been described in this chapter are known as voltage sources. A second type of source, known as a *current* source, will now be described. An *ideal independent current source* is a two-terminal device which maintains a given current through its terminals regardless of the voltage across them. Ideal current sources do not exist in practice any more than ideal voltage sources do, but some practical sources do approximate to the ideal; a solar cell, for example, acts like a current source under some conditions of operation.

The graphical symbol used in this book for an ideal current source is shown in fig. 1.21(*a*) and a practical source which has an output resistance R_0 is shown in fig. 1.21(*b*). What happens to the current I_0 of the ideal current source when the source is on open-circuit? The answer is that the current flows into the infinite resistance which constitutes this open-circuit and thereby produces an infinite voltage across the source terminals. Since the source has to deliver infinite power to meet this condition we recognize that it can never be established in practice. The question that has just been answered is similar to one that can be asked about ideal voltage sources. What happens to the voltage V_0 of an ideal voltage source when it is on short-circuit? The answer has the same form as before. The voltage V_0 drives an infinite current through the zero resistance of the short-circuit; infinite power is required to do it. Just as a practical voltage source has an output resistance which limits the short-circuit current of the source, so the practical current source has an output resistance R_0 which limits its open-circuit voltage. The limiting value is $R_0 I_0$.

Suppose a load resistance is connected to the terminals of a current source and draws a current I from it as shown in fig. 1.21(*b*). If V is the voltage developed across the load resistance by the current I, then the equation of the circuit is

$$I = I_0 - (V/R_0). \tag{1.13.1}$$

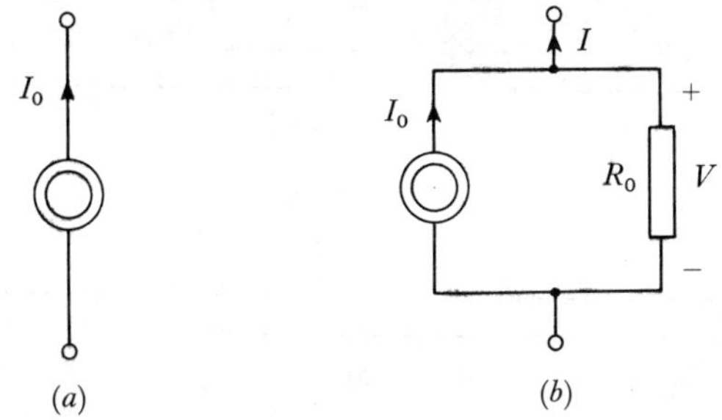

Fig. 1.21 Current sources: (*a*) graphical symbol for an ideal current source; (*b*) current source with output resistance R_0.

On making the substitution $V_0 = R_0 I_0$ the equation becomes

$$V = V_0 - R_0 I, \tag{1.13.2}$$

and since this is identical with (1.6.1) for a voltage source it follows that the terminal characteristics of the current source shown in fig. 1.21(*b*) are identical with the terminal characteristics of the voltage source shown in fig. 1.9. In other words any particular source can be represented either by a circuit of the form of fig. 1.9 or by that of fig. 1.21(*b*). It follows that a source given in one form may be converted to the other form via the equation $V_0 = R_0 I_0$. For example, a source which has a terminal voltage of 100 V on open circuit and which falls by 50 V per ampere of current drawn may be represented by an ideal source of 100 V (V_0) in series with a resistance of 50 Ω (R_0). Alternatively it may be represented by an ideal current source of 2 A ($I_0 = V_0/R_0$) in parallel with a resistance of 50 Ω (R_0). In the first model we start by viewing the source on open-circuit whereas in the second model we start by viewing it on short-circuit. The second model is the more natural if the information is given in the appropriate form. This would be that the source has a terminal current of 2 A on short circuit and falls by 1 A per

50 volts of applied voltage. The ability to convert a source from one form to the other is a very useful one.

Current sources exist not only in the independent forms described above but also in the ideal controlled forms shown in fig. 1.22(*a*) and (*b*) and in the practical controlled form of fig. 1.22(*c*).

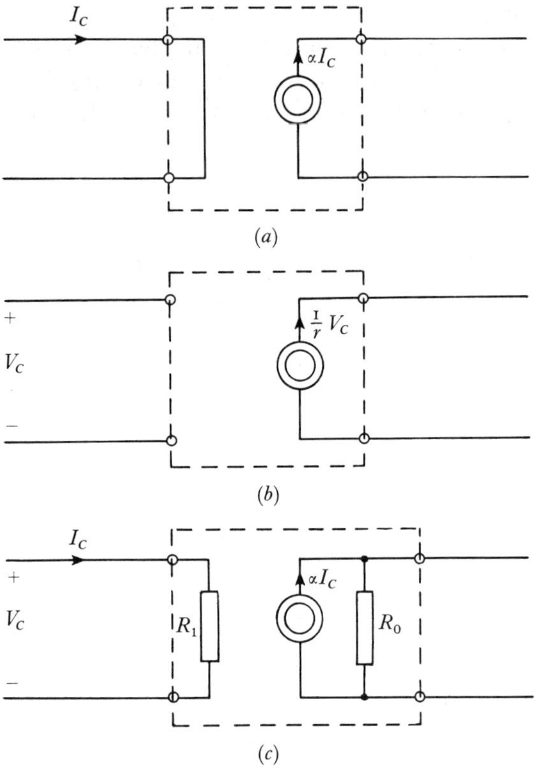

Fig. 1.22 Controlled current sources: (*a*) ideal current-controlled current source; (*b*) ideal voltage-controlled current source; (*c*) controlled current source with finite input and output resistances.

1.14 The node method of analysis

The node method of analysing electric circuits is an alternative to the loop method. The two are of equal importance and the reason for retaining both is that one gives a simpler solution than the other to circuits of a particular kind. The loop method usually gives the simpler solution to problems in electric machines whereas the node method often gives the simpler solution to the circuit configurations found in electronic engineering; electric power systems are also commonly found to be more susceptible to node analysis.

The simplicity of either method is related to the number of simultaneous equations that have to be solved, and since this number is usually different in the two methods one of the two is superior for the analysis of any particular circuit. This matter is considered in more detail in §1.15.

In node analysis the node voltages are postulated as the unknown quantities. One node is chosen as a reference point to which all other node voltages are referred. Simultaneous equations are set up by the successive application of Kirchhoff's current law to each node in turn and the equations are solved in terms of the unknown node voltages.

Consider the application of the method to fig. 1.23 in which I_1 and I_3 are known current sources and in which V_1, V_2, and V_3 are unknown voltages

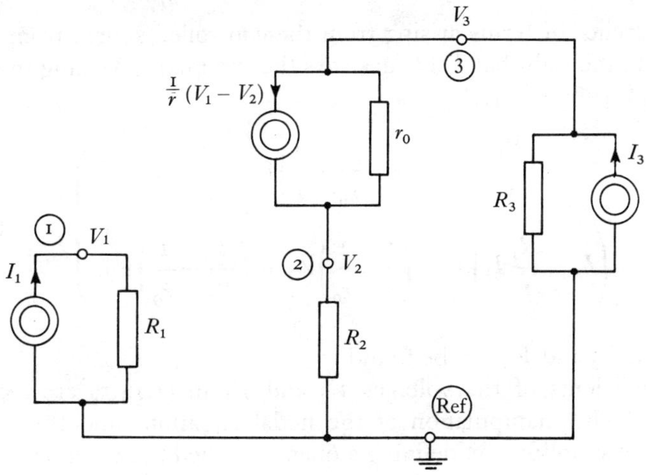

Fig. 1.23 Circuit arranged for nodal analysis.

defined with respect to earth. The circuit is in two parts and the equation for the left hand part is trivial. The current entering node 1 from the source is equal to the current leaving it via the passive branch R_1:

$$I_1 = \frac{1}{R_1} . V_1. \tag{1.14.1}$$

The right hand part of the circuit has two nodes other than the reference node (earth). Equating the source current entering each node to the current leaving it via the passive branches gives

$$\frac{1}{r}(V_1 - V_2) = \frac{1}{R_2} . V_2 + \frac{1}{r_0}(V_2 - V_3), \tag{1.14.2}$$

and

$$I_3 - \frac{1}{r}(V_1 - V_2) = \frac{1}{r_0}(V_3 - V_2) + \frac{1}{R_3}.V_3. \qquad (1.14.3)$$

Rearrangement of the right hand side of these two equations displays the symmetry about the principal diagonal which is a property of all such circuit equations (1.14.4),

$$\left.\begin{aligned} \frac{1}{r}(V_1 - V_2) &= \left(\frac{1}{R_2} + \frac{1}{r_0}\right)V_2 - \frac{1}{r_0}.V_3, \\ I_3 - \frac{1}{r}(V_1 - V_2) &= -\frac{1}{r_0}.V_2 + \left(\frac{1}{R_3} + \frac{1}{r_0}\right)V_3, \end{aligned}\right\} \qquad (1.14.4)$$

but transference of terms arising from the controlled source from the left hand side to the right hand side destroys the symmetry. Making this transference and using (1.14.1) gives

$$\left.\begin{aligned} \frac{R_1}{r}I_1 &= \left(\frac{1}{R_2} + \frac{1}{r_0} + \frac{1}{r}\right)V_2 - \frac{1}{r_0}V_3, \\ \left(I_3 - \frac{R_1}{r}I_1\right) &= -\left(\frac{1}{r} + \frac{1}{r_0}\right)V_2 + \left(\frac{1}{R_3} + \frac{1}{r_0}\right)V_3, \end{aligned}\right\} \qquad (1.14.5)$$

from which V_2 and V_3 can be found.

The coefficients of the voltages V_2 and V_3 in (1.14.5) are reciprocal resistances. The manipulation of the nodal equations and the resulting solutions are simplified by defining a quantity called the *conductance* G of a resistor which is the reciprocal of its resistance. All that is being done here is to rewrite Ohm's law in the form

$$I = GV \qquad (1.14.6)$$

where G is the conductance of the element measured in a unit which has in the past been called the *mho* (ohm spelt backwards) but which is now measured in a unit called the *siemens* (unit-symbol S) after the German-born family of electrical engineers of whom the most famous members are Werner von Siemens (1816–92), the founder of the electrical engineering firm Siemens and Halske, and his brother Sir William Siemens (1823–83). With this alteration (1.14.5) becomes

$$\left.\begin{aligned} \frac{g}{G_1}I_1 &= (G_2 + g_0 + g)V_2 - g_0 V_3, \\ I_3 - \frac{g}{G_1}I_1 &= -(g + g_0)V_2 + (G_3 + g_0)V_3, \end{aligned}\right\} \qquad (1.14.7)$$

and the voltage V_3, for example, is

$$V_3 = \frac{(G_2 + g_0 + g) I_3 - (gG_2/G_1) I_1}{(G_2 + g_0 + g) G_3 + g_0 G_2}. \tag{1.14.8}$$

1.15 Node analysis versus loop analysis

It might appear from the way in which the loop and the node methods of analysis have been presented that the loop method is applicable only to circuits which contain only voltage sources, and the node method only to circuits which contain only current sources. Fortunately this is not so. The presentation has been made in this way simply because the loop equations appear to be more systematic when only voltage sources are present, and the node equations appear to be more systematic when only current sources are present. But both methods may be used with either kind of source or with mixed sources. It is, however, sometimes convenient to convert a source of one kind to the other as outlined in §1.13, or to eliminate it by means of auxiliary equations. This is especially true of current-controlled sources in node analysis and of voltage-controlled sources in loop analysis.

If, then, either loop or node analysis may be used on any circuit, which is the better method to choose? In all but very simple circuits, the labour of analysis rests mainly in the solution of the equations rather than in their formulation or in the preparation of the circuit for analysis, and for this reason it is customary to choose that method of analysis which produces the smaller number of simultaneous equations. This number, in loop analysis, is equal to the number of independent loops in the circuit and, in node analysis, is equal to the number of independent node-pairs (a node voltage cannot exist across a single node but only across a node-pair). The number of independent loops and node-pairs in a circuit can be expressed in two formulae the derivation of which is outside our scope but which can easily be shown to be realistic. It is first necessary to define what is meant by a network which has *separate parts*. A network has P separate parts if it exists in P parts, none of which have any nodes in common but which may be mutually coupled, for example, by controlled sources (or, as will be seen later, by mutual inductance). It can then be shown that a network which has E elements, N_t nodes, P separate parts, V ideal voltage sources whether controlled or not, and C ideal current sources whether controlled or not, has N independent node-pairs and L independent loops where

$$N = N_t - P - V \tag{1.15.1}$$

and

$$L = E - N_t + P - C. \tag{1.15.2}$$

In these formulae, which will now be applied to several different circuits, an ideal source counts as an element, (but mutual inductance does not).

A Wheatstone bridge is shown in fig. 1.24 in which E = 7, $N_t = 5$, P = 1,

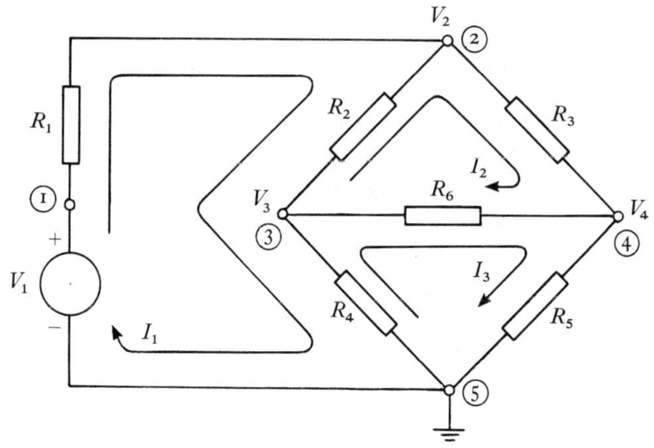

Fig. 1.24 A Wheatstone bridge.

V = 1, C = 0. Thus from (1.15.1) and (1.15.2) N = 3 and L = 3 so that both methods of analysis require the solution of three simultaneous equations. The loop equations, using the designated loop currents are

$$\left.\begin{aligned} V_1 &= (R_1+R_2+R_4)I_1 - R_2 I_2 - R_4 I_3, \\ 0 &= -R_2 I_1 + (R_2+R_3+R_6)I_2 - R_6 I_3, \\ 0 &= -R_4 I_1 - R_6 I_2 + (R_4+R_5+R_6)I_3, \end{aligned}\right\} \tag{1.15.3}$$

and the node equations, derived from nodes 2, 3, and 4 respectively are

$$\left.\begin{aligned} 0 &= -G_1 V_1 + (G_1+G_2+G_3)V_2 - G_2 V_3 - G_3 V_4, \\ 0 &= -G_2 V_2 + (G_2+G_4+G_6)V_3 - G_6 V_4, \\ 0 &= -G_3 V_2 - G_6 V_3 + (G_3+G_5+G_6)V_4. \end{aligned}\right\} \tag{1.15.4}$$

The choice between the two methods would, in this example, be made on grounds other than the labour of solving the equations.

A second example, shown in fig. 1.25, has the values E = 9, $N_t = 5$, P = 1, V = 1, C = 0, from which it follows that N = 3 and L = 5; so that whereas five equations are needed to determine the loop currents, only

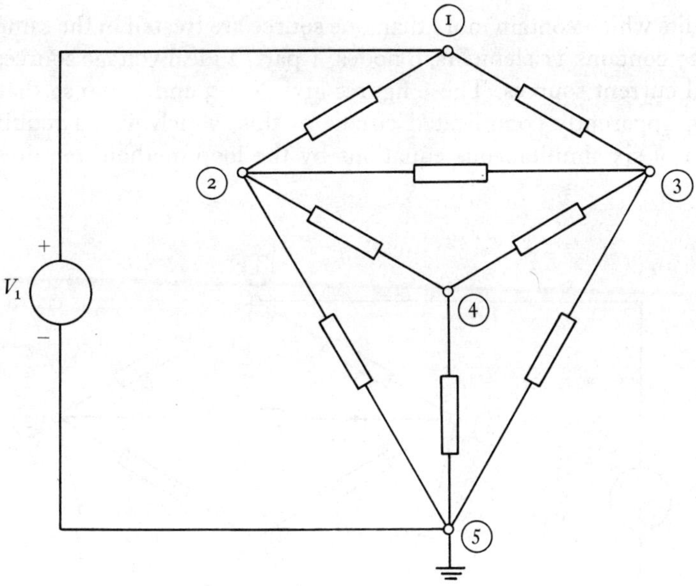

Fig. 1.25 Circuit suited to nodal analysis.

three (derived from nodes 2, 3, and 4) are needed to determine the node voltages. In this example the superiority of the node method is apparent. The reverse is true in fig. 1.26. For the values $E = 9$, $N_t = 6$, $P = 1$, $V = 0$ and $C = 1$, it follows that $N = 5$ and $L = 3$ showing that this circuit is better analysed by the loop method. Three suitable mesh currents I_1, I_2 and I_3 are shown which, together with the known mesh current I_0, will produce the three required equations.

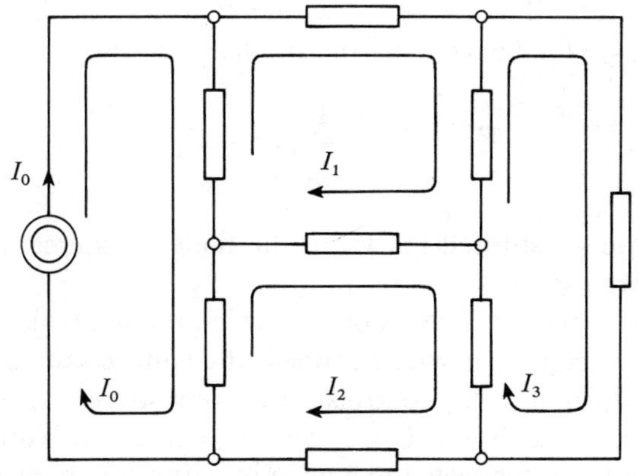

Fig. 1.26 Circuit suited to loop analysis.

Circuits which contain more than one source are treated in the same way. Fig. 1.27 contains 11 elements, 6 nodes, 1 part, 2 ideal voltage sources, and no ideal current sources. These figures give N = 3 and L = 6 so that even such an apparently complicated circuit as this, which would require the solution of six simultaneous equations by the loop method, requires only

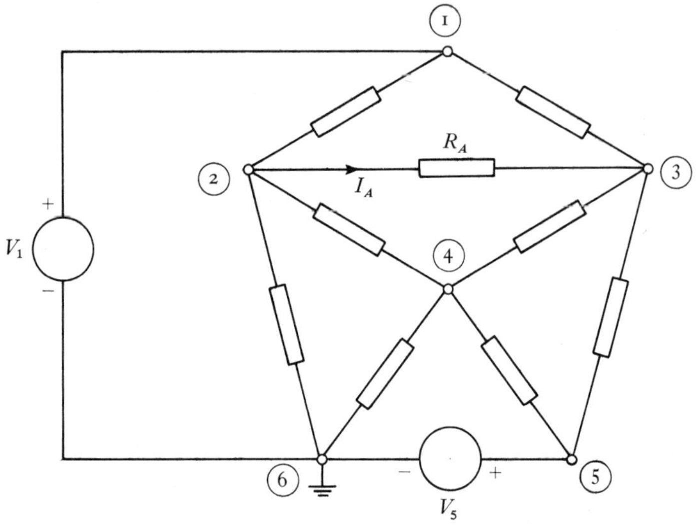

Fig. 1.27 Another circuit suited to nodal analysis.

three node equations for its solution. Nodes 2, 3, and 4 supply these equations; the voltages at nodes 1 and 5 are known quantities predetermined by the voltage sources. But suppose now that V_5 is a controlled source given perhaps by $V_5 = rI_A$. This creates the auxiliary equation

$$V_5 = rI_A = \frac{r}{R_A}(V_2 - V_3), \qquad (1.15.5)$$

which has to be substituted for V_5 in the node equations, but otherwise makes no difference.

The final example, fig. 1.28, is one in which the network has more than one part. The voltage V_3 in part 1 controls the voltage source μV_3 in part 2 whose current I_0 in branch 7 controls the current source αI_0 in part 1. Here E = 10, $N_t = 7$, P = 2, V = 2, C = 1; and N = 3, L = 4. Notice that if any one node in part 1 is connected to a node in part 2, the network becomes a one-part network but its operation is unchanged. N_t becomes 6 and P becomes 1, but N = 3 and L = 4 as before. When nodes 6 and 7 are tied together in this way, the node equations for nodes 2, 3 and 5 respectively are

$$\left.\begin{aligned} -\alpha I_0 &= -G_1 V_1 + (G_1+G_2+G_4) V_2 \quad -G_4 V_3, \\ 0 &= \qquad -G_4 V_2 \qquad +(G_3+G_4) V_3, \\ 0 &= \qquad\qquad -G_5 \mu V_3 \quad +(G_5+G_6+G_7) V_5, \end{aligned}\right\} \tag{1.15.6}$$

and the auxiliary equation $I_0 = G_7 V_5$ is available to eliminate I_0.

Equations (1.15.1) and (1.15.2) should be used with discrimination because the values of N and L can sometimes be changed significantly by

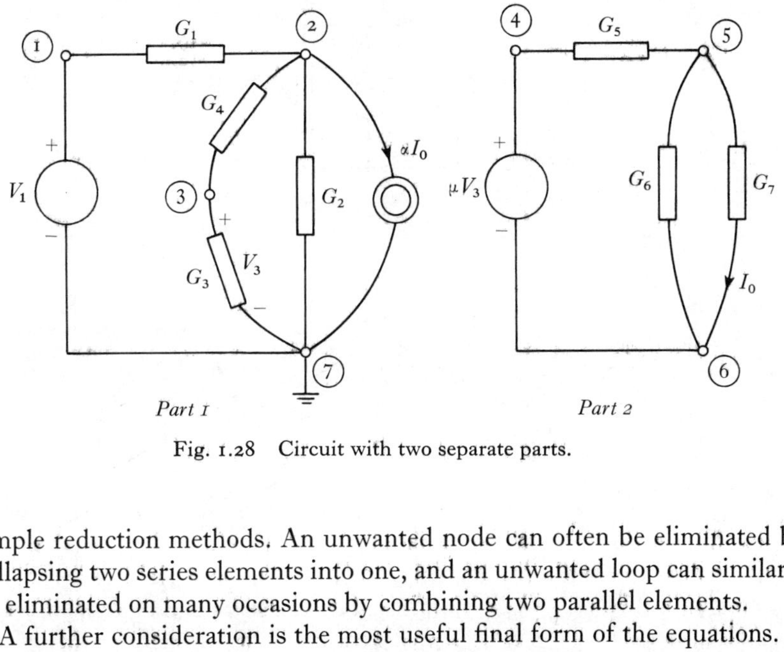

Fig. 1.28 Circuit with two separate parts.

simple reduction methods. An unwanted node can often be eliminated by collapsing two series elements into one, and an unwanted loop can similarly be eliminated on many occasions by combining two parallel elements.

A further consideration is the most useful final form of the equations. If shunt elements are likely to be added to the network between nodes, then the simplest method of accounting for these changes is to add new conductances to the existing ones. In other words, the equations are better expressed in the conductance (node) form. Conversely if series elements are likely to be inserted in the loops, the better representation is that of the loop equations.

The final choice of method is seen to depend on several factors which have to be weighted in accordance with the requirements of the problem. It is customary for unwanted nodes and loops to be eliminated from the network first of all, for N and L to be calculated, and a choice of method then made. If, however, it is apparent from the start that only two, or perhaps even three, equations will be needed in any one method, it may not be worth bothering with the calculation of N and L.

1.16 Generalized node equations

Just as it is possible to have a set of generalized loop equations for a circuit so it is also possible to have a set of generalized node equations. Fig. 1.29 shows what is supposed to be a typical node, node 2, in a circuit which

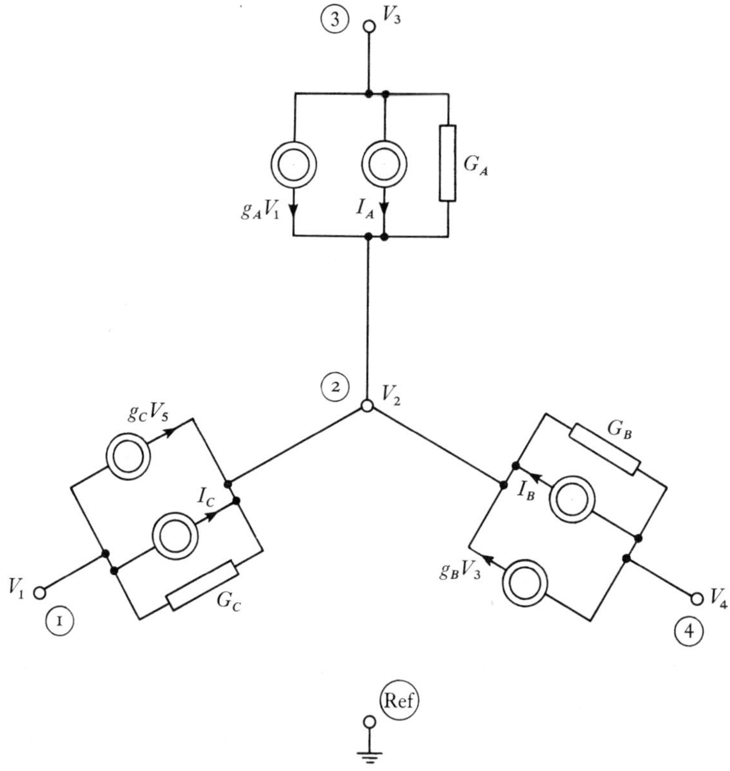

Fig. 1.29 A typical node.

contains many more nodes than those shown. The equation which results from the application of Kirchhoff's current law to this node is

$$I_A + I_B + I_C + g_A V_1 + g_B V_3 + g_C V_5$$
$$= - G_C V_1 + (G_A + G_B + G_C) V_2 - G_A V_3 - G_B V_4, \qquad (1.16.1)$$

and if there are n nodes other than the reference node, there will be a total of n equations similar to (1.16.1). These are shown in (1.16.2) in which the following systematic terminology is used:

I_j is the sum of the independent current sources attached to node j and flowing towards the node,

$g_{jk}V_k$ is the current directed towards node j by a voltage-controlled current source whose controlling voltage is V_k,

G_{jj} is the self conductance of node j, i.e. the sum of all conductance attached to node j,

G_{jk} is the conductance common to nodes j and k, i.e. the sum of the conductances between nodes j and k, taken as negative if V_j and V_k are defined in the same sense with respect to the reference terminal.

$$\left.\begin{aligned} I_1+g_{11}V_1+g_{12}V_2+\ldots+g_{1n}V_n &= G_{11}V_1+G_{12}V_2+\ldots+G_{1n}V_n, \\ I_2+g_{21}V_1+g_{22}V_2+\ldots+g_{2n}V_n &= G_{21}V_1+G_{22}V_2+\ldots+G_{2n}V_n, \\ \ldots\quad\ldots\quad\ldots\quad\ldots\quad\ldots & \quad\ldots\quad\ldots\quad\ldots\quad\ldots \\ I_n+g_{n1}V_1+g_{n2}V_2+\ldots+g_{nn}V_n &= G_{n1}V_1+G_{n2}V_2+\ldots+G_{nn}V_n. \end{aligned}\right\} \tag{1.16.2}$$

It can be seen from the way in which (1.16.2) is formed that $G_{jk}=G_{kj}$ and that the right hand side of these equations will be symmetrical about the principal diagonal. This symmetry is destroyed when the terms $g_{jk}V_k$, arising from the controlled sources, are transferred to the right hand side and are incorporated in new coefficients G'_{jk} where $G'_{jk}=G_{jk}-g_{jk}$. Since g_{jk} is not in general equal to g_{kj} it follows that the generalized node equations, (1.16.3), are not symmetrical about the principal diagonal:

$$\left.\begin{aligned} I_1 &= G'_{11}V_1+G'_{12}V_2+\ldots+G'_{1n}V_n, \\ I_2 &= G'_{21}V_1+G'_{22}V_2+\ldots+G'_{2n}V_n, \\ \ldots & \quad\ldots\quad\ldots\quad\ldots\quad\ldots \\ I_n &= G'_{n1}V_1+G'_{n2}V_2+\ldots+G'_{nn}V_n. \end{aligned}\right\} \tag{1.16.3}$$

The nodal equations, (1.16.2) and (1.16.3), are remarkably similar in form to the loop equations, (1.11.2) and (1.11.3), respectively. If currents are replaced by voltages and voltages by currents in one set of equations, and if resistances are replaced by conductances, or vice versa, the equations are transformed into the other set. The second set of equations is said to be the *dual* of the first set and vice versa. Since a set of (loop) equations has its counterpart in a dual set of (node) equations, the circuit which gave rise to the first set of (loop) equations has its counterpart in the circuit which gives rise to the dual set of (node) equations. The second circuit is called the dual of the first and vice versa. Thus figs. 1.19 and 1.29 represent dual circuits. Dual relationships will be explored more fully in chapter 6, but in the meantime it is interesting to observe that because of the duality of the equations, any circuit theorem which can be proved from the general loop equations has its dual in a theorem which can be proved in an identical fashion from the dual set of node equations. In particular, each of Kirchhoff's two laws is the dual of the other.

Examples on chapter 1

1.1 A moving-coil meter has a resistance of 50 Ω and gives a full-scale deflexion when passing a current of 15 mA. Find the value of the series resistor R_s and of the parallel resistor R_p which, separately, will give the meter a full-scale deflexion on the application of (*a*) a voltage of 100 V and (*b*) a current of 1 A.

1.2 Two 20 Ω resistors, in series, are connected to a 60 V battery. Across one of the 20 Ω resistors is connected a variable resistor R. What is the current through R, the voltage across it, and the power dissipated in it when it has a value of 20 Ω? What other value of R would give the same power dissipation in it?

1.3 A potential-divider network, consisting of three series-connected resistors R_1, R_2, and R_3, is placed across a 300 V supply with R_1 adjacent to the positive end A and R_3 adjacent to the negative end D. A current of 10 mA is to be drawn from the junction B of R_1 and R_2, and 20 mA from the junction C of R_2 and R_3, both currents being returned to the point D. Find the values of the resistors R_1, R_2, and R_3 which will supply these currents at voltages across BD and CD of 200 V and 150 V respectively, assuming that the 300 V supply delivers 50 mA to the network.

1.4 A source which has an open-circuit voltage of 1 V and an internal resistance of 1 Ω has terminals A and B across which are connected two pairs of resistors. The first pair consists of a 2 Ω resistor in series with a 3 Ω resistor, the 2 Ω one being adjacent to A, and the second pair consists of a 4 Ω resistor in series with a variable resistor R, the 4 Ω one being adjacent to A. Between the junction of the 2 Ω and 3 Ω resistors and the junction of the 4 Ω and variable resistors is connected a 1 Ω resistor which carries a current I. Derive an expression for this current I as a function of R and find the value of R for which I is zero.

1.5 Three d.c. sources are connected in a ring with nodes A, B, and C. The source across AB has an open-circuit voltage of 10 V directed so as to make A positive with respect to B, and an internal resistance of 4 Ω; the one across BC has an open-circuit voltage of 24 V directed so as to make C positive with respect to B, and an internal resistance of 8 Ω; the third, across AC, is a current source with a short-circuit current of 6 A directed so that current will flow through the source from C to A, and an internal conductance of 0.5 S. To nodes A, B, and C is additionally connected a star of resistors whose values are: to A, 3 Ω; to B, 6 Ω; and to C, 2 Ω. What is the current supplied by the 10 V source?

1.6 A 200 Ω potentiometer is connected across two series-aiding sources one of which has an open-circuit voltage of 10 V, the other of 12 V, and both of which have internal resistances of 100 Ω. The potentiometer slider is connected through a 100 Ω resistor to the junction point of the two sources. Find the two positions of the slider which will give a current of 10 mA through it, and the position for zero current.

1.7 Each of the four arms of a Wheatstone bridge is a 200 Ω resistor. The bridge is fed from a 2 V source which has negligible internal resistance. The detector is a 100 μA meter with an internal resistance of 50 Ω. By what percentage must one of the arms of the bridge be offset from balance in order to obtain full-scale deflexion on the meter?

1.8 Each of the four nodes, A, B, C, and D of a circuit is connected to every other node by resistors of values: AB, 3 Ω; AC, 3 Ω; AD, 3 Ω; BC, 3 Ω; BD, 1 Ω; and CD, 1 Ω. An ideal current source of value (5/3) A is connected between A and D (directed towards A) and a second of value 5 A is connected between C and D (directed towards C). What is the voltage between C and D?

1.9 A Wheatstone bridge with nodes A, B, C, and D is supplied at AC by a source whose open-circuit voltage is 4 V (making C positive with respect to A) and whose internal resistance is 1 Ω. In place of the detector a second source is connected across BD which acts like an ideal current source of 2 A (directed towards D) shunted by a 2 Ω resistor. What current is drawn from the 4 V source, and what voltage exists across the 2 A source when the arms AB, BC, CD, and DA are 6 Ω, 8 Ω, 4 Ω, and 8 Ω respectively?

1.10 A field-effect transistor which has terminals G (gate), S (source), and D (drain) may be represented internally by a voltage-controlled current source between D and S of magnitude $g_m V_{gs}$ (directed towards S) shunted by a resistor r_0, where V_{gs} is the voltage of terminal G with respect to that of S. G and S act internally like an open circuit. Resistors R_E and R_L are connected respectively between S and a common earth terminal E and between D and E, and a voltage source V_1 is applied to GE. What is the voltage between D and E if the component values are $R_E = 0.2$ kΩ, $R_L = 10$ kΩ, $r_0 = 50$ kΩ, $g_m = 1$ mS, and if $V_1 = 1$ V?

1.11 An active element is used to stabilize the output voltage of a primary source by operating in series with the source and the load. Between the terminals AB of the element it acts like a voltage-controlled voltage source of magnitude μV_{CB} (making B positive with respect to A) in series with a resistor r_0, where V_{CB} is the voltage of the control terminal C relative to B. The primary source, which is connected between A (positive) and terminal D (negative), has an open-circuit voltage V_s and an internal resistance R_s. A reference battery of voltage V_r is connected between C and D (C positive with respect to D); it supplies no current since C is essentially on open-circuit. For the values $V_s = 150$ V, $R_s = 500$ Ω, $V_r = 100$ V, $\mu = 20$, $r_0 = 1$ kΩ, find the voltage across a load resistor R_L connected between B and D (*a*) when $R_L = \infty$, and (*b*) when R_L draws a current of 0.1 A.

1.12 A bipolar junction transistor has three terminals B (base), C (collector), and E (emitter), and may be represented internally by a 0.8 kΩ resistor between B and E, and a current-controlled source shunted by a 10 kΩ resistor between C and E. The current of the controlled source is directed towards E and has a magnitude 100 times the current entering the terminal B. B is connected to the junction point of a 10 kΩ resistor and a 1.5 kΩ resistor which act as a potential divider across a 12 V supply PN (the 10 kΩ resistor being adjacent to the positive end of the supply P). Between E and N is a 0.4 kΩ stabilizing resistor and between C and P is a 1 kΩ load resistor. What current flows in the load resistor?

2 Ideal circuit elements

Direct current circuits can be analysed in terms of two concepts, the energy source and the energy sink. The energy source has two forms, the voltage source and the current source; the energy sink is the resistance element. But in circuits in which the stimulus is time-varying a further concept is necessary. This is the concept of an element which will *store* energy rather than produce it or consume it. In electromagnetic theory the storage of energy is associated with the existence of electric and magnetic fields, and each of these two forms of energy storage leads to a circuit element which will represent it. The names of the elements themselves are the *capacitor* and the *inductor*, and the properties which they display are called *capacitance* and *inductance* respectively.

This chapter will examine idealized versions of resistive, capacitive, and inductive elements in situations where the voltages and currents are varying with time.

2.1 Ideal circuit elements

The first process in the idealization of a circuit element is the isolation of its most significant property. The most significant property of a coil of wire, for example, may be its inductance (storage of energy in a magnetic field), but the coil may also contain the properties of resistance (energy consumption) and capacitance (storage of energy in an electric field). By visualizing an element which has only one property (e.g. inductance) and no other we conceive a *pure* element. A circuit element is pure if it is solely resistive or solely capacitive or solely inductive. Purity is the first of two conditions which must be satisfied in an *ideal* element; the second is *linearity*.

An element is said to be linear if the equation relating the current through it to the voltage across it is a mathematically linear one. The equation

$$v = f(i)$$

represents a linear equation if f symbolizes any mathematical operation or series of operations to be performed on i, and if

$$f(i_1 + i_2) = f(i_1) + f(i_2). \tag{2.1.1}$$

Put in words this means that the response to the stimulus $(i_1 + i_2)$ is the sum of the responses to i_1 and i_2 acting separately. Equations of the form $v = Ri$, $v = (1/C)\int i.\,dt$, and $v = L(di/dt)$ which satisfy condition (2.1.1) are linear ones, and elements which obey these equations are linear elements. Although the linearity of an element is formally defined in this mathematical way, it is more convenient from an engineering point of view to visualize the linearity in terms of a particular pair of physical quantities. For a resistor, this pair is voltage and current; for a capacitor it is voltage and charge; for an inductor it is current and flux-linkage. A linear relationship between these pairs of quantities ensures the linearity of the element as defined in the formal way.

Before continuing the discussion it is worth digressing on the coefficients R, C, and L which represent resistance, capacitance, and inductance and which arise when circuit equations are set up. It is customary to restrict these coefficients to *positive* and *constant* values. Positive values correspond to the state found in nature by, for example, a rod of compressed and bound carbon particles (which has positive resistance), or two metallic bodies (between which there is positive capacitance) or a coil of wire (which has positive inductance). The word *constant* is used to mean unchanging with time, or time-invariant. Neither of these two conditions need apply. Negative resistance, capacitance, and inductance can be made with the aid of controlled sources, and any element can be made to be time-variant like the variable resistor whose control is driven continuously up and down by a mechanical linkage. The justification for restricting the element parameters to positive constant values is not that such values are the only important ones but that an understanding of the behaviour of circuits containing only this restricted class of element forms a basis for a clear understanding of the more general class of circuits to which these restrictions are not applied. It is customary to take the positive and constant nature of an element for granted unless otherwise stated, and this practice will be followed hereafter.

An ideal element is one that is both pure and linear; it is assumed to exist at a point in space in a concentrated lump which occupies no volume. A long piece of wire, for example, whose resistance is distributed throughout its length would be represented by an ideal *lumped* resistor concentrated in a small volume and connected to the rest of the circuit by short resistanceless leads. Similarly, a coil of wire which has the properties of resistance and inductance inextricably mixed up in it would be represented by two quite separate lumped elements, one representing the resistive properties of the coil and one the inductive properties. The assumption that a practical element can be represented by lumped elements is valid when the physical dimensions of the whole circuit are sufficiently small and when the rate of change of the stimulus is not too great. The words *sufficiently small* and *not too great* are necessarily vague at this stage. What matters is that the effects of a stimulus should reach the most distant part of the circuit before the stimulus has had time to change significantly in value; it is then legitimate to assume that the electric effects occur simultaneously in all parts of the circuit. Under these conditions the concept of lumped elements is valid; the concept implies, in fact, that the velocity of propagation of electric effects is infinite.

In some circuits the approximation of lumped elements is not valid. The validity may be in doubt not only because the circuit is very large (e.g. a transmission line, hundreds of miles long) but because in smaller circuits the stimulus is changing so fast that electric effects at the end of the circuit are lagging behind effects in the circuit nearer the stimulus despite the high velocity of propagation of these effects. In the terminology of electromagnetic waves it would be said that the circuit dimensions are here significant with respect to the wavelength of the stimulus. It is under such conditions that the treatment of electric circuits by the concept of lumped elements has to be replaced by the laws of electromagnetism; but a useful bridge between the two approaches, of lumped-element circuit theory on the one hand and of electromagnetism on the other, is found in the notion of *distributed circuits*. An example of an object to which the distributed-circuit concept can usefully be applied is the coaxial cable. Such a cable has an inner rod-like conductor separated from a coaxial, cylindrical, outer conductor by an insulating medium. The predominant electrical effects in many situations are those of resistance and capacitance, and it is clear that both of these effects are distributed throughout the length of the cable. The full length of the cable is visualized as a succession of small lengths, and each small length is represented by a resistor and a capacitor. The resistor replaces the inner conductor of the small length of cable and represents the dissipative property not only of itself but also of the outer conductor, which can thereupon be represented by a resistanceless lead. The capacitor represents the capacitance between the inner and outer conductors of the small lengths of cable and is therefore placed between the mid-point of the resistor and the resistanceless lead. The whole cable consists of a cascade of such elemental circuits and the representation becomes more and more accurate as smaller and smaller elemental lengths are considered. When an infinite number of infinitesimal elements are assumed, the treatment is said to be that of the distributed circuit. Distributed circuits are treated in chapter 13.

2.2 Ideal resistors

An ideal resistor is defined by the equation

$$v(t) = Ri(t), \tag{2.2.1}$$

where the coefficient R is a positive constant and where the sign convention is shown in fig. 2.1. An alternative defining equation is

$$i(t) = Gv(t), \tag{2.2.2}$$

where G, the conductance of the resistor, is a positive constant. Equation (2.2.1) or (2.2.2) completely characterizes an ideal resistor because it

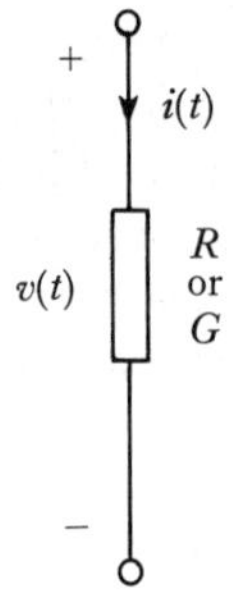

Fig. 2.1 Graphical symbol for an ideal resistor.

prescribes both linearity and purity. Linearity is established by the fact that the current i is proportional to the voltage v at all instants of time regardless of the magnitudes of v and i. The equation also prescribes purity, since it has no restriction on the rate at which v or i may change with time. If v changes instantaneously from a value v_1 to a value v_2 then i responds instantaneously from a corresponding value v_1/R to the new value v_2/R. Practical resistors display this property only within some margin of error and although it often suffices to represent a practical resistor by an ideal one, it sometimes happens that the practical element has to be synthesized by a conglomeration of ideal elements of different kinds. Ultimately, when dv/dt and di/dt become very large, the conglomeration of ideal elements required to represent the practical resistor becomes so great that the lumped treatment begins to break down and the approach involving distributed circuits has to be used.

The physical properties of an ideal resistor may be summarized as follows:

1. An ideal resistor absorbs energy. It does this by passing electric charge in such a manner that the resultant current is a linear function of the applied voltage. The instantaneous power developed in the element is

$$\left.\begin{aligned} p(t) &= v(t).i(t), \\ &= R[i(t)]^2, \\ &= G[v(t)]^2. \end{aligned}\right\} \tag{2.2.3}$$

2. The current changes *instantaneously* in accordance with the applied voltage and vice versa. Although this can never be exactly true in a practical resistor, it is often a good approximation.

2.3 Ideal capacitors

A capacitor is a circuit element which has the property of storing energy by the maintenance of an electric field. If q denotes the electric flux or charge in the element, and v the voltage across the element, both being functions of time, then the defining equation of an ideal capacitor is

$$q(t) = Cv(t), \tag{2.3.1}$$

where C is a positive constant called the capacitance of the element. C is measured in *farads* after the great experimentalist Michael Faraday (1791–1867). Equation (2.3.1) defines an element that is both linear and pure. The linearity is illustrated in fig. 2.2(*a*), but not all capacitors have this

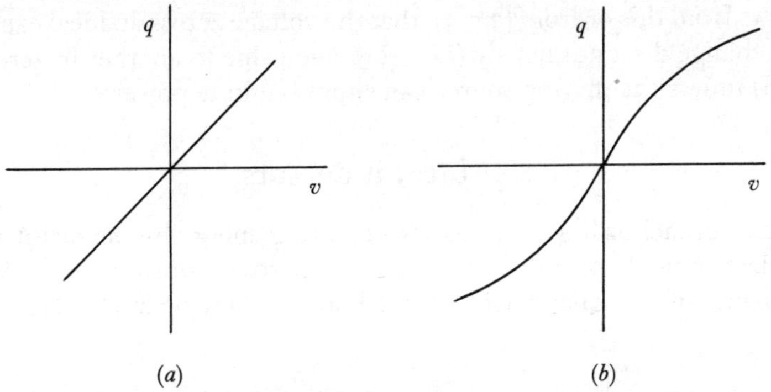

Fig. 2.2 Characteristics of capacitors: (*a*) linear; (*b*) non-linear.

characteristic; some, though not many, have the non-linear characteristic exemplified by fig. 2.2(*b*) in which the non-linearity is due to the characteristics of the dielectric material used in their construction. The purity of the element is entailed in (2.3.1) since no restriction is placed on the rates of change of q or v.

The current that flows in a capacitor is equal to the rate of change of stored charge and is given by

$$i = \frac{dq}{dt} = \frac{d}{dt}(Cv) = C\frac{dv}{dt}, \tag{2.3.2}$$

where the sign convention is defined by fig. 2.3.

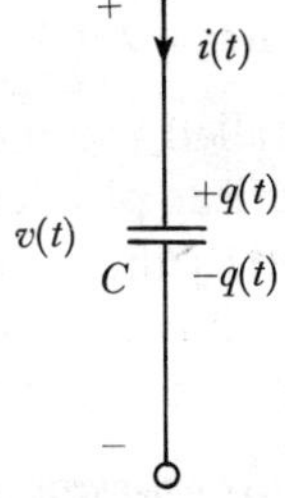

Fig. 2.3 Graphical symbol for an ideal capacitor.

The physical properties of an ideal capacitor may be summarized as follows:

1. An ideal capacitor stores energy by the maintenance of an electric field, and it is shown in textbooks on electromagnetism† that the energy stored is given by the equation

$$\text{Stored energy} = \tfrac{1}{2}Cv^2. \qquad (2.3.3)$$

2. The electric field manifests itself as a stored charge which responds instantaneously to changes in applied voltage in accordance with (2.3.1). It follows from this or from (2.3.3), that the voltage across an ideal capacitor cannot change discontinuously (i.e. from one value to another in zero time interval) unless the driving source can supply infinite power.

2.4 Ideal inductors

Whereas a capacitor has the property of storing energy by the maintenance of an electric field, an inductor has the property of storing energy by the maintenance of a magnetic field. If Λ denotes a magnetic flux-linkage in the

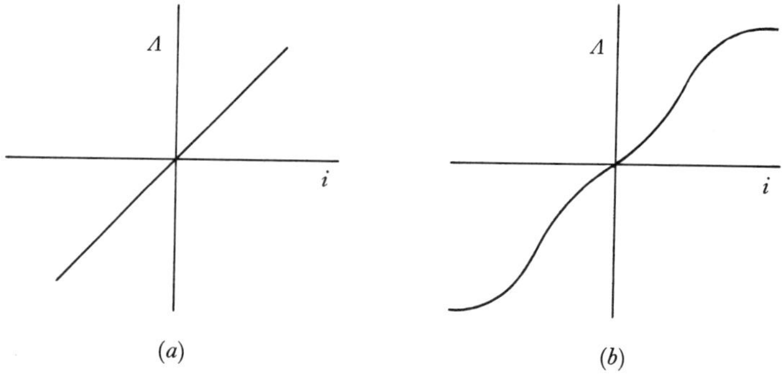

Fig. 2.4 Characteristics of inductors: (*a*) linear; (*b*) non-linear.

element and if i denotes the current flowing through the element, both being functions of time, then the defining equation of an ideal inductor is

$$\Lambda(t) = Li(t), \qquad (2.4.1)$$

where L is a positive constant called the inductance of the element. L is measured in *henrys* after the American physicist Joseph Henry (1797–1878).

A linear inductor is exemplified by the characteristic shown in fig. 2.4(*a*), but not all practical inductors are linear. Inductors made from coils of wire wound on non-magnetic materials are linear, but those wound on magnetic

† For example, G. W. Carter, *The electromagnetic field in its engineering aspects*, 2nd edition, p. 61, Longmans (1967).

materials such as iron are non-linear and have a characteristic which may be of the form shown in fig. 2.4(*b*). For such inductors the inductance L is not a constant and is a concept of little use.

If a current i flows through an ideal inductor and sets up a flux-linkage Λ in it then the voltage v across the inductor is given by the equation

$$v = \frac{d\Lambda}{dt} = \frac{d}{dt}(Li) = L\frac{di}{dt}. \tag{2.4.2}$$

The sign convention is defined by fig. 2.5.

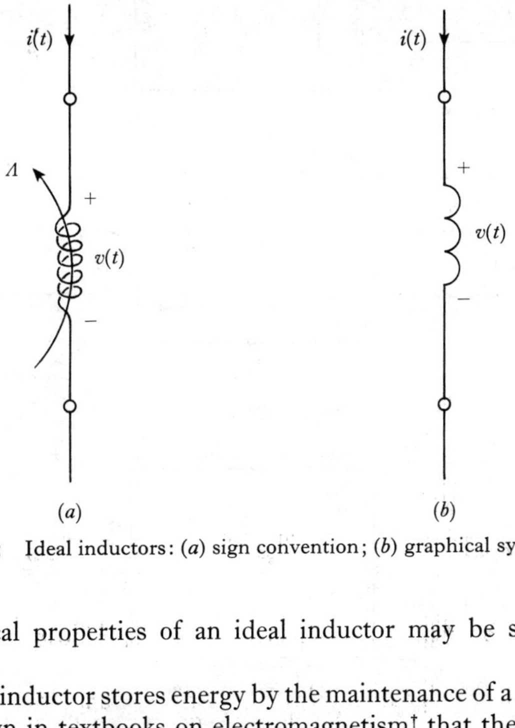

Fig. 2.5 Ideal inductors: (*a*) sign convention; (*b*) graphical symbol.

The physical properties of an ideal inductor may be summarized as follows:

1. An ideal inductor stores energy by the maintenance of a magnetic field, and it is shown in textbooks on electromagnetism† that the energy stored is given by the equation

$$\text{Stored energy} = \tfrac{1}{2}Li^2. \tag{2.4.3}$$

2. The magnetic field manifests itself as a magnetic flux-linkage, which responds instantaneously to changes of current in accordance with (2.4.1). It follows from this, or from (2.4.3), that the current through an ideal inductor cannot change discontinuously unless the driving source can supply infinite power.

† Ibid, p. 175.

2.5 Mutual inductance

When a current-carrying coil produces a magnetic flux that links its own turns, the coil is said to possess inductance as defined by (2.4.1). It sometimes happens however, either by accident or by design, that some part of this

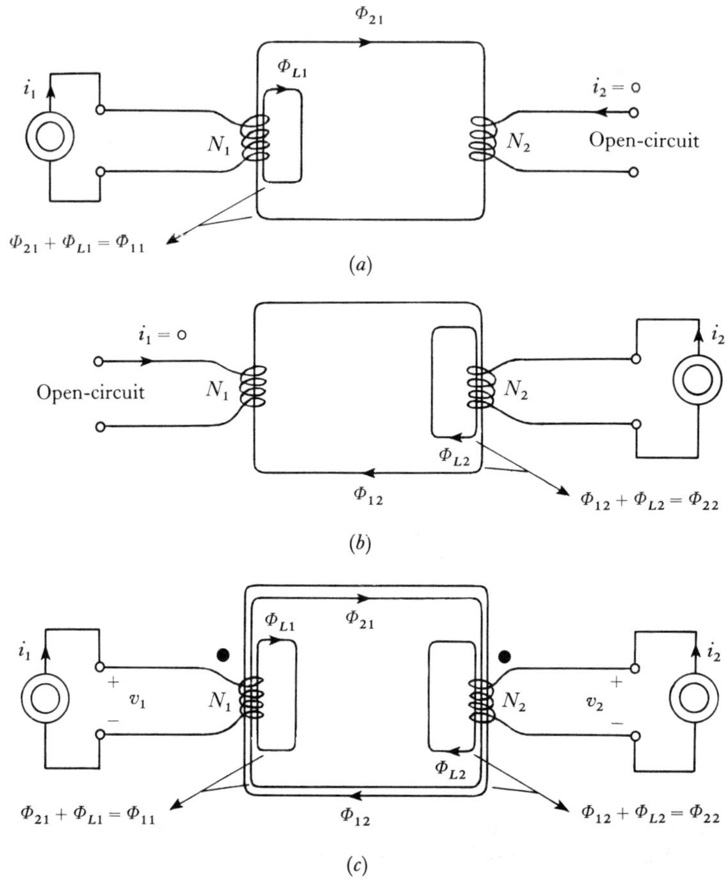

Fig. 2.6 Diagrammatic representation of fluxes linking two coils: (*a*) coil-1 energized; (*b*) coil-2 energized; (*c*) both coils energized.

flux also links the turns of a second nearby coil as shown diagrammatically in fig. 2.6(*a*). If this second coil is on open-circuit the situation is as follows. The *mutual flux*, Φ_{21}, is that part of the total flux which is common to both coils. The remaining part, Φ_{L1}, which links coil 1 only, is called the *leakage flux* of coil 1. The sum of these two fluxes constitutes the total flux linking coil 1; viz. $\Phi_{11} = \Phi_{21} + \Phi_{L1}$. The flux-linkage in coil 1 per unit current in this coil is called the inductance or, more strictly, the *self inductance* L_1 of coil 1 in accordance with the defining equation (2.4.1) which is repeated here in the terminology of fig. 2.6(*a*):

$$N_1\Phi_{11} = L_1 i_1. \tag{2.5.1}$$

The flux-linkage in coil 2 per unit current in coil 1 is called the *mutual inductance* M_{21} of the coils and is defined by the equation

$$N_2\Phi_{21} = M_{21} i_1; \tag{2.5.2}$$

M_{21} is measured in the same unit as L_1, the *henry*.

Suppose now that a current source is applied to the second coil and that the first coil is on open-circuit as shown in fig. 2.6(*b*). That component of the total flux which is common to both coils, Φ_{12}, is again called the mutual flux, and Φ_{L2} is called the leakage flux of coil 2. The total flux in coil 2 is $\Phi_{22} = \Phi_{12} + \Phi_{L2}$. The self inductance L_2 of coil 2 is given by

$$N_2\Phi_{22} = L_2 i_2, \tag{2.5.3}$$

and the flux-linkage produced in coil 1 per unit current in coil 2 defines the mutual inductance M_{12} of the two coils in accordance with the equation

$$N_1\Phi_{12} = M_{12} i_2. \tag{2.5.4}$$

It will be shown later that $M_{12} = M_{21}$.

For a linear system it is now possible to superimpose figs. 2.6(*a*) and (*b*) in order to find the total fluxes when i_1 and i_2 flow simultaneously. This application of the superposition theorem is made in fig. 2.6(*c*) from which it is seen that the total fluxes Φ_1 and Φ_2 in coils 1 and 2 respectively are

$$\left.\begin{aligned}\Phi_1 &= \Phi_{11} + \Phi_{12},\\ \Phi_2 &= \Phi_{21} + \Phi_{22},\end{aligned}\right\} \tag{2.5.5}$$

from which we obtain the flux-linkages

$$\left.\begin{aligned}\Lambda_1 &= N_1\Phi_1 = L_1 i_1 + M_{12} i_2,\\ \Lambda_2 &= N_2\Phi_2 = M_{21} i_1 + L_2 i_2.\end{aligned}\right\} \tag{2.5.6}$$

But the terminal voltage of an ideal inductive element is given by Faraday's law, $v = \mathrm{d}\Lambda/\mathrm{d}t$, and the application of this to (2.5.6) gives

$$\left.\begin{aligned}v_1 &= L_1\frac{\mathrm{d}i_1}{\mathrm{d}t} + M_{12}\frac{\mathrm{d}i_2}{\mathrm{d}t},\\ v_2 &= M_{21}\frac{\mathrm{d}i_1}{\mathrm{d}t} + L_2\frac{\mathrm{d}i_2}{\mathrm{d}t}.\end{aligned}\right\} \tag{2.5.7}$$

These are the equations of two mutually coupled, ideal, inductive elements; they refer to ideal elements because the elements have been assumed to be linear and because all effects other than inductive effects have been ignored. It remains to be shown however that $M_{12} = M_{21}$. This can be demonstrated simply if the fluxes are constrained to simple geometrical paths (as in fig. 2.6) and if it can be assumed that any component of flux is proportional to the ampere-turns of the coil producing it in accordance with the equation

$$\Phi = bNi, \tag{2.5.8}$$

where b is a constant depending upon the geometric nature of the path taken by the flux and on the magnetic properties of the material it traverses. If $\Phi_{21} = b_{21} N_1 i_1$, then

$$M_{21} = \frac{N_2 \Phi_{21}}{i_1} = b_{21} N_1 N_2. \tag{2.5.9}$$

Similarly,

$$M_{12} = \frac{N_1 \Phi_{12}}{i_2} = b_{12} N_1 N_2. \tag{2.5.10}$$

But since the paths of the fluxes Φ_{21} and Φ_{12} are the same, it follows that $b_{21} = b_{12}$. Hence $M_{21} = M_{12}$. A more rigorous treatment in terms of energy gives the same result without making any assumptions about the flux paths.

It should be observed in the treatment of mutual inductance just given that each flux has been assumed to link every turn of the appropriate coil in fig. 2.6. This will not be true in a practical situation; there will be some component of flux that will link only a fraction of the turns of a coil. This makes no essential difference to the analysis, and (2.5.7) remains valid, if the coefficients L_1, L_2, and M are defined by flux-linkages rather than by the product of the separate quantities flux and turns. It does follow, however, though it will not be shown here, that the concept of leakage flux is of limited value because of the difficulty of defining it in a practical situation. For this reason the coefficients L_1, L_2, and M are usually considered to be the primary coefficients of a pair of coupled coils; they are also readily available from simple measurements taken at the coil terminals.

In fig. 2.6 the fluxes Φ_{11} and Φ_{12} are shown in the same (aiding) sense for the given directions of i_1 and i_2. This makes the term $L_1(\mathrm{d}i_1/\mathrm{d}t)$ in (2.5.7) take the same sign as the term $M(\mathrm{d}i_2/\mathrm{d}t)$. But Φ_{11} always has the same sense as Φ_{21} (because they are both produced by the same current), and it follows that the terms $L_1(\mathrm{d}i_1/\mathrm{d}t)$ and $M(\mathrm{d}i_2/\mathrm{d}t)$ will have the same sign if Φ_{21} and Φ_{12} have the same sense. A similar argument applied to fluxes originating in the secondary winding shows that for the same condition, the term $L_2(\mathrm{d}i_2/\mathrm{d}t)$ takes the same sign as $M(\mathrm{d}i_1/\mathrm{d}t)$. If, however, the fluxes Φ_{21} and Φ_{12} are in opposition for the same directions of i_1 and i_2, then in each equation the term in M takes the opposite sign from that of the term in L. Whether the fluxes aid or oppose each other for any given directions of i_1 and i_2 depends on the way in which the turns of coil 2 are wound relative to the turns of coil 1. It is thus necessary to adopt a convention so as to introduce into the circuit diagram, and hence into (2.5.7), information on the relative directions of the windings. This is usually done by what is called the dot convention. A dot is placed at one end of each coil in the circuit diagram. The dots are so related to reality that if the currents i_1 and i_2 both flow into, or both flow out of, the dotted ends of their respective coils, then the mutual fluxes are aiding each other. A knowledge of the way in which the coils are wound enables the positions of the dots to be determined. Alternatively the following experiment establishes them. A dot is placed arbitrarily at one end of one coil. This dotted end is transiently made positive with respect to the other by connecting a battery with the correct polarity across the coil. That end of the second coil which goes transiently positive with respect to its other end should then be marked with a dot.

An examination of the way in which the coils in fig. 2.6(*c*) are wound shows that dots should be placed on the terminals as marked, or alternatively on both of the other two terminals. In any situation the dots simply tell us that if the currents i_1 and i_2 both flow from the external circuit towards the dots (or both flow out from them) then the mutual fluxes are aiding. If, therefore, they are both designated as flowing in towards the dots (or both out), then both terms in each equation have similar signs attached to them. A change in the designation of either current (but not both) will make the term involving M have a sign attached to it that is the opposite of that attached to the term involving L in the same equation.

The convention just described is sometimes utilized in conjunction with an equivalent circuit for two coupled inductors as an aid in setting up (2.5.7). Three different situations are shown in figs. 2.7, 2.8, and 2.9 together with the corresponding equivalent circuits. The corresponding equations are also given on the assumption that v_1 and v_2 are the designated voltages across the two pairs of terminals and that the upper terminal is designated as positive with respect to the lower one in each case. These designations are not marked on the diagram because its principal purpose is to relate the circuit to its equivalent, and this relationship is independent of the voltages v_1 and v_2. Fig. 2.7(*a*) shows the conventional diagram for two mutually coupled inductors in which their relative winding direction is given by the dot convention. Currents i_1 and i_2 are designated in an arbitrary manner and until this is done the sign of the term containing M is indeterminate. The equivalent circuit of fig. 2.7(*b*), in which no mutual coupling exists between the two inductors (because this is accounted for by means of the two ideal voltage sources), is now set up. This is done by recalling that if i_1 enters the dotted end of inductor 1 and is increasing ($\mathrm{d}i_1/\mathrm{d}t$ positive), then the dotted end of inductor 2 will be positive with respect to its other end.

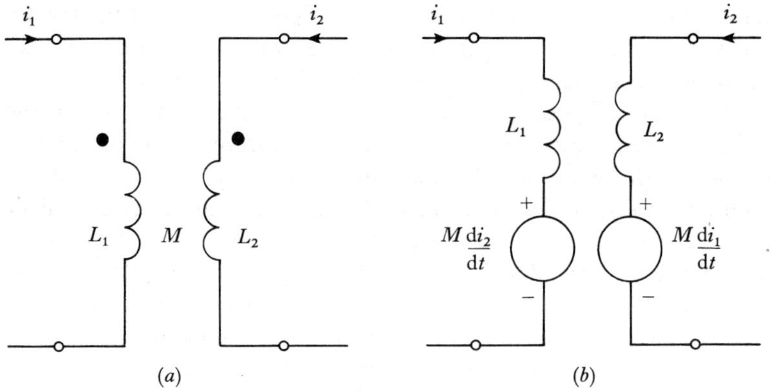

$$v_1 = L_1\frac{di_1}{dt} + M\frac{di_2}{dt}$$

$$v_2 = M\frac{di_1}{dt} + L_2\frac{di_2}{dt}$$

Fig. 2.7 (*a*) Circuit diagram of two mutually coupled inductors. (*b*) Circuit model.

The polarity of the ideal voltage source $M(di_1/dt)$ in inductor 2 is thereby established. An identical argument establishes the polarity of the ideal voltage source $M(di_2/dt)$ in inductor 1. The equations can be written down on inspection of the equivalent circuit. Similar arguments apply to figs. 2.8 and 2.9 which are shown because it is not always possible to arrange the circuit in the manner of fig. 2.7. The desire for a consistent set of loop

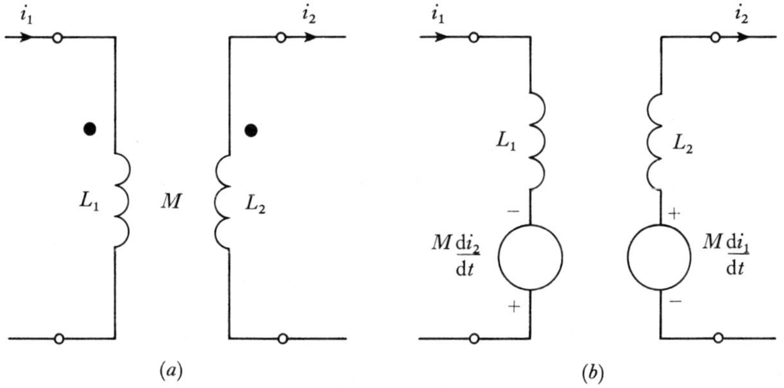

$$v_1 = L_1\frac{di_1}{dt} - M\frac{di_2}{dt}$$

$$v_2 = M\frac{di_1}{dt} - L_2\frac{di_2}{dt}$$

Fig. 2.8 (*a*) Mutually coupled inductors with differently assigned currents. (*b*) Circuit model.

equations, for example, may influence the directions in which the currents are designated.

Mutual inductance exists between two inductors not only when these two inductors comprise the whole system but also when they form merely a

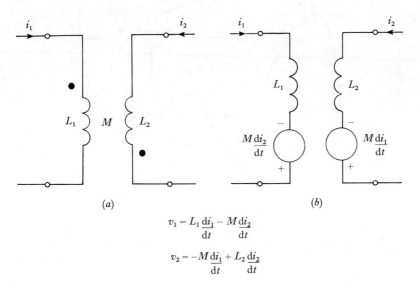

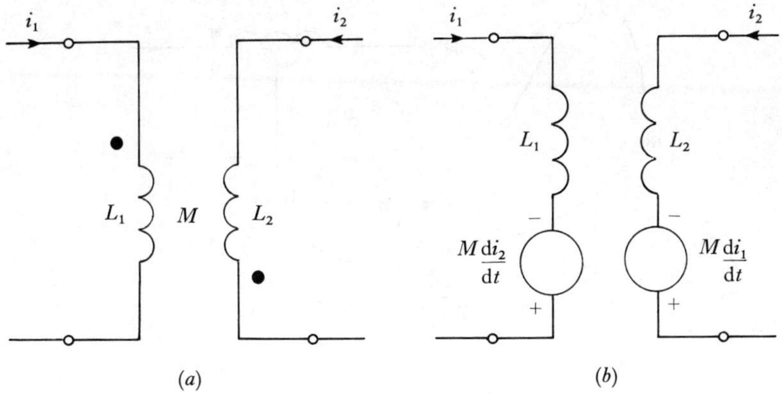

$$v_1 = L_1\frac{di_1}{dt} - M\frac{di_2}{dt}$$

$$v_2 = -M\frac{di_1}{dt} + L_2\frac{di_2}{dt}$$

Fig. 2.9 (*a*) Mutually coupled inductors with one coil reversed. (*b*) Circuit model.

part of a larger system of coupled inductors. Fig. 2.10(*a*) shows three coupled inductors randomly oriented. The mutual inductance coefficient (such as M_{23}) between any two of these inductors (such as inductors 2 and 3) represents the flux-linkage produced in one of them per unit current flowing in the other, when all of the remaining currents are zero. For example, the equation

$$N_2\Phi_{23} = M_{23}i_3 \tag{2.5.11}$$

defines the coefficient M_{23} ($=M_{32}$), where ($N_2\Phi_{23}$) is the flux-linkage in inductor 2 (due to current i_3) when i_1 and i_2 are zero. The total flux-linkage in any coil is given by the algebraic sum of the fluxes due to all the currents taken one at a time and, assuming for the moment that the fluxes are all in the same aiding direction, is given for inductor 2, for example, by the equation

$$\Lambda_2 = N_2\Phi_{21} + N_2\Phi_{22} + N_2\Phi_{23}, \tag{2.5.12}$$

$$= M_{21}i_1 + L_2i_2 + M_{23}i_3. \tag{2.5.13}$$

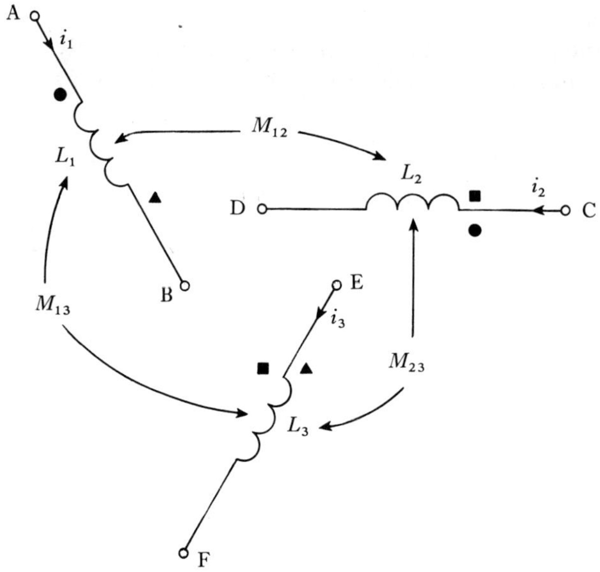

(*a*)

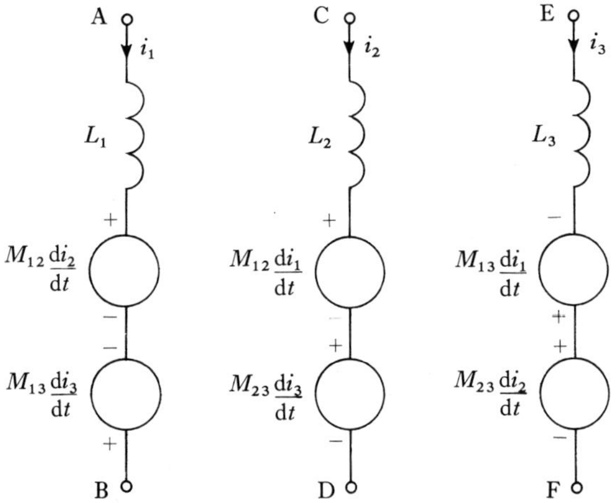

(*b*)

Fig. 2.10 (*a*) Three mutually coupled inductors. (*b*) Circuit model.

It follows that the voltage across inductor 2 is

$$v_2 = \mathrm{d}\Lambda_2/\mathrm{d}t,$$

$$= M_{21}\frac{\mathrm{d}i_1}{\mathrm{d}t} + L_2\frac{\mathrm{d}i_2}{\mathrm{d}t} + M_{23}\frac{\mathrm{d}i_3}{\mathrm{d}t}. \qquad (2.5.14)$$

Similar equations apply to inductors 1 and 3.

The relative direction of the two fluxes in any inductor due to its own current and the current in another inductor may be found experimentally from the system, and represented on the circuit diagram by the dot convention. In general a different kind of dot has to be used to relate each pair of coils, as shown in fig. 2.10(*a*), although some configurations can be adequately described with the aid of only one kind of dot. The equivalent circuit of fig. 2.10(*b*) may be drawn by taking the elements two at a time and representing the mutual inductance coupling by an appropriate ideal voltage source. Equations (2.5.15) follow immediately or, if preferred, directly from the fig. 2.10(*a*):

$$\left.\begin{aligned} v_1 &= L_1\frac{\mathrm{d}i_1}{\mathrm{d}t} + M_{12}\frac{\mathrm{d}i_2}{\mathrm{d}t} - M_{13}\frac{\mathrm{d}i_3}{\mathrm{d}t}, \\ v_2 &= M_{12}\frac{\mathrm{d}i_1}{\mathrm{d}t} + L_2\frac{\mathrm{d}i_2}{\mathrm{d}t} + M_{23}\frac{\mathrm{d}i_3}{\mathrm{d}t}, \\ v_3 &= -M_{13}\frac{\mathrm{d}i_1}{\mathrm{d}t} + M_{23}\frac{\mathrm{d}i_2}{\mathrm{d}t} + L_3\frac{\mathrm{d}i_3}{\mathrm{d}t}. \end{aligned}\right\} \qquad (2.5.15)$$

It is apparent that the dot convention becomes cumbersome when a large number of inductors are coupled together. An alternative method is to frame the equations with terms such as $M_{23}(\mathrm{d}i_3/\mathrm{d}t)$ of the same sign as the corresponding term $L_2(\mathrm{d}i_2/\mathrm{d}t)$ and to make the numerical value of the coefficient M_{23} positive or negative in accordance with the relative winding direction of the two inductors and the arbitrary designation of the two currents (i_2 and i_3).

2.6 The ideal transformer

A *transformer* consists of two or more coils inductively coupled by a specified amount. The required amount is determined by the particular application; often it is as great as can be achieved, sometimes it is very small. A measure of the amount of coupling between two coils is described by a coefficient k known as the *coefficient of coupling* which is defined by the equation

$$k = M/\surd(L_1 L_2). \qquad (2.6.1)$$

It is instructive to relate k to the fluxes in the transformer. Substitution of (2.5.1) to (2.5.4) into (2.6.1) gives

$$k = \sqrt{\left(\frac{M_{12}M_{21}}{L_1 L_2}\right)},$$

$$= \sqrt{\left(\frac{\Phi_{12}\Phi_{21}}{\Phi_{11}\Phi_{22}}\right)},$$

$$= 1 \bigg/ \sqrt{\left[\left(1+\frac{\Phi_{L1}}{\Phi_{21}}\right)\left(1+\frac{\Phi_{L2}}{\Phi_{12}}\right)\right]}, \tag{2.6.2}$$

and it is apparent that k is always less than unity. In the limit, when the leakage fluxes Φ_{L1} and Φ_{L2} of both coils are zero, $k = 1$ and the coils are said to be perfectly coupled or *unity coupled*. If k has a value close to unity, such as 0.9, the coils are said to be *closely coupled*, and if k is small, such as 0.01, the coils are said to be *loosely coupled*.

There are many applications in which it is desirable for a transformer to have coupling between its windings as close as possible; that is, to have a value of k as near to unity as can be achieved; and in applications of this kind it is obvious that the existence of leakage fluxes must be considered to be imperfections. The removal of these imperfections is the first stage in the visualization of an ideal transformer.

If the leakage fluxes are zero, the total flux linking coil 1, called the *primary winding* of the transformer, is $(\Phi_{21} + \Phi_{12})$, and this is the same as the total flux linking coil 2, known as the *secondary winding* of the transformer. Faraday's law then gives

$$v_1 = \frac{d}{dt}[N_1(\Phi_{21} + \Phi_{12})], \tag{2.6.3}$$

and

$$v_2 = \frac{d}{dt}[N_2(\Phi_{12} + \Phi_{21})], \tag{2.6.4}$$

from which it follows that

$$\frac{v_1}{v_2} = \frac{N_1}{N_2} = \frac{1}{n}, \tag{2.6.5}$$

where n is called the *turns ratio* of the transformer. Furthermore, (2.5.7) shows that v_1 can be expressed in terms of the currents i_1 and i_2 rather than the fluxes Φ_{21} and Φ_{12}:

$$v_1 = \frac{d}{dt}(L_1 i_1 + M i_2), \tag{2.6.6}$$

But $$\frac{M}{L_1} = \frac{Mi_1}{L_1 i_1} = \frac{N_2 \Phi_{21}}{N_1 \Phi_{11}} = \frac{N_2}{N_1} = n \text{ for } \Phi_{L1} = 0.$$

Thus $$v_1 = L_1 \frac{d}{dt}(i_1 + ni_2), \tag{2.6.7}$$

which shows that if a current $(i_1 + ni_2)$ flows through an isolated inductor L_1 it will produce across the inductor the voltage v_1, and vice versa. Equation

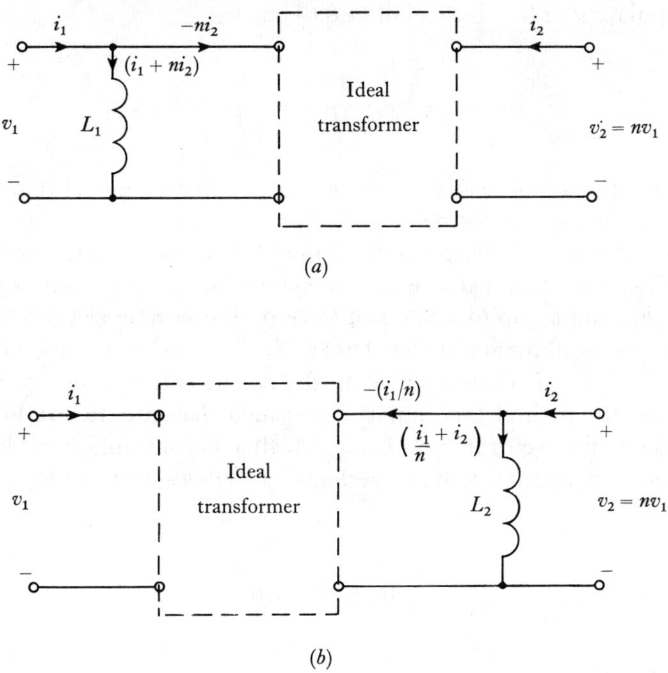

Fig. 2.11 (*a*) Circuit model of a unity coupled transformer. (*b*) Alternative model.

(2.6.7) is used in the left hand part of fig. 2.11(*a*) to construct a model of a unity coupled transformer; that is, one which has no leakage fluxes. The difference between the input current i_1 and the postulated current $(i_1 + ni_2)$ in L_1 is arranged in the model to flow into the input terminals of a box (shown dotted in the diagram) whose output current must be i_2, thereby necessitating that the input and output currents of the box be related by the factor $-n$ as shown. Also, to satisfy (2.6.5), it is necessary for the box to have its input and output voltages related by the factor $1/n$. In other words, that part of the model of a unity coupled transformer which is enclosed in the dotted box must be conceived to obey the equations

$$\left. \begin{aligned} v_1' &= \frac{1}{n} v_2, \\ i_1' &= -n i_2, \end{aligned} \right\} \tag{2.6.8}$$

where v_1', i_1' are the input quantities of the box. The box is known as an *ideal transformer*; and the model of a unity coupled transformer is an ideal transformer shunted on its primary side by the primary inductance L_1.

By manipulating (2.6.4) instead of (2.6.3) one gets

$$v_2 = L_2 \frac{\mathrm{d}}{\mathrm{d}t}\left(\frac{i_1}{n} + i_2\right),$$

which, in conjunction with (2.6.5) gives an alternative model (fig. 2.11(*b*)) of a unity coupled transformer.

The removal of the leakage fluxes from a transformer is the first stage in its conversion to an ideal one and is synonymous with making the coefficients b_{L1} and b_{L2} of (2.5.8) equal to zero. The second stage in the conversion is the elimination of the element L_1 from fig. 2.11(*a*), or L_2 from fig. 2.11(*b*), which is done by making the coefficients b_{12} and b_{21} infinite. This makes the path of the mutual fluxes such that infinite flux-linkage is produced by a finite current; L_1, L_2, and M then become infinite. If however v_1 is finite, an applied voltage perhaps, it follows from (2.6.7), for L_1 infinite, that

$$\frac{\mathrm{d}}{\mathrm{d}t}(i_1 + n i_2) = 0, \tag{2.6.9}$$

and hence that

$$i_1 = I_0 - n i_2, \tag{2.6.10}$$

where I_0 is a d.c. bias current in the primary winding which appears because of the constant of integration. It is customary to assume that $I_0 = 0$, so that

$$i_1 = -n i_2, \tag{2.6.11}$$

and this relation will then hold good however slow may be the variation of i_1 and i_2, down to the limit when i_1 and i_2 are D.C. The transformer is now an ideal one since its describing equations, (2.6.5) and (2.6.11), correspond to the basic definition of (2.6.8); it is usually represented in circuit diagrams in the manner shown in fig. 2.12.

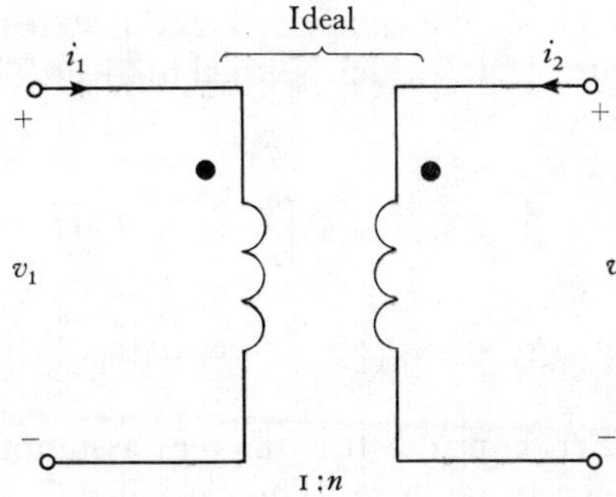

Fig. 2.12 Ideal transformer.

Notice that in this ideal device no power is absorbed. The total instantaneous power consumed by the ideal transformer is

$$\begin{aligned} p &= v_1 i_1 + v_2 i_2, \\ &= \frac{v_2}{n}(i_1 + n i_2), \\ &= 0. \end{aligned}$$

2.7 Paradoxes

Paradoxes sometimes arise from the process of idealizing circuit elements. Sometimes they arise because of a confusion between the real world and the idealized world of circuit theory. Of this kind is the one illustrated by fig. 2.13 in which an ideal voltage source of constant voltage V is applied at

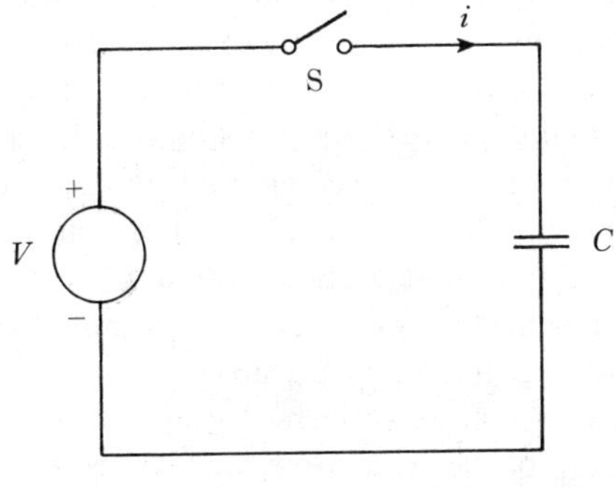

Fig. 2.13 Ideal voltage source supplying an ideal capacitor.

time $t = 0$ to an ideal capacitor C via resistanceless leads and an ideal switch, S. If the capacitor is initially uncharged its voltage is zero. On closing the switch its voltage rises instantaneously to V and it stores a charge Q. It now stores energy of amount $\frac{1}{2}CV^2$ which is equal to $\frac{1}{2}VQ$. The energy delivered by the battery, however, is

$$\int_0^\infty Vi.\mathrm{d}t = V\int_0^\infty i.\mathrm{d}t = VQ,$$

which is twice the energy stored by the capacitor. What has happened to the energy difference?

Before going further, consider the inherent assumptions in postulating the ideal circuit shown in fig. 2.13. The capacitor C dissipates no energy even though there exists in it an electric field which stresses its dielectric; the switch S changes its conductance from zero, when open, to infinity, when closed, in the infinitesimal time interval of its closure; the connecting wires have no resistance (nor inductance) even though infinite current will flow through them at an infinite rate of change with time; the battery supplies infinite power. When the matter is put in this way it is apparent that the idealized circuit cannot represent reality.

Nevertheless the reader may enquire what happens when a battery of very low internal resistance (approaching the ideal) is connected by a switch via very short thick leads to an almost ideal capacitor. To represent these conditions a finite resistance R would be inserted in the circuit to account for the total resistance, and when this is done it is found that R, whatever its value, accounts exactly for the missing energy. However, R can never tend to zero in practice, not only for the reasons set out above but also for the following reason. When a circuit carries a changing current it radiates electromagnetic waves and energy is transferred from the circuit to the surrounding space. The amount of energy radiated depends on the magnitude and rate of change of the current in the circuit and is normally negligible with respect to the energy consumed by the circuit in its lossy elements, unless the circuit is specifically designed to act as a radiator. But in fig. 2.13 the current changes from zero to infinity in zero time, and in these circumstances the radiated energy is not negligible. In fact, the element R, which accounts for the total energy consumed by the circuit, must include a term which corresponds to the radiated energy, and this term remains finite when the other constituent parts of R tend to zero. Fig. 2.13 is therefore not a valid representation of the actual circuit, though it may be a useful approximation to it for many purposes.

This treatment of the paradox appears to give a satisfactory solution to it, but conceals a further source of error which arises from the idealization process. A current flowing in a wire produces a magnetic field which links the current producing it. In other words a wire possesses inductance. The inductance of a conventional piece of wire depends on its dimensions and on the nature of the return path, but may be about 8 nH per centimetre of length, which is very small, and usually negligible; but if a current which changes with time at a very high rate flows through it (as in the circuit of fig. 2.13) the quantity $L(di/dt)$ becomes significant in the analysis; it may even dominate the situation. To account for it, the connecting leads may have to be represented not merely by resistance alone but by a series combination of resistance and inductance. It becomes apparent that the model used to represent a practical circuit must depend on the application of the circuit; in particular, it must depend on the nature or waveform of the driving source.

2.8 The circuit representation of practical elements

The paradox discussed in the previous section is connected with the accurate representation of practical elements by ideal ones. When a circuit diagram purports to represent a group of interconnected practical elements

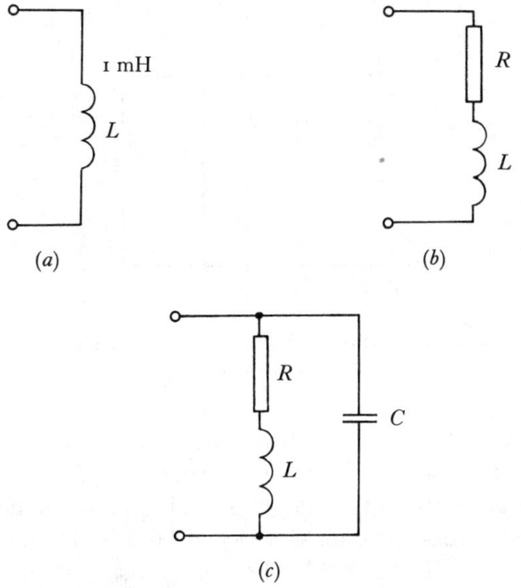

Fig. 2.14 Circuit models of an inductor: (*a*) first representation; (*b*) second representation; (*c*) third representation.

it attempts to incorporate all those features of each practical element which significantly affect the analysis of the circuit.

Suppose that we have a coil of wire labelled '1 mH inductor'. It will not normally suffice to assume either that the inductor has an inductance of exactly 1 mH, or that it is linear (unless wound on non-magnetic material), or that it is pure; the circuit model of fig. 2.14(*a*) will represent the element poorly. Upon examination it is sure to be found that the coil not only stores energy in the form of magnetic flux-linkage but also dissipates it. The dissipation may be assumed to arise from the resistivity of the material, probably copper, from which the coil is made, although this is by no means the full story. This resistivity or energy-dissipating property of the coil is distributed throughout the length of the wire, but it may be assumed that the coil can be adequately represented by an ideal lumped resistor R

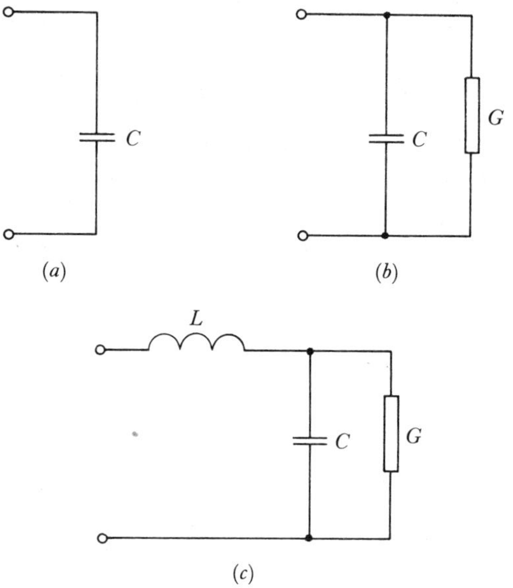

Fig. 2.15 Circuit models of a capacitor: (*a*) first representation; (*b*) second representation; (*c*) third representation.

connected in series with an ideal inductor L, as shown in fig. 2.14(*b*), which is an adequate representation for many purposes. If, however, the voltage that is to be applied to the coil is changing at a high rate, then a large voltage will be developed across the ends of the coil (and between adjacent turns) which will produce an electric field. The energy associated with this electric field will be stored in the corresponding space and it may be that this energy is significant with respect to the energy stored in the magnetic field. An approximate representation of this state of affairs is given in fig. 2.14(*c*) in which the whole of the energy stored in the electric field is represented by that appropriate to a capacitor placed across the ends of the coil.

A similar argument applies to the circuit representation of a capacitor. Fig. 2.15(*a*) is usually a much better representation of a practical capacitor than fig. 2.14(*a*) is of a practical inductor, and for many purposes may be adequate. Some capacitors, however, dissipate a significant amount of energy and may also produce a significant amount of magnetic flux due to the current flowing in them. The energy dissipation requires a conductance element G in the circuit model of this practical element as shown in fig. 2.15(*b*), and the magnetic flux storage requires an ideal inductor L as shown in fig. 2.15(*c*). The latter element is needed only when the rate of change of voltage across the practical element is sufficiently large for the resulting current $C(dv/dt)$ to set up a significant magnetic field.

It may be asked why the energy loss associated with a coil is represented by a *series* resistance R whereas the energy loss associated with a capacitor is represented by a *shunt* conductance G. This is purely a matter of convenience, of obtaining the simplest circuit that adequately represents what actually happens in practice. It is found from measurements on coils that in most circumstances the energy loss depends mainly on the current flowing through the coil; a constant resistor R in series with L gives a good approximation to what actually happens. In capacitors, however, the main part of the energy loss usually occurs in the dielectric and is dependent upon the voltage across the capacitor; a constant conductance G in shunt with C gives a good approximation to what actually happens.

There are many different ways of representing, by slightly different approximations, the same practical element. A more exact representation of a linear element always entails the addition of more ideal elements to the circuit model, and a balance has to be made between the faithfulness of the representation and the resultant complexity of the subsequent analysis. When the rates of change of voltage and current are large, it may be impossible to achieve by lumped elements an adequate approximation to the properties of a linear circuit, and the concept of lumped elements may have to be abandoned in favour of that of the distributed circuit.

A non-linear circuit can, of course, never be represented exactly by linear (ideal) elements however numerous; only in some limited (approximately linear) range is the representation of a non-linear circuit by linear elements a valid one.

Examples on chapter 2

2.1 An ideal 1 V d.c. voltage source is switched across an ideal 2 H inductor. What current flows and what energy is stored in the inductor 3 seconds later?

2.2 An ideal 0.1 A d.c. current source is connected in parallel with a closed switch and an ideal 1 μF capacitor. The switch opens at time $t = 0$. What is the voltage across the capacitor 1 ms later? If the source is not ideal but has an output resistance of 1 kΩ, what is the voltage across the capacitance at $t = 1$ ms and at $t = \infty$?

2.3 A linear rise in voltage v_c across a capacitor C is to be achieved by connecting the capacitor in series with two sources. The first has an internal resistance R_s and an open-circuit d.c. voltage V_s; the second has an internal resistance r_0 and an open-circuit controlled voltage μv_c. μv_c is directed round the loop comprising the two sources and C in opposition to the arbitrarily assigned direction of v_c. C is initially short-circuited by a switch. Show that, if $\mu = 1$, v_c changes linearly at the rate $V_s/C(R_s + r_0)$ V s^{-1} on opening the switch. What is the mathematical form of v_c if $\mu \neq 1$?

2.4 Two inductors have self inductances of 1 H and 2 H and a coefficient of coupling of 0.6. What is the inductance of the circuit formed by the two inductors connected (*a*) in series with the fluxes (i) aiding and (ii) opposing, and (*b*) in parallel with the fluxes (i) aiding and (ii) opposing?

2.5 Three inductors with terminals AB, CD, and EF have self inductances of 1 H, 2 H, and 4 H respectively and are magnetically coupled so that a current i_1, on flowing from A to B and increasing at the rate of 1 A s^{-1}, produces an open-circuit voltage at C relative to D of 0.6 V and at E relative to F of 0.8 V; and a current i_2, on flowing from C to D and increasing at the rate of 1 A s^{-1}, produces an open-circuit voltage at E relative to F of 1.2 V.

(*a*) What is the inductance between the terminals A and F when B is connected to C and D is connected to E?

(*b*) What is the rate of rise of current in a short-circuit across E and F when a 10 V d.c. voltage source is connected to the series connexion of AB and CD?

2.6 A 4 kΩ resistor is connected across the secondary of an ideal transformer which has a turns ratio of 2. What is the secondary voltage 1 ms after the application to the primary of an ideal voltage source $v_s = 10^3 t$ in series with a 1 kΩ resistor? What is it if the 4 kΩ and 1 kΩ resistors have previously been replaced by uncharged capacitors of 4 μF and 1 μF respectively?

2.7. A given source with an internal resistance R_s is to be connected to a given load R_L via an ideal transformer. Show that the transformer turns ratio required to give the maximum voltage across R_L is the same as that required to transfer maximum power from the source to the load and is given by $n = \sqrt{(R_L/R_s)}$.

2.8 An ideal three-terminal transformer with a turns ratio of 10 has input terminals AC and output terminals BC such that the output voltage V_{BC} is opposite in polarity to the input voltage V_{AC}. What is

(*a*) the input resistance of the circuit when ideal 1 kΩ resistors are connected across AB and BC?

(*b*) the input inductance of the circuit when ideal 1 H inductors are connected across AB and BC which have
  (i) no magnetic coupling between them?
  (ii) a mutual inductance of 0.5 H which is positive relative to currents flowing in the inductors from A to B and from C to B?

2.9 Two ideal capacitors C_1 and C_2 and an open switch are connected in series. The capacitors have initial voltages of V_1 and zero respectively. Observe that the charge lost by one capacitor on closing the switch is equal to the charge gained by the other, and thereby find the voltage across the capacitors when the switch is closed.

2.10 Two ideal inductors L_1 and L_2 and a closed switch are connected in parallel. There is mutual inductance M between the inductors of such a polarity that a current flowing round the loop comprising the two inductors would produce aiding fluxes in both. The inductors have initial currents flowing in them of I_1 and zero respectively. Observe that the decrease in flux linkages in one inductor on opening the switch is equal to the increase in flux linkages in the other, and thereby find the current through the inductors when the switch is open.

3 Alternating current circuits and the use of phasors

3.1 Definitions

An *alternating quantity* is a quantity which passes repeatedly through the same cycle of values, with a constant time-interval known as the *period* between each repetition. Fig. 3.1 shows three examples, namely a sine wave, a square wave and a saw-tooth wave; the use of the term 'wave' in this

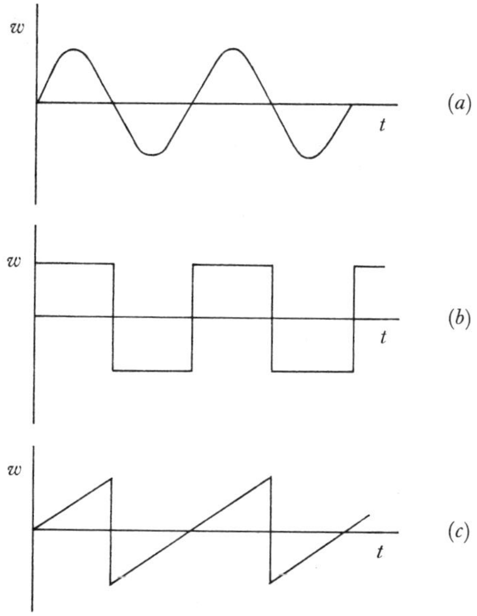

Fig. 3.1 Alternating waveforms.

connexion is natural, and the three descriptions are said to describe the respective *waveforms*. Fig. 3.2 shows another wave falling within the definition; unlike the quantities illustrated in fig. 3.1, this quantity has a mean value differing from zero and represented by the horizontal dotted line; it can indeed be obtained by adding a constant to the quantity illustrated in fig. 3.1(*a*). The waveform of fig. 3.2 is thus the sum of a *d.c. component* and a *pure a.c. component*. In the linear circuits which are the concern of this book, each component in a stimulus produces its own component response, the total response being the sum of the components. There is thus no need to consider a.c. quantities containing a d.c. component – the effect of the latter, if present, can be calculated separately; and the theory of a.c. circuits is developed entirely in terms of quantities for which the time-mean calculated over a cycle is zero – or, as we say, in which *the positive and negative half-cycles are equal in area.*

The number of repetitions of the cycle in unit time is known as the *frequency*, f, measured in hertz (Hz). Since the interval between corresponding points in successive cycles is the period, T,

$$fT = 1. \tag{3.1.1}$$

In this chapter we are solely concerned with circuits in which the frequency is constant. We shall furthermore be confined to the *quasi-steady state*; that

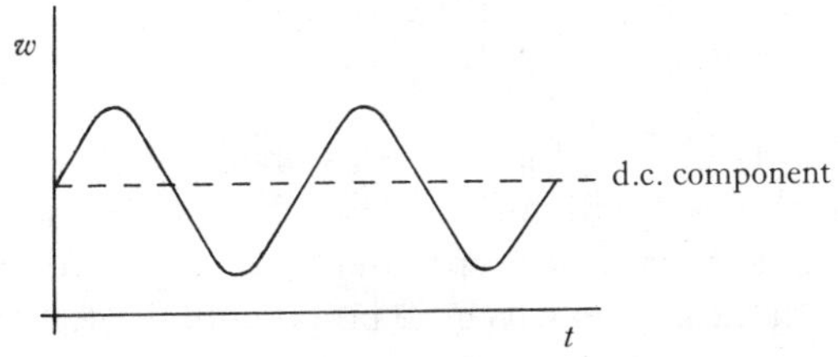

Fig. 3.2 Alternating waveform with d.c. component.

is, although the currents and voltages are varying in each cycle, successive cycles are identical. In practice this amounts to assuming the lapse of a sufficient time since the last switching operation; the transient phenomena which occur when any circuit, including an a.c. circuit, is switched to a new state, form the subject-matter of chapter 5.

Alternating current circuits occupy a central place in circuit theory, as do vibrations (in the broadest sense) in almost every field of physics. One is tempted to believe that Nature loves a sine wave as much as she abhors a vacuum. The reasons for the use of alternating currents in (say) a power supply system and a radio aerial are entirely different, yet – in each case – compelling; thus a.c. circuit theory is of equal importance to power and communication engineers, the only fundamental difference being that the power engineer can usually regard the frequency as constant, whereas the communication engineer cannot.

3.2 The root-mean-square value of an alternating quantity

The complete description of an alternating quantity must evidently include both the shape and size of the wave. The latter may be specified in a number of ways in accordance with the purpose envisaged. For example, if the breakdown of insulation by an alternating voltage is in question, it is the *maximum* or *peak* value of the voltage that is significant; if the maximum flux density in a transformer is required, a measurement of the *mean* voltage induced in a winding (averaged over a half cycle) is made. The most generally useful value, however, is the *effective* or *root-mean-square* (r.m.s.) value. This is chosen because it simplifies the formulae relating to power; in particular, if a current having the r.m.s. value I passes through a resistor R, the power loss is RI^2.

To find a formula for I in terms of $i(t)$, the instantaneous current expressed as a function of time, the energy loss in one cycle must be calculated; this is $\int_0^T R[i(t)]^2\,dt$. But if I is to satisfy the condition stated, the loss is RI^2T. Equating these, we obtain

$$I^2 = \frac{1}{T}\int_0^T [i(t)]^2\,dt, \tag{3.2.1}$$

and a similar formula obtains for the r.m.s. value of any other quantity. In this book, capital letters without a suffix will be used to denote the r.m.s. value of the a.c. quantities whose instantaneous values are given by the corresponding lower case letters. Where no possibility of confusion arises, the same notation may sometimes denote the maximum value attained during the cycle; but when both r.m.s. and maximum values occur in the same problem, the latter are distinguished by a suffix m (V_m, I_m).

If $w(t)$ is an alternating quantity in which the positive and negative half-cycles are similar, and if $t = 0$ at the instant when $w(t)$ is passing through zero in a positive direction, the mean value of $w(t)$ over the half-cycle is given by

$$\overline{W} = \frac{2}{T}\int_0^{T/2} w(t)\,dt. \tag{3.2.2}$$

For the same quantity, the r.m.s. value is given by W, where

$$W^2 = \frac{2}{T}\int_0^{T/2} [w(t)]^2\,dt, \tag{3.2.3}$$

since the contributions from the time $t = T/2$ to T are the same as from $t = 0$ to $T/2$. In these alternative methods of averaging, the second takes greater account of the high parts of the wave; thus, the more 'peaky' a wave is, the greater is the ratio $W/\overline{W}$ (known as the *form factor*). This may be illustrated by reference to fig. 3.1. The square wave has a form factor of unity; the sine wave, of 1.111; and the saw-tooth wave, of 1.155. The association of increasing form factor with increasing peakiness is clear.

3·3 Sinusoidal alternating quantities

Among all possible alternating waveforms, a sine wave occupies a unique place; so much so that, as will be seen, it is used as the element from which all other forms are built up. This uniqueness stems from the fact that, alone among alternating functions, a sine function is essentially unchanged by differentiation or integration. Thus, if $y = \sin\theta$, $dy/d\theta = \cos\theta = \sin(\theta + \pi/2)$; differentiation leads to the same function, but with the independent variable increased by $\pi/2$. For a sinusoidal function of time with period T, we may write

$$w(t) = W_m \sin\frac{2\pi t}{T} = W_m \sin 2\pi f t,$$

or

$$w(t) = W_m \sin\omega t, \text{ where } \omega = 2\pi f. \tag{3·3·1}$$

In this expression, ωt is called an *electrical angle*, and ω is the *angular frequency* (measured in electrical radians per unit time). Differentiation with respect to t now gives

$$\begin{aligned} w'(t) &= W_m \omega \cos\omega t, \\ &= W_m \omega \sin(\omega t + \pi/2). \end{aligned} \tag{3·3·2}$$

Differentiating this sinusoidal function of t therefore multiplies the maximum by ω, and increases the electrical angle by $\pi/2$; and the same is true when $w(t)$ is any similar function such as $W_m \sin(\omega t + \alpha)$ or $W_m \cos(\omega t + \alpha)$.

At any instant t_1, $\sin(\omega t + \pi/2)$ takes the value $\sin(\omega t_1 + \pi/2)$; but $\sin\omega t$ does not reach this value until $t = t_1 + \pi/2\omega$ – a later instant. $\sin\omega t$ is therefore said to *lag* $\sin(\omega t + \pi/2)$, and $\sin(\omega t + \pi/2)$ is said to *lead* $\sin\omega t$, by an electrical angle of $\pi/2$ or 90 electrical degrees. The relationship is shown in fig. 3·3. The electrical angle which separates corresponding points on two sine waves is called the *phase difference* between them. If the phase difference between two sinusoidally varying quantities is zero they are said

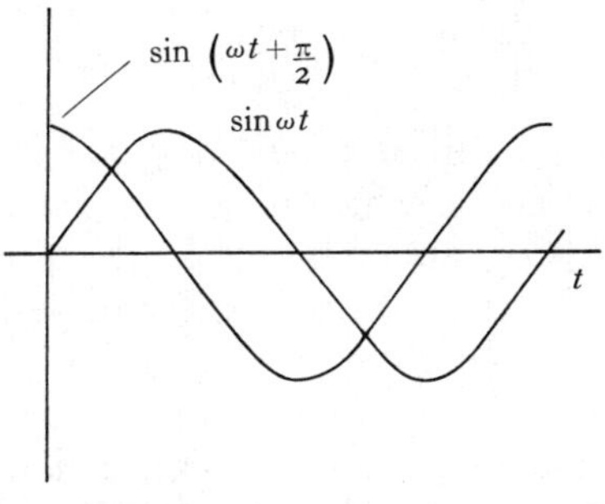

Fig. 3·3 Sinusoidal waveforms in quadrature.

to be *in phase*; if the phase difference is $\pi/2$ they are *in quadrature.* Often one of the quantities in a problem is treated as a starting-point for the description of the others; thus, if a network is energized from a single alternating voltage source, it may be convenient to denote its output by $V_m \sin \omega t$. The currents in various parts of the network then take the forms $I_{m1} \sin(\omega t + \alpha_1)$, $I_{m2} \sin(\omega t + \alpha_2)$, etc. Plainly each is fully described by I_{m1}, I_{m2}, ... (the magnitude, which is here the peak value of the wave) and α_1, α_2, ... (the phase difference from the source voltage); but as the source voltage is treated as a starting-point, this phase difference is simply described as the *phase* of the current in question.

It has already been pointed out that the magnitude of an alternating quantity is often conveniently specified by its r.m.s. value. In the case of a sinusoidal quantity $W_m \sin \omega t$, the r.m.s. value W is given by

$$W^2 = \frac{\omega}{2\pi} \int_0^{2\pi/\omega} W_m^2 \sin^2 \omega t \, dt,$$

$$= \frac{\omega W_m^2}{4\pi} \int_0^{2\pi/\omega} (1 - \cos 2\omega t) \, dt,$$

$$= \frac{W_m^2}{2}.$$

Thus
$$W = W_m/\sqrt{2}. \tag{3.3.3}$$

The mean value of $W_m \sin \omega t$ over a half-cycle is given by

$$\overline{W} = \frac{\omega}{\pi} \int_0^{\pi/\omega} W_m \sin \omega t \, dt,$$

$$= \frac{2}{\pi} W_m. \tag{3.3.4}$$

Hence the form factor:

$$\frac{W}{\overline{W}} = \frac{\pi}{2\sqrt{2}} = 1.111, \tag{3.3.5}$$

as has already been stated. Since W, $\overline{W}$, W_m are in fixed ratios for a sinusoidal wave, it is immaterial which is used to describe its magnitude; for example, we can speak of 'the ratio of voltage to current' in a circuit element without specifying whether r.m.s., mean or peak values are meant, provided the same measure is used for both quantities.

The theory of linear a.c. circuits is built on the assumption of sinusoidal variation of all voltages and currents. This assumption will be made henceforth, until the question of non-sinusoidal variation is reopened in chapter 9.

3.4 The geometrical representation of sinusoidally varying quantities

Consider a two-dimensional cartesian co-ordinate system in which a line OP, of fixed length r, rotates with angular velocity ω about its fixed end O (fig. 3.4(a)). If $t = 0$ when OP lies on Ox, the angle between OP and Ox at any other time is ωt, and the projection of OP on the x-axis is

$$\mathrm{OP}' = r \cos \omega t.$$

OP′ is thus an alternating quantity. Furthermore if OQ is a second line, of length R, fixed at an angle α to OP and therefore rotating with it,

$$\mathrm{OQ}' = R \cos(\omega t + \alpha).$$

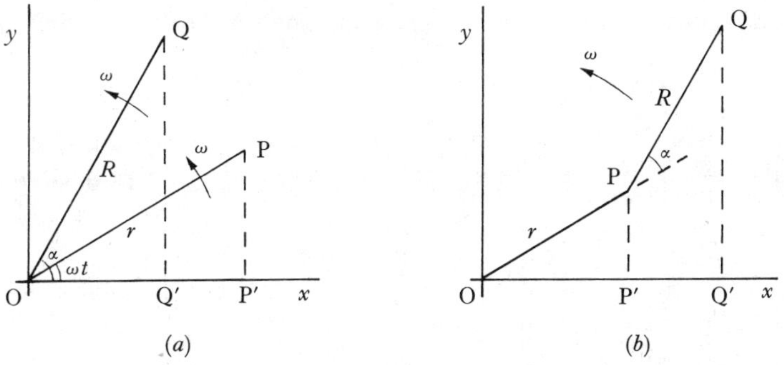

Fig. 3.4 Representation of alternating quantities by projection.

This is a second alternating quantity which leads OP′ (according to the definition in the last section) by an angle α. The geometrical implications of the words 'lead' and 'lag' are evident; if OP, OQ were two ranks of soldiers performing a left wheel, OQ would be the leading rank.

It is now apparent that any number of quantities which vary sinusoidally at the same frequency can be represented as the projections of rotating lines in the manner described. These lines need not be radii from the origin: in fig. 3.4(b), PQ is a line of length R fixed at an angle α to OP, and the projection P′Q′ is still equal to $R\cos(\omega t + \alpha)$. Had projections on the y-axis been used, the cosines would have been replaced by sines.

Imagine now a box with two terminals, having the property that, when an alternating current $I_m \cos \omega t$ flows through it, the voltage across the terminals is $Z_1 I_m \cos(\omega t + \phi_1)$. In words, the voltage has a magnitude Z_1 times that of the current, and leads the current by an angle ϕ_1; so that these two numbers Z_1, ϕ_1 characterize the properties of the box. Let this box be connected

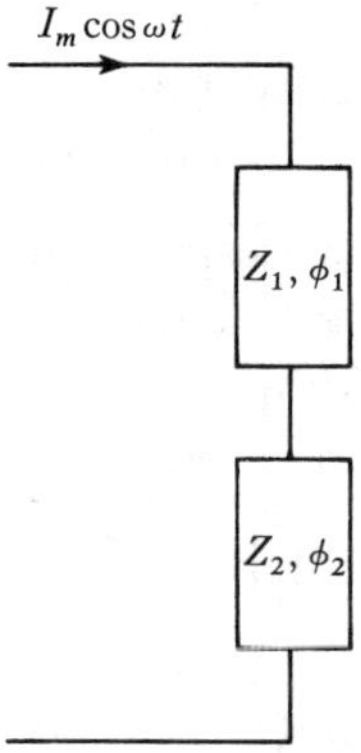

Fig. 3.5 A.C. components in series.

in series with a second, with characteristics Z_2, ϕ_2; and let the current $I_m \cos \omega t$ be passed through both (fig. 3.5). The total voltage is then

$$v(t) = Z_1 I_m \cos(\omega t + \phi_1) + Z_2 I_m \cos(\omega t + \phi_2). \qquad (3.4.1)$$

Fig. 3.6(*a*) is a diagram analogous to fig. 3.4(*b*), representing (3.4.1) at a particular instant of time. The current is represented by the projection of the

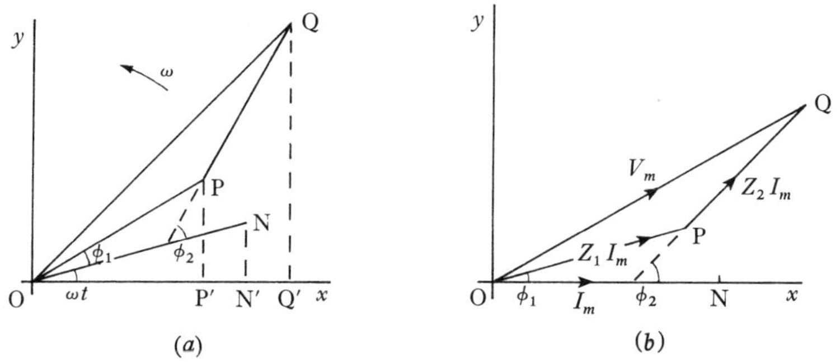

Fig. 3.6 Phasor diagrams for the circuit of fig. 3.5: (*a*) rotating diagram (ON $\propto I_m$, OP $\propto Z_1 I_m$, PQ $\propto Z_2 I_m$, OQ $\propto V_m$; ON′ $\propto I_m \cos \omega t$, etc.); (*b*) stationary diagram (phasors labelled with their magnitudes).

rotating line ON; the voltages across the two boxes by the projections of OP and PQ, which have lengths representing the magnitudes $Z_1 I_m$ and $Z_2 I_m$, and make angles ϕ_1, ϕ_2 with ON. Appropriate voltage and current scales must of course be defined. Since the sum of the projections OP′ and P′Q′ is OQ′, the projection of OQ, it is apparent that OQ represents the voltage $v(t)$ defined in (3.4.1); the length of OQ represents V_m, the maximum value of $v(t)$, to scale, and the angle NOQ is the electrical angle by which

$v(t)$ leads the current. Rotating lines used in this way to represent alternating quantities are called *phasors*.

In this diagram the lines are of constant lengths and at constant angles to each other, thus forming a pattern which rotates in one piece and does not vary. The question then arises, whether the repeated references to 'rotation' and 'projection' are really essential; the diagram is clearer if drawn as in fig. 3.6(*b*), with the line ON, which represents the current through the two boxes, laid along the x-axis. Since PQ and QP would represent alternating voltages having a phase difference of 180 degrees, arrow-heads are drawn on the lines to show which sense is intended. If it is desired to revert to the description of the quantities as functions of t, the rotation of the whole diagram and the projection of the lines in a fixed direction is easily visualized. But all the information is contained in the stationary diagram, and the rotating diagram has served its purpose in enabling us to prove that the sum of the quantities represented by OP and PQ is a quantity represented by OQ – in short, that *phasors are added vectorially*.

The lines in the stationary diagram are called *stationary phasors*, though the adjective is hardly ever necessary; the diagram itself is a *phasor diagram*. (The terms 'vector' and 'vector diagram' were formerly used, but these can lead to confusion in certain contexts outside the realm of circuit theory and are to be avoided.) The phasor ON which formed the starting point of the diagram is known as the *reference phasor*; from this the phase of all the other alternating quantities is measured.

The phasor diagram is the first example of a method which will be much used in this book – the replacement of a circuit problem by an *image problem* which contains the data of the circuit problem in a different form and may be amenable to a simpler solution. The phasor diagram is an image of a geometrical character. The circuit problem can be expressed as a set of linear equations (the Kirchhoff equations), relating currents and voltages all of which vary sinusoidally with time at the same frequency. In the image problem, each voltage and current is represented by a line of appropriate length and angular orientation, while the equations which relate them are represented by the positions of these lines – for example, the series connexion of the two elements in fig. 3.6 is represented by drawing the phasors OP and PQ end-to-end. Algebraic relations between time-varying quantities have thus been transformed into spatial relations between geometrical lines having no time-variation.

Phasors will be labelled in bold italic type; thus, the alternating quantity whose instantaneous value is $w(t)$, where

$$w(t) = W_m \cos(\omega t + \alpha) = W\sqrt{2}\cos(\omega t + \alpha),$$

is represented by a phasor labelled $\boldsymbol{W}$. It is often convenient so to scale the phasor diagram that the length of $\boldsymbol{W}$ gives a direct measure of the r.m.s.

value W. An equation such as $\boldsymbol{V} = R\boldsymbol{I}$ will carry the same meaning as in vector theory; the magnitudes of the phasors are related by $V = RI$, *and also* the directions of the phasors $\boldsymbol{V}$ and $\boldsymbol{I}$ are the same.

Before giving examples of the use of the phasor diagram, we must discuss the relations between the voltage and current phasors in the various ideal circuit elements.

3.5 Phasor diagrams of simple combinations of ideal circuit elements

The output of each source of alternating voltage or current is represented by a phasor; where there is only one such source, it may be convenient to take its output as the reference phasor. Frequently a system contains several sources whose outputs are in a fixed phase relationship.

The voltage and current in a resistor are represented by phasors whose directions are parallel and whose magnitudes are proportional ($\boldsymbol{V} = R\boldsymbol{I}$). The same is true for other elements in which there is direct proportionality between two quantities; for example, in a voltage-controlled voltage source,

$$\boldsymbol{V}_0 = \mu \boldsymbol{V}_c, \qquad \text{(see §1.6)}$$

and in an ideal transformer,

$$\boldsymbol{V}_2 = n\boldsymbol{V}_1, \qquad \boldsymbol{I}_2 = -\frac{1}{n}\boldsymbol{I}_1. \qquad \text{(cf. (2.6.8))}$$

There is little point in drawing diagrams to illustrate these simple relations.

In a capacitor the relation between the instantaneous values of voltage and current is

$$i = C\,\mathrm{d}v/\mathrm{d}t. \qquad \text{(equation (2.3.2))}$$

If $v = V_m \cos \omega t$,

$$i = -C\omega V_m \sin \omega t = C\omega V_m \cos(\omega t + \pi/2).$$

Thus the current leads the voltage by 90 degrees, and its magnitude is given by $I_m = C\omega V_m$, or $I = C\omega V$. In an inductor, for which

$$v = L\,\mathrm{d}i/\mathrm{d}t, \qquad \text{(equation (2.4.2))}$$

the roles of voltage and current are exchanged; the voltage now leads the current by 90 degrees, and its magnitude is given by $V_m = L\omega I_m$, or $V = L\omega I$. A corresponding relation obtains for mutual inductance. The phasor diagrams for a capacitor and an inductor are given in fig. 3.7; to emphasize the difference between them, the voltage is taken as reference phasor in each case.

As elements in a.c. circuits, capacitors and inductors resemble resistors in that the voltage and current are proportional in magnitude, but differ in that the voltage and current are in quadrature instead of being in phase, We are thus led to define, as an analogue of resistance, a quantity called *reactance* which is the ratio of the magnitudes of voltage and current in quadrature. This may be further particularized as *capacitive reactance* in which the current leads the voltage by 90 degrees, or *inductive reactance* in which the current lags the voltage by 90 degrees. A mutual inductor exhibits

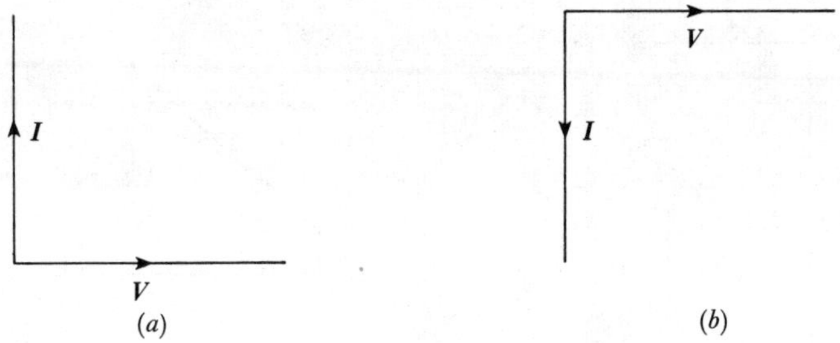

Fig. 3.7 (*a*) Phasor diagram for a capacitor ($I = C\omega V$). (*b*) Phasor diagram for an inductor ($V = L\omega I$).

mutual reactance in an a.c. circuit. The reciprocal of reactance, namely the ratio of the magnitudes of current and voltage when in quadrature, is called *susceptance*; this again may be capacitive or inductive. For a capacitor,

$$\text{reactance } X = 1/C\omega; \qquad \text{susceptance } B = C\omega; \tag{3.5.1}$$

while for an inductor,

$$\text{reactance } X = L\omega; \qquad \text{susceptance } B = 1/L\omega. \tag{3.5.2}$$

Figure 3.8(*a*) and (*c*) show simple combinations of resistance and inductance in series and in parallel respectively, supplied from a voltage source. The total voltage and current in each circuit are represented by phasors $\boldsymbol{V}$, $\boldsymbol{I}$, and the circuit diagrams are lettered with these names rather than with the corresponding functions of t. Fig. 3.8(*b*) and (*d*) show the phasor diagrams; here the phasors are labelled with their magnitudes only, the phases of the several quantities being denoted by the directions of the phasors. In fig. 3.8(*b*) the common current $\boldsymbol{I}$ is taken as the reference phasor, and the magnitude and phase of the voltage are given by

$$V^2 = (R^2 + L^2\omega^2)I^2, \tag{3.5.3}$$

and

$$\tan\phi = L\omega/R. \tag{3.5.4}$$

In fig. 3.8(*d*) the common voltage $\boldsymbol{V}$ is taken as the reference phasor, and the magnitude and phase of the current are given by

$$I^2 = \left(\frac{1}{R'^2} + \frac{1}{L'^2\omega^2}\right)V^2, \tag{3.5.5}$$

and

$$\tan\phi = R'/L'\omega. \tag{3.5.6}$$

In both circuits the phasor $\boldsymbol{V}$ leads the phasor $\boldsymbol{I}$ by an angle between 0 and 90 degrees, and it is possible so to choose R', L' that the ratio V/I and the phase difference ϕ is the same in the parallel circuit as in the series

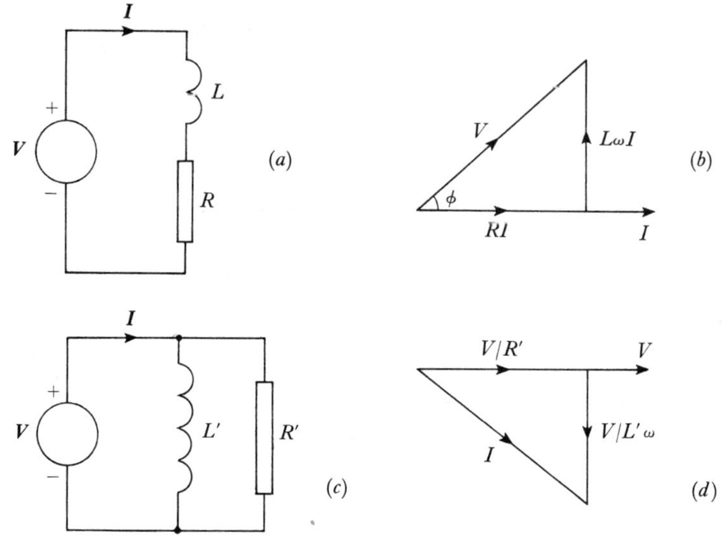

Fig. 3.8 Simple *RL* circuits and their phasor diagrams (phasors labelled with their magnitudes).

circuit. The characteristics of each resistance–inductance combination are then specified by the following pair of statements:

$$\left.\begin{aligned} &V = ZI, \\ &\boldsymbol{V} \text{ leads } \boldsymbol{I} \text{ by an angle } \phi. \end{aligned}\right\} \tag{3.5.7}$$

Z, the ratio of voltage to current without regard to phase difference, is called the *impedance* of the resistance–inductance combination. An alternative specification which interchanges the roles of current and voltage is

$$\left.\begin{aligned} &I = YV, \\ &\boldsymbol{I} \text{ leads } \boldsymbol{V} \text{ by an angle } -\phi. \end{aligned}\right\} \tag{3.5.8}$$

Y is then called the *admittance* of the combination.

It must be pointed out, however, that the terms 'impedance' and 'admittance' have become ambivalent, in that many writers regard them as giving information about the phase difference as well as the ratio. From this standpoint, impedance and admittance require for their specification a pair of numbers, which may be written thus: $Z\angle\phi$. The convention is adopted that ϕ is positive in an impedance in which the voltage leads the current and in an admittance in which the current leads the voltage, so that if the impedance of a network is $Z\angle\phi$, its admittance is $Z^{-1}\angle -\phi$. This wider meaning will reappear in the next chapter; for the present, 'impedance' and 'admittance' will denote the numerical ratios Z and Y.

From (3.5.3), for the combination of R and L in series,

$$Z^2 = R^2 + X^2, \tag{3.5.9}$$

X here standing for $L\omega$, the reactance of the inductor. For the parallel combination, (3.5.5) gives

$$Y^2 = G^2 + B^2, \tag{3.5.10}$$

where $G = 1/R'$, $B = 1/L'\omega$, the conductance of the resistor and the susceptance of the inductor respectively. It will generally be found advantageous to work in terms of impedance for series combinations of elements, admittance for combinations in parallel. If G, B in parallel are to be equivalent to R, X in series, the condition $Y = Z^{-1}$ gives

$$(R^2 + X^2)(G^2 + B^2) = 1 \tag{3.5.11}$$

while equality of the phase angles, as given by (3.5.4) and (3.5.6), requires the second condition

$$X/R = B/G \quad (= \tan\phi). \tag{3.5.12}$$

From these we readily obtain $RG = \cos^2\phi$, $XB = \sin^2\phi$; or

$$\left.\begin{aligned} G &= \frac{1}{R}\cos^2\phi = \frac{R}{R^2+X^2}, \\ B &= \frac{1}{X}\sin^2\phi = \frac{X}{R^2+X^2}. \end{aligned}\right\} \tag{3.5.13}$$

Since X and B contain the angular frequency ω, the equivalence described by the last equations is frequency-dependent. If a box with two terminals were found to have the characteristic $Z\angle\phi$ at a particular frequency, this would not determine whether it contained a series or parallel R–L combination; but, assuming it were known that the box contained one or the other, a measurement made at another frequency would determine the issue.

Series and parallel combinations of resistance and capacitance are dealt with in the same way as combinations of resistance and inductance, and lead to very similar results. Equations (3.5.9) to (3.5.13) are still valid, with X the reactance and B the susceptance of the capacitor as given by (3.5.1). The only difference is that ϕ is now the angle by which the current leads the voltage.

3.6 Power in a.c. circuits

The phase difference ϕ between $\boldsymbol{V}$ and $\boldsymbol{I}$ has a close association with the input of power from the voltage source to the network. In a resistor, which absorbs power, $\phi = 0$; in an inductor or capacitor, where energy is stored and unstored during each cycle but no net power is consumed, $\phi =$ 90 degrees. If any network is being supplied with current $I\sqrt{2}\cos\omega t$ from a generator having a terminal voltage of $V\sqrt{2}\cos(\omega t + \phi)$, the instantaneous power input is

$$2VI\cos\omega t\cos(\omega t + \phi) = VI[\cos\phi + \cos(2\omega t + \phi)].$$

The second term oscillates, but its mean value is zero. Thus

$$\text{Mean power input} = VI\cos\phi. \tag{3.6.1}$$

On the phasor diagram $VI\cos\phi$ is the length of one of the phasors $\boldsymbol{V}$, $\boldsymbol{I}$ multiplied by the projection of the other upon it. Frequently the voltage is taken as the reference phasor, and $I\cos\phi$ is then spoken of as the *inphase component* of the current. Cosϕ is called the *power factor* of the network. In power supply problems and other cases in which the transference or consumption of power is of prime importance, the symbol ϕ is reserved for the phase difference betweeen current and voltage; 'cosϕ' is then immediately understood to be a power factor and no special symbol is required. Cosϕ is positive for all angles ϕ between -90 and $+90$ degrees.

The product of the r.m.s. voltage and current, VI, is not a power and must not be measured in watts. It has its own significance, however; for example, a power transmission line is insulated for a given voltage and has conductors capable of carrying a given current, so VI measures its rating. The product of r.m.s. voltage and current without regard to phase is called *volt-amperes*, and the same name (abbreviated VA) is employed for the unit in which the product is measured. The power $VI\cos\phi$ is then one component of the volt-amperes; the other component, $VI\sin\phi$, is called the *reactive volt-amperes*. The unit of $VI\sin\phi$ is written VAr, and 'vars' is often used as a nickname for reactive volt-amperes.

3.7 The solution of circuit problems by phasor diagrams

For simple a.c. networks, phasor diagrams afford a rapid and often elegant mode of solution which is used less often than it deserves. Students tend to prefer the analytical method given in the next chapter, and will often get the wrong answer by making arithmetical errors when the simple use of a ruler and protractor would have given a close approximation to the right answer. Three examples will bring out the principal points of the method.

1. Certain appliances, notably induction motors, operate with a lagging power factor; and it is common practice to correct this by the parallel connexion of capacitors, in which the power factor is zero leading (that is, the current leads the voltage by 90 degrees). In this way the current drawn from the supply is reduced. As an example, we will calculate the capacitance required to bring the power factor of the supply current to 0.9 lagging, when a load of 60 kW at a power factor of 0.75 lagging is connected across a supply of 240 V, 50 Hz.

The circuit is shown in fig. 3.9(*a*) in which the load is represented by the conventional impedance symbol, namely a rectangle labelled *Z*. The phasor diagram is shown in fig. 3.9(*b*) with the supply voltage as the

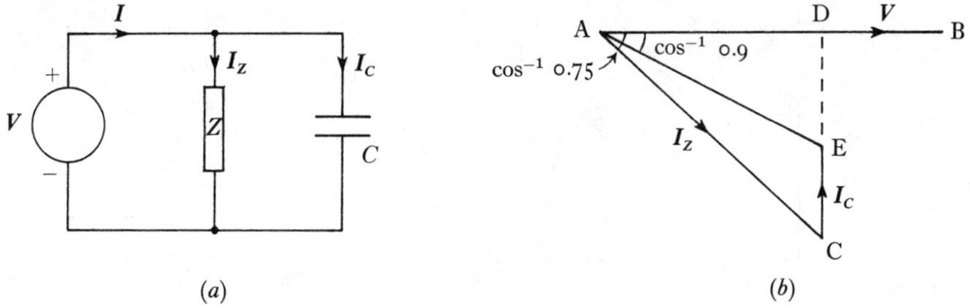

Fig. 3.9 Power factor correction by parallel connexion of capacitance: (*a*) circuit; (*b*) phasor diagram.

reference phasor AB. If $\cos\phi = 0.75$, $\phi = 41.4$ degrees; AC is therefore drawn at this angle to AB to give the direction of the current phasor, of such a length that its projection AD on AB represents the inphase current, which is 60 kW ÷ 240 V, or 250 A. CD, leading AB by 90 degrees, gives the direction of the capacitor current phasor $\boldsymbol{I}_c$; and if AE is drawn so that $\angle$ BAE $= \cos^{-1}(0.9)$ or 25.8 degrees, CE gives the magnitude of $\boldsymbol{I}_c$. It is found that $I_c = 99.5$ A; thus

$$C\omega V = 99.5,$$

where $V = 240$ and $\omega = 314$, from which

$$C = 1320\ \mu\text{F}.$$

The load current is 333 A, but the total current has been reduced to 278 A. By using a capacitor of 2920 μF the power factor could be corrected to unity and the current reduced to 250 A; but this comparatively small further reduction is won at a disproportionately great cost.

2. The principle of the Wheatstone bridge has been greatly developed for a.c. measurements, and the circuit of fig. 3.10(*a*) is the Hay bridge used for measuring the characteristics of an impure inductor represented by *L* in

series with R. The detector may take various forms, but all enable adjustments to be made until, when the bridge is in the 'balanced' condition, there is no current through the detector and no voltage across it. The balanced bridge is thus a comparatively simple problem with only two meshes and two currents $\boldsymbol{I}_1$, $\boldsymbol{I}_2$, and the conditions of balance are the conditions that there shall be no voltage between the points P and Q.

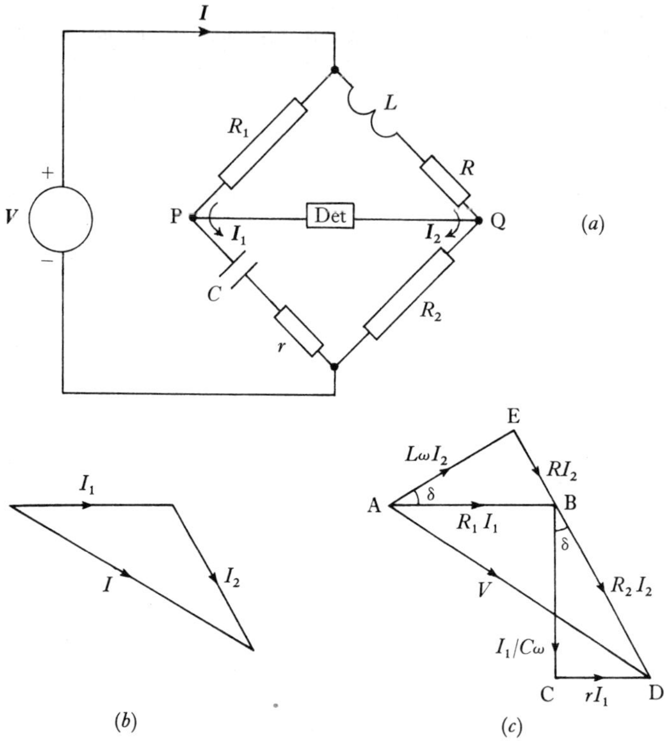

Fig. 3.10 Hay bridge for measurement of inductance combined with resistance: (*a*) circuit; (*b*) current phasor diagram; (*c*) voltage phasor diagram.

Figure 3.10(*b*) and (*c*) shows the phasor diagram with the voltage and current diagrams separated. It is developed by starting with $\boldsymbol{I}_1$ as reference phasor and drawing the phasors AB, BC, CD for the voltages across R_1, C, and r. Since the sum of the phasors BC and CD must equal the voltage across R_2, BD gives the direction of $\boldsymbol{I}_2$ and has the length $R_2 I_2$. If DB is produced to E and AE is drawn at right angles to it, AE and EB are voltages in quadrature and in phase with $\boldsymbol{I}_2$, whose sum is the voltage across R_1. Thus AE and EB represent the voltage across L and R respectively; it is to be noted that AE *leads* $\boldsymbol{I}_2$ by 90 degrees, the correct relationship for a pure inductor.

The total current is represented by $\boldsymbol{I}$, the sum of the phasors $\boldsymbol{I}_1$ and $\boldsymbol{I}_2$; and the total voltage is represented by the line AD, which is seen to be the phasor sum of the voltages across the elements in each branch.

The triangles AEB, BCD are similar, because ∠ CBD is the complement of ∠ ABE and is therefore equal to ∠ BAE. Writing δ for the two angles last named, we see that, from the two triangles,

$$\tan\delta = rC\omega = R/L\omega. \tag{3.7.1}$$

Also $I_1/C\omega = R_2 I_2 \cos\delta$, $L\omega I_2 = R_1 I_1 \cos\delta$; so

$$I_1/I_2 = R_2 C\omega \cos\delta = L\omega/R_1 \cos\delta.$$

Thus

$$L = R_1 R_2 C \cos^2\delta. \tag{3.7.2}$$

It is possible to eliminate δ between (3.7.1) and (3.7.2), but more convenient not to do so. The first part of (3.7.1) gives δ in terms of r and C, assuming that the value of ω is known; (3.7.2) then gives the unknown L in terms of known quantities, and the second part of (3.7.1) similarly gives R. If R is regarded as an undesirable impurity in the inductor, δ may be used as a measure of this impurity.

The need for two balance conditions to be satisfied in an a.c. bridge is of course associated with the fact that an a.c. quantity requires two numbers for its definition, namely the magnitude and the phase.

3. The Schering bridge, illustrated in fig. 3.11(*a*), is designed for the measurement of the characteristics of an impure capacitor, here denoted by pure elements C and R in parallel. The balance conditions will be determined.

As in the previous example, the current and voltage in the detector are zero; but the current phasor $\boldsymbol{I}_1$ for the left hand branch must be resolved into $\boldsymbol{I}_1'$ for the current through R_1 and $\boldsymbol{I}_1''$ for that through C_1. Since the voltages across R_1 and C_1 are identical, $\boldsymbol{I}_1'$ and $\boldsymbol{I}_1''$ must be in quadrature, with $R_1 I_1' = I_1''/C_1\omega$. The voltage across R_2 is the same; thus $\boldsymbol{I}_2$ is in phase with $\boldsymbol{I}_1'$. $\boldsymbol{I}_2$ is resolved into $\boldsymbol{I}_2'$ for the current through R and $\boldsymbol{I}_2''$ for the current through C; $\boldsymbol{I}_2''$ must be in phase with $\boldsymbol{I}_1$ in order that the voltages across C and C_0 may be in phase (as well as equal in magnitude). These relationships between the currents are shown in the current phasor diagram, fig. 3.11(*b*), which is so drawn that quantities in phase with each other are denoted by parallel rather than by coincident lines. The same procedure is followed in the voltage phasor diagram, fig. 3.11(*c*); and both diagrams are completed by phasors $\boldsymbol{I}$, $\boldsymbol{V}$, which give the total current and voltage.

If the angles ACB, CED are called δ, we have

$$\begin{aligned} \tan\delta &= I_1''/I_1', \\ &= R_1 C_1 \omega, \end{aligned} \tag{3.7.3}$$

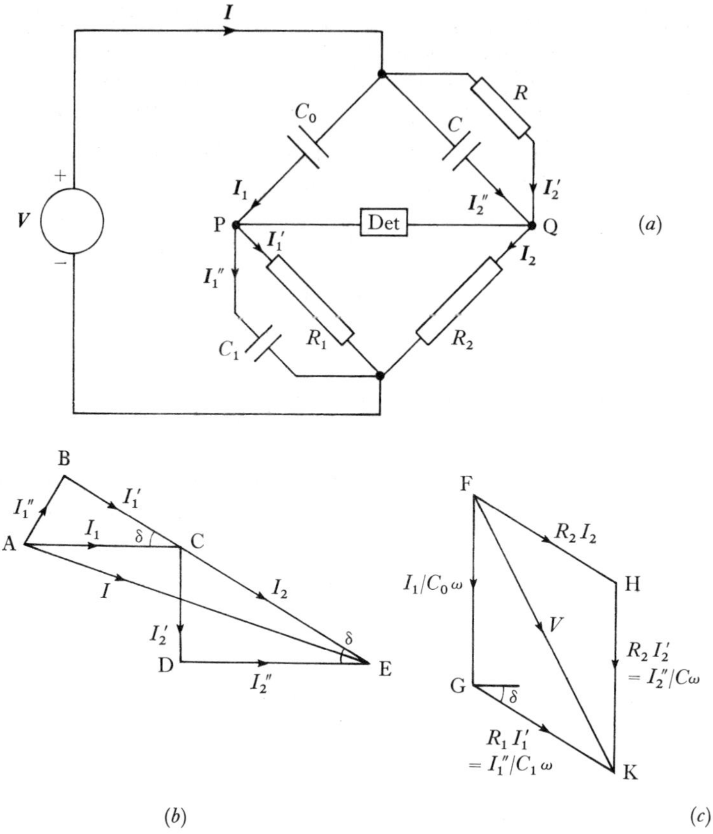

Fig. 3.11 Schering bridge for measurement of capacitance combined with resistance: (*a*) circuit; (*b*) current phasor diagram; (*c*) voltage phasor diagram.

since R_1 and C_1 are in parallel. Furthermore from the voltage diagram, FH = GK, or

$$R_2 I_2 = R_1 I_1' = R_1 I_1 \cos\delta. \tag{3.7.4}$$

Moreover FG = HK, or

$$I_1/C_0\omega = I_2''/C\omega = I_2 \cos\delta/C\omega. \tag{3.7.5}$$

Eliminating I_1/I_2 between these equations, we get

$$\frac{R_2}{R_1\cos\delta} = \frac{C_0\cos\delta}{C},$$

or

$$C = \frac{C_0 R_1}{R_2}\cos^2\delta. \tag{3.7.6}$$

Also, from the triangle CDE,

$$\tan\delta = I_2'/I_2'' = 1/RC\omega. \qquad (3.7.7)$$

Equations (3.7.6) and (3.7.7) give the unknown quantities C, R in terms of the other components C_0, C_1, R_1, R_2, together with ω; for $\tan\delta$ has already been obtained in terms of these in (3.7.3).

If the phasor diagram is drawn to scale for a set of numerically given quantities, a difficulty may well arise on account of the great disparity between the various phasors; in the example just given, for example, the ratio of the lengths FG and GK could exceed 1000. In such cases, parts of the diagram may have to be scaled up to enable lengths and angles to be accurately measured. In particular, in circuits containing transformers, the voltage and current scales on the primary and secondary sides are usually adjusted to give an apparent ratio of 1:1.

3.8 Geometrical properties of the phasor diagram

It has become apparent in the foregoing examples that each closed figure in the current phasor diagram stands for a group of currents whose sum (taken with appropriate signs) is zero. The closing of the figure is therefore the expression of a Kirchhoff node equation. For instance, in fig. 3.11(*b*) the triangle ABC stands for the phasor equation

$$\boldsymbol{I}_1' + \boldsymbol{I}_1'' = \boldsymbol{I}_1$$

and represents the summation of currents at the point P.

Each closed figure in the voltage diagram similarly represents a Kirchhoff loop equation. This diagram, however, has a further property of great interest. If several circuit elements are in series they are shown as connected end to end in the circuit diagram; and their phasors are drawn end to end in the voltage phasor diagram. Thus, provided the phasors are taken in the same order as the circuit elements in each current path, there is a geometrical affinity between the circuit diagram and the voltage phasor diagram, and corresponding connexion points may be lettered identically. This is illustrated in fig. 3.12, in which (for example) the element R, having terminals A, B, is associated with the voltage phasor AB. When the phasor diagram is drawn in this way, it contains the maximum of information about the circuit, in that the voltage between any two nodes (such as B, B′) in the circuit is given by the line joining the corresponding points in the voltage phasor diagram. The affinity between the two diagrams is made clearer if the circuit diagram is redrawn as in fig. 3.12(*d*).

The affinity between fig. 3.12(*a*) and (*c*) is of a subtler kind, and belongs to the realm of *topology*, the branch of geometry which explores the position relationships of geometrical elements such as points and lines, and which plays a considerable part in circuit theory. Fig. 3.12(*a*) and (*c*) are said to be *topologically equivalent*; fig. 3.12(*c*) and (*d*) may be called *geometrically similar*. It is often illuminating to draw a circuit diagram so that it is geometrically similar to the phasor diagram, but this may not necessarily be the clearest way to draw it. Again, it is desirable to make the phasor diagram

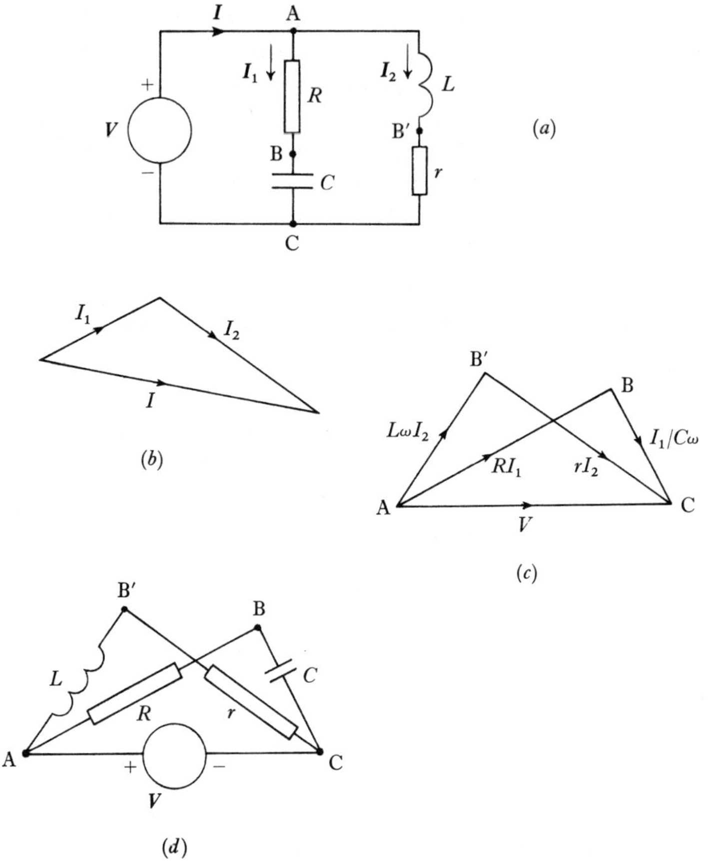

Fig. 3.12 Geometrical properties of phasor diagrams: (*a*) circuit; (*b*) current phasor diagram; (*c*) voltage phasor diagram; (*d*) circuit redrawn to emphasize topological equivalence with (*c*).

topologically equivalent to the circuit diagram, but not at the expense of clarity. Fig. 3.11(*c*) is an example of a phasor diagram which is not topologically equivalent to the circuit diagram; the phasors along the path FHK are taken in a different order from the corresponding circuit elements. Had they been taken in the same order, HK would have coincided with FG and FH with GK, and although the diagram would have been, in a sense, more 'correct', it would have been less clear.

3.9 Three-phase circuits

A *three-phase voltage source* consists of three voltage sources having equal magnitudes but phase differences of 120 degrees between them, and connected in a symmetrical manner. The two possible modes of connexion are the *star-connexion* (fig. 3.13(*a*)) and the *delta-connexion* (fig. 3.13(*b*)); these diagrams are drawn in a way which exhibits geometrical similarity to the voltage phasor diagrams. (It might be thought that the delta-connected sources would tend to circulate current round the delta, but this is not so,

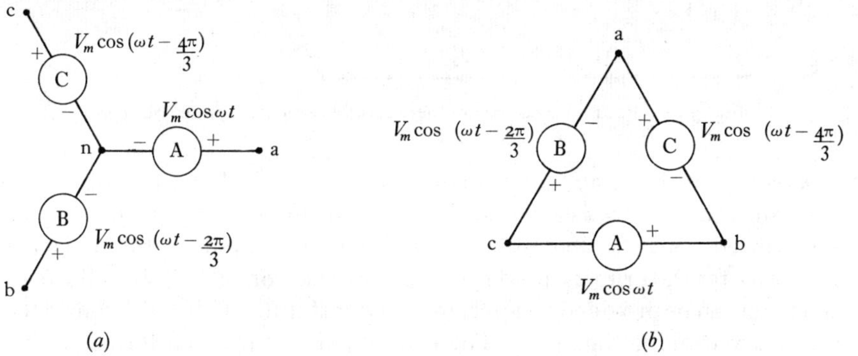

Fig. 3.13 (*a*) Star-connected three-phase voltage source. (*b*) Delta-connected three-phase voltage source.

because the sum of the voltage phasors of any two sources is equal and opposite to the voltage phasor of the third.) In practice the three sources are combined in a single device, and are described as its *phases*; but as a description for the purposes of circuit theory it is convenient to retain the diagram containing three separate sources A, B, C, distinguished as *single-phase* sources if necessary.

The order in which the three voltages reach their maxima is A, then B, then C; the three-phase source is said to have a *phase sequence* A, B, C. It is particularly to be noted, to avoid a possible source of confusion, that when the circuit diagram is drawn with geometrical similarity to the phasor diagram, the positive phase sequence is represented by a *clockwise* rotation from A to B to C. This is clearly seen in fig. 3.13(*a*).

The load connected to such a source may likewise be star- or delta-connected, each branch consisting of any combination of passive (non-power-producing) circuit elements. When the three branches are identical the load is said to be *balanced*. If both source and load are star-connected, the four wire connexion system may be used, the two *star-points* or *neutral terminals* being joined (fig. 3.14); we then effectively have three single-phase sources supplying single-phase loads and sharing the neutral connexion. If either source or load is delta-connected, the neutral connexion cannot be made, and the system is a three wire one with connexions between the *line*

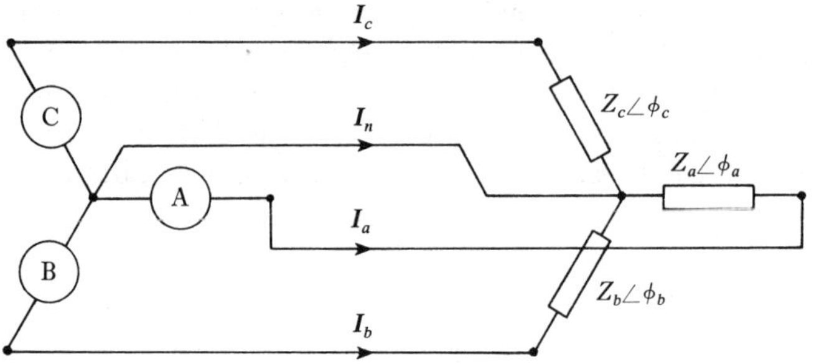

Fig. 3.14 Four wire connexion between three-phase source and load.

terminals only. The four wire system is essential if it is desired to maintain a constant voltage across each branch of a star-connected load which may be unbalanced, as in a domestic power distribution system; but it is not necessary for the primary power source to be star-connected, since the neutral point can be provided by interposing a transformer with a star-connected secondary winding (fig. 3.15). The neutral point of the transformer would

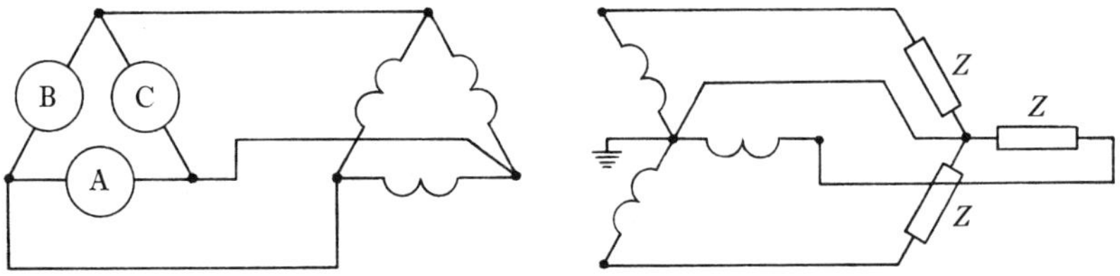

Fig. 3.15 Use of transformer to provide a neutral point.

commonly be earthed, so as to ensure that one end of each load was close to earth potential. Earthing considerations, however, form only one among several reasons which determine the choice between star- and delta-connexion in any given system, and it would not be appropriate to go into further detail here; the question will be discussed further in §9.5.

The three-phase transformer may be thought of as three single-phase transformers appropriately connected, though in practice they are likely to be combined in one unit. In the diagram, windings drawn with their axes in the same direction have their voltages in phase.

The intention in any three-phase system is that the load, like the source voltage, shall be balanced; and it is with balanced loads that we shall be chiefly concerned. If the star-connected impedances in fig. 3.14 are each equal to $Z\angle\phi$, the phasor diagram is as shown in fig. 3.16(*a*); the three currents are equal, and each lags its voltage by the same angle ϕ. The phasors representing the currents in the four wires must add up to zero, so the neutral current phasor is equal and opposite to the sum of the three line current phasors. Fig. 3.16(*b*) shows that, with a balanced load, that sum is itself zero; hence there is no current in the neutral connexion, and the wire

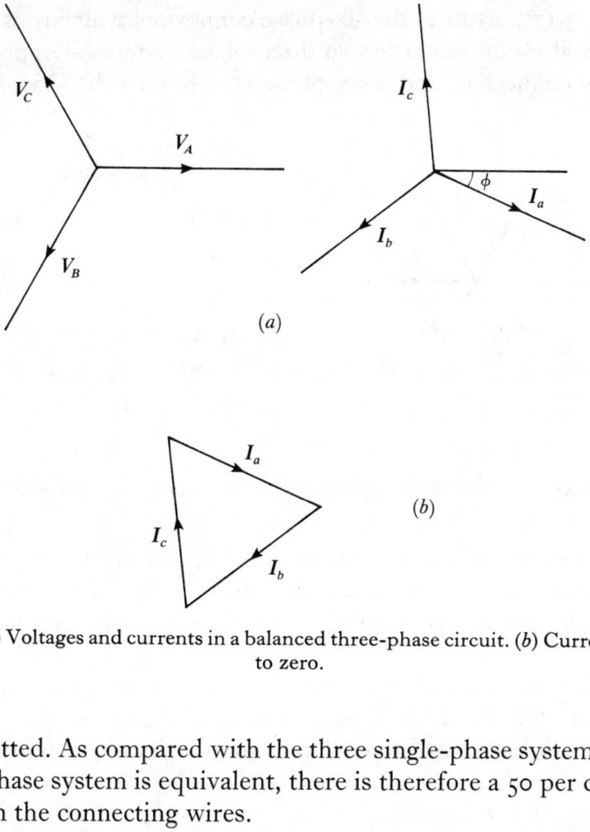

Fig. 3.16 (*a*) Voltages and currents in a balanced three-phase circuit. (*b*) Currents summed to zero.

may be omitted. As compared with the three single-phase systems to which the three-phase system is equivalent, there is therefore a 50 per cent saving of copper in the connecting wires.

In practice, to allow for some unbalance or for other purposes, a neutral connexion may be provided; nevertheless economy of copper remains one of the principal reasons for the use of three-phase generation and distribution of a.c. power in preference to single-phase. The other main reason is associated with the use of rotating machines, whether generators constituting the a.c. voltage source or motors forming a major part of the load on many supply systems; the phase sequence A, B, C forms a naturally rotating pattern which lends itself to the efficient design of both generators and motors.

An ideal three-phase current source is a feasible concept; but in view of the fact that three-phase circuits are invariably power circuits in which the voltage is as far as possible kept constant, the three-phase voltage source is a much more commonly used concept than the current source.

3.10 Other polyphase systems

Any number of identical voltage sources, successively separated by equal phase differences, may be connected symmetrically to form a *polyphase* source. Fig. 3.17 illustrates the six-phase connexion, which is used in rectifier systems; it is equivalent to two three-phase systems of opposite polarities suitably connected, and a six-phase supply may therefore be derived

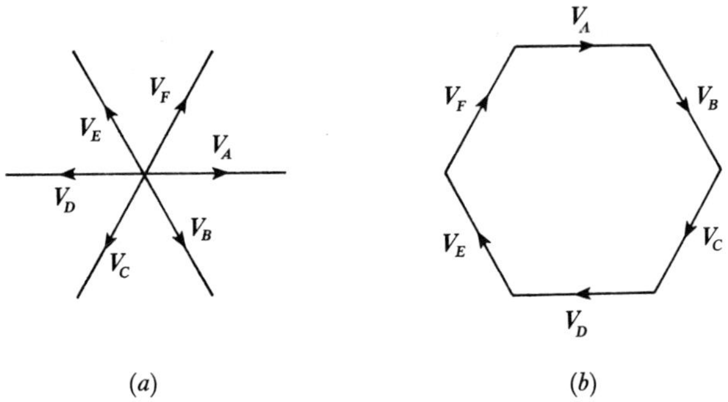

Fig. 3.17 Six-phase voltage phasor systems: (*a*) star-connexion; (*b*) mesh-connexion.

from a three-phase source by means of a transformer having two sets of secondary windings in appropriate connexion. The term *mesh*-connexion replaces 'delta' for describing the system in which all the sources are in series.

On this definition a 'two-phase source' should provide two voltage phasors in opposition – that is, with a phase difference of 180 degrees; a single-phase transformer with a centre-tapped secondary winding would be such a

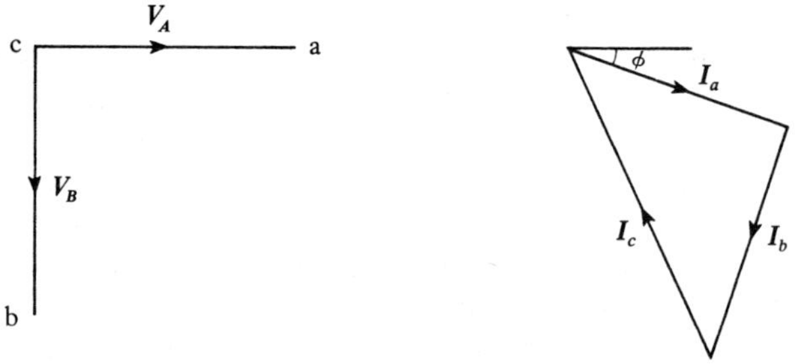

Fig. 3.18 Two-phase voltage and current phasors.

source. In point of fact the term *two-phase* is applied to the unsymmetrical system derived by suppressing two of the voltage phasors in a four-phase system. Now little used in practice, the two-phase system is of great theoretical importance in the analysis of rotating machines because these have two axes of symmetry, namely the centre-lines of the poles and of the interpolar spaces. Fig. 3.18 shows the voltage and current phasors for a two-phase source supplying a load of $Z\angle\phi$ between each of the line-pairs a, c and b, c; it is apparent that in this case the common conductor c carries a current $\sqrt{2}$ times as great as a and b. The voltage between the line terminals a, b is $\sqrt{2}$ times the phase voltage.

3.11 Balanced three-phase circuits

If the function of a star-connected three-phase voltage source ABC is regarded as the maintenance of the required a.c. voltages between each pair of the terminals a, b, c, it can be replaced as shown in fig. 3.19 by an equivalent delta-connected source A′B′C′. The diagram is drawn with geometrical

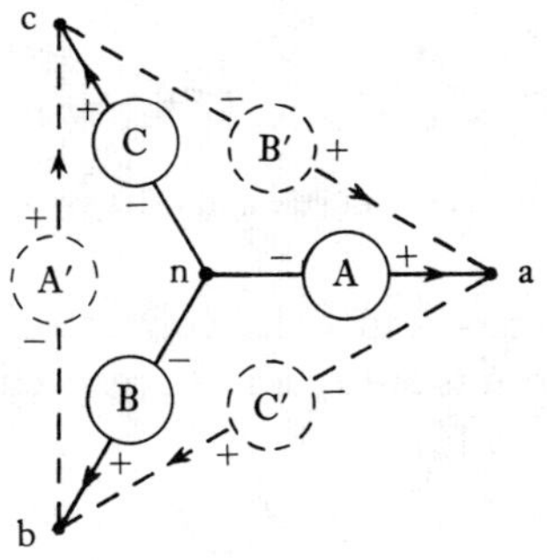

Fig. 3.19 Equivalent star- and delta-connected voltage sources.

similarity to the phasor diagram; the lengths of the branches containing the sources are therefore proportional to the voltages of the sources. If the magnitudes of the voltages in the star- and delta-connected sources are V_Y, V_Δ respectively, the geometry of the equilateral triangle shows that

$$V_\Delta = \sqrt{3}V_Y. \qquad (3.11.1)$$

Suppose that the three-phase source supplies a balanced load, star-connected across the terminals a, b, c, and taking currents denoted by phasors $\boldsymbol{I}_A, \boldsymbol{I}_B, \boldsymbol{I}_C$. When the source is star-connected, each current flows through the source which bears the same letter; but when the source is delta-connected, the current denoted by $\boldsymbol{I}_A$ is jointly supplied by the sources B′ and C′. The voltage and current polarity signs in fig. 3.19 show one of the possible systematic conventions; it is seen that, with this convention, the outflowing current from terminal a of the delta-connected source is given by the phasor $\boldsymbol{I}_{B'} - \boldsymbol{I}_{C'}$. In symbols, therefore

$$\left.\begin{aligned} \boldsymbol{I}_A &= \boldsymbol{I}_{B'} - \boldsymbol{I}_{C'}, \\ \boldsymbol{I}_B &= \boldsymbol{I}_{C'} - \boldsymbol{I}_{A'}, \\ \boldsymbol{I}_C &= \boldsymbol{I}_{A'} - \boldsymbol{I}_{B'}. \end{aligned}\right\} \quad (3.11.2)$$

In fig. 3.20 are drawn two three-phase sets of current phasors satisfying precisely these conditions; these then must represent the currents in the

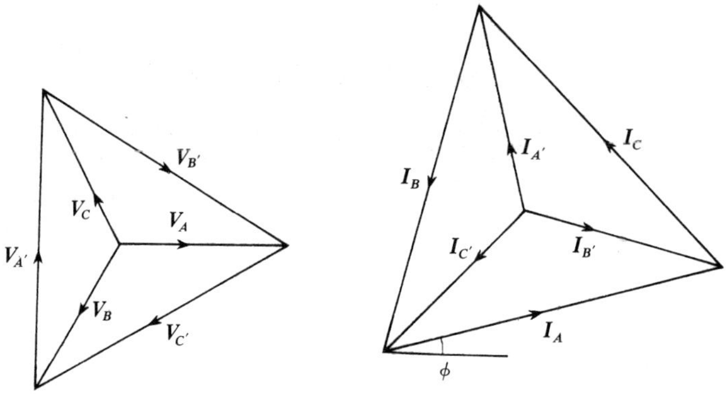

Fig. 3.20 Equivalence of star and delta systems of currents.

several phases of the star- and delta-connected current sources. If the magnitudes of these currents are I_Y and I_Δ respectively, the diagram shows that

$$I_Y = \sqrt{3} I_\Delta. \quad (3.11.3)$$

The voltage phasors for the two equivalent sources are also shown in fig. 3.20. It will be seen that each current makes the same angle with the associated voltage, this angle being ϕ, the phase angle of the load. Thus the total power output from the three phases of the star-connected source is $3V_Y I_Y \cos\phi$ (see (3.6.1)); from the delta-connected source, $3V_\Delta I_\Delta \cos\phi$. However, (3.11.1,3) show that $V_Y I_Y = V_\Delta I_\Delta$; so the power outputs of the two sources are equal – as they must be, if the sources are truly equivalent.

The same argument may be applied to a balanced load, which is essentially a negative power source. To any balanced star-connected load can be found an equivalent balanced delta-connected load, the voltages and currents being related by (3.11.1,3). Dividing one equation by the other, we find

$$Z_\Delta = 3Z_Y, \quad (3.11.4)$$

and inspection of fig. 3.20 shows that the phase angles are equal, so impedances $Z\angle\phi$ in star are equivalent to impedances $3Z\angle\phi$ in delta. It should be stated that an equivalence can also be established between unbalanced star- and delta-connected loads, but this problem is not our concern at the moment.

It has now become clear that the essential features of a three-phase system are those shown in fig. 3.21; the connexions inside the boxes are of secondary importance. For this reason the *line voltages* measured between each pair of the connexions aa′, bb′, cc′, and the *line currents* flowing in those connexions, are treated as the basic quantities descriptive of the state of the

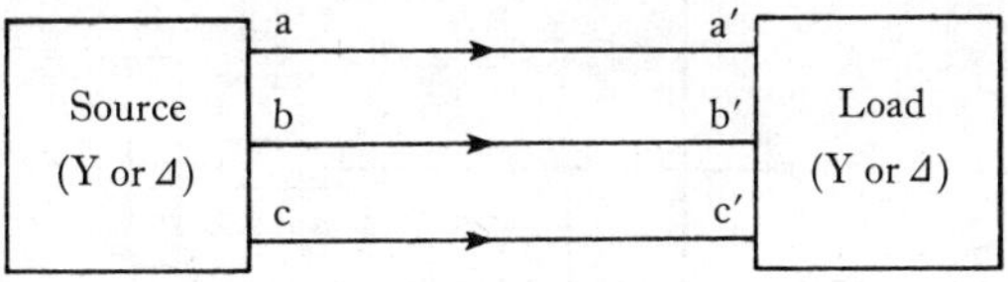

Fig. 3.21 Essentials of a three wire three-phase system.

network. For a balanced system, the line values V_L, I_L are related to the quantities already used by the equations

$$V_L = V_\Delta, \qquad I_L = I_Y. \tag{3.11.5}$$

Thus the total power flowing from the source to the load, shown above to be $3V_Y I_Y \cos\phi$ or $3V_\Delta I_\Delta \cos\phi$, is now written

$$\text{Power} = \sqrt{3}V_L I_L \cos\phi. \tag{3.11.6}$$

The terms *phase voltage* and *phase current* are used, primarily signifying the voltage and current in one phase of the source or load, and therefore dependent (unlike the line values) on the mode of connexion. The term 'phase voltage' is also used, however, in a transferred sense, to denote *phase voltage assuming star connexion*; this is because the reduction of a network to its equivalent star is commonly the first step in the analysis of it. If V_P is the phase voltage so defined, (3.11.1) shows

$$V_P = V_L/\sqrt{3}, \tag{3.11.7}$$

V_P being identical with the quantity V_Y already used.

3.12 Reduction and solution of balanced three-phase circuits

The possible connexions for a load are not exhausted by the simple star and delta. Fig. 3.22(*a*) shows a more complicated load network, in which a delta-connected source of line voltage V supplies star-connected impedances Z_1, and also delta-connected impedances Z_3 through series impedances Z_2. The deltas can however be replaced by equivalent stars as in fig. 3.22(*b*). On account of the symmetry of the problem, the three neutral points are all at the same potential, and can be imagined to be connected by a wire which carries no current. The addition of such a wire reduces the network to three single-phase networks, one of which is shown in fig. 3.22(*c*); this is solved by known methods when the impedances are given. It will be found that all balanced three-phase networks can be reduced to single-phase networks in this way.

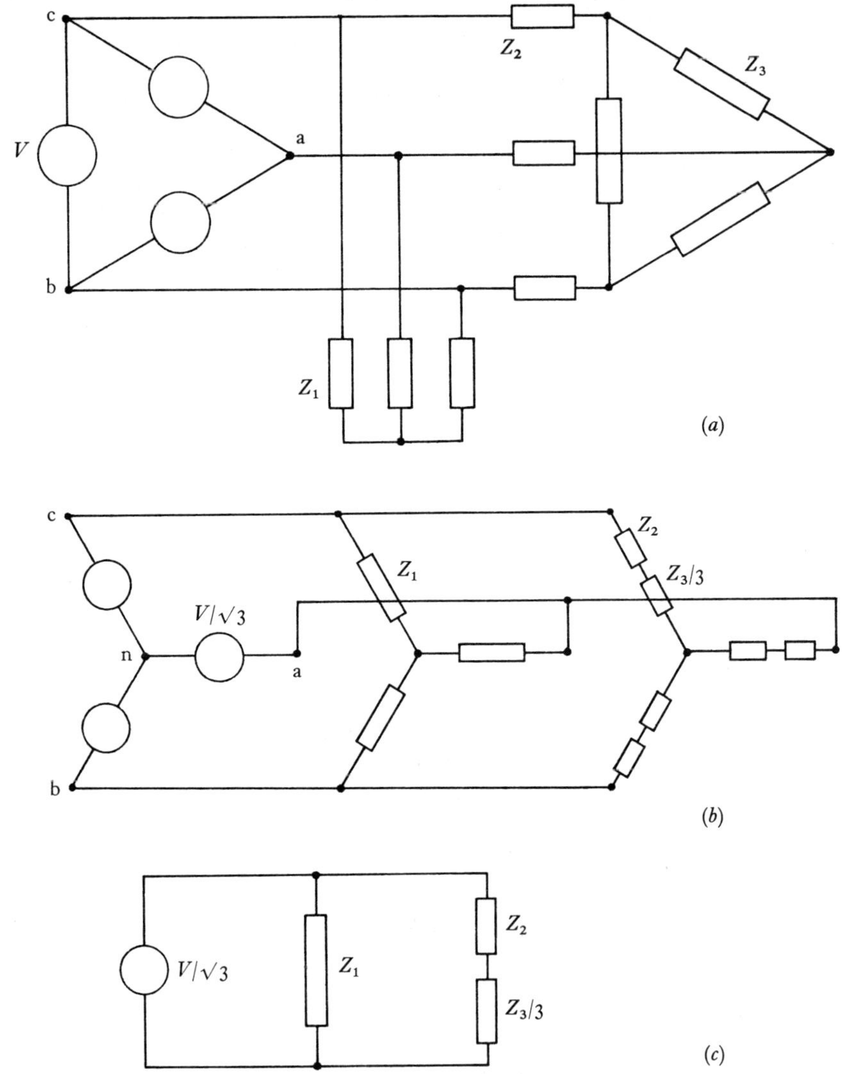

Fig. 3.22 (*a*) Complex load network. (*b*) Reduction of (*a*) by delta–star transformation. (*c*) Equivalent single-phase problem.

The connexion wires are themselves to be regarded as part of the load, which must include everything outside the terminals of the source. This raises a number of difficulties. If for the sake of definiteness we think of an open wire circuit such as an overhead transmission line, we can see that its resistance is distributed along the wires; with a little thought this is also seen to be true of the self inductance. The mutual inductances between the three loops in a three wire line must not be forgotten; it is, however, shown in books on electromagnetism that it is possible to ascribe to each wire an effective self inductance which includes the effects of mutual inductance.†

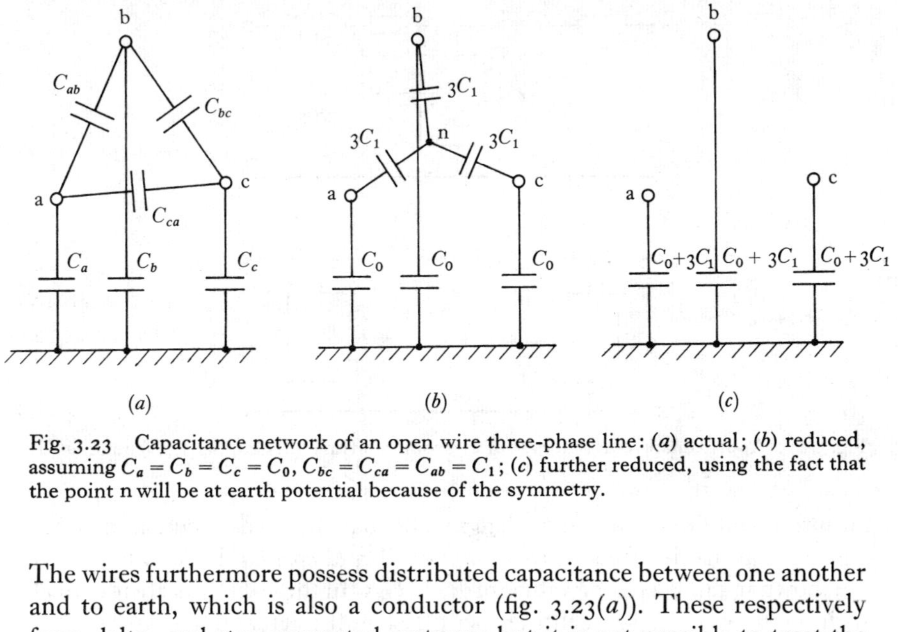

Fig. 3.23 Capacitance network of an open wire three-phase line: (*a*) actual; (*b*) reduced, assuming $C_a = C_b = C_c = C_0$, $C_{bc} = C_{ca} = C_{ab} = C_1$; (*c*) further reduced, using the fact that the point n will be at earth potential because of the symmetry.

The wires furthermore possess distributed capacitance between one another and to earth, which is also a conductor (fig. 3.23(*a*)). These respectively form delta- and star-connected systems; but it is not possible to treat the earth as a floating neutral point which will automatically take up the same potential as the neutral point of the supply, and if balanced conditions are to be approached the latter must be earthed. The diagram suggests that there will in any case be inequalities between the capacitances in each group, and the same is true of the inductances; however, these are eliminated by *transposition* of the conductors, which means that a given wire successively occupies each of the positions a, b, c. The three, effectively equal inter-wire capacitances can then be replaced by a star-connected system (fig. 3.23(*b*)) which may be regarded as being in parallel with the capacitance to earth (fig. 3.23(*c*)). (It is to be noted that the delta-connected capacitances C_1

† See, for example, G. W. Carter, *The electromagnetic field in its engineering aspects*, 2nd edition, §10.10, Longmans (1967).

are equivalent to star-connected capacitances $3C_1$, because the impedance of capacitance C_1 is $1/C_1\omega$, and (3.11.4) is given in terms of impedances.) In parallel with each of the capacitances in fig. 3.23(*c*) there may be leakage conductance, distributed, like the other quantities, along the line.

In sum, therefore, we have three wires, possessing effective resistance R and inductance L, across which, star-connected, are the effective conductance G and capacitance C to earth; but all these quantities are distributed along the line. Distributed circuits form the theme of chapter 13; in the

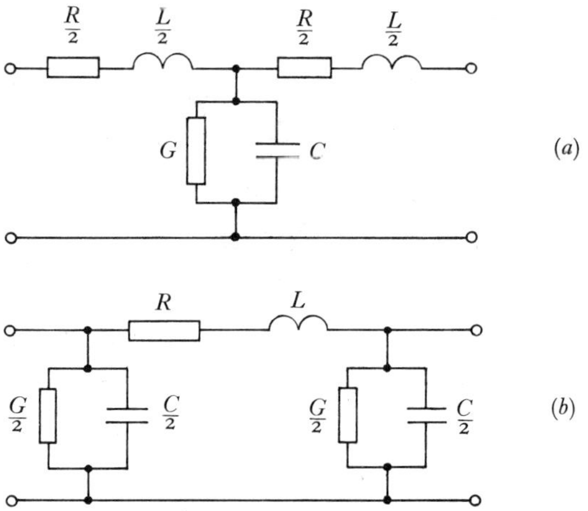

Fig. 3.24 Approximate representation of a transmission line: (*a*) nominal T; (*b*) nominal Π.

meantime, for lines of moderate length, adequate lumped-circuit approximations are available. These are the *nominal* T and *nominal* Π circuits, which are shown in fig. 3.24. The values of R, L, G, C in these circuits are identical with the totals of these quantities per phase in the actual line.

Figure 3.24 is an equivalent single-phase representation. Since the neutral conductor, if it exists, carries no current, there is no impedance in the lower wire. The same diagram might be used to represent a single-phase line, such as a telephone line, operated symmetrically with respect to earth; here, however, R and L must be the total resistance and inductance of the two wires, and it would be more correct (though analytically indistinguishable) to locate $R/2$ and $L/2$ in each wire.

Two examples of the solution of three-phase circuit problems by phasor diagrams will now be given.

1. A three-phase supply of 415 V, 50 Hz feeds a balanced load of 240 kVA and power factor 0.75 lagging. What value of capacitance, delta-connected, will raise the overall power factor to 0.9?

The phase voltage is $415/\sqrt{3}$, or 240; the equivalent single-phase problem is that in which 80 kVA are drawn at power factor 0.75 from a 240 V supply. This problem has already been solved in §3.7, where it was found that the capacitance required is 1320 μF per phase; this, however, would be the value in star-connexion, to which the equivalent value in delta-connexion is 1320/3, or 440 μF. From the standpoint of circuit theory the two sets of capacitors are exactly equivalent; but from the engineering standpoint, 3×440 μF of capacitors rated at 415 V (a.c.) might well be preferred, on grounds of lower cost, to 3×1320 μF at 240 V (a.c.). This illustrates the way in which an engineering decision is influenced by economic as well as technical considerations.

2. The following data refer to a 110 kV, three-phase, overhead transmission line, having its conductors transposed to achieve balanced conditions.

Resistance, each wire	0.081 Ω/km
Effective inductance, each wire	1.34 mH/km
Capacitance between each pair of wires	0.00135 μF/km
Capacitance between each wire and earth	0.00455 μF/km
Conductance between each wire and earth	negligible

The line connects a generator to a load at a distance of 100 km. The generator operates at a frequency of 60 Hz; its voltage is so adjusted that the voltage at the load is 110 kV, the load itself being 60000 kVA at a power factor of 0.9 lagging. It is required to find the voltage, current and power factor at the generator.

The problem is to be reduced to a star-connexion and hence to the equivalent single-phase problem. The conditions at the load are, from (3.11.7),

$$V_P = 110/\sqrt{3} = 63.5 \text{ kV},$$

and, as the load is 20000 kVA per phase,

$$I_L = \frac{20\,000}{63.5} = 315 \text{ A}.$$

These are the quantities labelled $\boldsymbol{V}_3$, $\boldsymbol{I}_3$ in the circuit diagram of fig. 3.25, in which the line is represented by its nominal T circuit. The numerical values are obtained thus from the line data:

$$r = \tfrac{1}{2} \times 0.081 \times 100 \qquad\qquad = 4.05\,\Omega,$$

$$X_L = \tfrac{1}{2} \times 1.34 \times 10^{-3} \times 100 \times 120\,\pi = 25.3\,\Omega,$$

$$X_C = \frac{10^6}{0.0086 \times 100 \times 120\,\pi} \qquad = 3090\,\Omega.$$

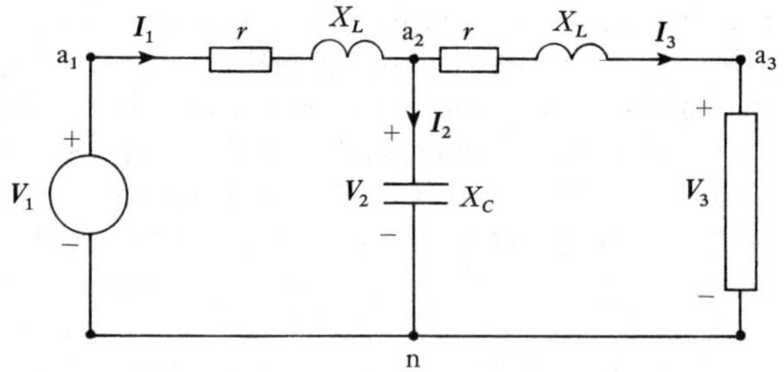

Fig. 3.25 Single-phase equivalent for a three-phase overhead line problem.

(The total capacitance of each line to neutral is made up of 0.00455 μF/km to earth, and 3 × 0.00135 μF/km as the star-connected capacitance equivalent to the capacitance between wires.)

From this point we may build up the phasor diagram of fig. 3.26 without having to evaluate the resistance and inductance of the load. $\boldsymbol{I}_3$ is taken as the

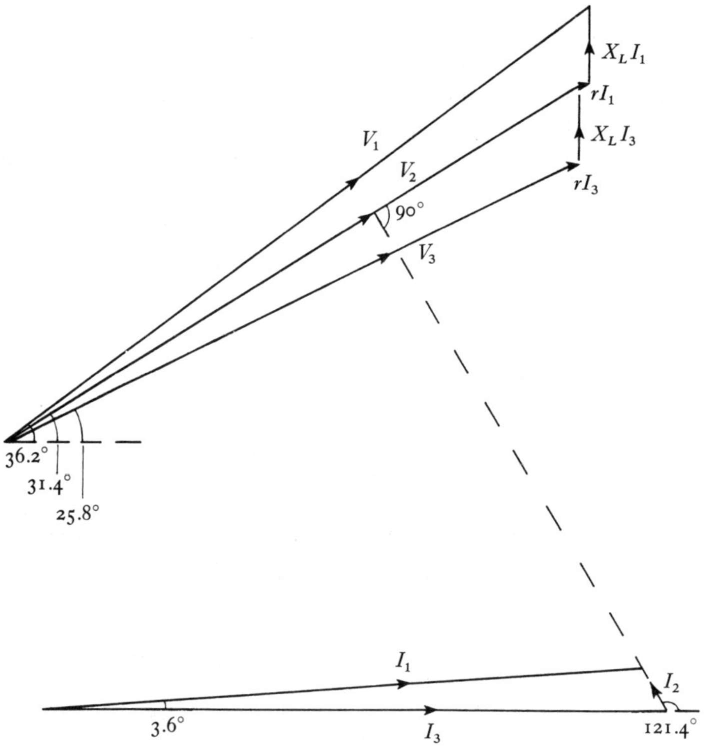

Fig. 3.26 Phasor diagrams for the line of fig. 3.25: $V_1 = 73.4$ kV, $V_2 = 68.4$ kV, $V_3 = 63.5$ kV ; $I_1 = 304$ A, $I_2 = 22$ A, $I_3 = 315$ A; $rI_1 = 1.2$ kV, $rI_3 = 1.3$ kV; $X_L I_1 = 7.7$ kV, $X_L I_3 = 8.0$ kV

reference phasor and $\boldsymbol{V}_3$ laid down at the angle $\cos^{-1}0.9$ (25.8 degrees) to it. The addition to $\boldsymbol{V}_3$ of the voltage drops $r\boldsymbol{I}_3$ and $X_L\boldsymbol{I}_3$ gives $\boldsymbol{V}_2$; at right angles to this is drawn the capacitance current $\boldsymbol{I}_2$. The sum of $\boldsymbol{I}_2$ and $\boldsymbol{I}_3$ is the generator current $\boldsymbol{I}_1$; its voltage $\boldsymbol{V}_1$ is obtained by adding $r\boldsymbol{I}_1$ and $X_L\boldsymbol{I}_1$ to $\boldsymbol{V}_2$. The numerical values are given in the diagram. The line voltage at the generator is seen to be $73.4\sqrt{3}$, or 127.1 kV, while the current is 304 A; the reason why the latter is somewhat less than the load current is because the capacitance current has a component which is in opposition to the load current. The phasor $\boldsymbol{V}_1$ leads $\boldsymbol{I}_1$ by 32.6 degrees, giving a power factor of $\cos 32.6°$, or 0.842. This may be verified by noting that the power input per phase is the sum of the output (18000 kW) and the ohmic loss in the line, which is $4.05\,(304^2+315^2)$ W, or 776 kW. Thus

$$\cos\phi = \frac{18776}{73.4\times 304} = 0.841.$$

It will be seen from this numerical example that the phasor diagram would have to be drawn large and accurately to get the best results; but even a small diagram will give a quick solution of modest accuracy, at the same time vividly illustrating the relative importance of the factors which influence the result.

Examples on chapter 3

3.1 Two sinusoidal voltage sources, of r.m.s. values 3 V and 4 V respectively, are connected in series. What is the r.m.s. value of the total voltage when the two voltages are (*a*) of the same frequency and in phase, (*b*) of the same frequency and in quadrature, (*c*) of different frequencies?

3.2 A coil has a resistance of 10 Ω and an inductance of 31.8 mH, and is connected across a voltage source having a complex waveform given by $v = 50 + 25\sin\omega t + 10\sin 3\omega t$. If the frequency is 50 Hz, find an expression for the current i that will flow in the circuit, and determine the r.m.s. values of v and i.

3.3 An inductive impedance, in series with a 10 Ω resistor, is connected to a 200 V, 50 Hz supply. If the voltage drops across the impedance and the resistor are 150 V and 100 V respectively, calculate the power in the impedance and its power factor. Determine also the value of the capacitor which, in parallel with the impedance, would raise the power factor of the circuit to unity.

3.4 A load consisting of a series combination of resistance R and inductive reactance X takes a power of 10 kW and a current of 52.1 A from a 240 V,

50 Hz source. Determine the values of R and X, and find the value of capacitance which should be connected in parallel with the load in order to raise the power factor to unity. What current is then drawn from the supply?

3.5 A two-terminal network consists of a resistor R in series with an inductor L. Show that the impedance of this network is equivalent to that of a resistor R' in parallel with an inductor L', where

$$R' = R[1 + (\omega L/R)^2],$$

and

$$L' = L[1 + (\omega L/R)^2]/(\omega L/R)^2.$$

The reactance of a 1 mH coil is 100 times as large as its series resistance, when measured at a frequency of 100 kHz. What are the values of an inductor and a resistor which, when connected in parallel, would represent the coil at this frequency?

3.6 A coil and a capacitor are connected in series between terminals A, B. The coil has inductance L and resistance R, and the capacitor may be represented by a pure capacitance C shunted by a conductance G. A sinusoidal voltage of angular frequency ω is applied to the terminals AB. Show that the voltage across the capacitor will be in quadrature with the applied voltage when

$$\omega = \sqrt{[(1 + RG)/LC]}.$$

3.7 A resistor of 10 Ω, a pure inductor of 100 mH and a capacitor of 1 μF are all connected in series in that order between terminals A, B. In parallel with the capacitor is a load impedance of $(500 \angle 30^\circ)$ Ω. By means of a phasor diagram find the voltage V_1 of a 50 Hz source which, when connected to the terminals AB, will produce a voltage of 1000 V across the load. What is the voltage drop across the resistor–inductor combination?

3.8 A circuit operating at 50 Hz consists of three branches:
(*a*) a 5 Ω resistor in series with a 0.02 H inductor;
(*b*) a 4 Ω resistor in series with a 398 μF capacitor;
(*c*) a 7 Ω resistor in series with a 0.032 H inductor.
Branches (*a*) and (*b*) are connected in parallel, and (*c*) is in series with the combination. Find the voltage across the entire network when a current of 10 A flows in the capacitor.

3.9 A bridge network PQRS is supplied at terminals PR from a 1000 Hz a.c. source, and the detector is across QS. The branch SP is an unknown air-cored coil. When the voltage and current in the detector are zero, the values of the components in the other branches are:

PQ, a resistor of 405 Ω;
QR, a capacitor of 0.55 μF shunted by a resistor of 1850 Ω;
RS, a resistor of 325 Ω.

Draw the phasor diagram and derive the inductance and resistance of the coil.

3.10 A branch PQ of a network consists of inductance L in series with resistance R (both unknown), and possesses known mutual inductance M with a secondary coil having terminals QR. Also connected across PQ is another network PSQ; between P and S are a resistor R_1 and a capacitor C_1 in parallel, while between S and Q is a resistor R_2. An a.c. supply of angular frequency ω is connected across PQ, and a detector across RS. Draw the phasor diagram for the condition in which the detector voltage and current are zero, and hence prove that

$$L = M\left(1 + \frac{R_1}{R_2}\cos^2\delta\right),$$

and

$$R = M\omega\frac{R_1}{R_2}\sin\delta\cos\delta,$$

where

$$\tan\delta = C_1\omega R_1.$$

3.11 A factory receives its power from a 415 V, 50 Hz, three-phase supply. The load comprises 100 kW of heating at unity power factor and 200 kW of induction motors operating at a lagging power factor of 0.7; each type of load is balanced as between the three phases. Calculate

(*a*) the line current and power factor for the whole factory;
(*b*) the value of capacitance, delta-connected, which will raise the overall power factor to 0.9 lagging;
(*c*) the power factor at which the factory will operate if the capacitance and the heating load are connected but the induction motors are shut down.

3.12 A three-phase line supplies a balanced load of 50 MVA, 0.8 power factor lagging, at a line voltage of 132 kV and a frequency of 50 Hz. The line is 100 miles long and has an inductance of 3 mH per mile for each phase and a total effective capacitance to neutral of 0.01 μF per mile for each phase; the conductor resistance and the conductance to neutral are neglected. Using the nominal T representation, draw the phasor diagram of this network (one phase only), and use it to estimate the line voltage, current and power factor at the input end of the line.

4 The use of complex numbers in sinusoidal analysis

The solution of network problems has been treated in chapter 3 by the phasor method. In this phasor method an image problem (the phasor diagram) is set up and is then solved (by trigonometrical means). The solution to the image problem is finally interpreted; that is to say, it is translated back into the original terminology. The present chapter deals with an alternative image problem, one which employs complex numbers. Like the phasor method, this alternative complex method is restricted to networks that are linear and to stimuli that are sinusoidal. Its usefulness lies primarily in its applicability to networks in which the stimulus may have any one of a wide range of frequencies for, whereas in the phasor method a new phasor diagram may have to be drawn for each frequency, in the complex method the frequency is included quite naturally in the solution. The complex method is therefore ideally suited to the analysis of electronic circuits.

4.1 Introduction to complex numbers

The first occasion on which one is likely to meet complex numbers is in the solution of a simple quadratic equation. Consider, for example, the equation

$$\boldsymbol{W}^2 - 4\boldsymbol{W} + 13 = 0. \tag{4.1.1}$$

The solution of (4.1.1) is

$$\boldsymbol{W} = 2 \pm \sqrt{(4 - 13)}, \tag{4.1.2}$$

but at this point there is an impasse if, as occurs here, the quantity under the square-root sign is negative. To resolve this impasse a quantity j is defined such that

$$\mathrm{j} = +\sqrt{(-1)}, \tag{4.1.3}$$

whence $\mathrm{j}^2 = -1.$

(Mathematicians normally use the symbol i instead of j in (4.1.3), but electrical engineers reserve the symbol i for current). Equation (4.1.2) now becomes

$$\boldsymbol{W} = 2 \pm \sqrt{[\mathrm{j}^2(13 - 4)]},$$
$$= 2 \pm \mathrm{j}3,$$

from which it is seen that the two solutions of the quadratic equation, (4.1.1), are $\boldsymbol{W}_1 = 2 + \mathrm{j}3$ and $\boldsymbol{W}_2 = 2 - \mathrm{j}3$. Such numbers $\boldsymbol{W}_1$ and $\boldsymbol{W}_2$ are called *complex* numbers and are denoted throughout this chapter by bold italic type. In general, if the symbols x and y denote *real* numbers, any number of the form $\boldsymbol{W} = x + \mathrm{j}y$ is a complex number. x is called the *real part* of $\boldsymbol{W}$ and y is called the *imaginary part* of $\boldsymbol{W}$. Symbolically this is written

$$x = \mathrm{Re}\,\boldsymbol{W},$$

and
$$y = \mathrm{Im}\,\boldsymbol{W}.$$

It will be noticed that the two solutions $\boldsymbol{W}_1$ and $\boldsymbol{W}_2$ of the quadratic equation (4.1.1) are related to each other in a particular fashion, namely that the real part of $\boldsymbol{W}_1$ is equal to the real part of $\boldsymbol{W}_2$, and that the imaginary part of $\boldsymbol{W}_1$ is equal to the negative of the imaginary part of $\boldsymbol{W}_2$. Two complex numbers which have this relationship are called *complex conjugates*. Thus if $\boldsymbol{W} = x + \mathrm{j}y$, and if $\boldsymbol{W}^*$ is the conjugate of $\boldsymbol{W}$, then $\boldsymbol{W}^* = x - \mathrm{j}y$.

4.2 The manipulation of complex numbers

Addition and subtraction of complex numbers

The sum (difference) of two complex numbers $\boldsymbol{A}$ and $\boldsymbol{B}$ is a complex number whose real part is the sum (difference) of the real parts of $\boldsymbol{A}$ and $\boldsymbol{B}$, and whose imaginary part is the sum (difference) of the imaginary parts of $\boldsymbol{A}$ and $\boldsymbol{B}$. Thus if

$$\boldsymbol{A} = a + \mathrm{j}b,$$

and if
$$\boldsymbol{B} = c + \mathrm{j}d,$$

then
$$\boldsymbol{A} + \boldsymbol{B} = \boldsymbol{C} = (a + c) + \mathrm{j}(b + d),$$

and
$$\boldsymbol{A} - \boldsymbol{B} = \boldsymbol{D} = (a - c) + \mathrm{j}(b - d).$$

Multiplication of complex numbers

Multiplication proceeds according to the ordinary rules of algebra; whenever j^2 appears, it may be replaced by -1. Thus

$$\boldsymbol{A} \times \boldsymbol{B} = (a + \mathrm{j}b)(c + \mathrm{j}d),$$
$$= (ac - bd) + \mathrm{j}(ad + bc).$$

Division of complex numbers

The division of a complex number $\boldsymbol{A}$ by another complex number $\boldsymbol{B}$ may be done by a process called *rationalization*. (An alternative method is described in §4.4). The imaginary term in the denominator is eliminated by multiplying both numerator and denominator by the conjugate of $\boldsymbol{B}$.

$$\begin{aligned}\frac{\boldsymbol{A}}{\boldsymbol{B}} &= \frac{a+\mathrm{j}b}{c+\mathrm{j}d},\\ &= \frac{(a+\mathrm{j}b)(c-\mathrm{j}d)}{(c+\mathrm{j}d)(c-\mathrm{j}d)},\\ &= \frac{ac+bd}{c^2+d^2}+\mathrm{j}\frac{bc-ad}{c^2+d^2}.\end{aligned}$$

4.3 Geometric interpretation of a complex number

Any complex number $\boldsymbol{W} = x + \mathrm{j}y$ can be described by the two real numbers x and y, respectively the real and imaginary parts of $\boldsymbol{W}$. A geometric interpretation of $\boldsymbol{W}$ becomes possible if $\boldsymbol{W}$ is plotted as a point on a plane described by two orthogonal axes, one axis representing a scale of real numbers and the other axis representing a scale of imaginary numbers. This plane is called the *complex plane*, and a diagram (fig. 4.1) representing this plane is often called an *Argand diagram* after the French mathematician

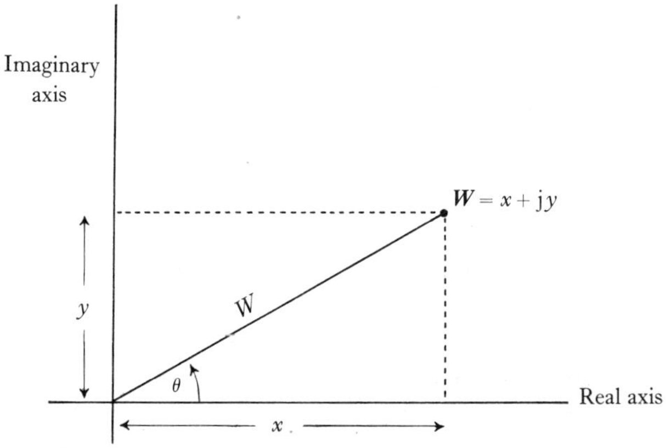

Fig. 4.1 Argand diagram.

(1768–1822). From the Argand diagram it is apparent that the complex number $\boldsymbol{W}$ may be represented not only by its cartesian co-ordinates x and y but equally well by its polar co-ordinates W and θ, where W is the length of the radius vector from the origin to the point $\boldsymbol{W}$ and where θ is the angle, measured in the direction shown, from the positive real axis to the radius vector. W is called the *magnitude* or *modulus* of the complex number, and is written $W = |\boldsymbol{W}|$, and θ is called the *phase* or *phase angle* of the complex number (θ is also known, by mathematicians, as the *argument* of the complex number, but this is a term rarely used by electrical engineers). A complex number $\boldsymbol{W}$ is often represented in polar co-ordinates by the notation

$$\boldsymbol{W} = |\boldsymbol{W}| \angle \theta = W \angle \theta,$$

and the equations relating the cartesian and polar co-ordinates are

$$\begin{aligned} W &= \surd(x^2 + y), \\ \tan\theta &= y/x, \end{aligned} \quad \text{and} \quad \begin{aligned} x &= W\cos\theta, \\ y &= W\sin\theta. \end{aligned} \qquad (4.3.1)$$

It is worth digressing for a moment to consider the ambiguity in θ as derived from the expression $\tan\theta = y/x$ in (4.3.1). If y/x is positive, θ will lie

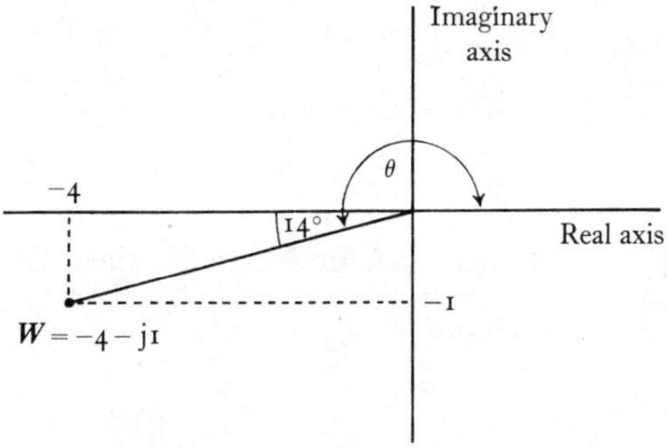

Fig. 4.2 Complex number with a phase angle lying in the third quadrant.

either in the first or in the third quadrant of the complex plane, and if y/x is negative θ will lie in the second or fourth quadrant. The ambiguity is resolved by the particular values of x and y. For example, in fig. 4.2 where $x = -4$ and $y = -1$, $\tan\theta = 0.25$ and $\theta = 194$ degrees. Thus although θ may lie in the first quadrant when $\tan\theta$ is positive, it must not be assumed to do so.

4.4 Exponential representation of a complex number

One of the most useful representations of a complex number is the *exponential* representation $\boldsymbol{W} = W e^{j\theta}$ which employs the polar co-ordinates W and θ and which may be considered to be a mathematical representation of the shorthand notation $W \angle \theta$. The exponential representation, which will now be derived, is the most convenient one when the operation of multiplication, division, differentiation, or integration has to be performed on $\boldsymbol{W}$. Consider (4.4.1), which is deducible from the definition of e^{α} as a function whose derivative is itself:

$$e^{\alpha} = 1 + (\alpha) + (\alpha^2/2!) + (\alpha^3/3!) + \dots. \qquad (4.4.1)$$

If α is replaced by $(j\theta)$, this equation becomes

$$e^{j\theta} = 1 + (j\theta) + (j\theta)^2/2! + (j\theta)^3/3! + \dots,$$

and since $j^2 = -1$, $j^3 = -j$, $j^4 = 1$, etc.,

$$\begin{aligned} e^{j\theta} &= 1 + (j\theta) - (\theta^2/2!) - j(\theta^3/3!) + (\theta^4/4!) + j(\theta^5/5!) + \dots \\ &= [1 - (\theta^2/2!) + (\theta^4/4!) + \dots] + j[\theta - (\theta^3/3!) + (\theta^5/5!) + \dots]. \end{aligned} \tag{4.4.2}$$

Now it can be deduced from the elementary trigonometrical definitions of $\cos\theta$ and $\sin\theta$ that

$$\left.\begin{aligned} \cos\theta &= 1 - (\theta^2/2!) + (\theta^4/4!) - (\theta^6/6!) + \dots, \\ \sin\theta &= \theta - (\theta^3/3!) + (\theta^5/5!) - (\theta^7/7!) + \dots. \end{aligned}\right\} \tag{4.4.3}$$

Substitution of (4.4.3) in (4.4.2) gives

$$e^{j\theta} = \cos\theta + j\sin\theta. \tag{4.4.4}$$

Similarly,

$$e^{-j\theta} = \cos\theta - j\sin\theta. \tag{4.4.5}$$

Any complex number may therefore be written in the form

$$\begin{aligned} \boldsymbol{W} &= x + jy \\ &= W\cos\theta + jW\sin\theta \\ &= We^{j\theta} \end{aligned}$$

where W and θ have the significance of magnitude and phase as shown in fig. 4.1.

Multiplication is easily carried out by means of the exponential representation, for the product of two numbers $W_1 e^{j\theta_1}$ and $W_2 e^{j\theta_2}$ is simply

$$W_1 e^{j\theta_1} \times W_2 e^{j\theta_2} = W_1 W_2 e^{j(\theta_1+\theta_2)}, \tag{4.4.6}$$

yielding the result that the magnitude of the product of two complex numbers is equal to the product of their individual magnitudes, and that the phase of the product is equal to the sum of their individual phases. A special case of particular importance is the multiplication of any complex number $We^{j\theta}$ by the particular complex number $e^{j\pi/2}$. The result, $We^{j(\theta+\pi/2)}$, shows that the phase of the original number has been increased by 90 degrees; but the particular complex number $e^{j\pi/2}$ is

$$e^{j\pi/2} = \cos(\pi/2) + j\sin(\pi/2) = j,$$

and it therefore follows that the multiplication of a number by j rotates the radius vector in the complex plane by 90 degrees in the anticlockwise direction. Fig. 4.3 illustrates successive multiplication of a number by j.

Note that multiplication by −j rotates the vector *backwards* by 90 degrees.

The quotient of two complex numbers is also easily obtained via the exponential representation. For example

$$\frac{W_1\,e^{j\theta_1}}{W_2\,e^{j\theta_2}} = \frac{W_1}{W_2}\,.\,e^{j(\theta_1-\theta_2)}. \tag{4.4.7}$$

The usefulness of (4.4.7) can scarcely be over-estimated because it is thereby possible to avoid the tedious process of rationalization described

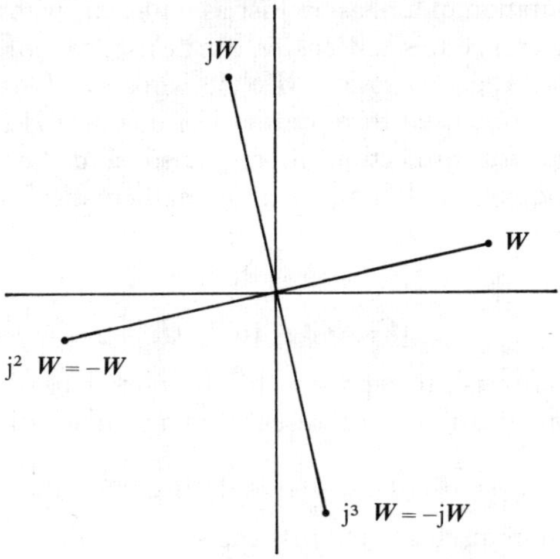

Fig. 4.3 Successive multiplication of a complex number by j.

on p. 103. For instance, in network theory, expressions of the form $(R+j\omega L)/(G+j\omega C)$ continually arise, and by means of (4.4.7) the magnitude and phase of such expressions may be written down directly

$$\frac{R+j\omega L}{G+j\omega C} = \sqrt{\frac{R^2+(\omega L)^2}{G^2+(\omega C)^2}}\,.\,\angle(\tan^{-1}(\omega L/R) - \tan^{-1}(\omega C/G)).$$

A further advantage of the exponential representation occurs when a complex number has to be raised to a given power. If $\boldsymbol{W} = We^{j\theta}$, then

$$\begin{aligned}\boldsymbol{W}^n &= W^n\,e^{jn\theta},\\ &= W^n \angle n\theta,\\ &= W^n(\cos n\theta + j\sin n\theta).\end{aligned}$$

4.5 $W\mathrm{e}^{\mathrm{j}\omega t}$ as a phasor

A connexion can now be established between complex numbers and phasors. It has been shown that the complex number $W\mathrm{e}^{\mathrm{j}\theta}$ represents a point in the complex plane at a distance W from the origin and making an angle θ with the positive real axis. Now if θ is the function of time, $\theta = \omega t$, where ω is a constant with the dimensions of radians per second, then the complex number $W\mathrm{e}^{\mathrm{j}\theta}$ becomes the complex function of time $W\mathrm{e}^{\mathrm{j}\omega t}$ and the point representing it moves in a circular path of radius W around the origin in the anticlockwise direction. The radius vector from the origin to the point thus represents a phasor, and the function $W\mathrm{e}^{\mathrm{j}\omega t}$ may be regarded as the mathematical representation of a phasor. Just as a phasor, whose horizontal projection on the reference axis is $W\cos\omega t$, can be used to represent the quantity $W\cos\omega t$, so also the expression $W\mathrm{e}^{\mathrm{j}\omega t}$, whose real part has the value $W\cos\omega t$, may also be used to represent the quantity $W\cos\omega t$. (The same is true of the *vertical* projection of the phasor and the *imaginary* part of $W\mathrm{e}^{\mathrm{j}\omega t}$ for the quantity $W\sin\omega t$.) The mathematical statement of this representation is

$$W\cos\omega t = \mathrm{Re}\, W\mathrm{e}^{\mathrm{j}\omega t},$$

and

$$W\sin\omega t = \mathrm{Im}\, W\mathrm{e}^{\mathrm{j}\omega t}.$$

If the quantity we wish to represent by the phasor has an arbitrary phase angle α, it is quite possible to represent it in a similar fashion,

$$W\cos(\omega t + \alpha) = \mathrm{Re}\, W\mathrm{e}^{\mathrm{j}(\omega t+\alpha)}, \tag{4.5.1}$$

but it is often more useful to express the same result in a slightly different form.

$$\begin{aligned} W\cos(\omega t + \alpha) &= \mathrm{Re}\, W\mathrm{e}^{\mathrm{j}(\omega t+\alpha)}, \\ &= \mathrm{Re}\, W\mathrm{e}^{\mathrm{j}\alpha}.\mathrm{e}^{\mathrm{j}\omega t}, \\ &= \mathrm{Re}\, \boldsymbol{W}\mathrm{e}^{\mathrm{j}\omega t}. \end{aligned} \tag{4.5.2}$$

Here, $\boldsymbol{W}$ is the complex number $W\mathrm{e}^{\mathrm{j}\alpha}$ (α = constant) and is called a *stationary phasor*. The stationary phasor $\boldsymbol{W}$ represents the position of the rotating phasor $\boldsymbol{W}\mathrm{e}^{\mathrm{j}\omega t}$ at time $t = 0$.

4.6 The image problem

Stationary phasors play an important part in the analysis of electric circuits; they take the place, in the image circuit, of the corresponding sinusoidal quantities of the actual circuit. The process of setting up the image problem from the actual problem is this: the time-varying voltage v and current i at any point in the actual circuit are replaced by corresponding stationary phasors $\boldsymbol{V}$ and $\boldsymbol{I}$. Kirchhoff's laws are applied to the image circuit, in terms of the stationary phasors, in conjunction with the rule that a complex current $\boldsymbol{I}$ on flowing through a resistor R, an inductor L, or a capacitor C produces a complex voltage drop $R\boldsymbol{I}$, $(\mathrm{j}\omega L)\boldsymbol{I}$, or $(1/\mathrm{j}\omega C)\boldsymbol{I}$ respectively. When these rules are applied it will be found that, in place of the *differential* equations in v and i applicable to the actual circuit, one obtains *algebraic* equations in $\boldsymbol{V}$ and $\boldsymbol{I}$ applicable to the image circuit. The solution of the algebraic equations is relatively simple and it then remains only to interpret the solution in terms of the original variables. This elegant technique was first introduced by C. P. Steinmetz (1865–1923), a famous electrical engineer who worked in America, and an example of it is given below.

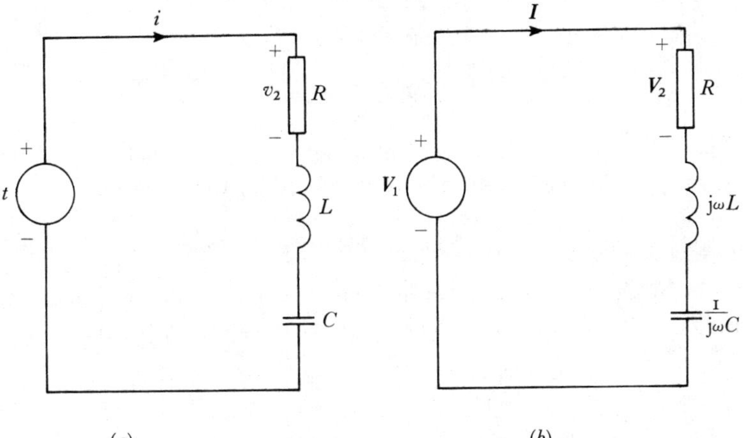

Fig. 4.4 Sinusoidal circuit: (*a*) actual circuit; (*b*) image circuit. ($R = 10\ \Omega$, $L = 10$ mH, $C = 2\ \mu$F, $V_1 = 100$ V, $\omega = 2\pi \times 10^3$ rad/s.)

Suppose that it is required to find the voltage v_2 across the resistor R in fig. 4.4(*a*). First the image circuit, fig. 4.4(*b*), is set up and then Kirchhoff's loop law is applied to it with the aid of the rules mentioned.

$$\boldsymbol{V}_1 = R\boldsymbol{I} + (\mathrm{j}\omega L)\boldsymbol{I} + (1/\mathrm{j}\omega C)\boldsymbol{I}, \tag{4.6.1}$$

$$= (R + \mathrm{j}\omega L + 1/\mathrm{j}\omega C)\boldsymbol{I}.$$

Thus
$$\boldsymbol{I} = \boldsymbol{V}_1/(R + \mathrm{j}\omega L + 1/\mathrm{j}\omega C). \tag{4.6.2}$$

Substitution of this expression for $\boldsymbol{I}$ into the equation $\boldsymbol{V}_2 = R\boldsymbol{I}$ gives

$$\boldsymbol{V}_2 = \left[\frac{R}{R + \mathrm{j}(\omega L - 1/\omega C)}\right]\boldsymbol{V}_1. \tag{4.6.3}$$

Equation (4.6.3) is the solution to the image problem, and by taking the magnitude of $\boldsymbol{V}_2$ and its phase relative to $\boldsymbol{V}_1$, the actual voltage v_2 can be constructed. The magnitude of $\boldsymbol{V}_2$ is given by

$$V_2 = [R/\sqrt{\{R^2 + (\omega L - 1/\omega C)^2\}}]\, V_1,$$
$$= [10/\sqrt{\{10^2 + (62.8 - 79.5)^2\}}] \times 100 \text{ V}$$
$$= 51.4 \text{ V}.$$

The phase ϕ of $\boldsymbol{V}_2$ relative to $\boldsymbol{V}_1$ is

$$\phi = -\tan^{-1}(\omega L - 1/\omega C)/R,$$
$$= -\tan^{-1}(62.8 - 79.5)/10,$$
$$= -\tan^{-1}(-1.67),$$
$$= +59.1 \text{ degrees}.$$

The voltage v_2 thus has a magnitude of 51.4 V and leads the applied voltage v_1 by 59.1 degrees; it may be represented by the equation

$$v_2 = 51.4 \cos(\omega t + 59.1^\circ). \tag{4.6.4}$$

So far, (4.6.1) has been set up by means of the rules outlined at the beginning of the section, and it is now necessary to justify them. In what follows,

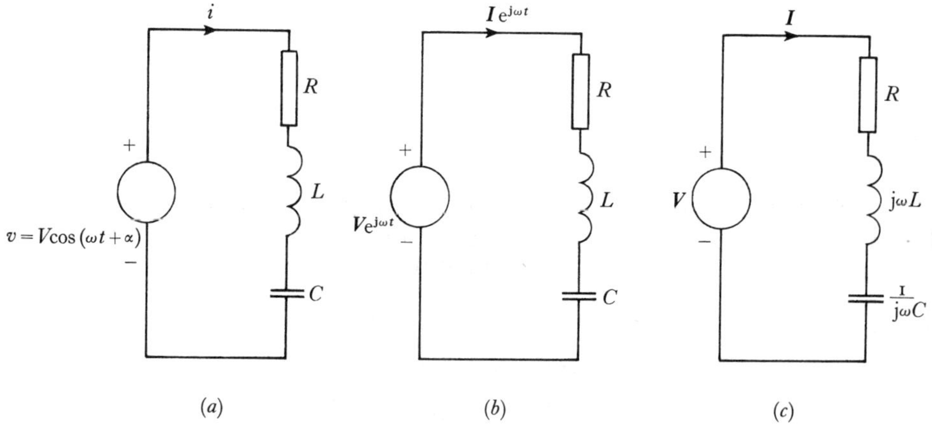

Fig. 4.5 *RLC* circuit: (*a*) actual circuit; (*b*) intermediate circuit; (*c*) image circuit.

a justification for the particular circuit shown in fig. 4.5(*a*) will be given; justification in more general terms is given in §4.9. It is supposed that the problem is to find the current i which flows as a result of the applied voltage v, and thereby to establish the validity of the image circuit (fig. 4.5(*c*)) and the loop equation, (4.6.1). A useful stepping stone in the argument is the intermediate circuit shown in fig. 4.5(*b*). Here the actual voltage $V\cos(\omega t+\alpha)$ has been replaced by the complex voltage $\boldsymbol{V}\mathrm{e}^{\mathrm{j}\omega t}$ where $\boldsymbol{V}$ is the stationary phasor $V\mathrm{e}^{\mathrm{j}\alpha}$ so chosen as to make the real part of the complex stimulus $\boldsymbol{V}\mathrm{e}^{\mathrm{j}\omega t}$ equal to the actual stimulus $V\cos(\omega t+\alpha)$. Similarly the actual current i is replaced in the intermediate circuit by the complex current $\boldsymbol{I}\mathrm{e}^{\mathrm{j}\omega t}$ where $\boldsymbol{I}$ is an unknown stationary phasor of the form $I\mathrm{e}^{\mathrm{j}\beta}$. The Kirchhoff loop equation for the intermediate circuit is then

$$\boldsymbol{V}\mathrm{e}^{\mathrm{j}\omega t} = R\boldsymbol{I}\mathrm{e}^{\mathrm{j}\omega t} + L\frac{\mathrm{d}}{\mathrm{d}t}(\boldsymbol{I}\mathrm{e}^{\mathrm{j}\omega t}) + \frac{1}{C}\int \boldsymbol{I}\mathrm{e}^{\mathrm{j}\omega t}\,.\,\mathrm{d}t. \tag{4.6.5}$$

But
$$L\frac{\mathrm{d}}{\mathrm{d}t}(\boldsymbol{I}\mathrm{e}^{\mathrm{j}\omega t}) = \mathrm{j}\omega L(\boldsymbol{I}\mathrm{e}^{\mathrm{j}\omega t}), \tag{4.6.6}$$

and
$$\frac{1}{C}\int \boldsymbol{I}\mathrm{e}^{\mathrm{j}\omega t}\,.\,\mathrm{d}t = \frac{1}{\mathrm{j}\omega C}(\boldsymbol{I}\mathrm{e}^{\mathrm{j}\omega t}). \tag{4.6.7}$$

The constant of integration which would normally appear in (4.6.7) is zero under the assumption, made throughout this chapter, that the sinusoidal stimulus was applied to the circuit an infinitely long time ago and that the circuit has settled down to its steady state. When this assumption is not true, the circuit has to be analysed by the methods described in chapter 5. Using (4.6.6) and (4.6.7), (4.6.5) becomes

$$\boldsymbol{V}\mathrm{e}^{\mathrm{j}\omega t} = (R+\mathrm{j}\omega L + 1/\mathrm{j}\omega C)\boldsymbol{I}\mathrm{e}^{\mathrm{j}\omega t}. \tag{4.6.8}$$

Dividing both sides of (4.6.8) by $\mathrm{e}^{\mathrm{j}\omega t}$, a term which is never equal to zero, reduces it to an equation containing only stationary phasors and illustrates the validity of the image circuit shown in fig. 4.5(*c*) and the rules which apply to it:

$$\boldsymbol{V} = (R+\mathrm{j}\omega L + 1/\mathrm{j}\omega C)\boldsymbol{I}. \tag{4.6.9}$$

It now remains only to relate the image current $\boldsymbol{I}$ to the actual current i. This is done by invoking a principle which states that if the actual stimulus v is the real (imaginary) part of the complex stimulus $\boldsymbol{V}\mathrm{e}^{\mathrm{j}\omega t}$ then the actual response i is the real (imaginary) part of the complex response $\boldsymbol{I}\mathrm{e}^{\mathrm{j}\omega t}$. This principle will be discussed more fully in §4.7. For the moment it suffices to say that an essential condition for its truth is that the mathematical operations performed on the complex variable in the process of obtaining the complex response must be linear ones. The application of this principle to (4.6.8) produces the solution for i. Since $\boldsymbol{V}e^{j\omega t}$ was so chosen as to satisfy the equation $v = \operatorname{Re}\boldsymbol{V}e^{j\omega t}$, then it follows that

$$i = \operatorname{Re}\boldsymbol{I}e^{j\omega t},$$

$$= \operatorname{Re}\frac{\boldsymbol{V}e^{j\omega t}}{R + j\omega L + 1/j\omega C},$$

$$= \operatorname{Re}\frac{V e^{j(\omega t+\alpha)}}{\sqrt{\{R^2 + (\omega L - 1/\omega C)^2\}}\, e^{j\phi}},$$

where $\tan\phi = (\omega L - 1/\omega C)/R$,

$$= \frac{V}{\sqrt{\{R^2 + (\omega L - 1/\omega C)^2\}}}\cdot\cos(\omega t + \alpha - \phi). \tag{4.6.10}$$

Before going any further it is useful to recall that the practical method of solving a circuit problem by complex algebra is that shown in the initial example at the start of this section and is not the method used above in

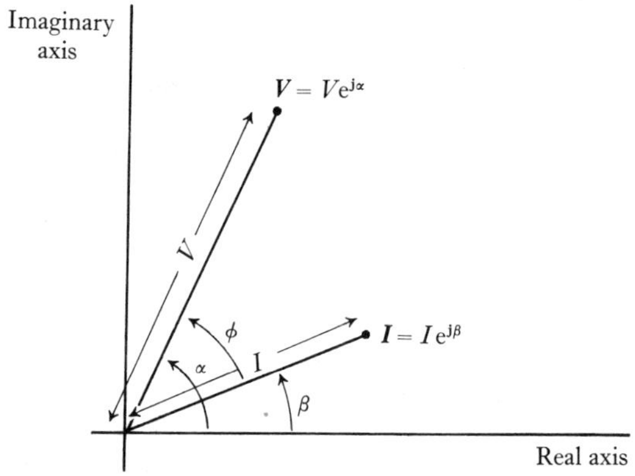

Fig. 4.6 Diagram of stationary phasors.

justification of it. Once the validity of the image circuit has been established in general, (4.6.9) may be written down by inspection or, in more complicated circuits, a series of algebraic equations may be written down from which the unknown image voltage or current may be obtained by simultaneous solution. It is then customary to translate the image quantity into the actual quantity by visualizing a diagram of stationary phasors. Such a diagram is shown in fig. 4.6 for the given problem; it shows that the geometrical treatment of a.c. circuits in chapter 3 and the analytical treatment in the present chapter are two facets of an essentially single method. If the magnitude I and phase β of the image current $\boldsymbol{I}$ are known, then the actual current i must take the form

$$i = I\cos(\omega t + \beta);$$

but from (4.6.9) and (4.4.7),

$$\text{magnitude of } \boldsymbol{I} = \frac{\text{magnitude of } \boldsymbol{V}}{\text{magnitude of } (R + \mathrm{j}\omega L + 1/\mathrm{j}\omega C)},$$

$$= \frac{V}{\sqrt{\{R^2 + (\omega L - 1/\omega C)^2\}}},$$

$$\text{and phase of } \boldsymbol{I} = \text{phase of } \boldsymbol{V} - \text{phase of } (R + \mathrm{j}\omega L - 1/\mathrm{j}\omega C),$$

$$= \alpha - \phi;$$

whence

$$i = \frac{V}{\sqrt{\{R^2 + (\omega L - 1/\omega C)^2\}}} \cdot \cos(\omega t + \alpha - \phi).$$

Once this process is grasped, the actual current i can be obtained from the image current $\boldsymbol{I}$ by inspection, and the whole solution occupies only a few lines.

4.7 Correspondence between real and imaginary parts of stimulus and response

It remains to be shown that, in the solution to the image problem, there is a direct correspondence between the real part of the complex stimulus and the real part of the complex response. Consider the specific problem of fig. 4.5. The differential equation of the actual circuit, fig. 4.5(*a*), is

$$V\cos(\omega t + \alpha) = Ri + L\frac{\mathrm{d}i}{\mathrm{d}t} + \frac{1}{C}\int i\,.\,\mathrm{d}t. \tag{4.7.1}$$

Now let the complex current shown in fig. 4.5(*b*) have real and imaginary parts i_R and i_I respectively so that

$$\boldsymbol{I}\,\mathrm{e}^{\mathrm{j}\omega t} = i_R + \mathrm{j}i_I.$$

Then the differential equation for fig. 4.5(*b*) is

$$\boldsymbol{V}\,\mathrm{e}^{\mathrm{j}\omega t} = R(i\ + \mathrm{j}i_I) + L\frac{\mathrm{d}}{\mathrm{d}t}(i_R + \mathrm{j}i_I) + \frac{1}{C}\int (i_R + \mathrm{j}i_I)\,.\,\mathrm{d}t,$$

and on expanding the left hand side and rearranging the right hand side, this becomes

$$V\cos(\omega t+\alpha)+\mathrm{j}V\sin(\omega t+\alpha)$$

$$=\left[Ri_R+L\frac{\mathrm{d}i_R}{\mathrm{d}t}+\frac{1}{C}\int i_R.\mathrm{d}t\right]+\mathrm{j}\left[Ri_I+L\frac{\mathrm{d}i_I}{\mathrm{d}t}+\frac{1}{C}\int i_I.\mathrm{d}t\right]. \quad (4.7.2)$$

Equating real and imaginary parts on both sides now splits (4.7.2) into two equations, one of which is identical with (4.7.1) provided that $i_R = i$. That is to say, the solution for i_R, the real part of the complex current $\boldsymbol{I}e^{j\omega t}$, is the solution for i, the actual current. Similarly, if the actual stimulus had been $V\sin(\omega t+\alpha)$ rather than $V\cos(\omega t+\alpha)$, the second equation contained in (4.7.2) would have corresponded to the solution of the actual circuit and the actual response i would then have been given by the *imaginary* part of the complex response $\boldsymbol{I}e^{j\omega t}$.

Although the validity of this principle has been demonstrated here only in a specific example, it may be shown by extension to apply to the steady state solution of any linear circuit, with sinusoidal stimuli, in which only linear operations (such as, for example, differentiation and integration) are performed on the variables. More will be said about non-linear operations when the calculation of power by the complex method is considered.

4.8 Complex impedance

In a d.c. circuit, the ratio of the voltage across a branch of the circuit, to the current flowing through that branch is called the resistance of the branch. In a.c. circuits this concept is extended, as was mentioned in §3.5; resistance becomes generalized into impedance and conductance into admittance. It is now useful to extend the concept further in order to include the use of complex numbers. The ratio of the stationary phasor (complex number) $\boldsymbol{V}$ to the corresponding stationary phasor (complex number) $\boldsymbol{I}$ is called the *complex impedance* $\boldsymbol{Z}$ of the branch to which these quantities refer. Thus the complex impedance of the circuit seen by the generator in fig. 4.5(*a*), and which consists of a resistor R, an inductor L and a capacitor C in series, is

$$\begin{aligned}\boldsymbol{Z} &= R+\mathrm{j}\omega L+1/\mathrm{j}\omega C,\\ &= R+\mathrm{j}(\omega L-1/\omega C),\\ &= \sqrt{\{R^2+(\omega L-1/\omega C)^2\}}.\angle[\tan^{-1}\{(\omega L-1/\omega C)/R\}],\\ &= Z\angle\phi. \end{aligned} \quad (4.8.1)$$

The complex impedance has a magnitude Z that is identical with the earlier definition of 'impedance' in chapter 3, but there is now associated with this magnitude a phase angle, ϕ. The complex impedance may therefore be displayed on an Argand diagram. The real part of $\boldsymbol{Z}$ corresponds to the resistive component of the impedance, and the imaginary part of $\boldsymbol{Z}$ corresponds to the reactive component of the impedance. Thus if

$$\boldsymbol{Z} = R + \mathrm{j}X, \tag{4.8.2}$$

then $$R = \mathrm{Re}\,\boldsymbol{Z},$$

and $$X = \mathrm{Im}\,\boldsymbol{Z}.$$

In general, both R and X will be functions of frequency.

Just as the complex impedance $\boldsymbol{Z}$ is defined by the equation

$$\boldsymbol{Z} = \boldsymbol{V}/\boldsymbol{I}$$

so also may the *complex admittance* of a branch be defined by

$$\boldsymbol{Y} = \boldsymbol{I}/\boldsymbol{V}$$

and if $\boldsymbol{Y}$ is expressed in terms of its real and imaginary parts, the real part corresponds to the conductive component of the admittance, and the imaginary part corresponds to the susceptive component of the admittance. As before, if

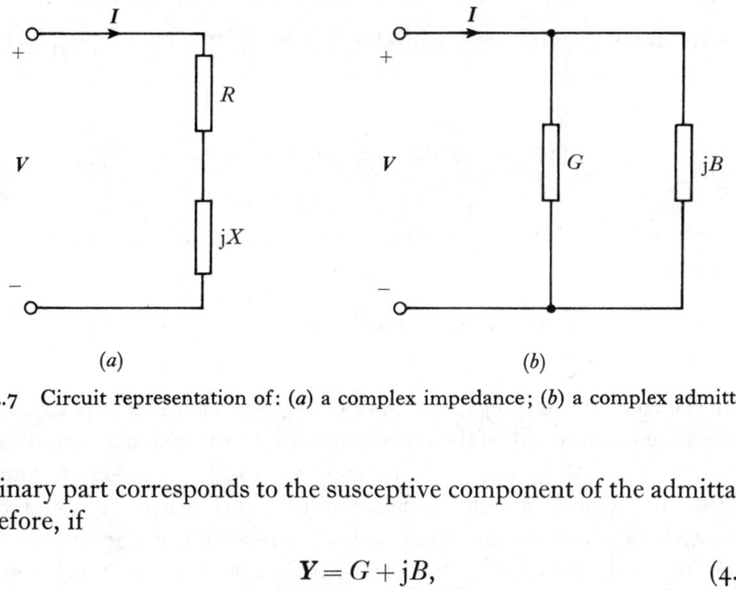

Fig. 4.7 Circuit representation of: (*a*) a complex impedance; (*b*) a complex admittance.

$$\boldsymbol{Y} = G + \mathrm{j}B, \tag{4.8.3}$$

then $$G = \mathrm{Re}\,\boldsymbol{Y},$$

and $$B = \mathrm{Im}\,\boldsymbol{Y}.$$

The image circuit representations of (4.8.2) and (4.8.3) are shown in figs. 4.7(*a*) and (*b*) respectively.

4.9 Generalization of the complex method

The solution to a particular circuit problem was obtained in §4.6 by means of the complex method. The object of the present section is to show that the same technique is valid for all problems in which it is required to find the steady-state response to a sinusoidal stimulus applied to a linear circuit. To do this consider a sinusoidal function of time $w(t) = W\cos(\omega t+\alpha)$, which may be pictured as the response of the circuit to a stimulus of similar form, and which may represent either a voltage or a current. Associated with $w(t)$ there is always a complex function whose real part is $w(t)$. Thus

$$\begin{aligned}\boldsymbol{W}\mathrm{e}^{\mathrm{j}\omega t} &= W\mathrm{e}^{\mathrm{j}\alpha}.\mathrm{e}^{\mathrm{j}\omega t},\\ &= W\mathrm{e}^{\mathrm{j}(\omega t+\alpha)},\\ &= W\cos(\omega t+\alpha)+\mathrm{j}W\sin(\omega t+\alpha),\\ &= w(t)+\mathrm{j}u(t).\end{aligned}$$

Now suppose we have a time-differential equation involving $w(t)$, of which a typical term is $A.\dfrac{\mathrm{d}^n}{\mathrm{d}t^n}[w(t)]$. If $w(t)$ is replaced in the time-differential equation by the associated complex function $\boldsymbol{W}\mathrm{e}^{\mathrm{j}\omega t}$, this typical term becomes

$$A.\frac{\mathrm{d}^n}{\mathrm{d}t^n}[\boldsymbol{W}\mathrm{e}^{\mathrm{j}\omega t}] = A\boldsymbol{W}.\frac{\mathrm{d}^n}{\mathrm{d}t^n}[\mathrm{e}^{\mathrm{j}\omega t}] = (\mathrm{j}\omega)^n.A\boldsymbol{W}\mathrm{e}^{\mathrm{j}\omega t}.$$

Similarly, if the term involves integration instead of differentiation the result is, for example

$$A\int\boldsymbol{W}\mathrm{e}^{\mathrm{j}\omega t}.\mathrm{d}t = A\boldsymbol{W}\int\mathrm{e}^{\mathrm{j}\omega t}.\mathrm{d}t = \frac{1}{\mathrm{j}\omega}.A\boldsymbol{W}\mathrm{e}^{\mathrm{j}\omega t},$$

where the constant of integration has been ignored because it is zero for the steady-state solution. All of the terms contain $\mathrm{e}^{\mathrm{j}\omega t}$ and, since the stimulus also contains this factor, it may be completely removed by division throughout the equation. The *differential* equation in $w(t)$ is thus shown to be reduced to an *algebraic* equation in $\boldsymbol{W}(\mathrm{j}\omega)$ and, after solving the algebraic equation for $\boldsymbol{W}$, the real part of $\boldsymbol{W}\mathrm{e}^{\mathrm{j}\omega t}$ is extracted to give the solution for w. The validity of this last step is rooted in the requirement that the mathematical processes performed on $\boldsymbol{W}\mathrm{e}^{\mathrm{j}\omega t}$ be linear ones, for it is only then that the separation of the real and imaginary parts of $\boldsymbol{W}\mathrm{e}^{\mathrm{j}\omega t}$ is maintained throughout the process. If the original equation in w demands some non-linear operation, such as would occur for example in the calculation of power, then the real and imaginary parts of $\boldsymbol{W}\mathrm{e}^{\mathrm{j}\omega t}$ become intermingled and the real part of the solution for $\boldsymbol{W}e^{j\omega t}$ is no longer the solution for w. This is consequent upon the fact that a linear operation on the real part of a complex variable gives the same result as that obtained by performing the same operation on the complex variable and then taking its real part. If L_0 symbolizes the mathematical instruction to perform a specific linear operation on $\boldsymbol{W}e^{j\omega t}$, then the above statement is expressed by the equation

$$\operatorname{Re} L_0(\boldsymbol{W}e^{j\omega t}) = L_0(\operatorname{Re} \boldsymbol{W}e^{j\omega t}). \tag{4.9.1}$$

Two examples make this clear. First consider the linear process of time-differentiation on $\boldsymbol{W}e^{j\omega t}$.

If
$$\boldsymbol{W}e^{j\omega t} = w + ju,$$

then
$$\frac{d}{dt}(\boldsymbol{W}e^{j\omega t}) = \frac{dw}{dt} + j\frac{du}{dt},$$

and
$$\operatorname{Re}\frac{d}{dt}(\boldsymbol{W}e^{j\omega t}) = \frac{dw}{dt} = \frac{d}{dt}(\operatorname{Re} \boldsymbol{W}e^{j\omega t}),$$

which is in accordance with (4.9.1). On the other hand if the operation to be performed on $\boldsymbol{W}e^{j\omega t}$ is the non-linear one of squaring, then

$$(\boldsymbol{W}e^{j\omega t})^2 = (w^2 - u^2) + 2jwu,$$

and whereas

$$\operatorname{Re}(\boldsymbol{W}e^{j\omega t})^2 = w^2 - u^2,$$
$$(\operatorname{Re} \boldsymbol{W}e^{j\omega t})^2 = w^2,$$

which is not the same. This is an important limitation to the complex method.

4.10 Examples of the complex method

1. A coil which, at a specific frequency, has a resistance of 30 Ω and a reactance of 100 Ω is connected in series with a capacitor which, at the same frequency, has a reactance of 60 Ω. Find the impedance and admittance of the circuit at the given frequency, and find the current that would flow under the application of 10 V at that frequency.

Solution:
$$\boldsymbol{Z} = R + j\omega L + \frac{1}{j\omega C},$$
$$= R + jX_L - jX_C,$$
$$= 30 + j(100 - 60)\,\Omega,$$
$$= 30 + j40\,\Omega,$$
$$= 50\angle 53.1^\circ\,\Omega;$$

and
$$\boldsymbol{Y} = 1/50\angle 53.1^\circ\ \text{S},$$
$$= 0.02\angle -53.1^\circ\ \text{S},$$
$$= (0.12 - \text{j}0.16)\ \text{S}.$$

Also
$$\boldsymbol{I} = \boldsymbol{YV},$$
$$= 0.02\angle -53.1^\circ\ \boldsymbol{V}.$$

Thus
$$I = 0.02\ V,$$
$$= 0.02 \times 10\ \text{A},$$
$$= 0.2\ \text{A},$$

and the current lags the applied voltage by 53.1 degrees.

2. Find the current i_2 in the circuit of fig. 4.8(*a*).

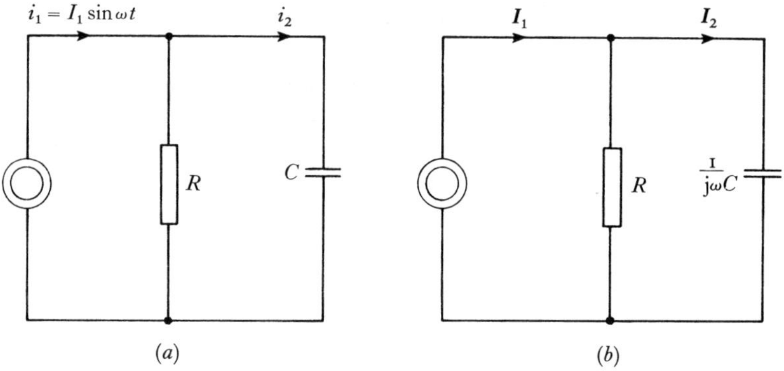

Fig. 4.8 *RC* circuit: (*a*) actual circuit; (*b*) image circuit. ($R = 100\ \Omega$, $C = 10\ \mu\text{F}$, $I_1 = 10$ A, $\omega = 2\pi \times 50$ rad/s.)

Solution: The current $\boldsymbol{I}_1$, in fig. 4.8(*b*), divides between the two branches in accordance with their relative complex impedances. Thus,

$$\boldsymbol{I}_2 = \frac{R}{R + 1/\text{j}\omega C}.\boldsymbol{I}_1,$$
$$= \frac{\text{j}\omega CR}{1 + \text{j}\omega CR}.\boldsymbol{I}_1;$$

and
$$I_2 = \frac{\omega CR}{\sqrt{\{1 + (\omega CR)^2\}}}.I_1,$$
$$= \frac{0.314}{\sqrt{\{1 + (0.314)^2\}}}\ 10\ \text{A},$$
$$= 3.0\ \text{A};$$

and $$\phi = \tan^{-1}\infty - \tan^{-1}\omega CR,$$
$$= 90° - 17.4°,$$
$$= 72.6°.$$

The current i_2 has a magnitude of 3.0 A and leads the source current by 72.6 degrees. It is therefore represented by the equation

$$i_2 = 3.0 \sin(\omega t + 72.6°).$$

3. A coil of inductance L_s and resistance R_s is used at a frequency which makes the reactance of the coil very large with respect to the resistance. Show that the coil may be represented at this frequency by an inductance L_p shunted by a resistor R_p and find the values of these components in terms of L_s, R_s, and ω.

Solution: The complex impedance of the coil is

$$\boldsymbol{Z} = R_s + \mathrm{j}\omega L_s,$$

and the complex admittance of the coil is

$$\boldsymbol{Y} = \frac{1}{\boldsymbol{Z}} = \frac{1}{R_s + \mathrm{j}\omega L_s},$$
$$= \frac{1}{R_s + \mathrm{j}\omega L_s} \cdot \frac{R_s - \mathrm{j}\omega L_s}{R_s - \mathrm{j}\omega L_s},$$
$$= \frac{R_s}{R_s^2 + (\omega L_s)^2} - \frac{\mathrm{j}\omega L_s}{R_s^2 + (\omega L_s)^2},$$
$$= \frac{1}{R_p} + \frac{1}{\mathrm{j}\omega L_p},$$

where $$R_p = \frac{R_s^2 + (\omega L_s)^2}{R_s} \approx \frac{(\omega L_s)^2}{R_s},$$

and $$L_p = \frac{R_s^2 + (\omega L_s)^2}{\omega^2 L_s} \approx L_s.$$

4.11 Complex power

It has been emphasized that the complex method of solving circuit problems is only valid for linear operations on the complex variables. It is clear that in the calculation of power, involving as it does the product of two variables, a non-linear operation is entailed. It therefore should not be expected that the product of the stationary phasors $\boldsymbol{V}$ and $\boldsymbol{I}$ will necessarily supply any useful result. This does not mean, however, that the application of complex numbers to the calculation of power is useless. On the contrary, the use of the stationary phasors $\boldsymbol{V}$ and $\boldsymbol{I}$ leads to an elegant notation for use in the calculation of power.

Figure 4.9(*a*) shows the phasor diagram of a linear circuit in which a sinusoidal voltage v drives a lagging current i. The corresponding image quantities are $\boldsymbol{V} = V\mathrm{e}^{\mathrm{j}\alpha}$ and $\boldsymbol{I} = I\mathrm{e}^{\mathrm{j}\beta}$. The product of $\boldsymbol{V}$ and $\boldsymbol{I}$ supplies no

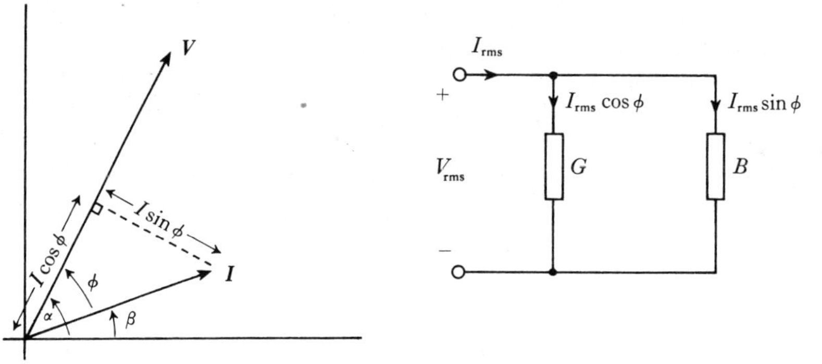

(*a*) (*b*)

Fig. 4.9 Power in a sinusoidal circuit: (*a*) phasor diagram ($\boldsymbol{V}$, $\boldsymbol{I}$ denote peak values); (*b*) circuit representation.

useful result, but the product of $\boldsymbol{V}$ and the *conjugate* of $\boldsymbol{I}$ leads to the required solution for the power in the circuit. If

$$\boldsymbol{I} = I\,\mathrm{e}^{\mathrm{j}\beta},$$

then

$$\boldsymbol{I}^* = I\,\mathrm{e}^{-\mathrm{j}\beta},$$

and thus

$$\begin{aligned}\tfrac{1}{2}\boldsymbol{VI}^* &= \tfrac{1}{2}VI\,\mathrm{e}^{\mathrm{j}(\alpha-\beta)},\\ &= \tfrac{1}{2}VI\,\mathrm{e}^{\mathrm{j}\phi},\\ &= \tfrac{1}{2}VI\cos\phi + \tfrac{1}{2}\mathrm{j}VI\sin\phi,\\ &= V_{\mathrm{rms}}I_{\mathrm{rms}}\cos\phi + \mathrm{j}V_{\mathrm{rms}}I_{\mathrm{rms}}\sin\phi, \qquad (4.11.1)\\ &= P + \mathrm{j}Q;\end{aligned}$$

where

$$P = \tfrac{1}{2}\,\mathrm{Re}\,\boldsymbol{VI}^*, \qquad (4.11.2)$$

and

$$Q = \tfrac{1}{2}\,\mathrm{Im}\,\boldsymbol{VI}^*.$$

The quantity $(P + \mathrm{j}Q)$ is called the *phasor power* or *complex power* of the circuit. P is the average power drawn by the circuit and, by analogy, Q is called the *reactive power* of the circuit. Both concepts are useful, and by employing them we are viewing the circuit in the manner represented by fig. 4.9(*b*) in which the purely conductive component G draws a current $I_{rms}\cos\phi$, and the purely susceptive component B draws a current $I_{rms}\sin\phi$. This representation emphasizes the fact that the generator must supply a current I_{rms} which is greater than the minimum value $I_{rms}\cos\phi$ necessary to dissipate the given power P. The excess current is associated with the energy-storing property of the susceptance B.

The significance of the product $V_{rms}I_{rms}$, as the *volt-ampere rating* of power apparatus, has already been pointed out in §3.6. In the terms there defined,

P is the power (W);
Q is the reactive volt-amperes (VAr);
$\surd(P^2+Q^2)$ is the volt-amperes (VA).

The complex notation conveniently handles the power and the reactive volt-amperes simultaneously.

4.12 The h-operator in three phase circuits

It was shown in §4.4 that the multiplication of a stationary phasor by $e^{j\theta}$ rotates the phasor in the anticlockwise direction by the angle θ, and that the special case of multiplication by $e^{j\pi/2}$, which is j, rotates the phasor by 90 degrees. It is helpful in this connexion to think of j as an *operator* so that $j\boldsymbol{V}$, for example, is pictured as the quantity j operating on $\boldsymbol{V}$, rotating it by 90 degrees in the forward direction. Operators which rotate a phasor by angles other than 90 degrees may also be useful, and an example is to be found in the analysis of three-phase systems.

It will be recalled from chapter 3 that a balanced three-phase source has voltages which are equal in magnitude but displaced from each other in phase by 120 degrees. If one of these three voltages is represented by the phasor $\boldsymbol{V}_A$, as in fig. 3.16(*a*), the other two may be written $\boldsymbol{V}_B=\boldsymbol{V}_A e^{j4\pi/3}$ and $\boldsymbol{V}_C=\boldsymbol{V}_A e^{j2\pi/3}$. If now we write

$$h=e^{j2\pi/3}, \tag{4.12.1}$$

the three phase-voltages may be expressed by the quantities $\boldsymbol{V}_A$, $h^2\boldsymbol{V}_A$, and $h\boldsymbol{V}_A$ respectively, in which h may be treated as a 120 degree operator. It is interesting to note that

$$h=-1/2+j\surd 3/2,$$

$$h^3=1,$$

and

$$1+h+h^2=0.$$

The first of these results is obtained by putting θ equal to 120 degrees in (4.4.4.), the second corresponds with the rotation of a phasor by 120 degrees three times in succession, and the last expresses the fact that the phasor sum of three balanced phasors is zero.

The h-operator finds its principal application in the analysis of unbalanced three-phase systems by the method of *symmetrical components*. In this method the unbalanced system is represented by the superposition of three sets of three phasors; two of the sets are balanced and the third has its three phasors in phase. A full treatment of this method will be found in textbooks on electric power systems.†

4.13 Notation and terminology

The actual variables v, i, etc., which are functions of time, and their image equivalents $\boldsymbol{V}$, $\boldsymbol{I}$, etc., which are functions of (jω), have been distinguished throughout this chapter by the use of small letters for the actual (time) variables and bold capital letters for their image equivalents. This distinction is not always maintained in practical work. For example, in the analysis of electronic circuits, in which a small letter is commonly used to denote an increment about a specific value of a variable, the same small letter is often used to denote not only the incremental quantity but also its corresponding image equivalent. Confusion is avoided simply by observing the type of equation in which the variable appears; equations in time denote the actual world whereas those in (jω) denote the image world.

A more important variant in notation arises when the stationary phasors are expressed in terms of the r.m.s. values rather than the peak values of the actual variables. Consider a phasor diagram in which the lengths of the individual phasors correspond to the peak values of the variables. If this diagram is now scaled down by the factor $\sqrt{2}$, the result is a phasor diagram unchanged except that the lengths of the phasors now correspond to the r.m.s. values of the variables. Each phasor can now be represented by an equation of the form $\boldsymbol{V} = V_{\mathrm{rms}}\mathrm{e}^{\mathrm{j}\alpha}$, and the modulus of $\boldsymbol{V}$ is now the r.m.s. value of the actual voltage v rather than its peak value. The manipulation of the complex numbers proceeds as before, but the moduli express r.m.s. values throughout. This approach is particularly useful in the calculation of power since it eliminates the factor $\frac{1}{2}$ which otherwise appears in equations like (4.11.2).

Examples on chapter 4

4.1 A Wheatstone bridge PQRS is supplied at PR from an ideal voltage source in series with a 1 Ω resistor. The four arms PQ, QR, RS, and SP respectively have impedances of 3j Ω, 4 Ω, −2j Ω, and 6 Ω at the source frequency. Find: (*a*) the impedance seen by the source; (*b*) the voltage between

† For example, A. E. Guile and W. Paterson, *Electrical power systems*, vol. 1, Oliver and Boyd (1969).

points S and Q and its phase relative to the source; (*c*) the power factor of the circuit seen by the source.

4.2 An ideal a.c. current source supplies a resistor R, an inductor L and a capacitor C, all connected in parallel. If (*a*) the capacitor is varied to give the maximum current in L and (*b*) the inductor is varied to give the maximum current in C, show that the ratio of these maxima is $\omega^2 L_1 C_1$, where L_1, C_1 are the initial values of L, C.

4.3 A resistor R_1 in series with a capacitor C_1 is connected across an ideal voltage source V_1; and across C_1 is connected a resistor R_2 in series with a capacitor C_2. Find the voltage ratio (V_2/V_1), where V_2 is the open-circuit voltage across C_2. For the special case $R_1 = R_2$, $C_1 = C_2$, sketch the magnitude and phase of (V_2/V_1) as a function of frequency.

4.4 Across the terminals of an ideal voltage source which has an open-circuit voltage of 1 V at a frequency f is connected an ideal inductor L in series with an ideal capacitor C. Across C is connected a second inductor (identical with the first) in series with a 1 kΩ resistor R. At a frequency of 5 kHz the inductors each have a reactance of 0.5 kΩ, and the capacitor has a reactance of 2 kΩ. What is the magnitude and phase, relative to that of the source, of the voltage across R when f is (*a*) 5 kHz and (*b*) 10 kHz?

4.5 A source has an open-circuit voltage of 100 V and an internal impedance of $(1178 + j444)$ Ω. It is shunted by two branches whose impedances are $(488 + j122)$ Ω and $(102 - j294)$ Ω. What is the current in the second branch?

4.6 In a two mesh electrical network the values of the total resistance, inductance, and capacitance in the first mesh are 4.9 Ω, 97.5 mH and 30 μF. In the second mesh there are resistors and inductors only, whose total values are 11 Ω and 0.386 H, whilst the values common to the two meshes are 0.74 Ω and 71.7 mH. Ideal voltage sources at a frequency of 60 Hz exist in the first and second meshes whose values are $75\angle 90°$ V and $115\angle -90°$ V respectively, acting clockwise round the meshes. What are the two mesh currents?

4.7 The primary and secondary resistances and inductances of a transformer are 2π Ω and 10 mH, and 3π Ω and 15 mH respectively, and its mutual inductance is 2 mH. Ideal voltage sources having a frequency of 1 kHz and values of $200\angle 0°$ V and $200\angle 90°$ V respectively are connected across the primary and secondary. Find the magnitudes and phase angles of the primary and secondary currents when the two windings are arranged so that the polarity-indicating dots are at the same end.

4.8 Consecutive elements round one mesh of a two mesh circuit are R_1, C, R_2 and L. An ideal voltage source in series with a resistor R_0 is applied to the junction of R_1 with C and of R_2 with L to form the second mesh. Find the conditions under which the impedance seen by the source is a real constant independent of frequency, and prove that under these conditions the voltage between the junctions of L with R_1 and of C with R_2 is zero at all frequencies.

4.9 The open-circuit primary and secondary inductances of a transformer and its mutual inductance are 0.6 mH, 0.9 mH and 0.1 mH respectively. Across the secondary is a 500 pF capacitor in series with a resistance of 30 Ω (which includes the resistance of the secondary winding). In series with the primary is a 1000 pF capacitor, a 10 Ω resistor (which includes the resistance of the primary winding), and an ideal voltage source which has a frequency of 200 kHz. What is the impedance seen by the source?

4.10 Two coils are connected in parallel across an ideal voltage source in such a way that when currents flow through the coils in the same direction their corresponding fluxes are aiding. The coil constants are $R_1 = 10$ Ω, $L_1 = 10$ mH, $R_2 = 20$ Ω, $L_2 = 10$ mH, $M = 3$ mH. What is the r.m.s. value of the total current drawn from the source if it has a value of 100 V r.m.s. at a frequency of 1 kHz?

4.11 Two three-phase, star/star transformers are connected with each phase winding of one transformer in parallel with the corresponding winding of the other, on both the primary and secondary side. The polarities are such that circulating currents are kept to a minimum; but they are not zero, for while the primary turns are equal in the two transformers, the secondary/primary turns ratios are 3.1 and 3.0 respectively. The 'leakage reactance' of each transformer may be represented by an inductive reactance of 100 Ω in each secondary phase; their resistance can be neglected. The primary windings are supplied at a line voltage of 11 kV; the secondaries jointly supply a star-connected load of $(1000 + j500)$ Ω per phase. Calculate the secondary terminal voltage and its phase angle relative to the secondary e.m.f.s. Hence calculate the current in each transformer.

4.12 A voltage, which may be represented by the complex number $(a + jb)$, is applied to a circuit which consists of two branches in parallel. The first draws a complex current $(d + je)$. The second consists of several elements in series, one of which, an ideal capacitor C, has a voltage $(f + jh)$ across it. What are the real and reactive powers drawn by the circuit?

5 Introduction to the theory of transients

5.1 Transients

The circuits discussed in the preceding chapters have all been in the condition known as a *steady state*; that is, the voltages and currents are either constant quantities, or sinusoidally varying quantities of constant amplitude and frequency. Such a state is attained in practice when a circuit has been left in operation for a sufficient time without any change in the connexions or in the impressed energy supplies. When these conditions are changed, however, – for example by the closing or opening of a switch – there generally ensues a period during which the circuit is adapting itself to the new conditions. It is then in a state of transition, and the phenomena occurring during this period are called *transients*. To begin the study of these is the purpose of this chapter.

5.2 The transient behaviour of a simple circuit

The basic features of the transient problem are best illustrated by considering an example which is as simple as possible. Consider then the series RC circuit, illustrated in fig. 5.1, into which a constant-voltage source V_1 (typified by a battery) can be connected by closing a switch. Except in the

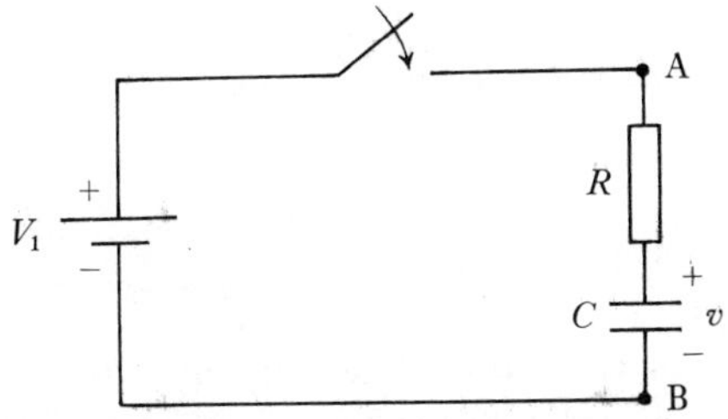

Fig. 5.1 Series RC circuit subjected to voltage stimulus suddenly applied.

particular case when the capacitor happens to be already charged to a voltage V_1, the closing is followed by a flow of current through the resistor, with consequent changes in the charge q and voltage v on the capacitor. Among these quantities attention will be directed at v, which is thus regarded as the *response* of the circuit to the shock, or *stimulus*, occasioned by the closure of the switch.

Let time t be measured from the instant of closure, and let the capacitor be initially charged to a voltage v_0. At subsequent instants the charges on its electrodes are $\pm q$, where $q = Cv$; and when q changes, the current through the resistor is given by

$$i = \mathrm{d}q/\mathrm{d}t = C\,.\,\mathrm{d}v/\mathrm{d}t.$$

Thus the Kirchhoff equation of the circuit, namely

$$Ri + v = V_1$$

may be written

$$RC\,.\frac{\mathrm{d}v}{\mathrm{d}t} + v = V_1,$$

or

$$\frac{\mathrm{d}v}{\mathrm{d}t} + \alpha v = \alpha V_1, \tag{5.2.1}$$

where

$$\alpha = 1/RC. \tag{5.2.2}$$

If both sides of (5.2.1) are multiplied by $\mathrm{e}^{\alpha t}$, the left hand member becomes equal to $\mathrm{d}/\mathrm{d}t(v\,\mathrm{e}^{\alpha t})$; thus the equation takes the form

$$\frac{\mathrm{d}}{\mathrm{d}t}(v\,\mathrm{e}^{\alpha t}) = \alpha V_1\,\mathrm{e}^{\alpha t}.$$

Integrating this equation over the range from $t = 0$ to $t = t_1$, we obtain

$$[v\,\mathrm{e}^{\alpha t}]_0^{t_1} = \alpha V_1 \int_0^{t_1} \mathrm{e}^{\alpha t}\,\mathrm{d}t,$$

$$= V_1[\mathrm{e}^{\alpha t}]_0^{t_1},$$

and if $v = v_1$ when $t = t_1$, this may be written

$$v_1\mathrm{e}^{\alpha t_1} - v_0 = V_1\,(\mathrm{e}^{\alpha t_1} - 1),$$

or

$$v_1 = V_1(1 - \mathrm{e}^{-\alpha t_1}) + v_0\,\mathrm{e}^{-\alpha t_1}.$$

Now t_1 can have any positive value. The suffixes 1 can therefore be omitted, and there is a continuing relation between v and t given by the equation

$$v = V_1(1 - \mathrm{e}^{-\alpha t}) + v_0\,\mathrm{e}^{-\alpha t}. \tag{5.2.3}$$

This solution presents features which are characteristic of transients in all linear circuits. It falls into two parts: the term $V_1(1 - \mathrm{e}^{-\alpha t})$, which depends upon the stimulus V_1 but is unaffected by the initial voltage v_0, and the term $v_0\mathrm{e}^{-\alpha t}$, in which the reverse is true. Textbooks on differential equations call the former term a *particular integral*, the latter a *complementary function*; but from the engineering standpoint it is more enlightening to observe that the two terms may be given a physical significance. They are indeed the solutions of two separate problems:

1. A capacitor, initially uncharged, is charged by connecting it through a resistor to a battery of voltage V_1.
2. A capacitor charged to a voltage v_0 is discharged through a resistor, there being no battery in the circuit.

Why can the problem be broken down into two simpler problems in this way? It is worth while to examine the question, for the answer introduces concepts which are fundamental in the theory of the transient behaviour of circuits.

5·3 The step function

Consider again the circuit of fig. 5.1, but suppose the capacitor to be initially uncharged. Closing the switch causes a sudden change of magnitude V_1 in the potential difference between the points A and B. When a quantity leaps in this way from one constant value to another, it is described as a *step function*. The standard step function is the function $u(t)$ defined by the following equations:

$$\left.\begin{aligned} u(t) &= 0 \text{ when } t < 0, \\ &= 1 \text{ when } t > 0. \end{aligned}\right\} \tag{5.3.1}$$

Oliver Heaviside (1850–1925), the great progenitor of the theory of transients in electric circuits and of much electrical theory besides, gave to $u(t)$ the name of *unit function*; but the term 'step' is so descriptive of its shape (fig. 5.2) that *unit step function* is preferable.

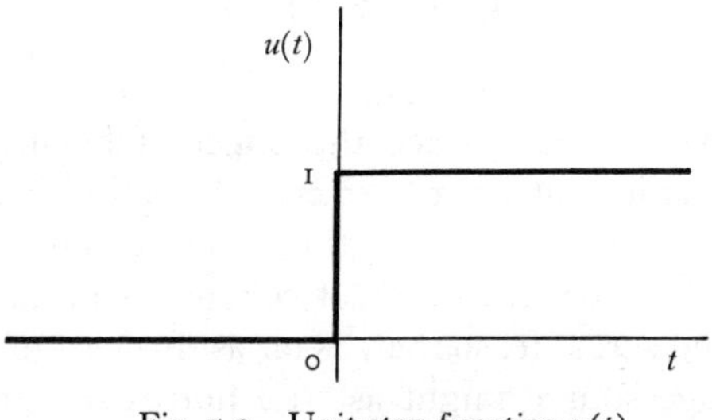

Fig. 5.2 Unit step function $u(t)$.

In addition to applying a step function of voltage between A and B, the closure of the switch effects a change in the circuit connexions, so that the impedance, infinite up to the instant $t = 0$, is finite thereafter. It will be seen that switching operations, whether closing or opening, will frequently possess this property of comprising two aspects in one. In the present case it is easy to see that the application of the step function voltage is significant, while the change in the circuit impedance is incidental; for the infinite impedance regime ends at time $t = 0$, and its end result is completely specified by citing the capacitor voltage at that instant. In other problems the change of impedance on switching will be significant; but there is always a possible ambiguity. Now electric circuit theory, like all other physical theories, consists in investigating the consequences of precisely defined assumptions, suggested by practice but not necessarily corresponding to it in every detail. Switching is not a precise concept; we therefore eliminate it by replacing the battery and switch of fig. 5.1 by an ideal voltage source generating the step function voltage $V_1 u(t)$. This substitution leads to the circuit shown in fig. 5.3, which is equivalent to the original circuit in every

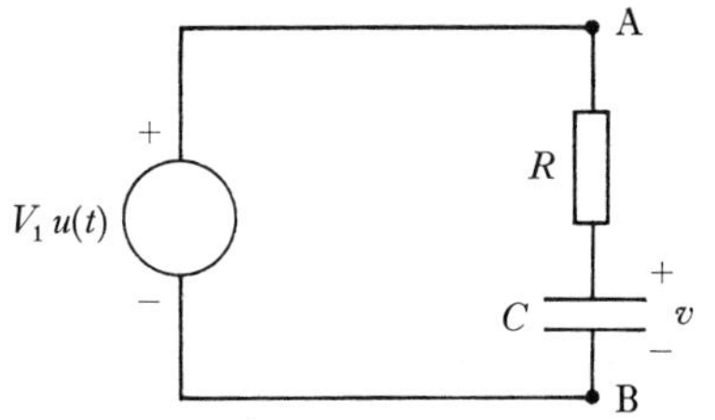

Fig. 5.3 Circuit of fig. 5.1 with step function source replacing battery and switch.

essential feature. The response of this circuit to the stimulus $V_1 u(t)$, given that $v = 0$ when $t = 0$, is already known from (5.2.3): namely $v = V_1(1 - e^{-\alpha t})$. As the circuit is quiescent up to the instant $t = 0$, a more precise description of its performance is given by

$$v = V_1(1 - e^{-\alpha t})u(t) \tag{5.3.2}$$

This is illustrated in fig. 5.4 (curve (1)).

It is now desirable to reintroduce the effect of having the capacitor initially charged so that its voltage at $t = 0$ is v_0. Equation (5.2.3) shows that this results in adding the term $v_0 e^{-\alpha t}$ to the response. This term, shown in fig. 5.4 as curve (2), describes the decay of voltage on a capacitor C when a resistor R is connected across it; so that, as far as this part of the response is concerned, the voltage source might as well not exist – it is completely quiescent and equivalent to a short circuit. There emerges the mental picture of two superimposed effects: first, the dying voltage (illustrated in curve (2*a*), which passes smoothly into curve (2)) on a capacitor whose charge is leaking away through the resistor and the voltage source; and second, the increasing voltage (curve (1)) due to the operation of the step function voltage source. The former effect takes no notice of the change in voltage at the terminals of the source; the latter is unaffected by the value of the initial charge on the capacitor. The fact that such a superposition of effects is possible is due to the linearity of the circuit elements, in consequence of which the total effect of several causes acting simultaneously is obtained by adding the separate effects obtained when each cause acts by itself.

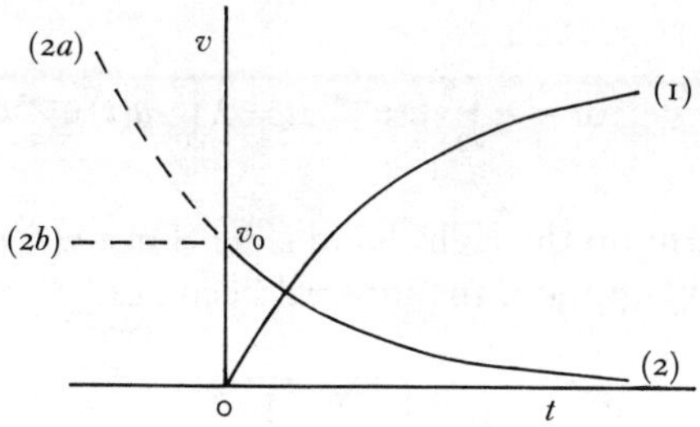

Fig. 5.4 Response of the circuit of fig. 5.3: (1) due to voltage source; (2) due to initial charge on capacitor: (2*a*) with no switch in circuit; (2*b*) with switch preventing loss of charge before $t = 0$.

When the circuit is energized by a battery and switch, the continuation of curve (2) into the region where $t < 0$ is represented by curve (2*b*) instead of (2*a*); but the difference has no effect after the instant $t = 0$.

5.4 Transformation of a problem into a more readily soluble image problem (the Laplace transformation)

We shall now approach the problem anew, as a means of introducing a powerful general method for solving transient problems in circuits. First, a slight generalization. Analysis of the growth of current in an inductor and resistor connected in series across a step function voltage source would lead to a differential equation very similar to (5.2.1), but with the current i as the dependent variable; we shall cover both voltage and current (and other possibilities) by employing a non-committal symbol w. Consider then the differential equation

$$\frac{dw}{dt} + \alpha w = \lambda u(t), \tag{5.4.1}$$

where α and λ are constants; and let $w = w_0$ when $t = 0$.

The reader must now be asked to follow a simple mathematical argument, without at present seeking to find any physical significance behind it. (In general the quest for physical meaning in mathematical manipulations comes naturally to the engineering mind, and indeed often helps to prevent the mathematics from going astray; but at the present juncture such a quest would be out of place.) Let s be an undefined complex number, having real and imaginary parts given by

$$\mathrm{s} = \sigma + \mathrm{j}\omega. \tag{5.4.2}$$

Let each term of the differential equation be multiplied by $\mathrm{e}^{-\mathrm{s}t}$, and let the products be integrated with respect to t from zero to infinity, so as to obtain

$$\int_0^\infty \frac{\mathrm{d}w}{\mathrm{d}t}\,\mathrm{e}^{-\mathrm{s}t}\,\mathrm{d}t + \alpha\int_0^\infty w\,\mathrm{e}^{-\mathrm{s}t}\,\mathrm{d}t = \lambda\int_0^\infty u(t)\,\mathrm{e}^{-\mathrm{s}t}\,\mathrm{d}t. \tag{5.4.3}$$

Consider first the term on the right hand side. Since $u(t)$ equals unity over the whole range of integration, the integral becomes

$$\int_0^\infty \mathrm{e}^{-\mathrm{s}t}\,\mathrm{d}t = \left[-\frac{\mathrm{e}^{-\mathrm{s}t}}{\mathrm{s}}\right]_0^\infty,$$

$$= \left[-\frac{\mathrm{e}^{-\sigma t}\,\mathrm{e}^{-\mathrm{j}\omega t}}{\sigma + \mathrm{j}\omega}\right]_0^\infty.$$

In order that this expression may have any meaning it is necessary for σ to be positive; for if σ is negative the integral tends to infinity with t, while if σ is zero the value for $t \to \infty$ is indeterminate. A limitation must therefore be placed on the undefined complex number s, namely that its real part must be positive. If this be granted, we obtain

$$\int_0^\infty u(t)\,\mathrm{e}^{-\mathrm{s}t}\,\mathrm{d}t = 1/\mathrm{s}. \tag{5.4.4}$$

The value of $\int_0^\infty w\mathrm{e}^{-\mathrm{s}t}\,\mathrm{d}t$ cannot be explicitly found until w is known. It can be seen, however, to be a function of s, so we write

$$\int_0^\infty w(t)\,\mathrm{e}^{-\mathrm{s}t}\,\mathrm{d}t = \boldsymbol{W}(\mathrm{s}). \tag{5.4.5}$$

Again it must be assumed that s is limited to values which make the integral meaningful; the condition $\sigma > 0$ is adequate for the type of problem at present in mind. The other integral is expressible in terms of $\boldsymbol{W}(\mathrm{s})$, for

$$\int_0^\infty \frac{\mathrm{d}w}{\mathrm{d}t}\,\mathrm{e}^{-\mathrm{s}t}\,\mathrm{d}t = \left[w\,\mathrm{e}^{-\mathrm{s}t}\right]_0^\infty + \mathrm{s}\int_0^\infty w\,\mathrm{e}^{-\mathrm{s}t}\,\mathrm{d}t, \quad \text{(integrating by parts)}$$

$$= -w_0 + \mathrm{s}\boldsymbol{W}(\mathrm{s}), \tag{5.4.6}$$

since $w = w_0$ when $t = 0$. Hence the whole differential equation becomes

$$(-w_0 + s\boldsymbol{W}) + \alpha \boldsymbol{W} = \lambda/s$$

or

$$(s + \alpha)\boldsymbol{W} = (\lambda/s) + w_0. \tag{5.4.7}$$

The significance of the operation now begins to appear. The variable t has been replaced by a new variable s; the function $w(t)$ has been associated (through (5.4.5)) with a function $\boldsymbol{W}(s)$, which will be called the *image* of $w(t)$; and the image of dw/dt is derived from that of w by the simple algebraical operation defined by (5.4.6). The process has resulted in a most elegant transformation of the differential equation into the algebraic equation (5.4.7), whose solution is

$$\boldsymbol{W} = \frac{\lambda + sw_0}{s(s+\alpha)} \tag{5.4.8}$$

– and all that is necessary to complete the solution is to find some means of getting back from the image world to the real or *original* one.

The process through which functions of a real variable t are associated with corresponding functions of a new variable s is known as a *Laplace transformation*. Equation (5.4.5) defines this transformation, and the new function $\boldsymbol{W}(s)$ is known as the *Laplace transform* of $w(t)$. The symbol $\mathscr{L}$ is used as an abbreviation for 'the Laplace transform of ...', so that we may write

$$\mathscr{L}w(t) = \int_0^\infty w(t)\,e^{-st}\,dt = \boldsymbol{W}(s). \tag{5.4.9}$$

The process of going back from the image world to the original world is called an *inverse Laplace transformation*, denoted by $\mathscr{L}^{-1}$; thus

$$\mathscr{L}^{-1}\boldsymbol{W}(s) = w(t). \tag{5.4.10}$$

While employing the symbols $\mathscr{L}$ and $\mathscr{L}^{-1}$, we shall generally prefer the word 'image' to 'Laplace transform', as being both briefer and more graphic. Functions of t will consistently be designated by lower case letters such as w, v, and i, so that the corresponding capital letters $\boldsymbol{W}$, $\boldsymbol{V}$, and $\boldsymbol{I}$ may be used to denote the corresponding image functions.

5.5 Properties of the Laplace transformation

In the preceding section, a differential equation, (5.4.1), relating a dependent variable w to an independent variable t, has been converted into the algebraic equation (5.4.7), which is easily solved so as to give the new dependent variable $\boldsymbol{W}$ in terms of the new independent variable s. The carrying out of the inverse process – namely, discovering $w(t)$ when $\boldsymbol{W}(s)$ is known –

requires a knowledge of certain general propositions which are conveniently gathered here.

1. If the image of $w(t)$ is $\boldsymbol{W}(s)$, and k is a constant, then the image of $kw(t)$ is $k\boldsymbol{W}(s)$. (5.5.1)

2. When an image consists of the sum of a number of functions of s, the original is given by adding together the corresponding functions of t. (5.5.2)

Both these follow at once from the definition of the transform; for example, if $\mathscr{L}w_1(t) = \boldsymbol{W}_1(s)$ and $\mathscr{L}w_2(t) = \boldsymbol{W}_2(s)$,

$$\mathscr{L}[w_1(t)+w_2(t)] = \int_0^\infty [w_1(t)+w_2(t)]\,e^{-st}\,dt,$$

$$= \boldsymbol{W}_1(s) + \boldsymbol{W}_2(s).$$

Therefore $\mathscr{L}^{-1}[\boldsymbol{W}_1(s)+\boldsymbol{W}_2(s)] = w_1(t)+w_2(t)$, which is proposition (5.5.2) in its simplest form. The reasoning can be extended to cover any number of terms, including an infinite number if the question of the convergence of the infinite series is watched.

3. If the image of $w(t)$ is $\boldsymbol{W}(s)$, the image of dw/dt is $s\boldsymbol{W}(s) - w_0$, where w_0 is the value of w when $t = 0$. (5.5.3)

This has already been proved—equation (5.4.6).

4. If the image of $w(t)$ is $\boldsymbol{W}(s)$,

the image of $\int_0^t w(\tau)\,d\tau$ is $\boldsymbol{W}(s)/s$. (5.5.4)

The integral is a function of t, its upper limit, and may therefore be written $f(t)$. (The variable under the sign of integration has been changed from t to τ to save possible confusion.) $f(t)$ is the function whose derivative is $w(t)$ and which is zero when $t = 0$, as may easily be verified from the definition of df/dt. Then

$$\mathscr{L}f(t) = \int_0^\infty f(t)\,e^{-st}\,dt,$$

$$= \left[-f(t)\frac{e^{-st}}{s}\right]_0^\infty + \frac{1}{s}\int_0^\infty f'(t)\,e^{-st}\,dt.$$

The bracketed term vanishes at both limits; at the lower because $f(0) = 0$, and at the upper by reason of the factor e^{-st}, assuming that the real part of s is sufficiently positive. Therefore

$$\mathscr{L}f(t) = \frac{1}{s}\int_0^\infty w(t)\,e^{-st}\,dt,$$

$$= \frac{\boldsymbol{W}(s)}{s},$$

which is (5.5.4).

5.6 Transition from the image system to the original system (inverse Laplace transformation)

The propositions set down in the last section can form the basis of a method for obtaining the solution of the original problem when that of the image problem is known. Though not universally applicable, this method covers a great range of circuit problems, and constitutes a general method for calculating transients in *lumped* circuits – that is, circuits which can be regarded as containing a finite number of ideal elements, such as pure resistors, inductors and capacitors. To go further would demand an acquaintance with the theory of functions of a complex variable; where a more elementary approach is available, there is much advantage in fully exploiting its possibilities.

It has already been shown, in (5.4.4), that $1/s$ is the image of the unit step function $u(t)$; thus $\mathscr{L}^{-1}(1/s) = u(t)$, or, since only values of t greater than zero are in question,

$$\mathscr{L}^{-1}\left(\frac{1}{s}\right) = 1. \tag{5.6.1}$$

Successive applications of (5.5.4) now give

$$\mathscr{L}^{-1}\left(\frac{1}{s^2}\right) = t,$$

$$\mathscr{L}^{-1}\left(\frac{1}{s^3}\right) = \frac{t^2}{2!},$$

and, in general,

$$\mathscr{L}^{-1}\left(\frac{1}{s^n}\right) = \frac{t^{n-1}}{(n-1)!}. \tag{5.6.2}$$

This formula makes it possible to determine the original, in the form of a series of positive powers of t, whenever the image can be expressed as a series of negative powers of s.

Consider now the image solution for the problem discussed in §5.4, namely

$$\boldsymbol{W}(s) = \frac{\lambda + s w_0}{s(s+\alpha)}. \qquad \text{(equation (5.4.8))}$$

This may be written

$$\boldsymbol{W}(s) = \frac{\lambda}{s(s+\alpha)} + \frac{w_0}{s+\alpha},$$

whence
$$w(t) = \lambda \mathscr{L}^{-1}\left(\frac{1}{s(s+\alpha)}\right) + w_0 \mathscr{L}^{-1}\left(\frac{1}{s+\alpha}\right), \tag{5.6.3}$$

on account of the propositions (5.5.1) and (5.5.2). Take first the second term; by dividing numerator and denominator by s, it is found that

$$\frac{1}{s+\alpha} = \frac{(1/s)}{1+(\alpha/s)},$$
$$= \frac{1}{s}\left(1+\frac{\alpha}{s}\right)^{-1},$$
$$= \frac{1}{s}\left(1-\frac{\alpha}{s}+\frac{\alpha^2}{s^2}-\dots\right),$$

(using the binomial expansion). Therefore, using (5.6.1) and (5.6.2),

$$\mathscr{L}^{-1}\left(\frac{1}{s+\alpha}\right) = 1 - \alpha t + \frac{\alpha^2 t^2}{2!} - \dots.$$

The series is recognizable as an exponential one, equal in fact to $e^{-\alpha t}$. It has thus been proved that

$$\mathscr{L}^{-1}\left(\frac{1}{s+\alpha}\right) = e^{-\alpha t} \tag{5.6.4}$$

– a result easily verified by substituting $e^{-\alpha t}$ for $w(t)$ in the Laplace transform equation, (5.4.5).

The first term may now be evaluated by writing $e^{-\alpha\tau}$ for $w(\tau)$ in (5.5.4). Since $1/(s+\alpha)$ is the image of $e^{-\alpha t}$, $1/s(s+\alpha)$ is the image of $\int_0^t e^{-\alpha\tau}\,d\tau$; thus

$$\mathscr{L}^{-1}\left(\frac{1}{s(s+\alpha)}\right) = \int_0^t e^{-\alpha\tau}d\tau,$$
$$= \frac{1-e^{-\alpha t}}{\alpha}. \tag{5.6.5}$$

Adding together the two terms as in (5.6.3), we find

$$w(t) = \lambda\left(\frac{1-e^{-\alpha t}}{\alpha}\right) + w_0\, e^{-\alpha t}. \tag{5.6.6}$$

The form of this solution is the same as was obtained by classical methods in (5.2.3). The first term arises from the stimulus, represented in this problem by the step function $\lambda u(t)$; the second is proportional to w_0, the initial value of the dependent variable w.

At first sight this method may seem clumsy and circuitous compared with the classical treatment. But even the Taj Mahal was not seen to be beautiful until the scaffolding was removed, and much of these last two sections has been mere scaffolding. Remove it: what remains? A few simple rules enable a differential equation to be replaced by an algebraic equation, its image; the solution of the image equation is child's-play; and a few more simple rules suffice to interpret the solution in terms of the original variables. A list of 'originals' with corresponding 'images', short enough to be easily memorized, gives the reader the power to carry out the transformations in all problems concerned with transients in lumped circuits. To establish the relation between 'original' and 'image' anew for each new problem is quite unnecessary; the thing is done once for all by methods like the one used above, and thenceforth is taken for granted.

A comprehensive list of Laplace transforms is given in appendix 1 (p. 517).

5.7 Circuits of the second order

The next possibility is conveniently illustrated by adding self inductance to the simple circuit already considered, so as to obtain the circuit of fig. 5.5.

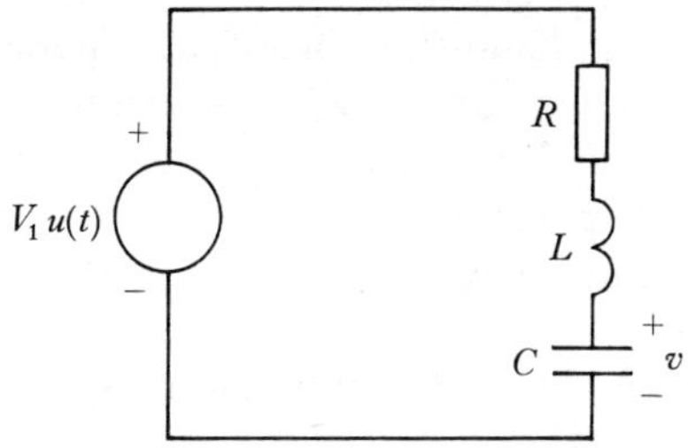

Fig. 5.5 Series *RLC* circuit stimulated by step function voltage.

There are two dependent variables, the voltage v on the capacitor and the current i; they are related by the differential equations

$$\left.\begin{aligned} L\frac{di}{dt} + Ri + v &= V_1\,u(t),\\ i - C\frac{dv}{dt} &= 0. \end{aligned}\right\} \qquad (5.7.1)$$

If the latter is used to eliminate i from the former, we obtain

$$LC\frac{d^2 v}{dt^2} + RC\frac{dv}{dt} + v = V_1 u(t), \tag{5.7.2}$$

which may be written

$$\frac{d^2 v}{dt^2} + a\frac{dv}{dt} + bv = \lambda u(t), \tag{5.7.3}$$

where $a = R/L$, $b = 1/LC$, $\lambda = V_1/LC$. Containing the second derivative d^2v/dt^2, this differential equation is described as being *of the second order*, and it is possible to specify two conditions at $t = 0$, namely v_0, i_0, the initial values of v, i.

From (5.5.3), the Laplace transforms of dv/dt, di/dt are $s\boldsymbol{V} - v_0$, $s\boldsymbol{I} - i_0$, where $\boldsymbol{V}$, $\boldsymbol{I}$ are the transforms of v, i. Thus the image equations corresponding to (5.7.1) are:

$$L(s\boldsymbol{I} - i_0) + R\boldsymbol{I} + \boldsymbol{V} = V_1/s,$$

$$\boldsymbol{I} - C(s\boldsymbol{V} - v_0) = 0,$$

or

$$\left.\begin{aligned}(Ls + R)\boldsymbol{I} + \boldsymbol{V} &= (V_1/s) + Li_0,\\ \boldsymbol{I} - Cs\boldsymbol{V} &= Cv_0.\end{aligned}\right\} \tag{5.7.4}$$

The three terms on the right hand side contribute separate elements to the solution. Its nature is best seen by considering one term at a time, and it will therefore now be assumed that v_0 and i_0 are zero, so that the circuit is dead up to the instant of application of the step function voltage. It will be found that the method can easily be extended to the more general case when v_0, i_0 are not zero.

With this simplifying condition, the elimination of $\boldsymbol{I}$ from the equations leads to

$$(LCs^2 + RCs + 1)\boldsymbol{V} = V_1/s,$$

or

$$(s^2 + as + b)\boldsymbol{V} = \lambda/s \tag{5.7.5}$$

in the notation used above. The problem is thus reduced to one of interpreting the image solution,

$$\boldsymbol{V} = \lambda/s(s^2 + as + b). \tag{5.7.6}$$

The quadratic expression has real, distinct factors if $a^2 > 4b$; complex factors if $a^2 < 4b$; and real, coincident factors if $a^2 = 4b$. Correspondingly there are three possible forms for the interpretation of (5.7.6).

(a) Real, distinct factors: $s^2 + as + b = (s+\gamma_1)(s+\gamma_2)$

The solution is divided into partial fractions of the following form:

$$\frac{\lambda}{s(s+\gamma_1)(s+\gamma_2)} \equiv \frac{A_0}{s} + \frac{A_1}{s+\gamma_1} + \frac{A_2}{s+\gamma_2}. \qquad (5{\cdot}7{\cdot}7)$$

The appropriate forms of fraction and the method of evaluating the constants are both well known in the integration of algebraic functions. It will be found that the identity (5·7·7) is satisfied by these values of A_0, A_1, A_2:

$$A_0 = \frac{\lambda}{\gamma_1\gamma_2}; \qquad A_1 = \frac{\lambda}{\gamma_1(\gamma_1-\gamma_2)}; \qquad A_2 = \frac{-\lambda}{\gamma_2(\gamma_1-\gamma_2)}$$

Thus
$$\boldsymbol{V} = \frac{\lambda}{\gamma_1\gamma_2}\left[\frac{1}{s} + \frac{1}{\gamma_1-\gamma_2}\left(\frac{\gamma_2}{s+\gamma_1} - \frac{\gamma_1}{s+\gamma_2}\right)\right], \qquad (5{\cdot}7{\cdot}8)$$

and
$$v = \frac{\lambda}{\gamma_1\gamma_2}\left[1 + \frac{\gamma_2 e^{-\gamma_1 t} - \gamma_1 e^{-\gamma_2 t}}{\gamma_1 - \gamma_2}\right],$$

the interpretation of the terms being known from (5.6.1) and (5.6.4). Noting that $\gamma_1\gamma_2 = b = 1/LC$, and that $\lambda = V_1/LC$, one may change the last equation to

$$v = V_1\left[1 + \frac{\gamma_2 e^{-\gamma_1 t} - \gamma_1 e^{-\gamma_2 t}}{\gamma_1 - \gamma_2}\right]. \qquad (5{\cdot}7{\cdot}9)$$

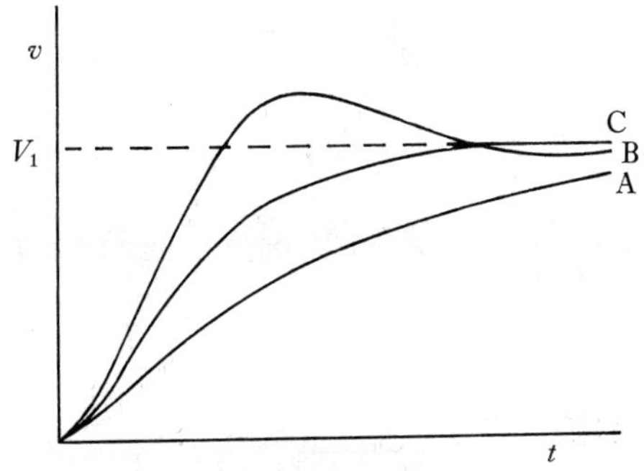

Fig. 5.6 Types of response in a second order circuit: A, non-oscillatory; B, oscillatory; C, critically damped.

For particular values of the constants, this result is illustrated in fig. 5.6 (curve A).

(b) Complex factors:
$$s^2 + as + b = (s+\alpha+j\beta)(s+\alpha-j\beta) \equiv (s+\alpha)^2+\beta^2$$

The solution here might be obtained by substituting $\alpha \pm j\beta$ for γ_1, γ_2 in (5.7.6)–(5.7.9); in numerical problems, however, it is preferable to test the reality of the factors at an early stage, and if they prove to be complex, to use the partial fractions

$$\frac{\lambda}{s[(s+\alpha)^2+\beta^2]} \equiv \frac{A_0}{s} + \frac{A_1 s + A_2}{(s+\alpha)^2+\beta^2}. \tag{5.7.10}$$

It is found that

$$A_0 = \frac{\lambda}{\alpha^2+\beta^2}; \qquad A_1 = \frac{-\lambda}{\alpha^2+\beta^2}; \qquad A_2 - \frac{-2\alpha\lambda}{\alpha^2+\beta^2};$$

so that

$$V \equiv \frac{\lambda}{s[(s+\alpha)^2+\beta^2]} \equiv \frac{\lambda}{\alpha^2+\beta^2}\left[\frac{1}{s} - \frac{s+2\alpha}{(s+\alpha)^2+\beta^2}\right]. \tag{5.7.11}$$

The second term in the bracket is of a form not hitherto mentioned. Reference to the table of transforms, however, brings to light the following (nos. 32, 33):

$$\frac{s+\alpha}{(s+\alpha)^2+\beta^2} = \mathscr{L}(e^{-\alpha t}\cos\beta t), \tag{5.7.12}$$

$$\frac{\beta}{(s+\alpha)^2+\beta^2} = \mathscr{L}(e^{-\alpha t}\sin\beta t). \tag{5.7.13}$$

From these it may be seen that

$$\frac{s+2\alpha}{(s+\alpha)^2+\beta^2} = \mathscr{L}[e^{-\alpha t}\{\cos\beta t + (\alpha/\beta)\sin\beta t\}]$$

so that

$$v = \mathscr{L}^{-1}\left(\frac{\lambda}{s[(s+\alpha)^2+\beta^2]}\right) = \frac{\lambda}{\alpha^2+\beta^2}[1 - e^{-\alpha t}\{\cos\beta t + (\alpha/\beta)\sin\beta t\}].$$

But $\lambda = V_1/LC$, and $(\alpha^2+\beta^2) = b = 1/LC$. Hence

$$v = V_1[1 - e^{-\alpha t}\{\cos\beta t + (\alpha/\beta)\sin\beta t\}]. \tag{5.7.14}$$

This type of response is illustrated in fig. 5.6, curve B.

The two new transforms which have been required, (5.7.12) and (5.7.13), are quite easily derived by writing $(\alpha + j\beta)$ for α in (5.6.4), so as to obtain

$$\mathscr{L}^{-1}\left(\frac{1}{s+\alpha+j\beta}\right) = e^{-(\alpha+j\beta)t},$$

$$= e^{-\alpha t}(\cos\beta t - j\sin\beta t).$$

The left hand side may be rewritten as

$$\mathscr{L}^{-1}\left(\frac{s+\alpha-j\beta}{(s+\alpha)^2+\beta^2}\right);$$

upon equating separately the real and imaginary parts in this equation, both results are obtained.

(c) Real, coincident factors: $s^2 + as + b = (s+\alpha)^2$

This condition, transitional between the two already discussed, is attained when $a^2 = 4b$, or (in the present circuit problem) when $R^2 = 4L/C$. Increasing R, so that the circuit passes from the oscillatory to the non-oscillatory condition, is said to increase the *damping* of the circuit; in the transitional state it is said to be *critically damped*, and the appropriate partial fractions are

$$\boldsymbol{V} = \frac{\lambda}{s(s+\alpha)^2} \equiv \frac{A_0}{s} + \frac{A_1}{(s+\alpha)} + \frac{A_2}{(s+\alpha)^2}. \tag{5.7.15}$$

The values of A_0, A_1, A_2 are found to be such that

$$\boldsymbol{V} = \frac{\lambda}{\alpha^2}\left[\frac{1}{s} - \frac{1}{(s+\alpha)} - \frac{\alpha}{(s+\alpha)^2}\right].$$

The third term, a new form, is found as no. 26 in the table of transforms:

$$\mathscr{L}\left(\frac{1}{(s+\alpha)^2}\right) = t\,e^{-\alpha t}. \tag{5.7.16}$$

From this, and from others already known,

$$v = \frac{\lambda}{\alpha^2}[1 - e^{-\alpha t} - \alpha t\, e^{-\alpha t}],$$

or, substituting for λ/α^2 in the manner already familiar,

$$v = V_1[1 - (1+\alpha t)\,e^{-\alpha t}]. \tag{5.7.17}$$

This solution is illustrated with the others in fig. 5.6 (curve C).

The transform (5.7.16) could be obtained by expanding $(s+\alpha)^{-2}$ in powers of $1/s$, but a more elegant way of obtaining it will be given in §10.7.

5.8 Excitation of a circuit by a stimulus which is not a step function

The last section has opened the door to the generalization required for more complex circuits, but another generalization is still needed – that of the stimulus. For example, a series *RLC* circuit might be connected, by closing a switch at $t=0$, to an a.c. voltage generator; this would be represented by a voltage source whose e.m.f., zero until the instant $t=0$, was equal to $V_1 \cos(\omega_1 t+\epsilon)$ thereafter. Reference to (5.7.3) shows that the differential equation of a second order circuit excited by a voltage source will in general take the form

$$\frac{d^2 v}{dt^2}+a\frac{dv}{dt}+bv=f(t), \tag{5.8.1}$$

where $f(t)$ is proportional to the e.m.f. of the source.

In principle, a problem of this kind is attacked by the method already given. The image of each term is written down by the known rules, and the term appearing on the right hand side is $\boldsymbol{F}(s)$, the image of $f(t)$. In the example already cited, for instance, the equation takes the form

$$\begin{aligned}\frac{d^2 v}{dt^2}+a\frac{dv}{dt}+bv &= \lambda_1 \cos(\omega_1 t+\epsilon)\,u(t),\\ &= \lambda_1(\cos\omega_1 t\cos\epsilon-\sin\omega_1 t\sin\epsilon)\,u(t).\end{aligned}$$

In the table of transforms are found, as nos. 28 and 29,

$$\cos\omega_1 t=\mathscr{L}^{-1}\left(\frac{s}{s^2+\omega_1^2}\right), \tag{5.8.2}$$

and

$$\sin\omega_1 t=\mathscr{L}^{-1}\left(\frac{\omega_1}{s^2+\omega_1^2}\right) \tag{5.8.3}$$

– the factor $u(t)$ being assumed in each function of t. Therefore the image equation, analogous to (5.7.5), is

$$(s^2+as+b)\,\boldsymbol{V}=\lambda_1\left(\frac{s\cos\epsilon-\omega_1\sin\epsilon}{s^2+\omega_1^2}\right). \tag{5.8.4}$$

The solution proceeds on the same lines as before, the two terms in the formula for $\boldsymbol{V}$ being split into partial fractions.

It is at once evident that the success of this method is dependent upon $\boldsymbol{F}(\mathrm{s})$ taking the form of a ratio of two polynomials in s. If $\boldsymbol{F}(\mathrm{s})$ were equal to $\mathrm{e}^{-k\mathrm{s}}$, for example, the resolution into partial fractions would be impossible. This imposes a limitation upon the form of the stimulus $f(t)$; it must be expressible as a finite number of powers of t, or of sinusoidal or exponential functions of t. Fortunately such stimuli are of very common occurrence. Methods given later will enable the response to more sophisticated stimuli to be calculated.

5.9 Image circuits in transient analysis

The use of the symbols $\boldsymbol{V}, \boldsymbol{I}$ for the images of voltage and current, in this chapter as in the preceding one, will have suggested to the reader that the 'image circuit' approach described in §4.6 may be capable of adaptation to transient problems. It will now be shown that this can be done in a very simple manner, provided that the circuit is initially dead.

The passive elements of a linear circuit are of two types. The first is characterized by proportionality, at every instant, between two variable quantities; examples of this type are resistance, in which $v = Ri$, and ideal transformers, in which $v_2 = nv_1$, $i_1 = -ni_2$. Controlled sources as defined in §1.6, having an output voltage or current proportional to an input voltage or current, are to be regarded as elements in this category. The second type consists of elements in which one variable is proportional at every instant to the rate of change of another; such are inductance ($v = L\,\mathrm{d}i/\mathrm{d}t$), mutual inductance ($v_1 = M\,\mathrm{d}i_2/\mathrm{d}t, v_2 = M\,\mathrm{d}i_1/\mathrm{d}t$), and capacitance ($i = C\,\mathrm{d}v/\mathrm{d}t$). Taking resistance as typical of the first type, the Laplace transformation of $v = Ri$ yields the relation

$$\boldsymbol{V} = R\boldsymbol{I}, \tag{5.9.1}$$

so that $\boldsymbol{V}, \boldsymbol{I}$ can be regarded as the voltage and current in an 'image element' of impedance R. As for the second type, the transformation of the inductance equation leads to the image equation

$$\boldsymbol{V} = L\mathrm{s}\boldsymbol{I}, \tag{5.9.2}$$

provided that $i = 0$ when $t = 0$, with similar forms for mutual inductance; and the capacitance equation becomes

$$\boldsymbol{I} = C\mathrm{s}\boldsymbol{V}, \tag{5.9.3}$$

provided $v = 0$ when $t = 0$. These two equations show that, if $\boldsymbol{V}$ is regarded as a voltage and $\boldsymbol{I}$ as a current, impedances of $L\mathrm{s}$ and $1/C\mathrm{s}$ must be ascribed to inductance and capacitance elements respectively. There is an exact correspondence with the image treatment of a.c. circuits developed in §4.6, s here replacing $j\omega$.

The complete network will also contain sources, which may be either independent or controlled. The outputs of the former are functions of t; in an image network, these must be replaced by sources of the same type, having outputs which are the corresponding function of s. For example, a voltage source generating the step function $V_1 u(t)$ becomes transformed into a voltage source generating V_1/s. Controlled sources are characterized by proportionality between the output and input variable; they are therefore treated in the same way as the passive elements described above, and are transformed into controlled sources in the image network.

The procedure is so simple, and so closely parallel to that employed with a.c. circuits, that elaborate illustration is superfluous. Fig. 5.7 shows the

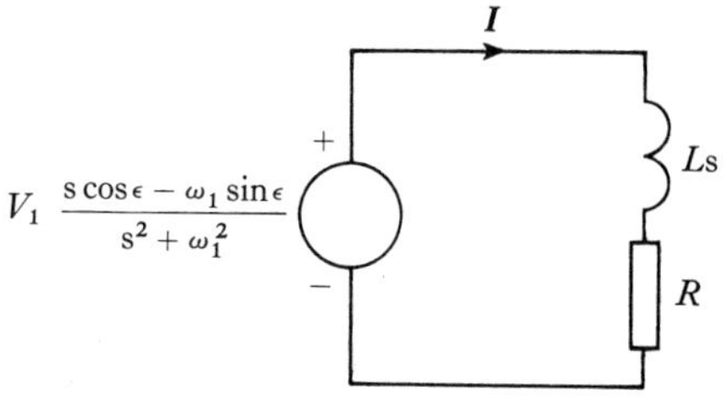

Fig. 5.7 Image circuit for a coil suddenly connected to an a.c. voltage source.

image circuit which describes the sudden connecting of a resistive inductor to an a.c. voltage supply $V_1 \cos(\omega_1 t + \epsilon)$, represented by an image voltage source having an e.m.f. of $V_1(s\cos\epsilon - \omega_1 \sin\epsilon)/(s^2 + \omega_1^2)$. The image current is seen to be given by

$$\boldsymbol{I} = \frac{V_1(s\cos\epsilon - \omega_1 \sin\epsilon)}{(Ls + R)(s^2 + \omega_1^2)}, \tag{5.9.4}$$

which is interpreted by the usual method.

When a circuit has been replaced in this way by an equivalent image circuit, it becomes natural to think of the image currents and voltages $\boldsymbol{I}$, $\boldsymbol{V}$ as 'currents' and 'voltages' pure and simple, and of their ratios as 'impedances' or 'admittances'. This practice does not lead to confusion, for the quantities in the image circuit obey the Kirchhoff and other circuit laws. The image circuit shown in fig. 5.8(*a*) represents the application of a step function voltage to an inductance–capacitance network, in which it is desired to calculate the current in the inductance; the generated 'voltage' V_1/s, the 'impedances', and the required 'current' $\boldsymbol{I}$, are all functions of s.

Inverted commas will not in future be employed in describing quantities in the image circuit.

The current $\boldsymbol{I}$ sets up in the inductor a voltage $Ls\boldsymbol{I}$; applied to the parallel capacitor, this voltage produces a current $LCs^2\boldsymbol{I}$. The total current in the circuit is thus $(LCs^2+1)\boldsymbol{I}$, and the Kirchhoff loop equation is

$$Ls\boldsymbol{I} + \frac{1}{C_1 s}(LCs^2+1)\boldsymbol{I} = \frac{V_1}{s},$$

whence
$$\boldsymbol{I} = \left(\frac{C_1 s}{L(C+C_1)s^2+1}\right)\frac{V_1}{s}. \qquad (5.9.5)$$

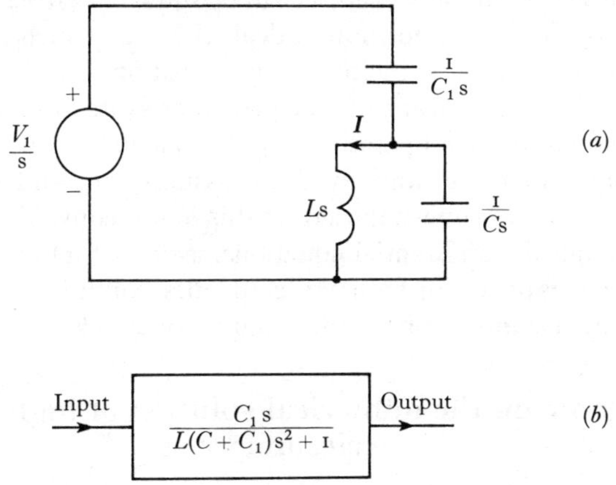

Fig. 5.8 (*a*) Image circuit for LC network stimulated by step function voltage. (*b*) Transfer function for (*a*).

The quantity in brackets, by which the 'input' V_1/s must be multiplied in order to derive the 'output' $\boldsymbol{I}$, is called the *transfer function* of the circuit.

It is reasonable to use the term *input* for the stimulus (whether voltage or current) which activates the circuit, but the *output* is in this instance a particular current, arbitrarily chosen, and to this extent the transfer function is also arbitrary. In most cases the purpose for which the circuit is intended determines which quantity shall be regarded as its output. The transfer function then summarizes the operation of the circuit; it is a very general concept, and may be defined for mechanical and electromechanical as well as for purely electrical systems. If the step function input (in the example just discussed) had been replaced by some other stimulus, the response or output would have been derived by multiplying the new input by the same transfer function; hence the performance of the circuit may be graphically represented by a diagram (fig. 5.8(*b*)), consisting of a rectangle on which the transfer function is inscribed. When there are several circuits, in which the output of one is the input of another, several interconnected rectangles can be drawn; but it is necessary to point out that this is not done by simply joining up the rectangles which would represent the several circuits in isolation, for the transfer function of a circuit is affected by connecting another circuit across its output terminals. Nevertheless such a *block diagram* can very clearly illustrate the function of each portion of a complicated system.

It is to be noted that the concept of a transfer function has been developed and defined only for circuits which are initially dead, and which contain only one independent energy source.

The 'image circuit' approach is capable of being generalized to include circuits which are not initially dead, as will be shown in §§10.2–4. Even now, however, the reader is in a position to deal with many such circuits by the image method through the principle of superposition. If a circuit is brought to its final condition by means of a sequence of switching operations, the *change* introduced by each operation may be calculated with the aid of an image network under the 'initially dead' assumption, and the results of the successive changes superimposed. If this is not convenient, the formal approach through the differential equations, as in (5.7.1) to (5.7.4), is still available as a means of dealing with non-zero initial conditions. The examples in §5.11 include instances of both these modes of attack.

5.10 A note on the numerical solution of higher order circuits

It requires no great effort of imagination to see that the image solution of a more complicated circuit, subjected to a stimulus of the types discussed, must take the form

$$\boldsymbol{W} = \mathrm{P(s)/Q(s)}, \tag{5.10.1}$$

where P(s), Q(s) are polynomials in s. The subsequent resolution into partial fractions requires that Q(s) shall be factorized. Since the coefficients in Q(s) are necessarily real, any complex factors must occur in conjugate pairs, which may be multiplied to form a real quadratic factor; thus Q(s) can be resolved into real factors, either linear or quadratic. Unfortunately no useful general formula of factorization exists for polynomials of degree higher than 2. Nevertheless the factorization can be performed in any numerical case; and when the degree of Q(s) is not greater than 5, the method may be quite elementary.

Taking first the case $n = 3$, it is evident that a cubic polynomial must have at least one real factor – or, what is the same thing, the equation

$$\mathrm{s}^2 + a\mathrm{s}^2 + b\mathrm{s} + c = 0 \tag{5.10.2}$$

must have at least one real root. Any real root may be determined approximately by plotting the function, or by a simple process of trial and error; Q(s) may then be divided by the corresponding factor, leaving a quadratic expression which can immediately be factorized in all cases. So, too, a fifth degree polynomial can be reduced to one of the fourth degree by finding a real root and dividing by the corresponding factor. The resolution of a quartic (fourth degree) polynomial into factors may, however, introduce a new difficulty, since it need have no real roots – two conjugate pairs of complex roots are quite possible, and these can hardly be found by trial and error. If, however, the quartic polynomial is written in this form:

$$s^4 + as^3 + bs^2 + cs + d \equiv (s^2 + a's + b')^2 - (c's + d')^2, \quad (5.10.3)$$

and corresponding coefficients on the two sides are equated, it is not difficult to eliminate a', c', and d' from the resulting four equations, leaving an equation for b' which is found to be a cubic; this may be solved as already described. The reason why it is a cubic is that the four linear factors of the original quartic expression may be paired off in three ways; if the factors are all imaginary, one of the values of b' will lead to real values for all the coefficients a', b', c', d', with the result that the right hand side of (5.10.3) is resolved into the real quadratic factors

$$[s^2 + (a' + c')s + (b' + d')][s^2 + (a' - c')s + (b' - d')].$$

Thus the factorization of a quartic or quintic polynomial is reducible to the solution of a cubic equation, and more sophisticated methods are not required until expressions of the sixth degree are reached.

The polynomial in the denominator on the right hand side of (5.10.1) being thus factorized into real factors, the resolution into partial fractions is carried out in the usual way. Thus, if Q(s) has one or more real factors typified by $(s + \gamma)$, and complex factors typified by $[(s + \alpha)^2 + \beta^2]$,

$$\frac{P(s)}{Q(s)} \equiv \frac{C}{s+\gamma} + \ldots + \frac{As + B}{(s+\alpha)^2 + \beta^2} + \ldots. \quad (5.10.4)$$

A squared linear factor $(s + \gamma)^2$ requires the assumption of fractions

$$\frac{C}{s+\gamma} + \frac{C'}{(s+\gamma)^2};$$

a squared quadratic factor $[(s + \alpha)^2 + \beta^2]^2$, fractions

$$\frac{As + B}{(s+\alpha)^2 + \beta^2} + \frac{A's + B'}{[(s+\alpha)^2 + \beta^2]^2}.$$

Factors raised to higher powers than the square are most unlikely. The coefficients A, B, C, etc. are determined in general by multiplying both sides of (5.10.4) by $Q(s)$ and equating corresponding coefficients of s, though the C coefficients are frequently found more quickly by giving s the particular values $-\gamma$; for, if two functions of s are identically equal, they are equal for all values of s.

$W(s)$ having been thus expressed as a sum of several fractions, the original quantity $w(t)$ is found by reference to the table of transforms. No transforms apart from those already used in §5.7 are required, with the exception of those corresponding to the squared quadratic s-terms; these are found in the appendix, being derived by a method to be given in §10.7.

5.11 Some examples

The methods just described are, in essence, sufficient for calculating the response of any linear, lumped circuits to any of the stimuli most commonly encountered. It appears useful at this juncture to demonstrate the application of the image method to a number of specific problems.

1. The circuit of fig. 5.9(a), initially dead, is energized by closing the switch at an instant when the e.m.f. of the voltage generator is a maximum. It is required to find the secondary current i_2.

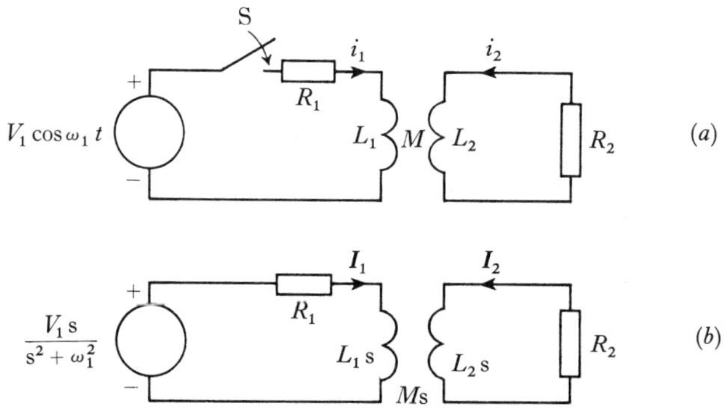

Fig. 5.9 (a) Circuit containing mutual inductance stimulated by a.c. voltage suddenly applied. (b) Image circuit for (a).

As in the problem of §5.3, the switch can be eliminated if the generator is replaced by a voltage source $(V_1 \cos \omega_1 t)u(t)$. In the image circuit (fig. 5.9(b)), this becomes $V_1 s/(s^2 + \omega_1^2)$. The relative directions of the currents are so chosen that a positive secondary current sets up flux through the primary winding in the same sense as a positive primary current, and vice versa. With this convention the Kirchhoff loop equations for the image circuit are

$$\left.\begin{aligned}(L_1 s + R_1)\boldsymbol{I}_1 + \quad M s \boldsymbol{I}_2 &= \frac{V_1 s}{s^2+\omega_1^2},\\ M s \boldsymbol{I}_1 + (L_2 s + R_2)\boldsymbol{I}_2 &= 0.\end{aligned}\right\} \tag{5.11.1}$$

Eliminating $\boldsymbol{I}_1$, we obtain

$$\{(L_1 s + R_1)(L_2 s + R_2) - M^2 s^2\}\boldsymbol{I}_2 = -MV_1 \frac{s^2}{s^2+\omega_1^2},$$

which may be written

$$(L_1 L_2 - M^2)(s+\alpha)(s+\beta)\boldsymbol{I}_2 = -MV_1 \frac{s^2}{s^2+\omega_1^2},$$

where $-\alpha$, $-\beta$ are the roots of the equation

$$(L_1 L_2 - M^2) s^2 + (L_1 R_2 + L_2 R_1) s + R_1 R_2 = 0. \tag{5.11.2}$$

(It may be verified that the roots are real.) Thus the image solution is

$$\boldsymbol{I}_2 = \frac{-MV_1}{L_1 L_2 - M^2} \frac{s^2}{(s+\alpha)(s+\beta)(s^2+\omega_1^2)}. \tag{5.11.3}$$

The partial fraction form is

$$\boldsymbol{I}_2 = \frac{-MV_1}{L_1 L_2 - M^2}\left\{\frac{A}{s+\alpha} + \frac{B}{s+\beta} + \frac{Cs+D}{s^2+\omega_1^2}\right\}, \tag{5.11.4}$$

where, as the reader may verify,

$$\left.\begin{aligned}A &= \frac{-\alpha^2}{(\alpha-\beta)(\alpha^2+\omega_1^2)}; & B &= \frac{\beta^2}{(\alpha-\beta)(\beta^2+\omega_1^2)};\\ C &= \frac{(\alpha+\beta)\,\omega_1^2}{(\alpha^2+\omega_1^2)(\beta^2+\omega_1^2)}; & D &= \frac{\omega_1^2(\omega_1^2-\alpha\beta)}{(\alpha^2+\omega_1^2)(\beta^2+\omega_1^2)}.\end{aligned}\right. \tag{5.11.5}$$

When (5.11.4) is converted back from the image to the original it gives

$$i_2 = \frac{-MV_1}{L_1 L_2 - M^2}\left\{A\,e^{-\alpha t} + B\,e^{-\beta t} + C\cos\omega_1 t + \frac{D}{\omega_1}\sin\omega_1 t\right\}. \tag{5.11.6}$$

The form of this current with particular values of the constants, along with that of the applied primary voltage, is shown in fig. 5.10. It will be noted that the exponential terms in (5.11.6) represent the transient part of the solution, while the terms in $\cos\omega_1 t$ and $\sin\omega_1 t$ give the ultimate, steady state current.

2. In the circuit shown in fig. 5.11(*a*), the switch S is initially closed, and a generator of negligible impedance passes a constant current I_0 through L and R. At time $t = 0$ the switch is opened. It is required to find the subsequent voltage across the switch contacts.

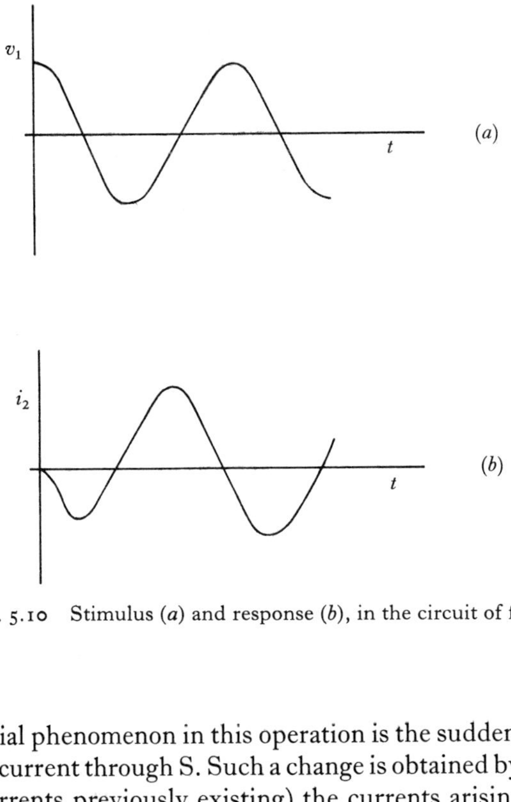

Fig. 5.10 Stimulus (*a*) and response (*b*), in the circuit of fig. 5.9.

The essential phenomenon in this operation is the sudden change from I_0 to zero in the current through S. Such a change is obtained by superimposing (upon the currents previously existing) the currents arising when an ideal current source injects a step function current $I_0 u(t)$ at the switch terminals in opposition to I_0, so as to annul I_0 in the switch branch from $t = 0$ onwards. The superimposed currents start from zero, so that the image circuit by which they are calculated may be drawn as in fig. 5.11(*b*). The 'generator of negligible impedance' plays no part, and is replaced by a short circuit. The voltage required is the sum of the initial voltage across the switch branch (namely zero) and the superimposed voltage, of which $\boldsymbol{V}$, in fig. 5.11(*b*), is the image; thus, when the image problem has been solved, $\mathscr{L}^{-1}\boldsymbol{V}$ gives the complete solution of the original problem.

Taking the current $\boldsymbol{I}$, as shown, as a second variable, we obtain for the image circuit the equations

$$\left.\begin{aligned} \boldsymbol{I}+\left(Cs+\frac{1}{r}\right)\boldsymbol{V}&=\frac{I_0}{s}, \\ (Ls+R)\boldsymbol{I}-\boldsymbol{V}&=0. \end{aligned}\right\} \tag{5.11.7}$$

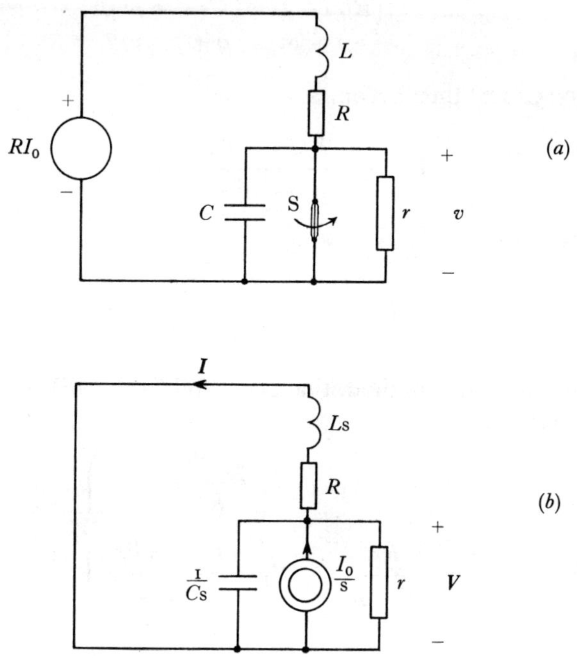

Fig. 5.11 (*a*) Circuit stimulated by opening a switch S. (*b*) Image circuit with the switch replaced by a step function current source.

Eliminating $\boldsymbol{I}$,

$$\left[(Ls+R)\left(Cs+\frac{1}{r}\right)+1\right]\boldsymbol{V}=\frac{I_0(Ls+R)}{s}. \tag{5.11.8}$$

The roots of the quadratic expression multiplying $\boldsymbol{V}$ may be either real or complex. Consider, then, this numerical case: $L=1$ H, $C=0.05$ μF, $R=1000\ \Omega$, $r=2500\ \Omega$, $I_0=100$ mA. Equation (5.11.8) becomes

$$(5\times10^{-8}s^2+4.5\times10^{-4}s+1.4)\boldsymbol{V}=10^{-1}\left(\frac{s+10^3}{s}\right),$$

or
$$(s^2 + 9 \times 10^3 s + 2.8 \times 10^7)\boldsymbol{V} = 2 \times 10^6\left(\frac{s + 10^3}{s}\right).$$

The quadratic has complex factors, and may be written $(s+4500)^2 + (2780)^2$. To avoid writing four-figure numerals, this will be written as $(s+\alpha)^2+\beta^2$; it is to be noted that, since

$$(Ls + R)(Cs + 1/r) + 1 \equiv LC\{(s+\alpha)^2 + \beta^2\},$$
$$(R/r) + 1 = LC(\alpha^2 + \beta^2),$$

or
$$\alpha^2 + \beta^2 = (R+r)/rLC \tag{5.11.9}$$

The equation (5.11.8) then becomes

$$\boldsymbol{V} = \frac{I_0(Ls + R)}{LCs[(s+\alpha)^2 + \beta^2]}, \tag{5.11.10}$$

which may be written

$$\boldsymbol{V} = I_0\left[\frac{A_0}{s} + \frac{A_1 s + B_1}{(s+\alpha)^2 + \beta^2}\right]. \tag{5.11.11}$$

It is found, by the usual method, that $A_0 = R/LC(\alpha^2+\beta^2)$; but the use of (5.11.9) reduces this to

$$\left.\begin{aligned} &A_0 = \frac{Rr}{R+r}. \\ \text{Also}\quad A_1 = -\frac{Rr}{R+r}; \quad &B_1 = \frac{1}{C} - \frac{2\alpha Rr}{R+r}. \end{aligned}\right\} \tag{5.11.12}$$

Thus

$$\boldsymbol{V} = \frac{RrI_0}{R+r}\left[\frac{1}{s} - \frac{s+2\alpha}{(s+\alpha)^2+\beta^2}\right] + \frac{I_0}{C}\cdot\frac{1}{(s+\alpha)^2+\beta^2}. \tag{5.11.13}$$

The transforms required for interpreting this formula have already been employed in §5.7, and enable us to write the following equation for v:

$$v = \frac{RrI_0}{R+r}\left[1 - e^{-\alpha t}\left(\cos\beta t + \frac{\alpha}{\beta}\sin\beta t\right)\right] + \frac{I_0}{C\beta}e^{-\alpha t}\sin\beta t. \tag{5.11.14}$$

In numerical terms, for the values discussed, this is

$$\left.\begin{aligned} &v = 71\left[1 - e^{-\alpha t}\left(\cos\beta t + \frac{\alpha}{\beta}\sin\beta t\right)\right] + 720\,e^{-\alpha t}\sin\beta t, \\ \text{where}\quad &\alpha = 4500\,s^{-1}; \qquad \beta = 2780\,s^{-1}. \end{aligned}\right\} \tag{5.11.15}$$

This solution is illustrated in fig. 5.12.

It will be noted that the time scale is conveniently measured in milliseconds rather than in seconds. The reader may verify that if L is measured in millihenrys and C in millifarads – but R and r are still given in ohms – (5.11.15) is still obtained but with t in milliseconds. The stated values of α and β are now divided by 1000.

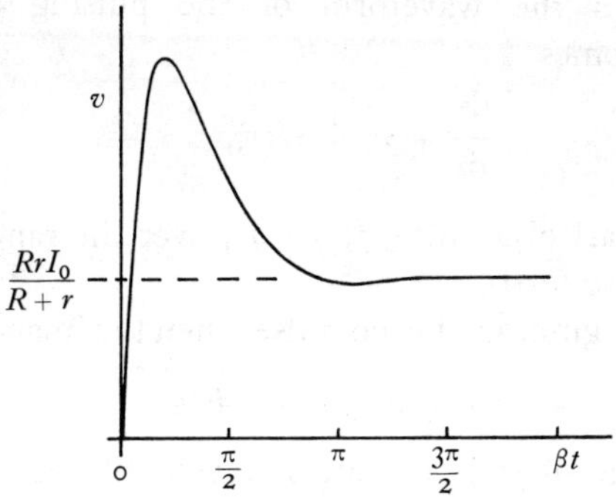

Fig. 5.12 Voltage appearing across the switch in fig. 5.11(*a*).

3. A simple series *RC* circuit is excited from a voltage generator which generates a succession of square pulses of magnitude V_1 and duration t_1, the voltage during the intervening intervals t_2 being zero (fig. 5.13; the

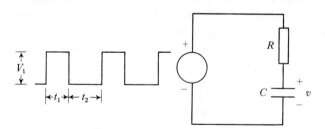

Fig. 5.13 Series *RC* circuit subjected to recurrent square pulses of voltage.

ratio $t_1 : t_2$ is called the *mark-space ratio*). Assuming that the circuit has been excited for a time so long that events in successive cycles exactly recur, it is required to discover the voltage v across the capacitor at any instant of the cycle.

A circuit in which the voltages and currents are undergoing a cyclic variation is said to be in a *quasi-steady state*. Strictly speaking the state of an alternating current circuit, operating sinusoidally at a constant frequency, is quasi-steady; however, the standard method of a.c. circuit analysis, as set forth in chapter 4, amounts to replacing the original circuit by a steady state image. For this reason the term 'steady state' is commonly (if somewhat loosely) applied to describe the condition of the original circuit also.

The differential equation of the circuit, obtained as in §5.2, is

$$RC\frac{dv}{dt} + v = f(t),$$

where $f(t)$ describes the waveform of the pulsing source. Writing $(1/RC) = \alpha$, this becomes

$$\frac{dv}{dt} + \alpha v = \alpha f(t), \tag{5.11.16}$$

and if $t = 0$ at the start of a pulse, $f(t) = V_1$ over the range $0 < t < t_1$, $= 0$ over $t_1 < t < t_2$, and so forth.

Let $v = v_0$ at the beginning of each pulse; then the image equation is

$$(s + \alpha)\boldsymbol{V} = v_0 + \alpha \boldsymbol{F}(s), \tag{5.11.17}$$

$\boldsymbol{F}(s)$ being the image of $f(t)$. We are not yet able to derive $\boldsymbol{F}(s)$. The following device must therefore be invoked to break the problem into manageable parts:

(*a*) At the beginning of a pulse, when the hangover of previous pulses has left the capacitor charged to voltage v_0, a step function voltage $V_1 u(t)$ is applied.

(*b*) When $t = t_1$, v has attained a new value v_1, which is the starting-point for a new period of duration t_2, during which the source is quiescent and C discharges freely through it. At the end of this period, if the circuit is in a quasi-steady state, v should again be v_0.

During the interval $0 \leqslant t \leqslant t_1$, therefore, $f(t)$ is replaced by $V_1 u(t)$, and $\boldsymbol{F}(s)$ by V_1/s. Consequently

$$\boldsymbol{V} = \frac{sv_0 + \alpha V_1}{s(s+\alpha)},$$

$$= \frac{v_0}{s+\alpha} + V_1\left[\frac{1}{s} - \frac{1}{s+\alpha}\right].$$

Hence
$$v = v_0 e^{-\alpha t} + V_1(1 - e^{-\alpha t}), \tag{5.11.18}$$

and at $t = t_1$,
$$v = v_1 = v_0 e^{-\alpha t_1} + V_1(1 - e^{-\alpha t_1}), \tag{5.11.19}$$

Now let the origin of time be transferred to $t = t_1$. During the next interval, $f(t) = 0$, and

$$\boldsymbol{V} = \frac{v_1}{s+\alpha},$$

so that
$$v = v_1 e^{-\alpha t}.$$

At the end of the interval, which is the start of the next pulse, $v = v_1 e^{-\alpha t_2}$; this must equal v_0. Therefore

$$v_0 = v_1 e^{-\alpha t_2} = v_0 e^{-\alpha(t_1+t_2)} + V_1 e^{-\alpha t_2}(1 - e^{-\alpha t_1})$$

from (5.11.19); whence

$$v_0 = V_1 \left[\frac{e^{-\alpha t_2}(1 - e^{-\alpha t_1})}{1 - e^{-\alpha(t_1+t_2)}}\right]. \tag{5.11.20}$$

When this is substituted into (5.11.18), it is found that v may be expressed in the form

$$v = V_1 \left[1 - \left(\frac{1 - e^{-\alpha t_2}}{1 - e^{-\alpha(t_1+t_2)}}\right) e^{-\alpha t}\right]. \tag{5.11.21}$$

This holds good from $t = 0$ to t_1; at the latter instant,

$$v = v_1 = V_1 \left[\frac{1 - e^{-\alpha t_1}}{1 - e^{-\alpha(t_1+t_2)}}\right]. \tag{5.11.22}$$

For the remaining period t_2,

$$v = V_1 \left[\frac{1 - e^{-\alpha t_1}}{1 - e^{-\alpha(t_1+t_2)}}\right] e^{-\alpha t}$$

relatively to a new time-origin at the end of the pulse; or, using the same time-origin as in (5.11.21),

$$v = V_1 \left[\frac{1 - e^{-\alpha t_1}}{1 - e^{-\alpha(t_1+t_2)}}\right] e^{-\alpha(t-t_1)}. \tag{5.11.23}$$

Equations (5.11.21) and (5.11.23) completely describe the variations of voltage on the capacitor; the nature of the response is indicated in fig. 5.14(*a*). The minimum and maximum values of v are v_0 and v_1, given by (5.11.20) and (5.11.22).

This problem could also have been solved by a direct application of the method of superposition. Upon the response to a positive step function, impressed upon the circuit at $t = 0$ with v initially equal to v_0, would be superimposed the response to a negative step function impressed at $t = t_1$. Fig. 5.14(*b*) indicates how the response can be resolved into two components in this way. Physically this is equivalent to replacing the pulse generator by a large number of step function sources connected in series, the steps being alternately $\pm V_1$ and applied at intervals alternately t_1 and t_2. It should be noted that, in calculating the response to the second or negative step, the

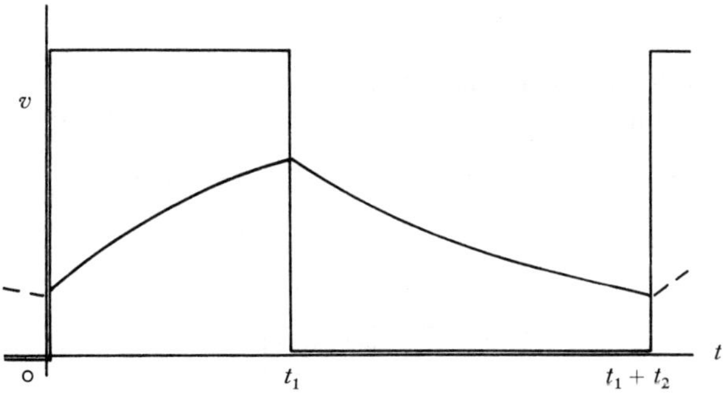

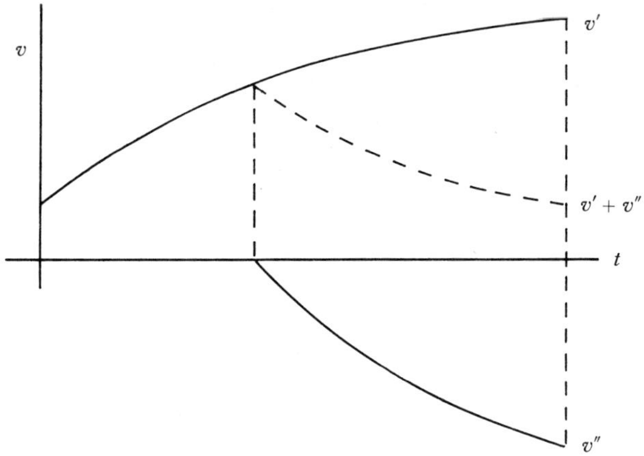

Fig. 5.14 (*a*) Response of the circuit of fig. 5.13. (*b*) Derivation of (*a*) by superposition of responses to positive step (v') and negative step (v'').

value of v must be assumed to be initially zero; in fact, $v = v_1$ at this instant, but this is part of the response to the first step, and the *extra* effect of the second step is that which would be associated with a zero initial value of v.

Examples on chapter 5

5.1. Find the functions of t corresponding to the following functions of s:

(*a*) $\dfrac{16s + 20}{4s^2 + 12s + 5}$; (*b*) $\dfrac{5s^3 + 35s}{s^4 + 13s^2 + 36}$; (*c*) $\dfrac{s + 8}{s^2 + 4s + 13}$;

(*d*) $\dfrac{5s^2 + 17s + 22}{s^3 + 5s^2 + 12s + 8}$; (*e*) $\dfrac{2s^2 + 9s + 9}{s^3 + 5s^2 + 8s + 4}$.

5.2 A resistor R is in parallel with a pure inductor L; in series with this combination is a second resistor R, and a voltage $V_1 u(t)$ is applied across the whole. Find the current in each component of the circuit.

5.3 A resistor of 1 MΩ and a capacitor of 1 μF are connected in series across terminals A, B. In parallel with the capacitor is a branch consisting of a second 1 MΩ resistor in series with a second 1 μF capacitor. The capacitors are initially without charge, and a voltage $V_1 u(t)$ is applied across AB. Calculate the voltage on the second capacitor as a function of time.

5.4 A circuit consists of two subsidiary networks connected in series:
(*a*) a pure inductor L shunted by a resistor R_1;
(*b*) a pure capacitor C shunted by a resistor R_2.
If this network is initially dead and a constant current is suddenly applied to it, find the conditions under which the voltage developed across the network will be independent of time.

5.5 Two coils, each of inductance 30 mH, have a mutual inductance of 20 mH. A 5 Ω resistor is connected across the terminals of the one; the other is connected, in series with a 3 Ω resistor, to a voltage source $10u(t)$ V. The circuit is initially dead. Calculate the current in the 5 Ω resistor, and find the time that elapses before it reaches its maximum.

5.6 A circuit consists of an inductor L, a resistor R and a capacitor C, all in series. It is initially dead, and a voltage $V_1 \cos \omega t$ is applied to its terminals from $t = 0$. Find the current, given that R^2 is less than $4L/C$ and $1/LC = \omega^2$.

5.7 A 0.01 F capacitor is in parallel with a 10 Ω resistor; in series with this combination is a coil of inductance 1 H and resistance 10 Ω. The circuit is quiescent at $t = 0$, when a voltage $500 \sin 10t$ volts is applied to it. Find the subsequent charge on the capacitor as a function of t.

5.8 A bridge network ABCD consists of a capacitor C in the branch AB, and equal resistors R in each of BC, CD, DA. The input terminals are A, C; the output terminals, open-circuited, are B, D. Derive the transfer function relating output to input voltage. Hence show that the response to a step function input voltage is an output whose initial and final values are equal and opposite. Find also the response to a suddenly applied sinusoidal voltage $(V_1 \sin \omega t) u(t)$.

5.9 A coil of inductance L and resistance R is in parallel with a capacitor C; the combination can be connected through a switch to a d.c. voltage source V_1. The switch is opened at $t = 0$, the circuit having previously been in a steady state. Find the subsequent current in the coil, assuming that R^2 is greater than $4L/C$.

5.10 A 1 μF capacitor is connected in parallel with a 100 Ω resistor, and an alternating current of 5 A (r.m.s), 50 Hz is passed into the combination from a generator of negligible impedance. A switch in series with the generator is opened at $t = 0$, when the current in the generator is instantaneously zero. Find the voltage which appears across the terminals of the switch.

5.11 The primary and secondary windings of a transformer have resistance and inductance R_1, L_1 and R_2, L_2, and the mutual inductance between them is $\sqrt{(L_1 L_2)}$. At some instant when the short-circuited secondary winding carries a current i_{20} and the primary winding is open-circuited, an ideal voltage source V_1 is connected across the primary terminals, of such magnitude that the secondary current is instantaneously reduced to zero. Find V_1, and derive an expression for the primary current as a function of time.

5.12 A pure inductor L is connected in series with a capacitor C, and a voltage $V_1 \cos \omega t$ is applied to the terminals of the circuit from $t = 0$ onwards, $LC\omega^2$ being equal to 1. Find the current in the circuit.

If the circuit also contains a rectifier which extinguishes the current when its value comes back to zero, calculate the charge left on the capacitor.

6 Reduction methods and network theorems

6.1 Introduction

Network theorems are formal statements about the properties of electric networks; they either help to reduce the labour of analysis, or enable complicated circuits to be visualized in a helpful way, or both. The theorems themselves are explicit statements of facts which are implicit in the loop or nodal equations of the circuits to which they apply, and the formal proof of any theorem often takes one of these two sets of equations as its starting point. In so far as these equations are valid not only in the d.c. world of chapter 1 but also in the $j\omega$-world of chapter 4 and in the s-world of chapter 5, the consequent theorems are also valid in the corresponding worlds.

In addition to these theorems there are a number of analytical techniques, often of circuit manipulation rather than of mathematical manipulation, which are helpful in reducing the labour of analysis. These are called reduction methods.

The aim of this chapter is to present some of the more important theorems and reduction methods of circuit theory.

6.2 Millman's theorem

'A useful network theorem' is how Millman† describes his theorem; and indeed it is. In its simplest application it gives the open-circuit voltage V_c of any number of generators connected in parallel. Three such generators are shown in fig. 6.1. The upward current through the admittance Y_1 is $(V_1 - V_c)Y_1$, and similar expressions apply to the currents through Y_2 and Y_3. From Kirchhoff's current law the sum of these currents must be zero, and it follows that the open-circuit voltage across the generators is given by

$$V_c = \frac{V_1 Y_1 + V_2 Y_2 + V_3 Y_3}{Y_1 + Y_2 + Y_3}. \tag{6.2.1}$$

When the circuit has n generators in parallel the open-circuit voltage is given by

$$V_c = \frac{V_1 Y_1 + V_2 Y_2 + \ldots + V_n Y_n}{Y_1 + Y_2 + \ldots + Y_n}, \tag{6.2.2}$$

† Jacob Millman, *Proc. I.R.E.* **28**, 413 (Sept. 1940).

which may be written in the mnemonic form

$$V_c = \frac{\sum (VY)}{\sum Y}. \tag{6.2.3}$$

In accordance with fig. 6.1, Millman's theorem is sometimes called the theorem of parallel generators, but this is really a misnomer because the theorem achieves its greatest successes in circuits which have nothing at all

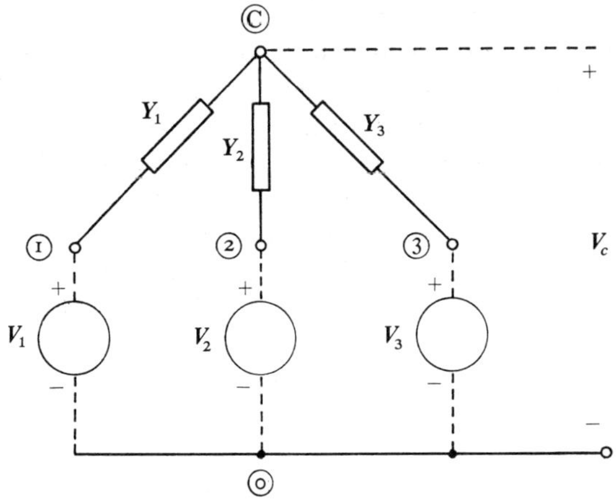

Fig. 6.1 Three generators in parallel; an application of Millman's theorem.

to do with generators in parallel. So long as the admittance elements terminate on a common point C, and so long as the actual voltages V_1, V_2, etc. at the other ends 1, 2, etc. of the admittance elements are known with respect to an arbitrary reference point 0, it is not necessary that the voltages V_1, V_2, etc. arise from ideal voltage sources actually connected between the specific points 1,0; 2,0; etc. This will be illustrated by two examples.

1. What is the a.c. component of the output voltage V_c in the circuit of fig. 6.2(*a*)?

By accepting the model shown in fig. 6.2(*b*), which is valid for the alternating components V_1 and V_c of the input and output voltages, we see that Millman's theorem can be applied directly. The point 0 is chosen as the reference terminal, and the points 1, 2, and 3 are the points at which we require to know the voltages with respect to the reference. These are V_1, $-\mu V_1$, and zero respectively. Then from (6.2.3),

$$V_c = \frac{(j\omega C_2)\, V_1 - (g_a)\, \mu V_1}{j\omega C_2 + g_a + (Y_L + j\omega C_3)}. \tag{6.2.4}$$

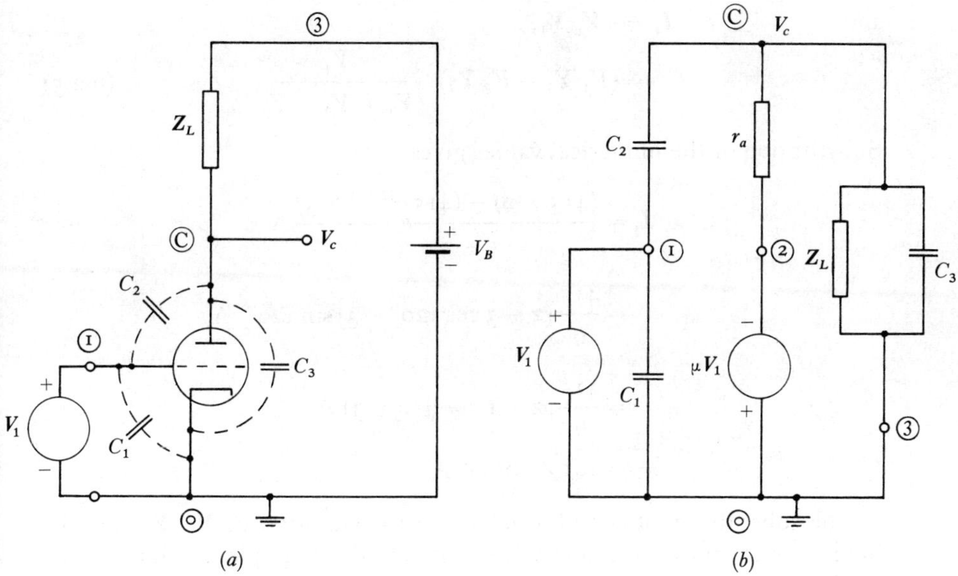

Fig. 6.2 (*a*) Thermionic valve amplifier. (*b*) Circuit model.

2. A 415 V, three-phase, delta-connected source feeds an unsymmetrical star-connected load as shown in fig. 6.3. What is the line current I_1?

Since it is required to find the current flowing into terminal 1, it is convenient to choose the reference point 0 so as to make it coincident with 1. The voltage across Y_1 is thereby made equal to V_c from which the required current can be calculated.

$$V_c = \frac{-V_1 Y_2 + V_2 Y_3}{Y_1 + Y_2 + Y_3},$$

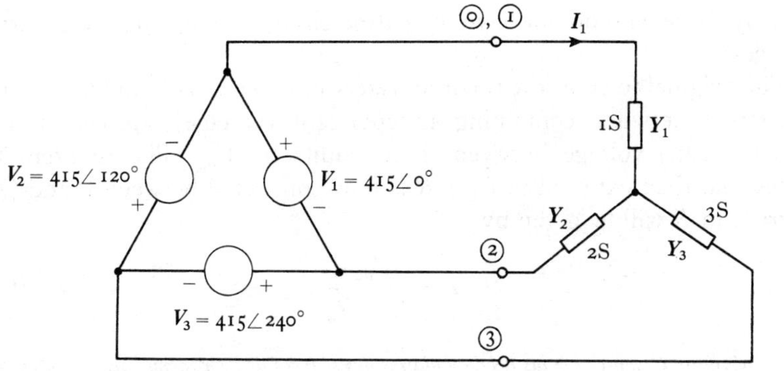

Fig. 6.3 Delta-connected source feeding an unsymmetrical star-connected load.

and
$$\boldsymbol{I}_1 = -\boldsymbol{V}_c \boldsymbol{Y}_1,$$
$$= (\boldsymbol{V}_1 \boldsymbol{Y}_2 - \boldsymbol{V}_2 \boldsymbol{Y}_3) \cdot \frac{\boldsymbol{Y}_1}{\boldsymbol{Y}_1 + \boldsymbol{Y}_2 + \boldsymbol{Y}_3}. \tag{6.2.5}$$

Substitution of the numerical values gives

$$\boldsymbol{I}_1 = \frac{(415 \times 2) - (415\,\mathrm{e}^{\mathrm{j}2\pi/3} \times 3)}{1 + 2 + 3}\,\mathrm{A},$$
$$= \frac{415}{6}(2 - 3\cos 120^\circ - 3\mathrm{j}\sin 120^\circ)\,\mathrm{A},$$
$$= \frac{415}{6}(2 + 1.5 - \mathrm{j}1.5\sqrt{3})\,\mathrm{A},$$
$$= 301\angle{-36.6^\circ}\,\mathrm{A}.$$

The substitution of complex values for the admittances $\boldsymbol{Y}_1$, $\boldsymbol{Y}_2$, and $\boldsymbol{Y}_3$ in place of the purely conductive ones used in this example would make the algebra slightly longer but would not affect the method.

6.3 Thevenin's theorem

In 1853, the German scientist H. Helmholtz (1821–94) described a network theorem which was subsequently forgotten; it now usually bears the name of L. Thévenin† (1857–1926), a French physicist, who published it independently thirty years later. In its original form it applied to linear d.c. circuits in which there were no controlled sources, but in the course of time it has become extended to circuits which contain controlled sources and also to a.c. circuits and transient circuits by means of the jω- and s-image worlds. The extension to controlled sources necessitates a formulation of the theorem which is slightly different from the original, and although the difference in wording is small it is important; the original form of wording is now retained only by those who do not have to analyse circuits which contain controlled sources.

The original form of the theorem refers to two points P and Q in a linear (resistive) network containing independent (voltage) sources. Let the (open-circuit) voltage between these points be V_{oc}. The theorem then states that if an external resistor R is now connected between P and Q the current in R will be given by

$$I = \frac{V_{oc}}{R + R_0}, \tag{6.3.1}$$

† L. Thévenin, *Comptes rendus hebdomadaires des séances de l'Académie des Sciences*, XCVII, 159 (1883).

where R_0 is the resistance of the original network, seen from the points PQ, when all the internal sources are turned down to zero. This phrase means that the sources are replaced by passive elements, namely their own internal resistances. The concept behind this, illustrated in fig. 6.4, is that any linear resistive network, viewed from two chosen terminals, may be represented by an ideal voltage source V_{oc} in series with a resistor R_0, and that the new representation is indistinguishable from the old at the terminals PQ. R_0 is called the output resistance of the circuit at the terminals PQ.

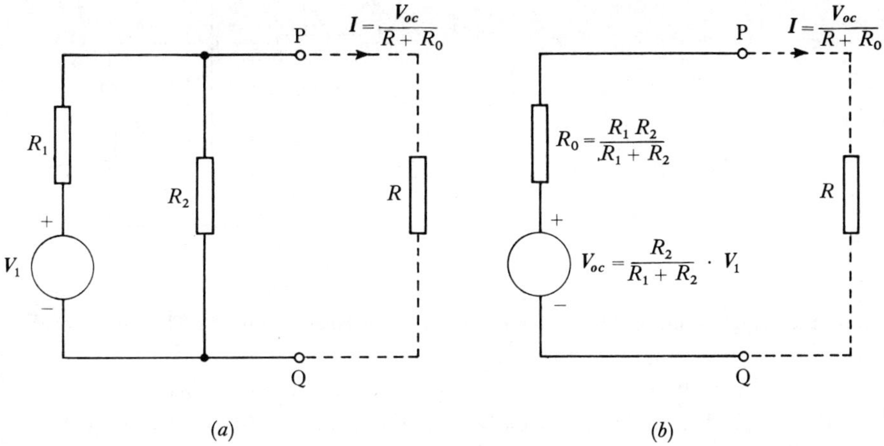

Fig. 6.4 Illustration of Thevenin's theorem: (*a*) original circuit; (*b*) Thevenin model.

The modification that must be made to the theorem to accommodate controlled sources concerns the evaluation of R_0. This must be evaluated with only the *independent* sources turned down to zero; controlled sources must be left operating. The values of V_{oc} and R_0 are, however, given in all situations by the following rules:

1. V_{oc} is the open-circuit voltage across the terminals PQ of the original circuit.

2. R_0 is the ratio of the open-circuit voltage V_{oc} to the current I_{sc} that would flow through a short-circuit placed across the terminals PQ.

The theorem is readily extended to alternating conditions by replacing R and R_0 by the complex impedances $\boldsymbol{Z}$ and $\boldsymbol{Z}_0$ and by letting $\boldsymbol{I}$ and $\boldsymbol{V}_{oc}$ stand for the complex values of the corresponding d.c. quantities.

The theorem will be demonstrated by two examples before a proof is given; first, in an a.c. circuit containing only independent sources; secondly, in a resistive circuit with a controlled source.

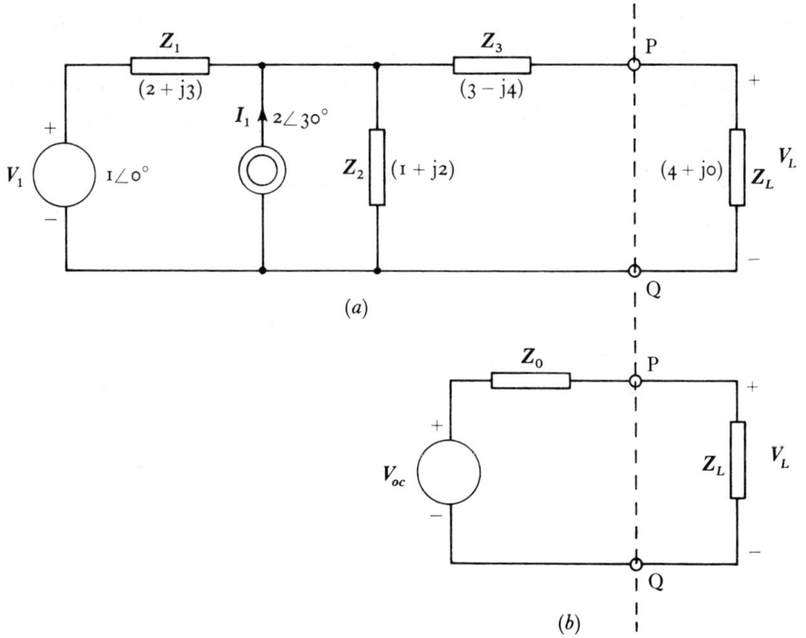

Fig. 6.5 Application of Thevenin's theorem: (*a*) original circuit; (*b*) Thevenin model.

1. Find the voltage $\boldsymbol{V}_L$ across the impedance $\boldsymbol{Z}_L$ in fig. 6.5(*a*).

Terminals P and Q are visualized at the ends of $\boldsymbol{Z}_L$. The voltage $\boldsymbol{V}_{oc}$, with $\boldsymbol{Z}_L$ removed, can be found by applying the superposition principle to the two sources. Thus

$$\boldsymbol{V}_{oc} = \frac{\boldsymbol{Z}_2}{\boldsymbol{Z}_1 + \boldsymbol{Z}_2} . \boldsymbol{V}_1 + \frac{\boldsymbol{Z}_1 \boldsymbol{Z}_2}{\boldsymbol{Z}_1 + \boldsymbol{Z}_2} . \boldsymbol{I}_1,$$

$$= \frac{(1 + j2)(1) + (1 + j2)(2 + j3)(1.73 + j1)}{3 + j5} \text{ V},$$

$$= \frac{-12.92 + j10.11}{3 + j5} \text{ V},$$

$$= 2.82 \angle 82.9^\circ \text{ V}.$$

The output impedance $\boldsymbol{Z}_0$ is equal to the impedance seen from the terminals PQ, looking back into the circuit, with the sources turned down to zero; this is true because the circuit contains only independent sources. Thus

$$\boldsymbol{Z}_0 = \boldsymbol{Z}_3 + \frac{\boldsymbol{Z}_1 \boldsymbol{Z}_2}{\boldsymbol{Z}_1 + \boldsymbol{Z}_2},$$

and the quantity $(\boldsymbol{Z}_0 + \boldsymbol{Z}_L)$, which will be needed in a moment, is

$$\boldsymbol{Z}_0 + \boldsymbol{Z}_L = \boldsymbol{Z}_L + \boldsymbol{Z}_3 + \frac{\boldsymbol{Z}_1 \boldsymbol{Z}_2}{\boldsymbol{Z}_1 + \boldsymbol{Z}_2},$$

$$= (4 + \mathrm{j}0) + (3 - \mathrm{j}4) + \frac{(1 + \mathrm{j}2)(2 + \mathrm{j}3)}{3 + \mathrm{j}5}\,\Omega,$$

$$= \frac{37 + \mathrm{j}30}{3 + \mathrm{j}5}\,\Omega,$$

$$= 8.16\angle{-20^\circ}\,\Omega.$$

The current flowing in the load is $\boldsymbol{V}_{oc}/(\boldsymbol{Z}_0 + \boldsymbol{Z}_L)$, and the voltage $\boldsymbol{V}_L$ is given by

$$\boldsymbol{V}_L = \frac{\boldsymbol{Z}_L}{\boldsymbol{Z}_0 + \boldsymbol{Z}_L}.\boldsymbol{V}_{oc},$$

$$= \frac{(4)(2.82\angle 82.9^\circ)}{8.16\angle{-20^\circ}}\,\mathrm{V},$$

$$= 1.38\angle 103^\circ\,\mathrm{V}.$$

Although the theorem has hitherto been used for finding an equivalent to the whole of a network feeding a load at terminals PQ, it can equally apply to feeding another network, as the following example shows.

2. What is the current through the resistor R_2 in fig. 6.6(*a*)?

The source $\boldsymbol{V}_2$ feeds the circuit in two places, so it is replaced by two sources in fig. 6.6(*b*), where we also mark the Thevenin section PQ at which the final calculation is required (this can be as shown – to use the actual terminals of R_2 is unnecessary) and another section P′Q′ which we propose to use *en route*. We first replace all to the left of P′Q′ by $\boldsymbol{V}'_{oc}$ in series with R'_0, then all to the left of PQ by $\boldsymbol{V}_{oc}$ in series with R_0. The first replacement is shown in part (*c*) where

$$\boldsymbol{V}'_{oc} = \frac{R_y}{R_x + R_y}.\boldsymbol{V}_2 \tag{6.3.2}$$

$$R'_0 = r_1 + \frac{R_x R_y}{R_x + R_y}, \tag{6.3.3}$$

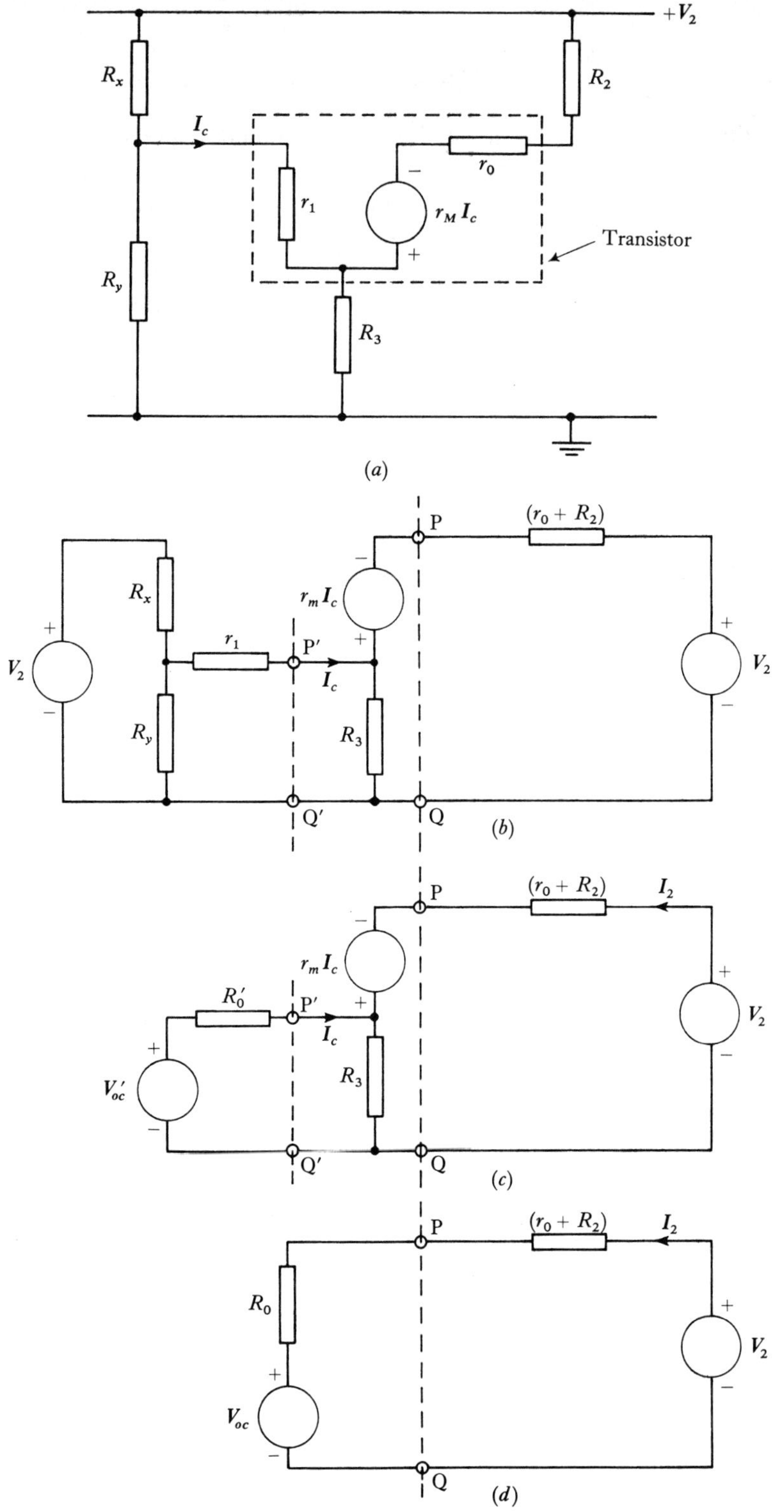

Fig. 6.6 Successive reduction by Thevenin's theorem: (*a*) original circuit; (*b*) circuit prepared for Thevenin's theorem; (*c*) first reduction by Thevenin's theorem; (*d*) second reduction by Thevenin's theorem.

and the second replacement is shown in part (*d*). There is no reason, of course, why the theorem should not be applied once only at the terminals PQ, but successive applications in simple stages are often preferable.

Most of the solution consists in getting the values of $\boldsymbol{V}_{oc}$ and R_0 from part (*c*). If $\boldsymbol{I}_{c1}$ is the value of the current $\boldsymbol{I}_c$ when PQ are open-circuited, then $\boldsymbol{V}_{oc}$, the open-circuit voltage at these terminals, is

$$\boldsymbol{V}_{oc} = -r_m \boldsymbol{I}_{c1} + R_3 \boldsymbol{I}_{c1},$$

$$= \frac{(R_3 - r_m)}{R_3 + R_0'} . \boldsymbol{V}_{oc}'. \tag{6.3.4}$$

The current $\boldsymbol{I}_{sc}$ through a short-circuit across PQ can be evaluated in the following way: under short-circuit conditions the voltage across R_3 is $r_m \boldsymbol{I}_{c2}$, where $\boldsymbol{I}_{c2}$ is the value of $\boldsymbol{I}_c$ when PQ are short-circuited, and the current through R_3 is $(\boldsymbol{I}_{c2} - \boldsymbol{I}_{sc})$. These two quantities must be related by Ohm's law,

$$r_m \boldsymbol{I}_{c2} = R_3(\boldsymbol{I}_{c2} - \boldsymbol{I}_{sc}). \tag{6.3.5}$$

$\boldsymbol{I}_{c2}$ can be found from the fact that the voltage across R_0' is $(\boldsymbol{V}_{oc}' - r_m \boldsymbol{I}_{c2})$ and that this is equal to $R_0' \boldsymbol{I}_{c2}$, giving

$$R_0' \boldsymbol{I}_{c2} = \boldsymbol{V}_{oc}' - r_m \boldsymbol{I}_{c2}. \tag{6.3.6}$$

The solution for $\boldsymbol{I}_{sc}$ from (6.3.5) and (6.3.6) is

$$\boldsymbol{I}_{sc} = \frac{R_3 - r_m}{R_3(R_0' + r_m)} . \boldsymbol{V}_{oc}', \tag{6.3.7}$$

and the output resistance R_0 at the terminals PQ is therefore

$$R_0 = \boldsymbol{V}_{oc}/\boldsymbol{I}_{sc},$$

$$= R_3(R_0' + r_m)/(R_3 + R_0'). \tag{6.3.8}$$

Notice that because of the presence of the controlled source this value of R_0 is not the same as that which would be obtained by turning all the sources down to zero and looking back into the terminals PQ. The result of this latter procedure would give the false value $R_0' R_3/(R_0' + R_3)$. It would, however, be possible to evaluate R_0 correctly by turning down to zero only the independent sources and evaluating the effective resistance seen on looking back into PQ. This could be done by imagining a voltage $\boldsymbol{V}$ applied to these terminals, calculating the current $\boldsymbol{I}$ that flows as a result of $\boldsymbol{V}$, and taking the ratio of these quantities. This would give the same result as shown in (6.3.8).

The wanted current $\boldsymbol{I}_2$ is now given by

$$\boldsymbol{I}_2 = \frac{(\boldsymbol{V}_2 - \boldsymbol{V}_{oc})}{R_0 + r_0 + R_2},$$

$$= \frac{\boldsymbol{V}_2(R_3 + R_0') - \boldsymbol{V}_{oc}'(R_3 - r_m)}{R_3(R_0' + r_m) + (r_0 + R_2)(R_3 + R_0')}. \tag{6.3.9}$$

Thevenin's theorem affords a useful means of making approximations in the analysis of a circuit. For example, (6.3.3) shows that if r_1 is small with respect to the parallel combination of R_x and R_y it has an insignificant effect on the circuit and may be omitted from the analysis. This is apparent to the user of Thevenin's theorem even at stage (*a*) of fig. 6.6.

The proof of the theorem depends on the general loop equations of a linear circuit and will be given in a manner directly applicable to steady-state

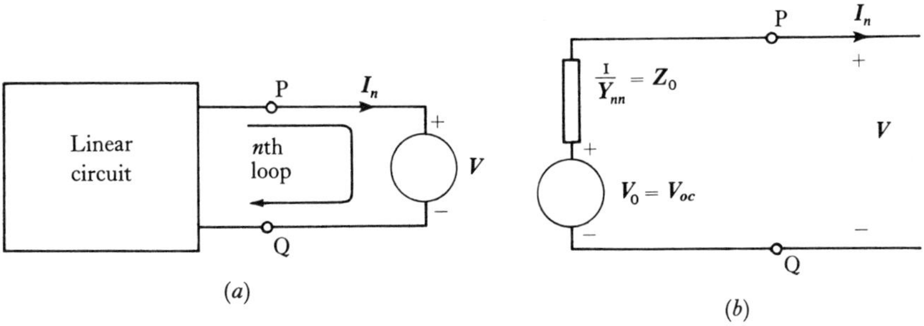

Fig. 6.7 Proof of Thevenin's theorem: (*a*) insertion of a voltage source in the *n*th loop of a network; (*b*) equivalent network.

d.c. and a.c. circuits, though capable of immediate extension to the transient case. Consider a circuit, containing both independent and controlled sources, which has a voltage source $\boldsymbol{V}$ inserted in its *n*th loop as illustrated in fig. 6.7(*a*). Its loop equations are

$$\left.\begin{aligned} \boldsymbol{V}_1 &= \boldsymbol{Z}_{11}'\boldsymbol{I}_1 + \boldsymbol{Z}_{12}'\boldsymbol{I}_2 + \ldots + \boldsymbol{Z}_{1n}'\boldsymbol{I}_n, \\ \boldsymbol{V}_2 &= \boldsymbol{Z}_{21}'\boldsymbol{I}_1 + \boldsymbol{Z}_{22}'\boldsymbol{I}_2 + \ldots + \boldsymbol{Z}_{2n}'\boldsymbol{I}_n, \\ \ldots &\quad \ldots \quad \ldots \quad \ldots \quad \ldots \\ \boldsymbol{V}_n - \boldsymbol{V} &= \boldsymbol{Z}_{n1}'\boldsymbol{I}_1 + \boldsymbol{Z}_{n2}'\boldsymbol{I}_2 + \ldots + \boldsymbol{Z}_{nn}'\boldsymbol{I}_n, \end{aligned}\right\} \tag{6.3.10}$$

where $\boldsymbol{V}_1$, $\boldsymbol{V}_2$, etc. are the sums of the clockwise independent voltages in their respective loops. Since (6.3.10) is a set of linear equations, the solution for $\boldsymbol{I}_n$ has the form

$$\boldsymbol{I}_n = \boldsymbol{Y}_{n1}\boldsymbol{V}_1 + \boldsymbol{Y}_{n2}\boldsymbol{V}_2 + \ldots + \boldsymbol{Y}_{nn}(\boldsymbol{V}_n - \boldsymbol{V}), \tag{6.3.11}$$

where the coefficients $\boldsymbol{Y}_{n1}$, etc. are constants, ratios of determinants of which the terms are impedances $\boldsymbol{Z}_{jk}$. Now if $\boldsymbol{I}_n$ is zero when $\boldsymbol{V}$ has the value $\boldsymbol{V}_0$, it follows from (6.3.11) that

$$\boldsymbol{V}_0 = \frac{\boldsymbol{Y}_{n1}}{\boldsymbol{Y}_{nn}}.\boldsymbol{V}_1 + \frac{\boldsymbol{Y}_{n2}}{\boldsymbol{Y}_{nn}}.\boldsymbol{V}_2 + \ldots + \frac{\boldsymbol{Y}_{nn}}{\boldsymbol{Y}_{nn}}.\boldsymbol{V}_n, \qquad (6.3.12)$$

and the substitution of (6.3.12) in (6.3.11) produces

$$\boldsymbol{I}_n = \boldsymbol{Y}_{nn}\boldsymbol{V}_0 - \boldsymbol{Y}_{nn}\boldsymbol{V}.$$

This is the equation of the circuit shown in fig. 6.7(*b*). The voltage $\boldsymbol{V}_0$ is the open-circuit voltage $\boldsymbol{V}_{oc}$ of the network because it is that value of $\boldsymbol{V}$ at which $\boldsymbol{I}_n$ equals zero, and the reciprocal of $\boldsymbol{Y}_{nn}$ can be shown to equal the complex ratio of the open-circuit voltage and the short-circuit current as follows. Let $\boldsymbol{I}_{sc}$ be the short-circuit value of $\boldsymbol{I}_n$, i.e. the value of $\boldsymbol{I}_n$ for the condition $\boldsymbol{V} = 0$. Putting $\boldsymbol{V}$ equal to zero in (6.3.11) and combining this with (6.3.12) produces

$$\begin{aligned}\boldsymbol{I}_{sc} &= \boldsymbol{Y}_{nn}\boldsymbol{V}_0,\\ &= \boldsymbol{Y}_{nn}\boldsymbol{V}_{oc}.\end{aligned}$$

Thus,

$$\frac{1}{\boldsymbol{Y}_{nn}} = \boldsymbol{Z}_0 = \frac{\boldsymbol{V}_{oc}}{\boldsymbol{I}_{sc}} = \frac{\text{open-circuit voltage}}{\text{short-circuit current}}. \qquad (6.3.13)$$

6.4 Rosen's theorem and the star–delta transformation

The equivalence which exists between a symmetrical star-connected load and a symmetrical delta-connected load is mentioned in chapter 3. This is a special case of a more general equivalence between what are called star-networks and mesh-networks (fig. 6.8) which was described by A. Rosen in

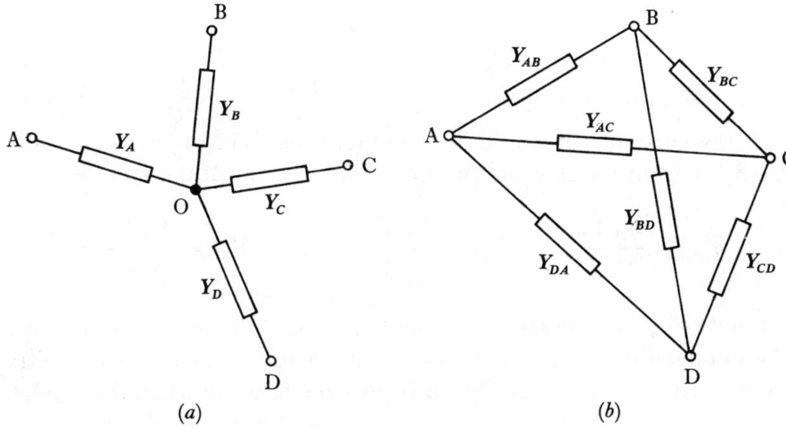

Fig. 6.8 Rosen's theorem: (*a*) star network; (*b*) equivalent mesh network.

1924.† Rosen's theorem states that in any network, a star of N elements OA, OB, ..., ON (fig. 6.8(a)) may be replaced by a mesh of $N(N-1)/2$ elements (fig. 6.8(b)) in which every pair of terminals A, B, ..., N is joined (O being eliminated) without affecting the rest of the network. It goes on to say that if the admittances of the star elements are $\boldsymbol{Y}_A$, $\boldsymbol{Y}_B$, ..., $\boldsymbol{Y}_N$, then the elements of the equivalent mesh are

$$\left.\begin{aligned} \boldsymbol{Y}_{AB} &= \frac{\boldsymbol{Y}_A \boldsymbol{Y}_B}{\sum \boldsymbol{Y}}, \qquad \boldsymbol{Y}_{BC} = \frac{\boldsymbol{Y}_B \boldsymbol{Y}_C}{\sum \boldsymbol{Y}}, \text{ etc.,} \\ \text{where} \qquad \sum \boldsymbol{Y} &= \boldsymbol{Y}_A + \boldsymbol{Y}_B + \ldots + \boldsymbol{Y}_N. \end{aligned}\right\} \tag{6.4.1}$$

The original proof depends on the mathematical structure of the equations which are necessary to satisfy the equivalence, but a direct proof can be obtained by Millman's theorem. Let voltage sources be connected between A, R; B, R; etc. where R is a reference point outside the network. The open-circuit voltage between the star point O and the reference point R is then

$$\boldsymbol{V}_{OR} = \frac{\boldsymbol{V}_{AR} \boldsymbol{Y}_A + \boldsymbol{V}_{BR} \boldsymbol{Y}_B + \ldots + \boldsymbol{V}_{NR} \boldsymbol{Y}_N}{\sum \boldsymbol{Y}}, \tag{6.4.2}$$

where $\sum \boldsymbol{Y}$ has the value defined by (6.4.1). The current in any star element, such as $\boldsymbol{Y}_B$ for example, flowing into the terminal B towards O is

$$\boldsymbol{I}_B = (\boldsymbol{V}_{BR} - \boldsymbol{V}_{OR}) \boldsymbol{Y}_B, \tag{6.4.3}$$

which, on substitution for $\boldsymbol{V}_{OR}$, becomes

$$\begin{aligned} \boldsymbol{I}_B = (\boldsymbol{V}_{BR} - \boldsymbol{V}_{AR}) \left(\frac{\boldsymbol{Y}_A \boldsymbol{Y}_B}{\sum \boldsymbol{Y}}\right) + (\boldsymbol{V}_{BR} - \boldsymbol{V}_{CR}) \left(\frac{\boldsymbol{Y}_B \boldsymbol{Y}_C}{\sum \boldsymbol{Y}}\right) \\ + \ldots + (\boldsymbol{V}_{BR} - \boldsymbol{V}_{NR}) \left(\frac{\boldsymbol{Y}_B \boldsymbol{Y}_N}{\sum \boldsymbol{Y}}\right). \end{aligned} \tag{6.4.4}$$

This is the current that would flow into terminal B of the equivalent mesh (fig. 6.8(b)), from the external circuit, if the elements of the mesh were

$$\boldsymbol{Y}_{AB} = \frac{\boldsymbol{Y}_A \boldsymbol{Y}_B}{\sum \boldsymbol{Y}}, \qquad \boldsymbol{Y}_{BC} = \frac{\boldsymbol{Y}_B \boldsymbol{Y}_C}{\sum \boldsymbol{Y}}, \ldots, \qquad \boldsymbol{Y}_{BN} = \frac{\boldsymbol{Y}_B \boldsymbol{Y}_N}{\sum \boldsymbol{Y}},$$

as given by (6.4.1). The same argument applies to all the other terminals.

It is not possible, in general, to reverse the theorem, to convert an arbitrary mesh into an equivalent star, but it is possible to do so when the mesh has

† A. Rosen, 'A new network theorem', *J.I.E.E.* **62**, 916 (1924).

only three elements; the conversion of a three-element mesh into a three-element star, or vice versa, is known as the star–delta transformation. Equations (6.4.1) give the elements of a delta (a three-element mesh) in terms of the given star admittances, and (6.4.5) give the elements of a star in terms of the given delta elements.

$$\left.\begin{aligned} \boldsymbol{Z}_A = \frac{\boldsymbol{Z}_{AB}\boldsymbol{Z}_{AC}}{\sum \boldsymbol{Z}}, \quad \boldsymbol{Z}_B = \frac{\boldsymbol{Z}_{BA}\boldsymbol{Z}_{BC}}{\sum \boldsymbol{Z}}, \quad \boldsymbol{Z}_C = \frac{\boldsymbol{Z}_{CA}\boldsymbol{Z}_{CB}}{\sum \boldsymbol{Z}}, \\ \text{where} \quad \sum \boldsymbol{Z} = \boldsymbol{Z}_{AB} + \boldsymbol{Z}_{BC} + \boldsymbol{Z}_{CA}. \end{aligned}\right\} \quad (6.4.5)$$

As seen from fig. 6.9 any unknown element is the product of the two adjacent elements divided by the sum of all three elements. In going from a star

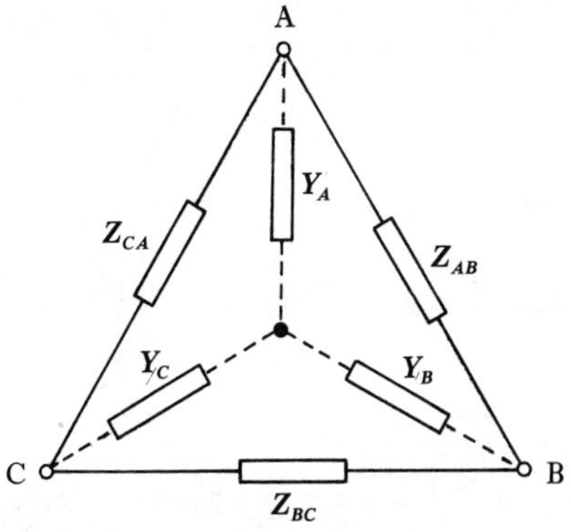

Fig. 6.9 Star–delta transformation.

to a delta this statement refers to admittances; in going from a delta to a star it refers to impedances. In the special case of equal elements which occurs in chapter 3,

$$\boldsymbol{Y}_\Delta = \frac{\boldsymbol{Y}_Y^2}{3\boldsymbol{Y}_Y} = \frac{\boldsymbol{Y}_Y}{3},$$

or

$$\boldsymbol{Z}_\Delta = 3\boldsymbol{Z}_Y.$$

Figure 6.10 shows the application of the star–delta transformation to a simple d.c. Wheatstone bridge in which it is required to find the current drawn from the voltage source. Transformation of the delta made up of the 2 Ω, 4 Ω, and 6 Ω resistors to the corresponding star shown in fig. 6.10(*b*) enables the total load on the source to be calculated from the simple rules of elements in series and parallel; the source is connected to an effective load of 4.2 Ω and therefore supplies a current of 2.1 A.

If a star or mesh network contains a mixture of *R*, *L*, and *C* elements, the elements in the corresponding transformed network will in general be

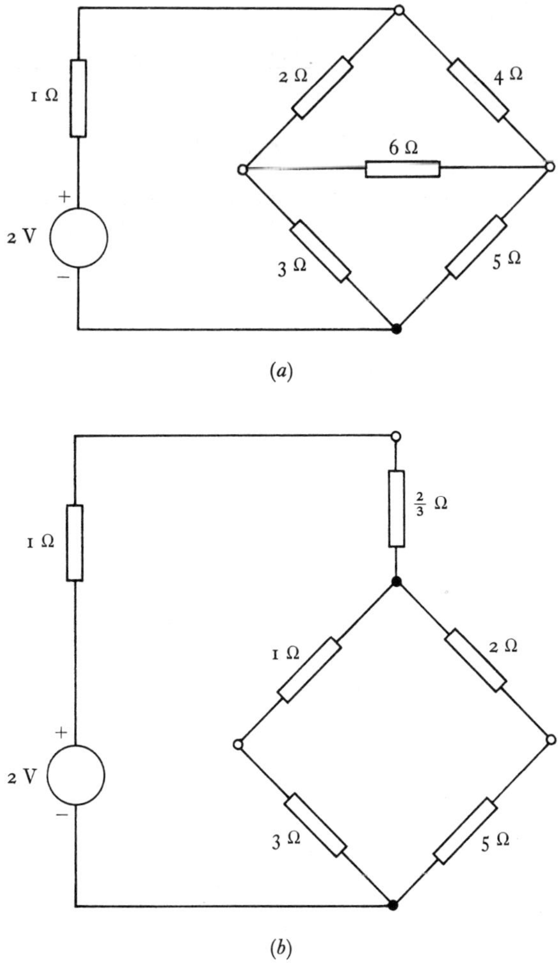

Fig. 6.10 Application of the star–delta transformation: (*a*) Wheatstone bridge; (*b*) Wheatstone bridge after transformation.

functions of frequency. This normally restricts the usefulness of the transformation either to networks which are operated at a fixed frequency or to networks composed entirely of elements of the same kind. It should also be observed that the transformation sometimes leads, quite correctly, to elements containing negative resistance.

6.5 The substitution theorem

If a passive circuit element $\boldsymbol{Z}$ passes a current $\boldsymbol{I}$ and thereby develops a voltage $\boldsymbol{V}$, it may be replaced by an ideal voltage source of value $\boldsymbol{V}$ without affecting the rest of the circuit. This is the substitution theorem; and it is such an obvious consequence of the Kirchhoff loop equations that its significance tends to be overlooked. The significance is that the substitution is

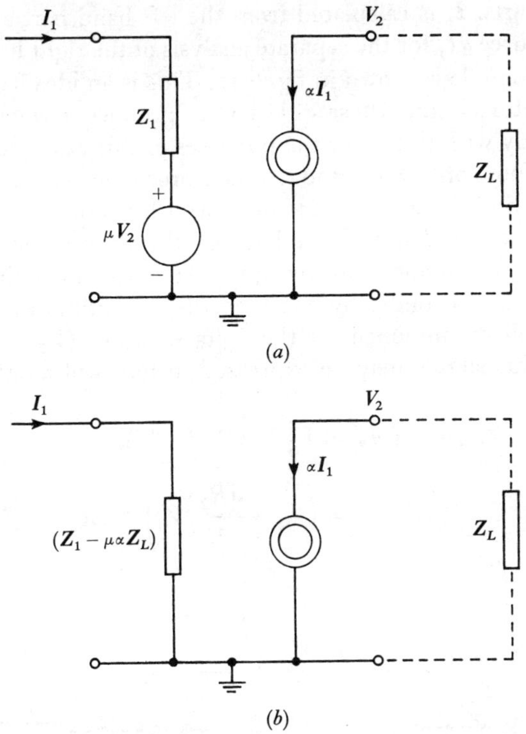

Fig. 6.11 Application of the substitution theorem: (*a*) original network; (*b*) network after application of the substitution theorem.

valid both ways round; not only can a passive element $\boldsymbol{Z}$ be replaced by a (current-controlled) voltage source $\boldsymbol{ZI}$, but a current-controlled voltage source $\boldsymbol{ZI}$ can be replaced by a passive element $\boldsymbol{Z}$. The theorem is used, in favourable circumstances, to eliminate current-controlled voltage sources, as will be shown in the following two examples.

The circuit of fig. 6.11(*a*) should be thought of as a small part of a larger circuit which extends on both sides of the diagram; the part on the right hand side is symbolized by the impedance $\boldsymbol{Z}_L$. The analysis of the full circuit can be simplified by eliminating the feedback effect of the source $\mu\boldsymbol{V}_2$ and this can be done by means of the substitution theorem. The voltage $\boldsymbol{V}_2$ measured with respect to the earth terminal is $(-\alpha\boldsymbol{I}_1\boldsymbol{Z}_L)$, and the controlled source $\mu\boldsymbol{V}_2$ is

$$\mu\boldsymbol{V}_2 = -(\mu\alpha\boldsymbol{Z}_L)\,\boldsymbol{I}_1,$$

a current-controlled voltage source through which the current $\boldsymbol{I}_1$ flows. This source may therefore be replaced by an impedance $-(\mu\alpha\boldsymbol{Z}_L)$ as shown in part (*b*), which consequently splits the overall circuit into two more manageable parts. $\boldsymbol{I}_1$ is calculated from the left hand part and its value is used in the source $\alpha\boldsymbol{I}_1$ for the separate analysis of the right hand part.

A second example is shown in fig. 6.12. This is an idealized version of a Miller-integrator circuit whose object is to produce a voltage (v_0) which changes linearly with time. For the purposes of this example we may treat V_1 as a step function $V_1 u(t)$ and V_2 as a preconnected d.c. source. The problem is to find the actual value of v_0 as a function of time, given that $v_0 = V_2$ when $t = 0$. Parts (*b*) and (*c*) of the diagram show successive reductions of the circuit, first by applying Thevenin's theorem to the terminals PQ and secondly by applying the substitution theorem to the current-controlled component of the voltage source $(V_2 - \beta R_2 i_B)$ shown in part (*b*). This source may be expressed in terms of i rather than i_B as follows:

$$(V_2 - \beta R_2 i_B) = V_2 - \beta R_2(I_1 - i),$$

$$= \left(V_2 - \frac{\beta R_2}{R_1}.V_1\right) + \beta R_2 i,$$

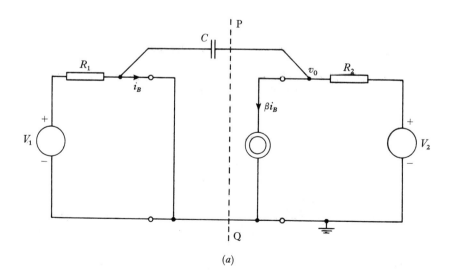

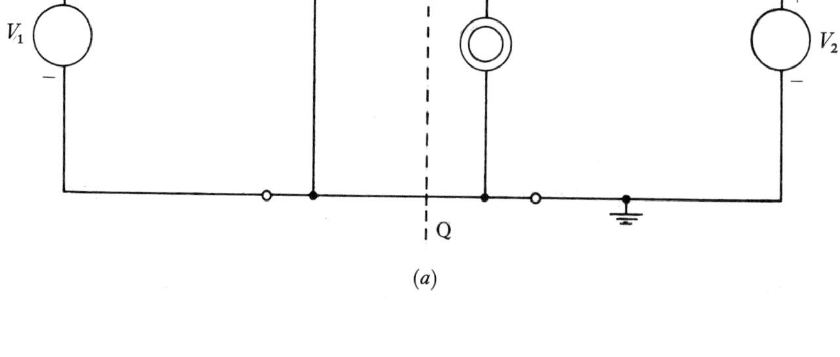

(*a*)

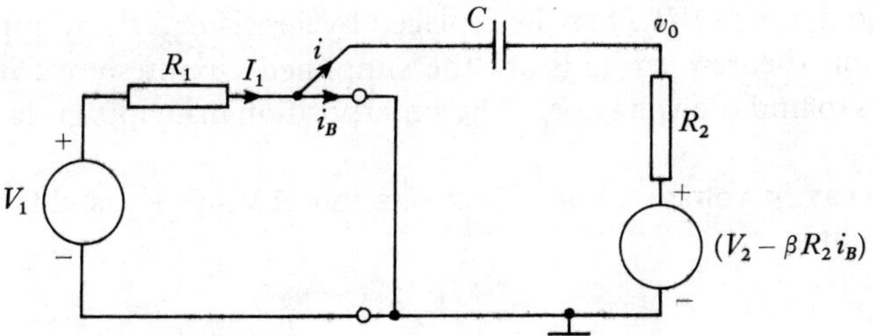

(b)

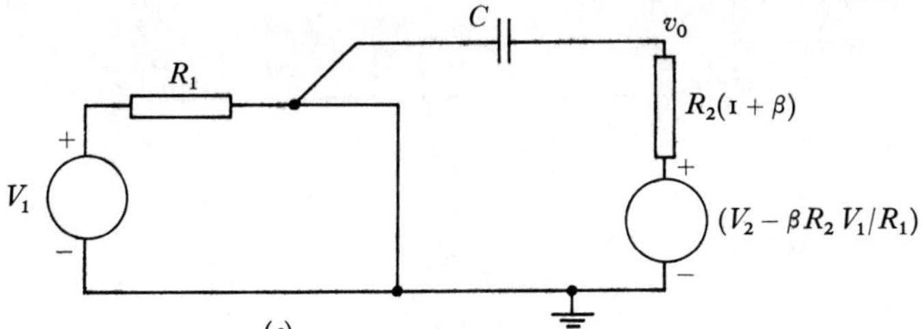

(c)

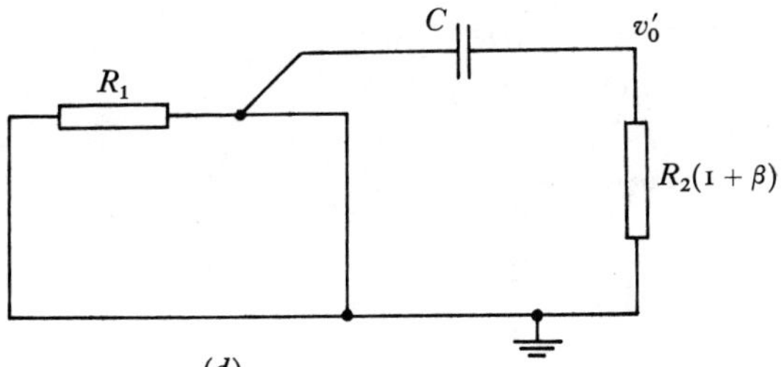

(d)

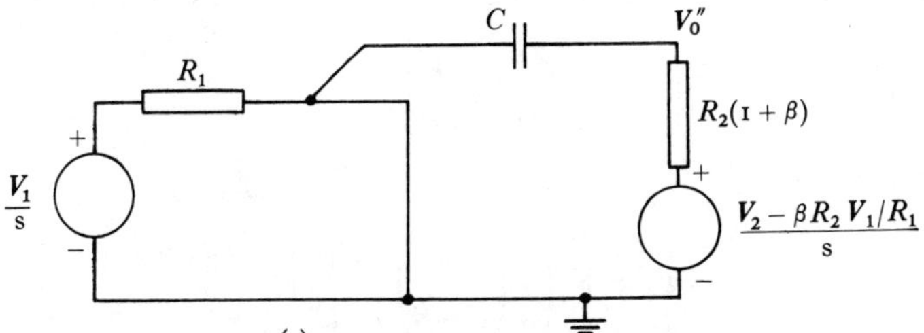

(e)

Fig. 6.12 The Miller-integrator circuit: (*a*) idealized Miller-integrator circuit; (*b*) after reduction by Thevenin's theorem; (*c*) after additional reduction by substitution theorem; (*d*) circuit controlling decay of voltage v_0' on C from its initial value V_2; (*e*) image circuit giving voltage component v_0'' on C due to operation of voltage sources from $t = 0$ onwards.

and the component $\beta R_2 i$ may be replaced by a resistor βR_2 by applying the substitution theorem to it. From the simplified circuit shown in part (*c*) it remains to find the voltage v_0. The superposition principle gives a suitable method.

The decay of voltage v_0' on C from its initial value V_2 is obtained from part (*d*) and is

$$v_0' = V_2 \mathrm{e}^{-t/\tau},$$

where $\tau = CR_2(1+\beta)$. The voltage component v_0'' on C due to the operation of the voltage sources from $t = 0$ onwards is obtained from part (*e*). The image voltage $\boldsymbol{V}_0''$ is given by

$$\boldsymbol{V}_0'' = \frac{1/C\mathrm{s}}{R_2(1+\beta)+1/C\mathrm{s}} \cdot \frac{V_2 - \beta R_2 V_1/R_1}{\mathrm{s}},$$

$$= \frac{V_2 - \beta R_2 V_1/R_1}{\tau \mathrm{s}(\mathrm{s}+1/\tau)},$$

$$= \left(V_2 - \frac{\beta R_2}{R_1}.V_1\right)\left(\frac{1}{\mathrm{s}} - \frac{1}{\mathrm{s}+1/\tau}\right).$$

Hence $$v_0'' = \left(V_2 - \frac{\beta R_2}{R_1}.V_1\right)(1 - \mathrm{e}^{-t/\tau}),$$

and $$v_0 = v_0' + v_0'',$$

$$= V_2 - \frac{\beta R_2}{R_1}.V_1(1 - \mathrm{e}^{-t/\tau}). \qquad (6.5.1)$$

In practice, (6.5.1) would not be valid for negative values of v_0 because circuit (*a*) would not be a valid representation of the actual circuit in this region. Putting this restriction on v_0 we find that for the condition

$$\frac{\beta R_2}{R_1} \gg \frac{V_2}{V_1} \qquad (6.5.2)$$

(6.5.1) becomes

$$v_0 \approx V_2 - \left(\frac{\beta}{1+\beta}\right)\frac{V_1}{R_1 C}.t. \qquad (6.5.3)$$

The substitution theorem has so far been described in only one of its two possible forms. It will become apparent from §6.8 on the duality of networks that not only may a current-controlled voltage source $\boldsymbol{ZI}$ which has a current $\boldsymbol{I}$ flowing through it be replaced by an impedance $\boldsymbol{Z}$, but a voltage-controlled current source $\boldsymbol{YV}$ which has a voltage $\boldsymbol{V}$ across it may be replaced by an admittance $\boldsymbol{Y}$. The second form is of equal value.

6.6 The manipulation of current sources

If a circuit contains a current source between two nodes A and B (fig. 6.13(*a*)), the source can be replaced by a string of identical current sources which starts at A, finishes at B, and includes in its path as many other nodes

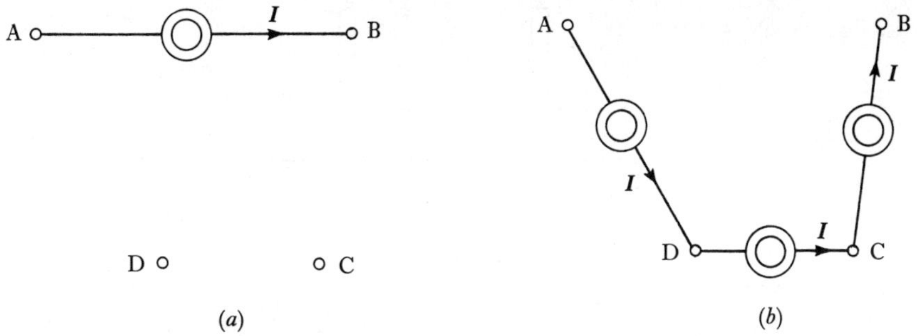

Fig. 6.13 Manipulation of current sources: (*a*) network with a current source; (*b*) changed network with unchanged node voltages.

C, D, etc. as may be desired (fig. 6.13(*b*)), without affecting any of the node voltages. This is true because the Kirchhoff current law is undisturbed at each node. For example, the node D has an extra current pushed into it by the source AD and an equal current withdrawn from it by the source DC, leaving the current balance unchanged.

At first sight it appears that this process complicates rather than simplifies the subsequent analysis, but this is not so if the substitution theorem can be applied simultaneously. Two situations of this kind are shown in figs. 6.14 and 6.15. In the first, the controlled source $\boldsymbol{Y}_m\boldsymbol{V}_1$ is replaced by two identical sources, one across AE and the other across EB. The source across AE, driving from E to A, is then replaced by an admittance $-\boldsymbol{Y}_m$ by means of the substitution theorem, and this is combined with the element $\boldsymbol{Y}_1$ to produce $(\boldsymbol{Y}_1 - \boldsymbol{Y}_m)$ as shown in part (*b*). The final arrangement, like the original, contains only one source, but at a different place. Whether or not this is an advantage depends on the circumstances. If the response at the terminals BE to a stimulus at AE is required, the rearrangement is advantageous because the controlled source, now directly across the output terminals, is likely to produce the primary component of the response. Moreover, if $\boldsymbol{Y}_3$ is small enough to be neglected in an approximate analysis, the circuit can be analysed in two separate parts; its effect upon the network connected to terminals AE is merely that of an admittance ($\boldsymbol{Y}_1 - \boldsymbol{Y}_m$).

Similar arguments apply to fig. 6.15(*a*). Here the substitution theorem is first used to convert the element $\boldsymbol{Y}_f$ to the current source shown in part (*b*)

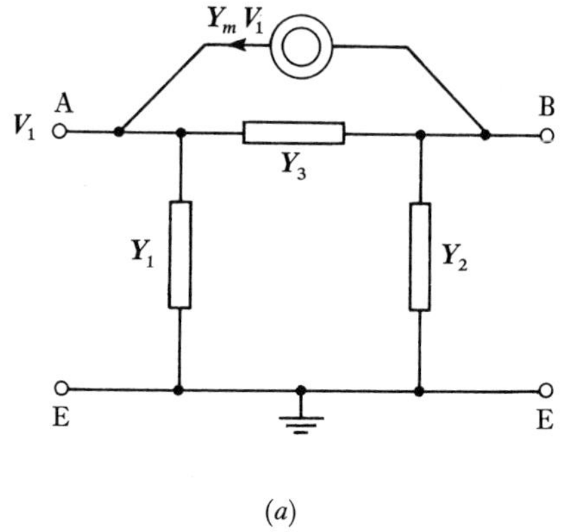

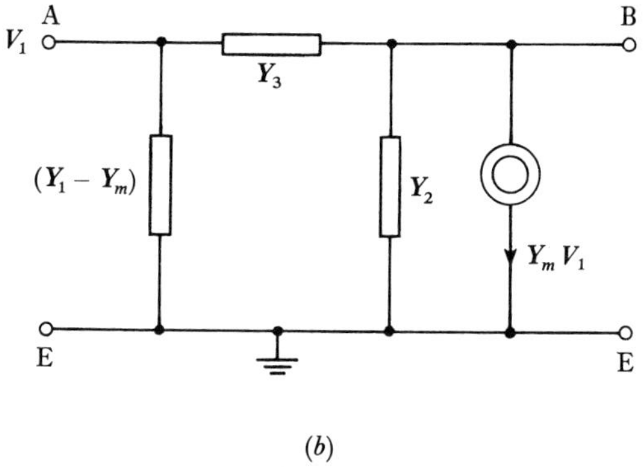

Fig. 6.14 Movement of a current source: (*a*) original circuit; (*b*) equivalent circuit.

and, after replacing this source by two current sources (part (*c*)), is then used again to convert the current sources back into admittance elements (part (*d*)). The conversion of the left hand current source in (*c*) to the corresponding admittance element in (*d*) is made by dividing the source current $\boldsymbol{Y}_f \boldsymbol{V}_1(1 - \mu)$ by the voltage $\boldsymbol{V}_1$ that exists across the source. The right hand current source is converted in a similar manner; the source current flowing from B to E, namely $\boldsymbol{Y}_f \boldsymbol{V}_1(\mu - 1)$, is divided by the voltage $\mu \boldsymbol{V}_1$ across the

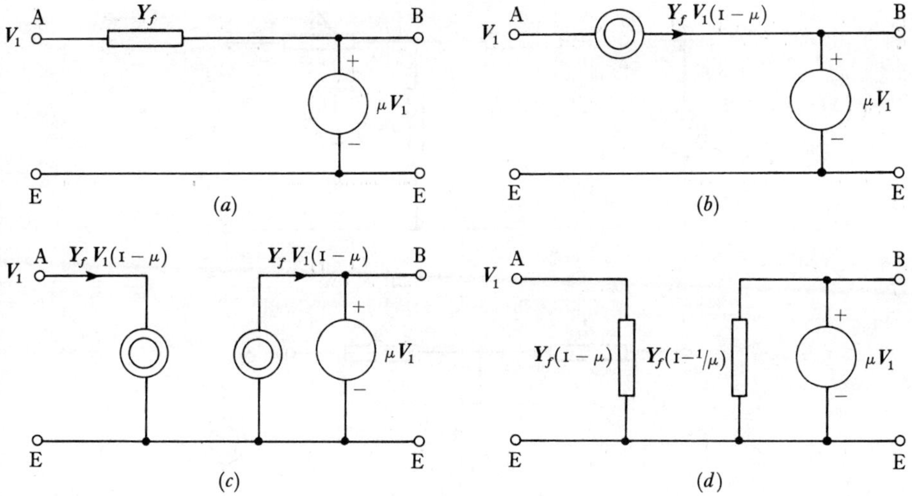

Fig. 6.15 Removal of an element: (*a*) portion of an active circuit; (*b*) substitution of a current source for the element $\boldsymbol{Y}_f$; (*c*) replacement of the current source by two identical sources; (*d*) substitution of an admittance element for each current source.

source to produce the admittance element $\boldsymbol{Y}_f(1 - 1/\mu)$. Part (*d*) shows that the terminals A and B have been isolated by the removal of the original element $\boldsymbol{Y}_f$ and that the circuit can now be analysed in two separate parts.

6.7 Reciprocal and non-reciprocal networks

Figure 6.16 shows a simple reciprocal network. The relevant property is that the voltage response to a current stimulus is independent of an exchange of positions of stimulus and response; i.e. the ratio of stimulus to response in circuit (*a*) is the same as in circuit (*b*):

$$\left(\frac{\boldsymbol{V}_2}{\boldsymbol{I}_1}\right)_{I_2=0} = \left(\frac{\boldsymbol{V}_1}{\boldsymbol{I}_2}\right)_{I_1=0}. \tag{6.7.1}$$

This property applies to a large class of networks, and a statement of it is called the *reciprocity theorem*. Networks which display this property are called *reciprocal networks*; those that do not are called *non-reciprocal networks*. All networks which employ only linear *R*, *C*, *L* and *M* elements are reciprocal networks; those which employ controlled sources are likely to be non-reciprocal. What matters is the equality of the coefficients $\boldsymbol{Y}_{jk}$ and $\boldsymbol{Y}_{kj}$ in the node equations, and a reciprocal network is *defined* as one in which these coefficients are equal, thereby producing symmetry about the principal diagonal of their array of coefficients, as in fig. 1.14. Equation (6.7.1) follows from this equality.

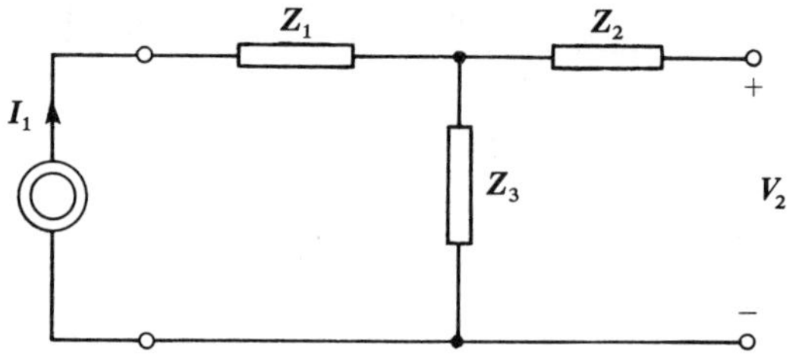

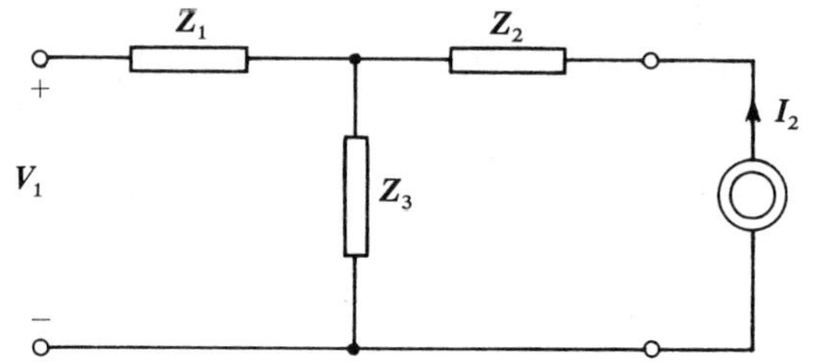

Fig. 6.16 Node version of the reciprocity theorem: (*a*) current stimulus producing a voltage response; (*b*) positions of stimulus and response exchanged.

An alternative form of the reciprocity theorem, in which the stimulus is a voltage and the response a current, is given by (6.7.2) which applies to the test conditions shown in fig. 6.17.

$$\left(\frac{\boldsymbol{I}_2}{\boldsymbol{V}_1}\right)_{V_2=0} = \left(\frac{\boldsymbol{I}_1}{\boldsymbol{V}_2}\right)_{V_1=0}. \tag{6.7.2}$$

This is a consequence of an equality of the impedance coefficients $\boldsymbol{Z}_{jk}$ and $\boldsymbol{Z}_{kj}$ in the loop equations of a reciprocal network which is an inevitable consequence of the corresponding equality of its admittance coefficients. It is, however, important to observe that the reciprocity theorem is not valid when the stimulus and response are of the same kind thereby making the transfer function dimensionless. This is because an interchange of stimulus and response will, in this situation, necessitate a change in the circuit state at the terminals. To see this clearly it is helpful to view a current source as a generalized open-circuit and a voltage source as a generalized short-circuit. It can be seen, for example, that the exchange of an ammeter (a short-circuit) for a current source (an open-circuit) at a pair of terminals changes the circuit configuration and renders the theorem invalid. Similarly, an exchange of a voltmeter (an open-circuit) for a voltage source (a short-circuit) renders the theorem invalid.

A device which violates the reciprocity theorem by virtue of its conception is the *gyrator*. This is a network which has two pairs of terminals, and at

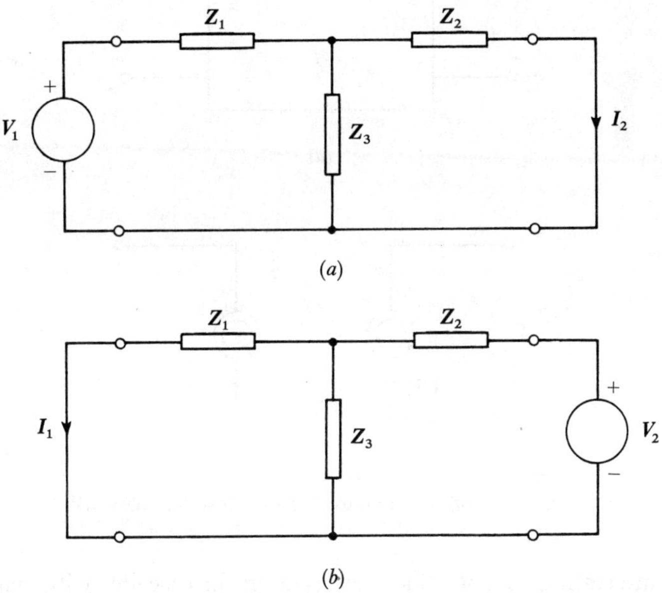

Fig. 6.17 Loop version of the reciprocity theorem: (*a*) voltage stimulus producing a current response; (*b*) positions of stimulus and response exchanged.

each pair the current is dependent only on the voltage at the other pair. For the terminology of fig. 6.18(*a*) the defining equations are

$$\left.\begin{aligned} I_1 &= GV_2, \\ I_2 &= -GV_1, \end{aligned}\right\} \tag{6.7.3}$$

where G is a constant; the equivalent circuit is shown in part (*b*). The instantaneous power supplied to a gyrator is

$$\begin{aligned} p &= v_1 i_1 + v_2 i_2, \\ &= v_1 G v_2 - v_2 G v_1, \\ &= 0, \end{aligned}$$

so that, like the ideal transformer, it consumes no power but transmits all the power supplied to it. It is therefore passive in addition to being linear and non-reciprocal. Its justification as a network element lies in the existence of some passive microwave components which possess gyrator properties, but networks with these properties can also be constructed from active non-reciprocal devices like transistors.

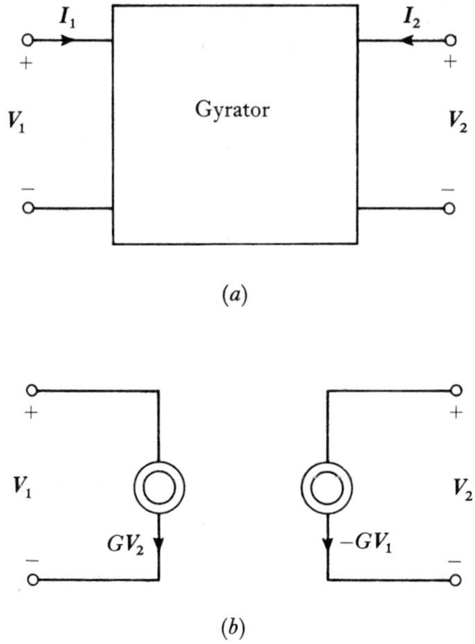

Fig. 6.18 (a) The gyrator. (b) An equivalent network.

It is interesting to note that two gyrators in cascade, with parameters G_a and G_b respectively, form an ideal transformer of turns ratio (G_a/G_b).

6.8 Duality

A glance at the pairs of equations which relate the voltage and the current in ideal R, L, and C elements reveals a striking similarity in their mathematical forms.

$$\left.\begin{array}{ll} v = Ri; & i = Gv; \\ v = L\dfrac{di}{dt}; & i = C\dfrac{dv}{dt}; \\ v = \dfrac{1}{C}\displaystyle\int i\,dt; & i = \dfrac{1}{L}\displaystyle\int v\,dt. \end{array}\right\} \quad (6.8.1)$$

This likeness is described by the word *duality*, and the elements that correspond to the coefficients in corresponding equations are called the *duals* of one another. A similar likeness can exist not only between specific pairs of elements but also between specific pairs of complete circuits, and when this is so the *circuits* are said to be the duals of one another. Fig. 6.19 shows such a pair of dual circuits, and it can be seen from (6.8.2) that the equation of either circuit can be produced from the equation of the other by interchanging v and i, R and G, L and C, and C and L:

$$\left.\begin{aligned} v &= Ri + L\frac{di}{dt} + \frac{1}{C}\int i\,dt, \\ i &= Gv + C\frac{dv}{dt} + \frac{1}{L}\int v\,dt. \end{aligned}\right\} \tag{6.8.2}$$

These circuits are duals because in addition to the duality of the corresponding elements there is a duality between the Kirchhoff loop equation, $\sum v = 0$, of the series circuit and the Kirchhoff node equation, $\sum i = 0$, of

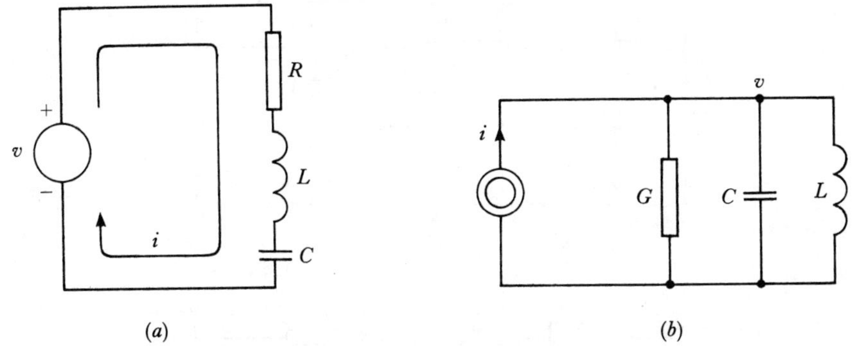

Fig. 6.19 Dual circuits: (*a*) series circuit fed from a voltage source; (*b*) parallel circuit fed from a current source.

the parallel circuit. A table of some dual quantities is given here. These dualities are mutual; that is, the quantity in either column is the dual of the corresponding quantity in the other column.

Quantity	Dual quantity
Resistance	Conductance
Inductance	Capacitance
Impedance	Admittance
Voltage source	Current source
Series connexion	Parallel connexion
Mesh current	Node voltage
Open-circuit voltage	Short-circuit current

Any linear mappable circuit that does not contain mutual inductance has a dual, and one method of deriving the dual is shown in fig. 6.20. A node is placed in each mesh of the original circuit and a reference node is placed outside the circuit. The voltages $\boldsymbol{V}_A$, $\boldsymbol{V}_B$, etc. of these new nodes, measured with respect to the reference node R, form the dual quantities of the loop

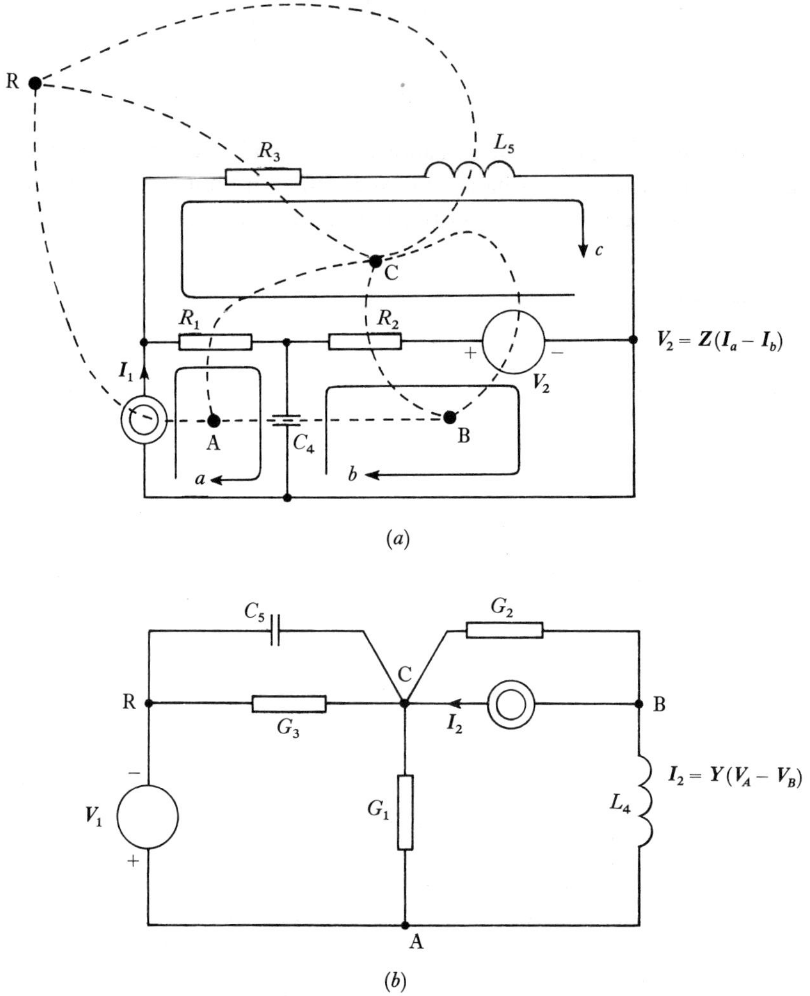

Fig. 6.20 The construction of a dual circuit: (*a*) original; (*b*) dual.

currents $\boldsymbol{I}_a$, $\boldsymbol{I}_b$, etc. of the original circuit. Lines are drawn from the new nodes to each other so as to cross the elements of the original circuit, and in these lines are inserted elements which are the duals of the corresponding originals. Such lines are shown dotted in fig. 6.20(*a*) and are arranged in a more conventional way in part (*b*). Equations (6.8.3) and (6.8.4) which are respectively the Kirchhoff mesh equations for the meshes b, c in the original circuit and the node equations for the nodes B, C in the dual, show that duality exists between the mesh currents of the original and the node voltages of the dual.

$$\left.\begin{aligned} \boldsymbol{I}_1\left(\frac{1}{j\omega C_4}-\boldsymbol{Z}\right) &= \left(R_2+\frac{1}{j\omega C_4}-\boldsymbol{Z}\right)\boldsymbol{I}_b - R_2\boldsymbol{I}_c, \\ \boldsymbol{I}_1(\boldsymbol{Z}+R_1) &= (\boldsymbol{Z}-R_2)\boldsymbol{I}_b + (R_1+R_2+R_3+j\omega L_5)\boldsymbol{I}_c. \end{aligned}\right\} \quad (6.8.3)$$

$$\left.\begin{aligned} \boldsymbol{V}_1\left(\frac{1}{j\omega L_4}-\boldsymbol{Y}\right) &= \left(G_2+\frac{1}{j\omega L_4}-\boldsymbol{Y}\right)\boldsymbol{V}_B - G_2\boldsymbol{V}_C, \\ \boldsymbol{V}_1(\boldsymbol{Y}+G_1) &= (\boldsymbol{Y}-G_2)\boldsymbol{V}_B + (G_1+G_2+G_3+j\omega C_5)\boldsymbol{V}_C. \end{aligned}\right\} \quad (6.8.4)$$

In the construction described, care must be taken with the polarity of the dual sources. Assume that clockwise mesh currents in the original are to correspond with positive node voltages in the dual. Then it follows that a source which tends to produce a clockwise mesh current $\boldsymbol{I}_a$ in the original must be replaced by a source (the dual of the original) which tends to produce a positive node voltage $\boldsymbol{V}_A$ in the dual. This is shown in fig. 6.20. Notice also the duality between the coefficients of the controlled sources $\boldsymbol{V}_2$ and $\boldsymbol{I}_2$.

The concept of duality is useful because it enables us to draw conclusions about a circuit from a knowledge of its dual, and because it may enable us to extend a circuit theorem by the inclusion of a dual version. Thevenin's theorem is a good example. The dual version, described by E. L. Norton of the Bell Telephone Laboratories in 1926, runs as follows. Any linear two-terminal network may be replaced by an ideal current source $\boldsymbol{I}_{sc}$ in parallel with an admittance $\boldsymbol{Y}_0$, where $\boldsymbol{I}_{sc}$ is the short-circuit current at the terminals and $\boldsymbol{Y}_0$ is the ratio of the short-circuit current to the open-circuit voltage at

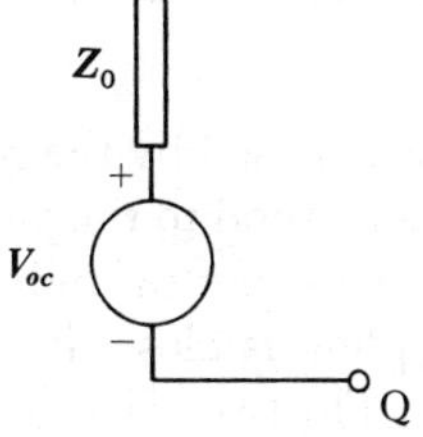

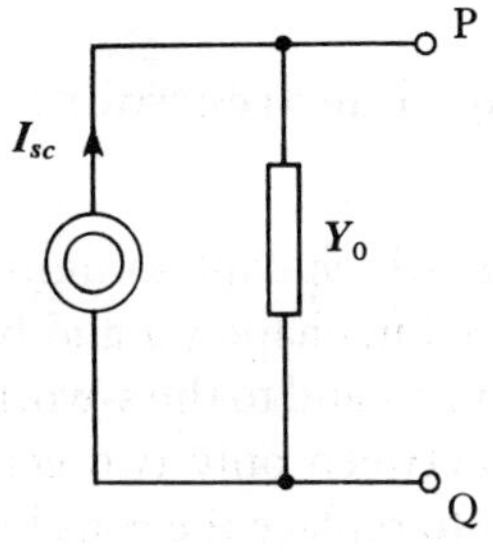

Fig. 6.21 Dual versions of Thevenin's theorem: (*a*) Thevenin representation; (*b*) Norton representation.

the terminals. The equivalent circuits produced by Thevenin's theorem and Norton's theorem are shown in fig. 6.21; the relationship between them is given by the equations

$$\left.\begin{aligned} \boldsymbol{Y}_0 &= 1/\boldsymbol{Z}_0, \\ \boldsymbol{I}_{sc} &= \boldsymbol{V}_{oc}/\boldsymbol{Z}_0. \end{aligned}\right\} \tag{6.8.5}$$

These may be obtained by equating the open-circuit voltage of each version and by equating the short-circuit current of each version.

A second example concerns the manipulation of current sources described in §6.6. The dual version of this, illustrated in fig. 6.22, is the manipulation

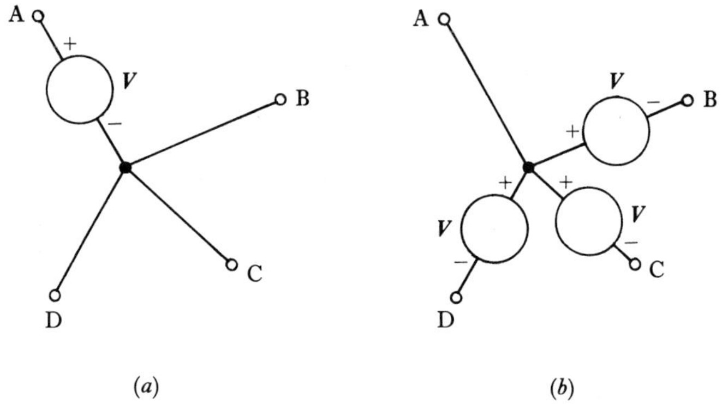

Fig. 6.22 Manipulation of voltage sources: (*a*) network with a voltage source; (*b*) changed network with unchanged loop currents.

of voltage sources. The change from part (*a*) to part (*b*), which has been described as 'pushing the voltage source through the node', leaves the net loop voltages, and hence the loop currents unchanged.

6.9 The treatment of the linear transformer in circuit analysis

Circuits which contain mutual inductance may be analysed by the method described in chapter 2 and by the extension of this method to the jω-world of chapter 4 and to the s-world of chapter 5, but when the mutual inductance exists between only two coils, and when the coupling is close, it is often helpful to replace the coils by an equivalent circuit. The two coils constitute a transformer and since the most useful equivalent circuit of a transformer includes an ideal one, this section begins with ideal transformers and then goes on to describe some equivalent circuits of non-ideal transformers.

An ideal transformer of turns ratio n is shown in fig. 6.23(*a*) loaded by a series impedance $\boldsymbol{Z}_2$ on the output side. The equations relating the terminal voltages and currents of the dotted box are

$$\left.\begin{aligned} \boldsymbol{I}_1 &= -n\boldsymbol{I}_2, \\ \boldsymbol{V}_1 &= \boldsymbol{V}_2'/n = (\boldsymbol{V}_2/n) + (\boldsymbol{Z}_2\boldsymbol{I}_1)/n^2. \end{aligned}\right\} \tag{6.9.1}$$

An examination of the terminal voltages and currents of the dotted box in part (*b*) shows that they too are related by (6.9.1); and since the two boxes,

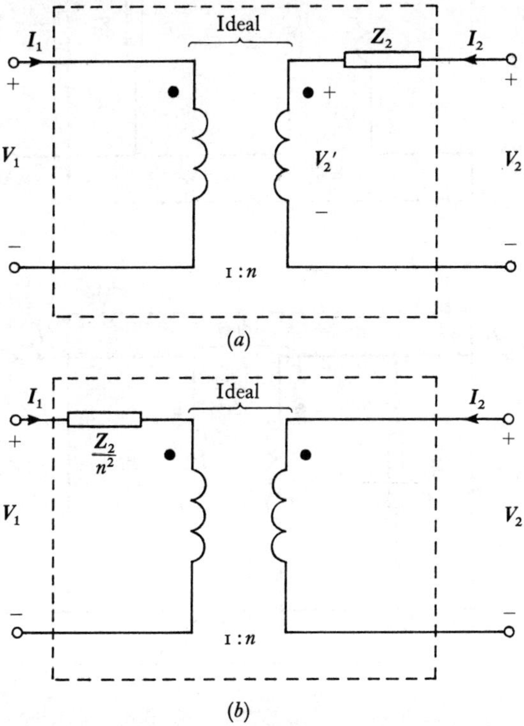

Fig. 6.23 Transfer of a series impedance across an ideal transformer: (*a*) series impedance on output side; (*b*) equivalent impedance on input side.

as seen from their terminals, are indistinguishable, a series impedance on one side of an ideal transformer may be transferred to the other without affecting the rest of the circuit, provided that it is modified by the factor n^2. A similar equivalence exists between the two parts of fig. 6.24, in which the transformers have shunt loading, and between the two parts of fig. 6.25 in which part (*a*) shows a three-terminal ideal transformer with mutual loading and part (*b*) shows a similar transformer with shunt loading.

The use of an ideal transformer to represent a real one may be justified when the coupling is close, the leakage inductances small, the coupling inductances large, and the energy loss negligible; but it is often necessary to account for the imperfections due to these quantities. This is done by adding

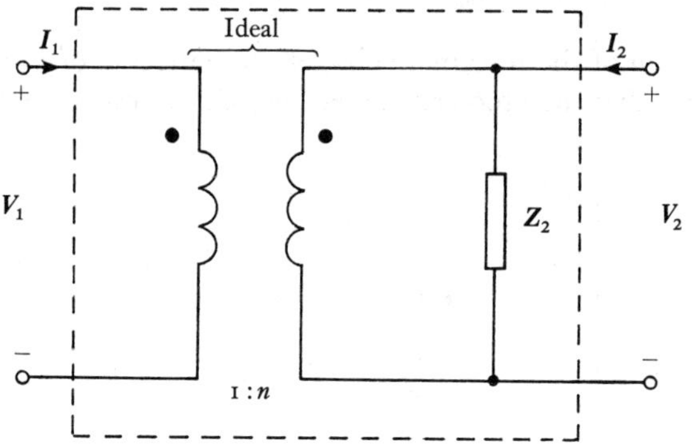

(*a*)

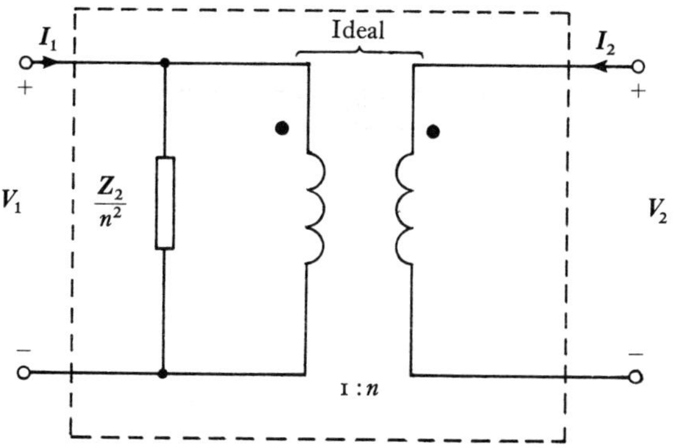

(*b*)

Fig. 6.24 Transfer of shunt impedance across an ideal transformer: (*a*) shunt impedance on output side; (*b*) equivalent impedance on input side.

elements to the ideal transformer. Fig. 6.26(*a*) shows the imperfections that are accounted for in the equivalent circuit of part (*b*). The most noteworthy omissions are the winding capacitances, which may be important at frequencies high with respect to 50 Hz, and the energy loss in the core material which may be important in power transformers. No restriction is placed on the coefficient of coupling k, but the representation that follows is of most value for transformers in which k is close to unity.

The derivation of the equivalent circuit starts from (6.9.2) which apply to part (*a*):

$$\left.\begin{aligned} \boldsymbol{V}_1 &= (R_1 + \mathrm{j}\omega L_1)\boldsymbol{I}_1 + \mathrm{j}\omega M\boldsymbol{I}_2, \\ \boldsymbol{V}_2 &= \mathrm{j}\omega M\boldsymbol{I}_1 + (R_2 + \mathrm{j}\omega L_2)\boldsymbol{I}_2. \end{aligned}\right\} \tag{6.9.2}$$

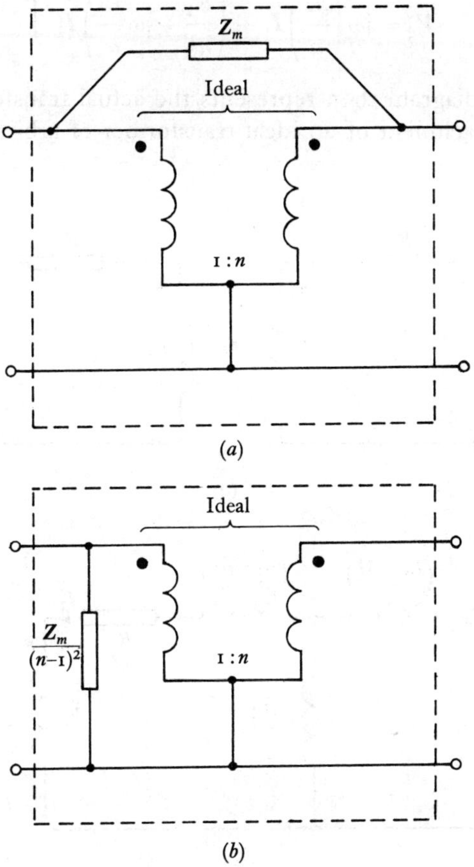

Fig. 6.25 (*a*) Three-terminal ideal transformer with mutual load. (*b*) Transformer with equivalent shunt load.

These equations are manipulated to introduce an ideal transformer at the output, which is done by changing the variables $\boldsymbol{V}_2$ and $\boldsymbol{I}_2$ in accordance with (6.9.3). Let

$$\left.\begin{aligned} \boldsymbol{V}_2' &= \boldsymbol{V}_2/a, \\ \boldsymbol{I}_2' &= -a\boldsymbol{I}_2, \end{aligned}\right\} \tag{6.9.3}$$

and

where a is an arbitrary constant. The substitution of the new variables in (6.9.2) produces (6.9.4), and it can be seen that (6.9.4) corresponds to the left hand part of fig. 6.26(*b*) up to the ideal transformer:

$$\left.\begin{aligned} \boldsymbol{V}_1 &= (R_1 + \mathrm{j}\omega L_1)\boldsymbol{I}_1 + \mathrm{j}\omega\left(\frac{M}{a}\right)\boldsymbol{I}_2', \\ \boldsymbol{V}_2' &= \mathrm{j}\omega\left(\frac{M}{a}\right)\boldsymbol{I}_1 + \left(\frac{R_2}{a^2} + \mathrm{j}\omega\frac{L_2}{a^2}\right)\boldsymbol{I}_2'. \end{aligned}\right\} \tag{6.9.4}$$

The complete diagram then represents the actual transformer of part (*a*) because the attachment of an ideal transformer of arbitrary turns ratio a

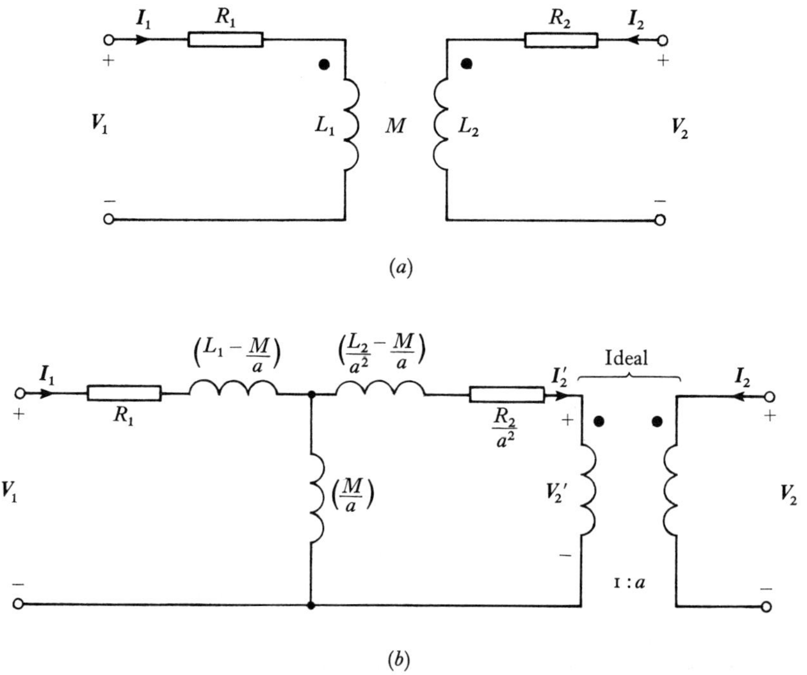

Fig. 6.26 Equivalent circuit of a transformer: (*a*) linear transformer; (*b*) equivalent circuit.

establishes the required terminal quantities $\boldsymbol{V}_2$ and $\boldsymbol{I}_2$ in accordance with (6.9.3).

Two useful choices of a will now be considered. The first is

$$a_1 = \sqrt{(L_2/L_1)}. \tag{6.9.5}$$

Remembering that the coefficient of coupling, k, is equal to $\sqrt{(L_1 L_2)}$, we find that

$$\left.\begin{aligned} M/a_1 &= kL_1, \\ L_1 - (M/a_1) &= (1-k)\,L_1, \\ \text{and} \qquad (L_2/a_1^2) - (M/a_1) &= (1-k)\,L_1, \end{aligned}\right\} \tag{6.9.6}$$

so the equivalent circuit reduces to that shown in fig. 6.27(*a*). In so far as k tends to unity, the two elements $(1-k)L_1$, tend to zero and the circuit

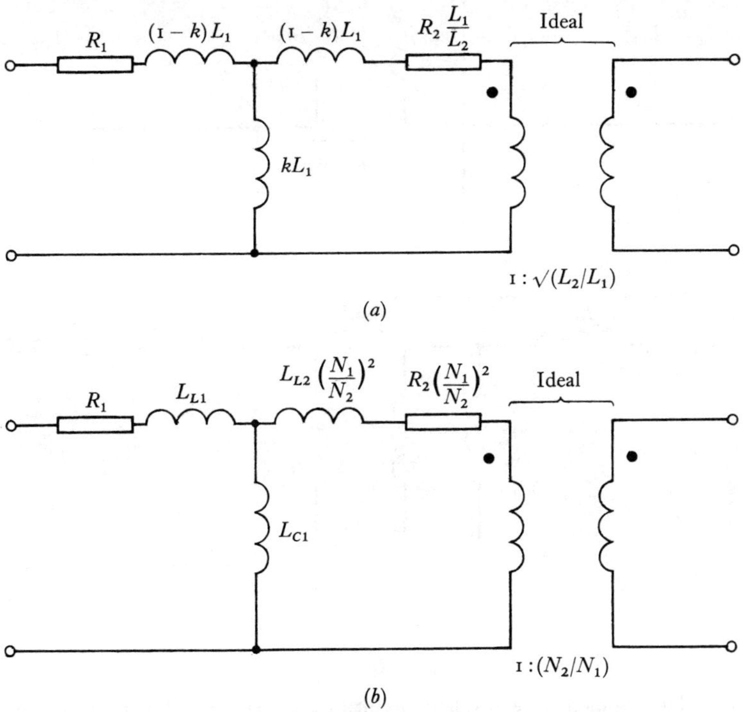

Fig. 6.27 More equivalent circuits of a transformer: (*a*) transformer with turns ratio $\sqrt{(L_2/L_1)}$; (*b*) transformer with turns ratio (N_2/N_1).

reduces to that of an ideal transformer of turns ratio $\sqrt{(L_2/L_1)}$ shunted on the primary side by the inductance L_1 and having in addition the elements R_1 and $R_2 L_1/L_2$ which represent the energy loss. It is common practice to identify the two elements $(1-k)L_1$ with the primary leakage inductance and the secondary leakage inductance as transferred to the primary winding, but this is an approximation. It is, however, a good one for values of k close to unity.

The second choice of a to be considered here is

$$a_2 = N_2/N_1, \tag{6.9.7}$$

where N_1 is the number of turns on the primary winding and N_2 is the number of turns on the secondary winding. This choice leads to the circuit

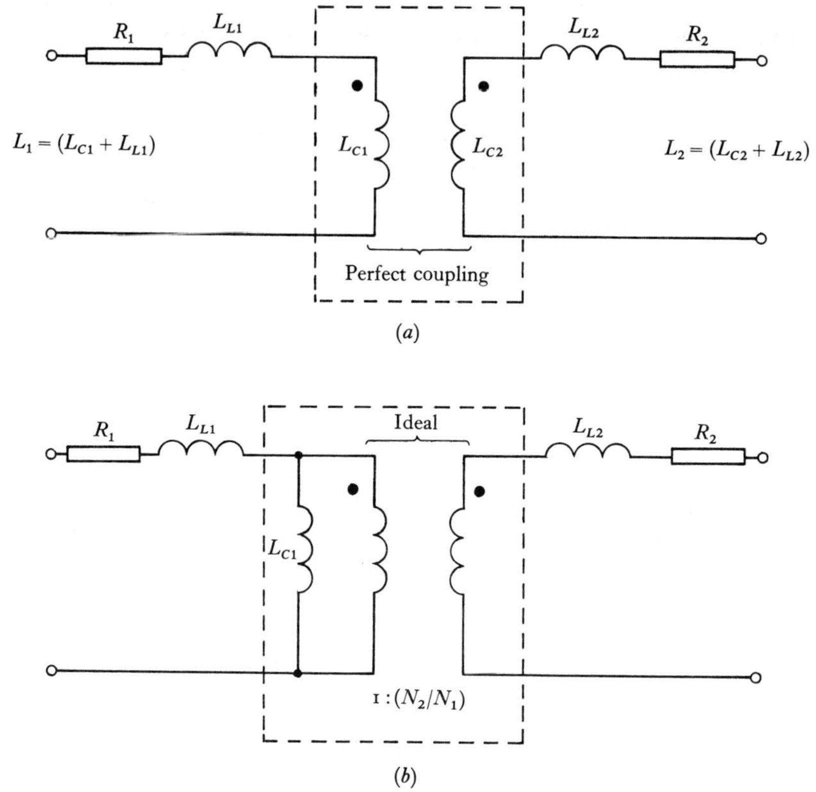

Fig. 6.28 Development of equivalent circuit of a transformer: (*a*) windings split into leakage and coupling components; (*b*) replacement of perfectly coupled coils by ideal transformer shunted by inductance.

shown in fig. 6.27(*b*) in which L_{L1} and L_{L2} represent the leakage inductances and L_{C1} (equal to the difference between L_1 and L_{L1}) is the primary coupling inductance, but the same result can be achieved more directly by the following argument. If the total self inductance of each winding is split into two components, the leakage inductance (which has no coupling to the other winding) and the coupling inductance (which is perfectly coupled to the other winding), then the two perfectly coupled elements L_{C1} and L_{C2} may be represented by an ideal transformer shunted by L_{C1} as outlined in chapter 2. The turns ratio of this ideal transformer is given by

$$L_{C1}/L_{C2} = (N_1/N_2)^2, \tag{6.9.8}$$

since the turns ratio of the perfectly coupled elements is the same as that of the transformer as a whole. The elements L_{L2} and R_2 may now be transferred across the ideal transformer to produce fig. 6.27(*b*). The earlier stages of the argument are illustrated by fig. 6.28.

Both of the representations of fig. 6.27 are in common use in the analysis of circuits which contain close-coupled transformers.

6.10 The power-transfer theorem

All of the previous theorems in this chapter are concerned with the linear quantities voltage and current; the following theorem is concerned with the non-linear quantity power. Picture a source, of internal impedance $\boldsymbol{Z}_s$, feeding a load $\boldsymbol{Z}_L$ which can be varied. What value of $\boldsymbol{Z}_L$ will produce the maximum transfer of power from the source to the load, and what is the

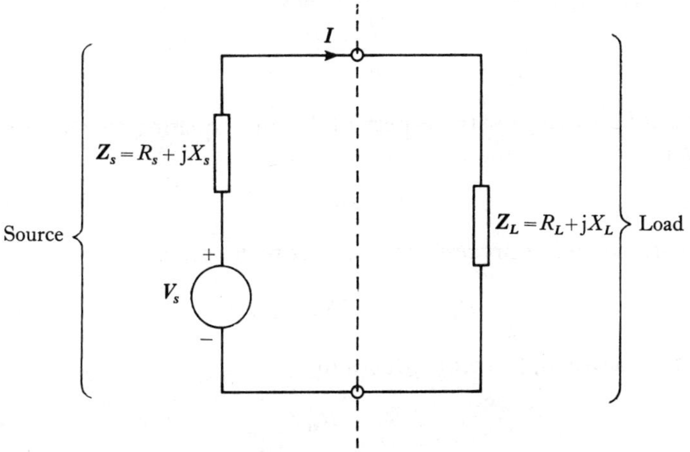

Fig. 6.29 Transfer of power from a source to a load.

value of this power? The power-transfer theorem gives the answers to these questions.

The current flowing in the load of fig. 6.29 is

$$\boldsymbol{I} = \frac{\boldsymbol{V}_s}{(R_s + R_L) + \mathrm{j}(X_s + X_L)}, \tag{6.10.1}$$

and its r.m.s. value is

$$I = \frac{V_s}{\sqrt{[(R_s + R_L)^2 + (X_s + X_L)^2]}}, \tag{6.10.2}$$

where V_s is the r.m.s. value of the source voltage. The power delivered to the load is

$$P = \frac{R_L V_s^2}{(R_s + R_L)^2 + (X_s + X_L)^2}, \tag{6.10.3}$$

and if the components R_L and X_L of the load can be varied independently there will be a specific value of each which, together, will maximize P. This maximum value P_{as} is called the *available power* of the source because under no conditions of loading can a greater power be drawn from it. When the load draws this maximum power it is said to be *power-matched* to the source.

Since P is a function of the two independent variables R_L and X_L, the required conditions for a maximum are

$$\frac{\partial P}{\partial X_L} = 0,$$

and

$$\frac{\partial P}{\partial R_L} = 0.$$

Differentiating (6.10.3) with respect to X_L and equating to zero gives the condition

$$X_L = -X_s, \tag{6.10.4}$$

and the corresponding process with respect to R_L gives

$$R_L^2 = R_s^2 + (X_s + X_L)^2. \tag{6.10.5}$$

If (6.10.4) is satisfied, (6.10.5) reduccs to

$$R_L = R_s, \tag{6.10.6}$$

and it is established that for maximum power-transfer the load impedance must be the conjugate of the source impedance; i.e.

$$\boldsymbol{Z}_L = \boldsymbol{Z}_s^*. \tag{6.10.7}$$

Moreover, if this conjugate matching occurs, the power delivered to the load is

$$P_{as} = \frac{V_s^2}{4R_s}. \tag{6.10.8}$$

A similar analysis shows that the dual source, which comprises an ideal current source $\boldsymbol{I}_s$ shunted by an admittance $\boldsymbol{Y}_s(=G_s+\mathrm{j}B_s)$, delivers maximum power to a load admittance $\boldsymbol{Y}_L$ which is the conjugate of $\boldsymbol{Y}_s$, and that the power available from the source is

$$P_{as} = \frac{I_s^2}{4G_s}. \tag{6.10.9}$$

The argument that led to (6.10.7) and (6.10.8) depends on the assumption that R_L and X_L can be varied independently. There are, however, many situations in which the load $\boldsymbol{Z}_L$ cannot be changed. In these circumstances a

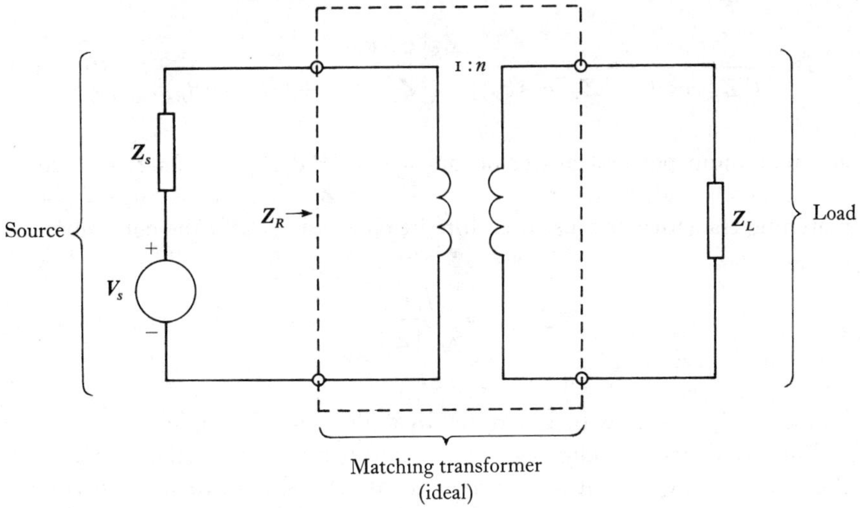

Fig. 6.30 Use of a matching transformer between source and load.

close-coupled transformer, inserted between the source and the load, may produce a greater transfer of power than would otherwise be possible. The problem here is to find the turns ratio, n, of the matching transformer, shown in fig. 6.30, which will produce the greatest transfer of power. This power P_{max} will not be as great as P_{as} but will be the best we can achieve in the circumstances.

The problem is simplified by assuming the transformer to be ideal. The source then sees an apparent load $\boldsymbol{Z}_L/n^2$ to which it delivers power, and the power delivered to this apparent load is equal to the power delivered to the real load because the ideal transformer absorbs none. The problem is thus reduced to one of matching a fixed source $\boldsymbol{Z}_s$ to a load $\boldsymbol{Z}_R$ whose phase angle

θ_R is fixed but whose magnitude $|\boldsymbol{Z}_R|$ is variable. This is apparent from (6.10.10) which shows that $\theta_R = \theta_L$ and $|\boldsymbol{Z}_R| = |\boldsymbol{Z}_L|/n^2$:

$$|\boldsymbol{Z}_R| \angle \theta_R = \frac{|\boldsymbol{Z}_L|}{n^2} \angle \theta_L. \tag{6.10.10}$$

Now let

$$\left.\begin{aligned} \boldsymbol{Z}_s &= (R_s + \mathrm{j}X_s) = |\boldsymbol{Z}_s| \angle \theta_s = |\boldsymbol{Z}_s| \cos\theta_s + \mathrm{j}|\boldsymbol{Z}_s| \sin\theta_s, \\ \boldsymbol{Z}_R &= (R_R + \mathrm{j}X_R) = |\boldsymbol{Z}_R| \angle \theta_R = |\boldsymbol{Z}_R| \cos\theta_R + \mathrm{j}|\boldsymbol{Z}_R| \sin\theta_R. \end{aligned}\right\} \tag{6.10.11}$$

Then the power transferred to the load is

$$P = \frac{V_s^2 |\boldsymbol{Z}_R| \cos\theta_R}{(|\boldsymbol{Z}_s| \cos\theta_s + |\boldsymbol{Z}_R| \cos\theta_R)^2 + (|\boldsymbol{Z}_s| \sin\theta_s + |\boldsymbol{Z}_R| \sin\theta_R)^2}, \tag{6.10.12}$$

and maximum power-transfer occurs when $\mathrm{d}P/\mathrm{d}|\boldsymbol{Z}_R|$ equals zero, a condition which leads to the result that $|\boldsymbol{Z}_R| = |\boldsymbol{Z}_s|$. The turns ratio n of the matching transformer must therefore be chosen to satisfy the condition

$$n = \sqrt{\frac{|\boldsymbol{Z}_L|}{|\boldsymbol{Z}_s|}}. \tag{6.10.13}$$

These results will now be illustrated by a numerical example.

The circuit shown in fig. 6.31(*a*) is to supply a resistive load of 100 Ω from the terminals PQ. What transformer turns ratio should be used to couple these terminals to the load so as to produce maximum power in it, what is the value of this power, and what is the available power at the terminals PQ?

The circuit is first converted to its Thevenin equivalent as shown in part (*b*) from which the equivalent source impedance $\boldsymbol{Z}_s$ is calculated.

$$\begin{aligned} \boldsymbol{Z}_s &= R_2 + \frac{\mathrm{j}\omega L_3 R_1}{R_1 + \mathrm{j}\omega L_3 R_1}, \\ &= \left[R_2 + \frac{(\omega L_3)^2 R_1}{R_1^2 + (\omega L_3)^2}\right] + \mathrm{j}\left[\frac{(\omega L_3) R_1^2}{R_1^2 + (\omega L_3)^2}\right], \\ &= R_s + \mathrm{j}X_s, \\ &= (2.6 + \mathrm{j}0.8)\ \Omega. \end{aligned}$$

The r.m.s. value of the voltage $\boldsymbol{V}_{oc}$ is given by

$$V_{oc} = \frac{\omega L_3}{\sqrt{(R_1^2 + \omega^2 L_3^2)}} \,.\, V_1,$$

$$= 8.95 \text{ V}.$$

The power available from the source is

$$P_{as} = \frac{V_{oc}^2}{4R_s},$$

$$= 7.7 \text{ W},$$

and this power would be drawn by an impedance (2.6 – j0.8) Ω connected to the terminals PQ. For a resistive load R_L of 100 Ω, however, it will not be possible to draw 7.7 W from the source. With this load, most power will

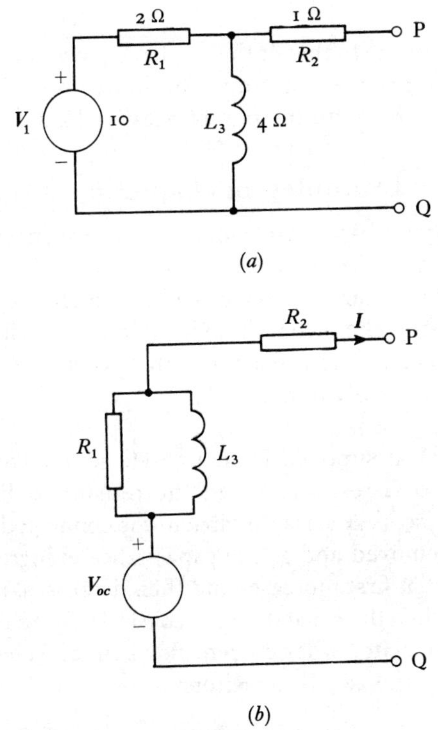

Fig. 6.31 Calculation of power in a load: (*a*) circuit supplying power at the terminals PQ; (*b*) equivalent Thevenin circuit.

be drawn when an ideal transformer of turns ratio n is inserted between it and the terminals PQ, where

$$n^2 = R_L/|\boldsymbol{Z}_s|,$$
$$= 100/\sqrt{(2.6^2+0.8^2)},$$

and

$$n = 6.1.$$

With this transformer, the terminals PQ see a resistive load of R_L/n^2 which is 2.71 Ω, and the power delivered to the load is

$$P_L = \left(\frac{R_L}{n^2}\right).I^2,$$

$$= \left(\frac{R_L}{n^2}\right).\frac{V_{oc}^2}{(R_s+R_L/n^2)^2+X_s^2},$$

$$= \frac{2.71\times 8.95^2}{5.31^2+0.8^2}\,\text{W},$$

$$= 7.5\ \text{W}.$$

The closeness of this figure to that of the available power of the source is due to the predominantly resistive nature of the source impedance which can therefore be closely matched to the purely resistive load.

Examples on chapter 6

6.1 A balanced, three-phase, star-connected source in which the phase voltages are 100 V r.m.s. (i.e. a source like that shown in fig. 3.13(*a*) with $V_m = 100\sqrt{2}$) supplies a star-connected load in which one arm of the star is an ideal 100 Ω resistor, the second is an ideal inductor with a reactance of 50 Ω, and the third is an ideal capacitor with a reactance of 100 Ω. What power does the source supply to the load?

6.2 A network which is supplied from a 50 Hz source contains a branch AB consisting of a 20 Ω resistor. When the resistor is disconnected the voltage across AB is 300 V. A variable capacitor is connected between A and B with the resistor removed and as the capacitance is increased from zero the current through it at first increases and then decreases, passing through a maximum of 10 A when the capacitance is 20 μF. If the 20 Ω resistor is now substituted for the capacitor what current flows in it? What current flows in it if it is shunted by the 20 μF capacitor?

6.3 Three impedances of values (178 + j344) Ω, (102 − j294) Ω, and (488 + j122) Ω are connected respectively to terminals A, B, and C so as to form a star network. When a source whose open-circuit voltage is 200 V and whose internal impedance is $(1000 + j100)\ \Omega$ is connected to AC a current $\boldsymbol{I}$ flows through an impedance $\boldsymbol{Z}_L$ connected to BC. What are the characteristics of a second source which will supply the same current $\boldsymbol{I}$ to the same impedance $\boldsymbol{Z}_L$ when $\boldsymbol{Z}_L$ is connected directly across it? What is the value of $\boldsymbol{I}$ when $\boldsymbol{Z}_L = 0$?

6.4 The four arms AB, BC, CD, and DA of an unbalanced Wheatstone bridge are resistors with values of 250 Ω, 260 Ω, 220 Ω, and 200 Ω respectively. The supply voltage at AC is 30 V at a frequency of 50 Hz and the detector at BD is replaced by a 50 Ω resistor in series with a 10 μF capacitor. What is the magnitude of the current through the capacitor, and what is its phase with respect to the supply voltage?

6.5 A star network with terminals A, B, and C consists of three 200 Ω resistors. It is bridged at AB by a 600 Ω resistor and at BC by a second 600 Ω resistor. What resistance would be seen by a source connected to AC? If 9 μF and 3 μF capacitors are connected across each of the 200 Ω resistors and 600 Ω resistors respectively, what is the input impedance at the terminals AC at 50 Hz?

6.6 Three 20 Ω resistors are connected in series between terminals A and B of a three-terminal network. From each of the two junctions of these resistors is connected to the third terminal C a 50 Ω resistor, and a 100 Ω load resistor is connected between B and C. By means of the star–delta transformation find the voltage and current needed at the terminals AC to produce a power of 10 W in the load.

6.7 A three-terminal active element is represented by a current source of value $(I_0 + g_m V_g)$ between two terminals A and B across which is connected a resistor r_0; the source is directed towards B. V_g is the potential of an open-circuited control terminal C relative to that of B. Show that when a resistor R is connected between B and a fourth terminal D, the circuit acts between terminals A, C, and D like an active element of the same kind as before but with $(I_0 + g_m V_g)$ replaced by $(I_0 + g_m V_g)/[1 + R(g_0 + g_m)]$ and r_0 replaced by $[r_0 + R(1 + r_0 g_m)]$.

6.8 Find a Wheatstone bridge which contains at least one non-pure reactive element in its arms, which has a purely resistive detector, which is fed from a purely resistive source, and which has for its dual the same bridge circuit. Find the dual of two such bridges fed in parallel from the same resistive source and show that the dual of this circuit is not the same as the original.

6.9 An impedance $\boldsymbol{Z}_L = (80 + j60)\ \Omega$ is to be power-matched by means of an ideal transformer to a source which produces 50 V on open-circuit and has an internal impedance $\boldsymbol{Z}_s = (3 + j4)\ \Omega$. Find the turns ratio needed in the transformer to transfer maximum power from the source to the load. What is the value of this power, and what is the available power of the source?

6.10 A source has an open-circuit voltage V_s and an internal impedance which is purely resistive and which is 1 Ω in value. Across the terminals of the source is connected an ideal 1 H inductor, and across that is connected a second ideal 1 H inductor in series with an ideal 0.5 F capacitor. Across the second inductor is connected the load which is a 1 Ω resistor. Show that the ratio of the power developed in the load to the available power of the source is

$$\omega^6/(1 + \omega^6),$$

where ω is the angular frequency of the source. If the source and load resistances are raised to 50 Ω, how should the circuit be modified to obtain the same ratio of power in the load to available power?

6.11 An ideal capacitor C is shunted by an ideal inductor L in series with a second identical capacitor C to form a matching network which is to be used between a load R_L (across the second capacitor) and a source of internal resistance R_s and open-circuit voltage V_s (across the first). Derive an expression for the input admittance of the matching network when terminated by R_L, and show that it is purely real and of value $R_L C/L$ at the frequency $\omega_0 = 1/\sqrt{(LC)}$. Thus find suitable values of L and C in the matching network for transferring power from a 1 kΩ source to a 16 kΩ load at an angular frequency of 10^5 rad/s.

7 The frequency characteristics of elementary circuits

7.1 The magnitude and phase response of a circuit

The transfer function of a circuit is the mathematical expression relating its response to its stimulus. When the stimulus is sinusoidal of angular frequency ω, the transfer function, displayed in terms of ω, is called the *frequency response* of the circuit. It has two parts: the magnitude, displayed as a function of ω, is called the *amplitude response* of the circuit: the phase angle, similarly displayed, is called the *phase* response.

A graphical display of the frequency response of a circuit leads to its description in terms of its *critical frequencies*. The first part of this chapter describes these critical frequencies in elementary circuits. The second part establishes some of the properties of resonant circuits and shows how their responses can be treated approximately by the critical frequency method. The final part shows that by extending the concept of critical frequency to include complex as well as real values, these frequencies themselves completely define the shape of the frequency response, and a display of these frequencies, called a pole–zero diagram, can have more significance than the response itself.

7.2 Circuits with single time-constants

The parallel RC circuit shown in fig. 7.1 is a circuit with a single time-constant. Others of this kind are the series RC circuit, the series RL circuit,

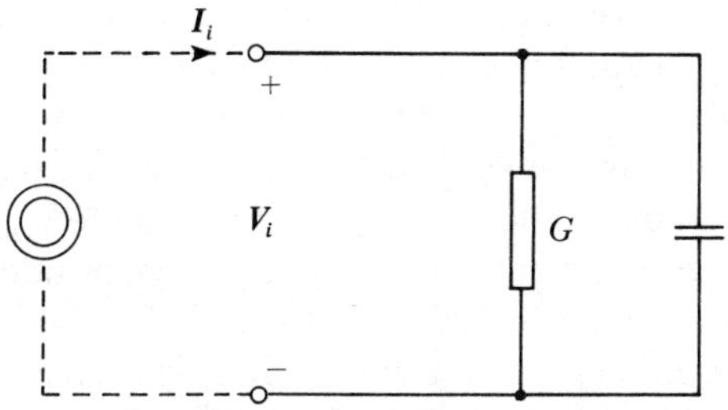

Fig. 7.1 Circuit with only one energy-storing element.

and the parallel RL circuit. The common feature of these four structures is the presence of only one energy-storing element, and this feature gives to them transfer functions of a particularly simple kind. For example, the transfer function $\boldsymbol{V}_i/\boldsymbol{I}_i$, better known as the impedance of the circuit of fig. 7.1 is

$$\boldsymbol{Z} = \frac{1}{G + j\omega C},$$

$$= \frac{1}{G(1 + j\omega/\omega_a)}, \tag{7.2.1}$$

where the angular frequency ω_a, equal to (G/C), is the reciprocal of the time constant and is called the *corner frequency*, the *break frequency*, or the *critical frequency*.

The impedance may be *normalized* by dividing both sides of (7.2.1) by $(1/G)$.

$$\boldsymbol{Z}G = \frac{1}{1 + j\omega/\omega_a},$$

$$= \frac{1}{1 + jx},$$

$$= \frac{1}{\sqrt{(1 + x^2)}} \angle -\tan^{-1} x. \tag{7.2.2}$$

This process of normalization in which $\boldsymbol{Z}(\omega)$ is divided by $\boldsymbol{Z}(0)$, the d.c. value of $\boldsymbol{Z}(\omega)$, or by any other convenient value, produces the dimensionless quantity $(\boldsymbol{Z}G)$ which is called the normalized transfer function. For similar reasons, the quantity $x(=\omega/\omega_a)$ is called the normalized angular frequency, though the word *normalized* is often omitted when the context is clear. The object of normalizing is to produce only one equation or one diagram that will refer to *all* parallel *RC* circuits like fig. 7.1 no matter what component values they possess.

The magnitude and phase angle of $(\boldsymbol{Z}G)$ are sketched in fig. 7.2 which is worth examining in detail since expressions of the form $(1 + jx)$, or its reciprocal, occur not only in electric circuits of the single time-constant type but also as parts of more complicated functions. Sketches of this kind are called Bode diagrams.† The magnitude is plotted to log-log scales; the quantity plotted on the vertical axis is $20 \log_{10}$ (magnitude of normalized transfer function), and is labelled *decibels*, for which the abbreviation is db. The origin of this term and of the factor 20 in it is of historic interest (the unit is named after Alexander Graham Bell (1847–1922), the inventor of the telephone), but need not concern us here. The dotted lines on the magnitude diagram represent the equations $|\boldsymbol{Z}G| = 1$ (zero db), and

† H. W. Bode, *Network analysis and feedback amplifier design*, Van Nostrand (1945).

$|ZG| = 1/x$, the values to which the magnitude tends as x tends to zero and infinity respectively. These are represented by straight lines on a graph with logarithmic scales. They intersect at $x = 1$, and the slope of the line $|ZG| = 1/x$ is such that a doubling of x causes $|ZG|$ to decrease by $20 \log_{10} 2$ or, for practical purposes, by 6 db. Since two frequencies related by a factor of two are said to be separated by an octave, the slope of the line is said to be

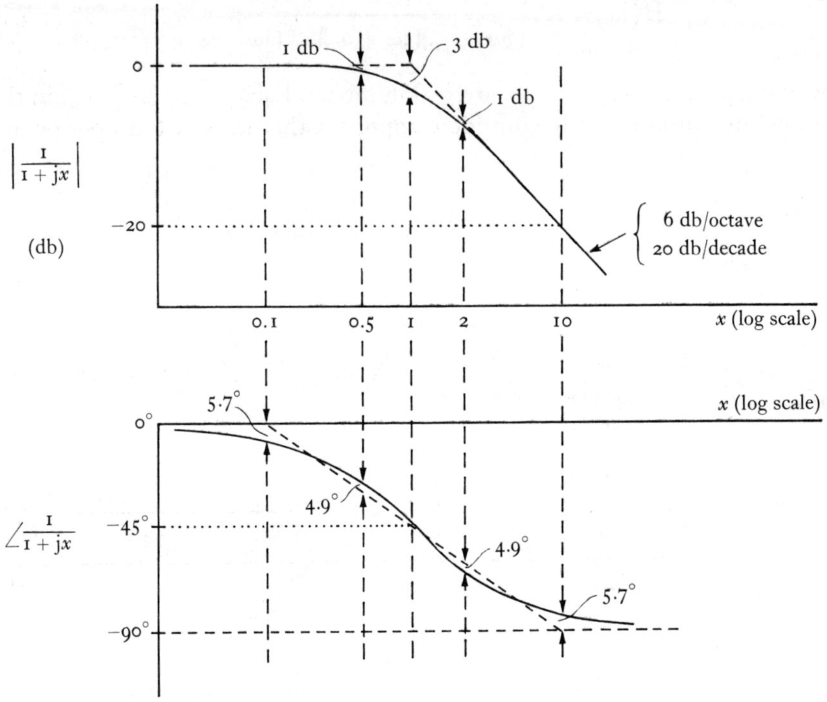

Fig. 7.2 Magnitude and phase angle of the function $1/(1 + jx)$.

−6 db per octave. The unfortunate use of the word octave to signify a factor of two arises from its use in music. It is, however, well established in circuit theory, though some people prefer the alternative description, −20 db per decade, in which the word decade stands for a frequency ratio of ten. The function $|1/(1 + jx)|$ departs from its asymptotes by 3 db at $x = 1$ and by 1 db at frequencies an octave above and below this point.

The phase angle of (ZG), $-\tan^{-1} x$, is also shown in fig. 7.2. It closely approaches a straight line drawn through the points 0 and −90 degrees at frequencies respectively a decade below and above the value $x = 1$. If the phase angle is assumed to be zero at frequencies below $x = 0.1$, and 90 degrees at frequencies above $x = 10$, the error in the straight line approximation is nowhere greater than 6 degrees.

7·3 Transfer functions with more than one corner frequency

Transfer functions more complicated than (7.2.2) can be plotted in the same way provided that they can be expressed in the form

$$\boldsymbol{H}(\mathrm{j}\omega) = K.\frac{(\mathrm{j}\omega+\omega_1)(\mathrm{j}\omega+\omega_2)\ldots(\mathrm{j}\omega+\omega_i)}{(\mathrm{j}\omega+\omega_a)(\mathrm{j}\omega+\omega_b)\ldots(\mathrm{j}\omega+\omega_j)}, \tag{7.3.1}$$

where ω_a, ω_b, ..., ω_1, ω_2, etc. are real numbers. Each term $(\mathrm{j}\omega + \omega_m)$ in the numerator produces a magnitude component that rises at 6 db per octave

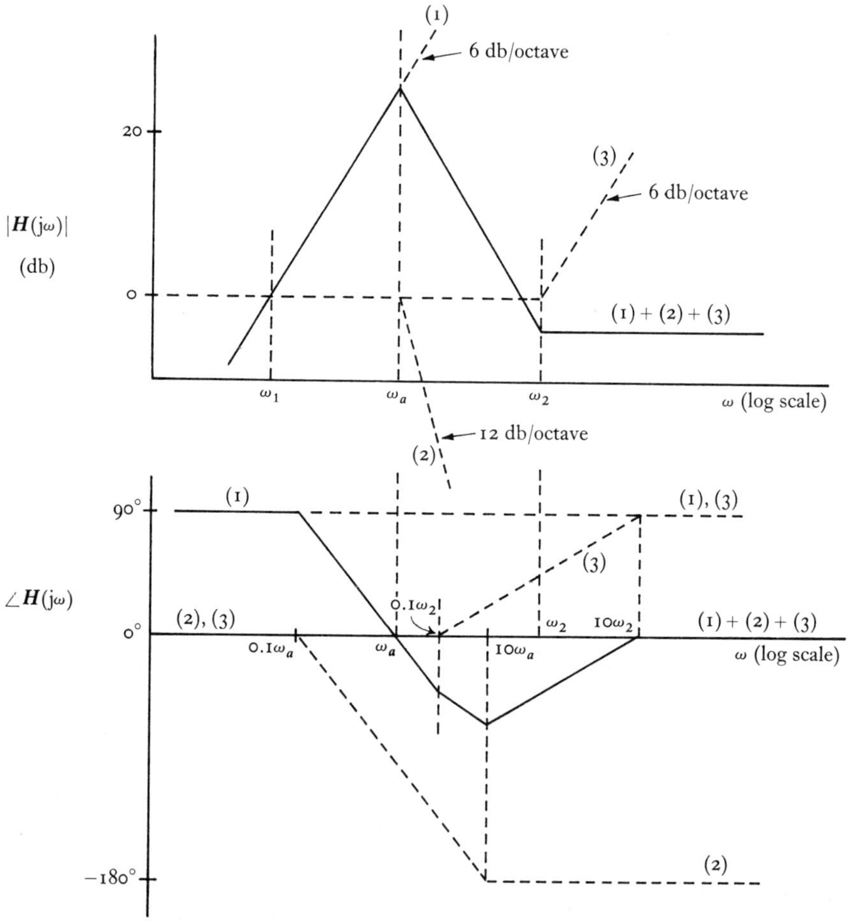

Fig. 7·3 Magnitude and phase response of $(\mathrm{j}\omega/\omega_1)\,(1+\mathrm{j}\omega/\omega_2)/(1+\mathrm{j}\omega/\omega_a)^2$.

above the corner frequency ω_m and produces a phase angle component that is positive; each term $(j\omega + \omega_n)$ in the denominator produces a magnitude component that falls at 6 db per octave above the corner frequency ω_n and produces a phase angle component that is negative. Each term is plotted separately, being represented by its asymptotes. The overall response, whether in magnitude or phase angle, is obtained by addition of all the components. Corrections to the asymptotic approximations can best be made after this addition.

For example, consider the transfer function

$$\boldsymbol{H}(j\omega) = \frac{(j\omega/\omega_1)(1 + j\omega/\omega_2)}{(1 + j\omega/\omega_a)^2}, \tag{7.3.2}$$

where $\omega_2 > \omega_a > \omega_1$. The magnitude, in decibels, is given by

$$|\boldsymbol{H}(j\omega)| = 20 \log_{10}|j\omega/\omega_1| + 20 \log_{10}|1 + j\omega/\omega_2| - 40 \log_{10}|1 + j\omega/\omega_a|, \tag{7.3.3}$$

and the phase angle is given by

$$\angle \boldsymbol{H}(j\omega) = \angle j\omega/\omega_1 + \angle(1 + j\omega/\omega_2) - 2\angle(1 + j\omega/\omega_a). \tag{7.3.4}$$

The individual components of these two quantities, approximated by straight line segments, are shown dotted in fig. 7.3. The term $(j\omega/\omega_1)$ gives, in magnitude, a straight line of slope 6 db per octave passing through the frequency ω_1 at zero db, and a constant phase angle of 90 degrees. This component is labelled (1) in fig. 7.3. The other two components are labelled (3) and (2), and the sum of the three components is shown by the solid line.

7.4 Resonance

Resonance is a property displayed by innumerable physical systems. Such systems exhibit relatively large responses to stimuli of specific frequencies f_0, f_0', f_0'', etc., and relatively small responses to stimuli of other frequencies. The values of f_0, f_0', f_0'', etc. are called the *resonant frequencies* of the system. This property depends, in its simplest form, upon a free exchange of energy between two kinds of energy-storing elements, and in electric systems these elements are the inductor and the capacitor. (More complicated types of resonant circuits may employ only one kind of energy-storing element but they must then also contain controlled sources).

Figure 7.4 shows a *parallel-resonant circuit*. This consists of two energy-storing elements in parallel together with a conductance which accounts for the energy loss. Its frequency response, (fig. 7.5), shows that it has a single resonance at the frequency f_0. More than one resonant frequency is possible if the circuit contains more inductors and capacitors, and the response

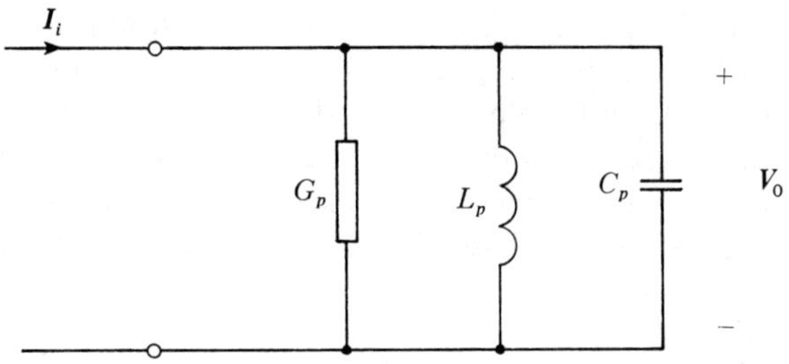

Fig. 7.4 Parallel-resonant circuit.

would then have a peak like that of fig. 7.5 at each resonant frequency.

The transfer function V_o/I_i of the parallel-resonant circuit is

$$\frac{V_o}{I_i} = \frac{1}{G_p + j\omega C_p + 1/j\omega L_p}$$
$$= \frac{1}{\sqrt{[G_p^2 + (\omega C_p - 1/\omega L_p)^2]}} \angle -\tan^{-1}(\omega C_p - 1/\omega L_p)/G_p. \qquad (7.4.1)$$

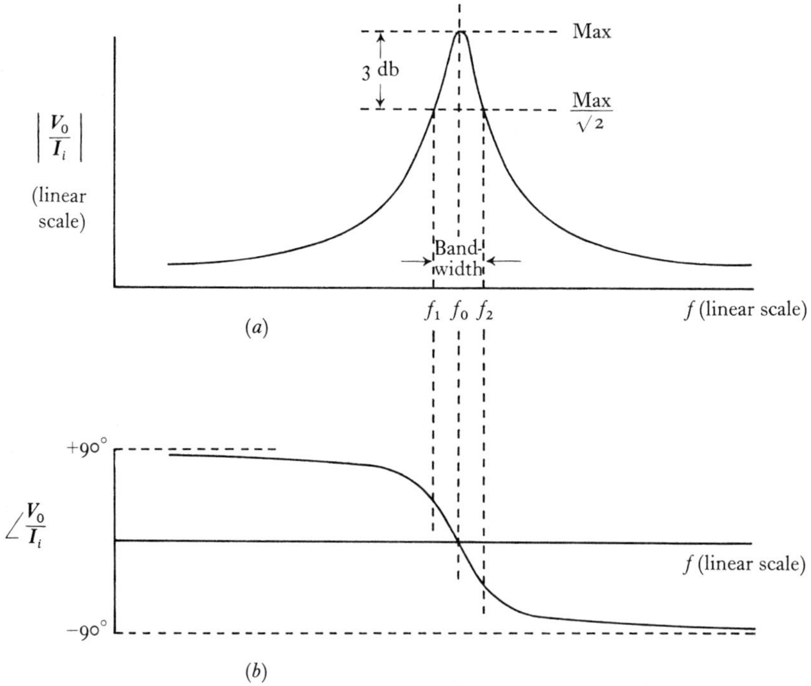

Fig. 7.5 Magnitude response (a) and phase response (b) of a parallel-resonant circuit.

At the angular frequency ω_{0p} such that

$$\omega_{0p} C_p = 1/\omega_{0p} L_p, \tag{7.4.2}$$

the magnitude of the transfer function is a maximum, $1/G_p$, and the phase angle of the transfer function is zero; the frequency $\omega_{0p}/2\pi$ is the resonant frequency of the circuit. Not all resonant circuits, however, have their maximum response and zero phase angle at the same frequency. The difference, when it occurs, is usually negligible from an engineering point of view, but for the purpose of definition the resonant frequency is attached to the frequency of zero phase angle.

The physical significance of a resonant circuit lies in its ability to discriminate between a stimulus which has a frequency in the neighbourhood of f_0 and others which do not. The circuit responds to a *band* of frequencies centred on f_0, and very little to frequencies outside this band. The *bandwidth* is arbitrarily defined by the limits f_1 and f_2 at which the response is $1/\sqrt{2}$ of the maximum response and is given by

$$\text{Bandwidth} = f_2 - f_1. \tag{7.4.3}$$

On a decibel scale, the response at f_1 or f_2 is 3 db lower than at the maximum, and for this reason the bandwidth $(f_2 - f_1)$ is sometimes called the 3 db *bandwidth* of the circuit. The frequencies f_1 and f_2 are called the 3 db frequencies or the 'half-power' frequencies.

Another measure of the sharpness of the response curve is given by the *quality-factor* or *Q-factor* of the circuit. The Q-factor of any physical resonant system is defined by the equation

$$Q = 2\pi . \frac{\text{Maximum energy stored in the system}}{\text{Energy dissipated by the system in one cycle}}, \tag{7.4.4}$$

in which it is assumed that a sinusoidal forcing function is applied to the system at the resonant frequency. If a stimulus of frequency ω_{0p}, applied to the circuit of fig. 7.4, produces a response voltage

$$v_o = V_{om} \cos \omega_{0p} t,$$

then the energy stored at any instant t by the capacitor is

$$W_C = \tfrac{1}{2} C_p V_{om}^2 \cos^2 \omega_{0p} t, \tag{7.4.5}$$

and that stored by the inductor at the same instant is

$$W_L = \tfrac{1}{2} L_p \left(\frac{V_{om}}{\omega_{0p} L_p} \right)^2 . \sin^2 \omega_{0p} t. \tag{7.4.6}$$

Since $\omega_{0p} C_p = 1/\omega_{0p} L_p$, (7.4.6) can be written

$$W_L = \tfrac{1}{2} C_p V_{om}^2 \sin^2 \omega_{0p} t. \tag{7.4.7}$$

The total energy stored by the circuit at time t is the sum of the two components (7.4.5) and (7.4.7) which is

$$\begin{aligned} W_T &= W_C + W_L, \\ &= \tfrac{1}{2} C_p V_{om}^2 (\cos^2 \omega_{0p} t + \sin^2 \omega_{0p} t), \\ &= \tfrac{1}{2} C_p V_{om}^2. \end{aligned} \tag{7.4.8}$$

W_T is, in fact, a constant, independent of time; the stored energy is shuttled back and forth between the capacitor and the inductor, the total remaining constant. Energy is, however, dissipated in the circuit by the element G_p. The mean rate of energy dissipation is G_p times the r.m.s. value of the voltage v_o. The energy dissipated in one cycle of the waveform is therefore

$$\begin{aligned} W_R &= G_p \left(\frac{V_{om}}{\sqrt{2}} \right)^2 \cdot \frac{2\pi}{\omega_{0p}}, \\ &= \frac{\pi G_p V_{om}^2}{\omega_{0p}}. \end{aligned} \tag{7.4.9}$$

The ratio of (7.4.8) and (7.4.9), multiplied by 2π, gives the Q-factor of the parallel-resonant circuit

$$Q_p = \frac{\omega_{0p} C_p}{G_p}. \tag{7.4.10}$$

It is useful to remember the alternative forms of this expression

$$Q_p = \frac{\text{susceptance}}{\text{conductance}} = \frac{\omega_{0p} C_p}{G_p} = \frac{1}{\omega_{0p} L_p G_p} = \frac{1}{G_p} \sqrt{\frac{C_p}{L_p}}. \tag{7.4.11}$$

Circuits which have numerical values of Q greater than about 10 or 20 are called high-Q circuits; those which have values less than about 2 or 3 are called low-Q circuits. It is convenient to take the arbitrary value of 10 as the dividing line between these two classes.

7.5 The parallel-resonant circuit

In this section the transfer function (V_o/I_i) of the parallel-resonant circuit is examined further. In particular it is shown that, by means of a simple frequency-transformation, the frequency response of a parallel-resonant circuit may be represented by a diagram of the same form as the Bode diagram shown in fig. 7.2. Finally, the bandwidth of the circuit is expressed in terms of its Q-factor.

The impedance $\boldsymbol{Z}_p$ of the circuit shown in fig. 7.4 is given by the transfer function $(\boldsymbol{V}_o/\boldsymbol{I}_i)$. Thus

$$\boldsymbol{Z}_p = 1/G_p + \mathrm{j}(\omega C_p - 1/\omega L_p),$$

$$= 1/G_p\left[1 + \mathrm{j}\left(\frac{\omega C_p}{G_p} - \frac{1}{\omega L_p G_p}\right)\right],$$

$$= 1/G_p\left[1 + \mathrm{j}Q_p\left(\frac{\omega}{\omega_{0p}} - \frac{\omega_{0p}}{\omega}\right)\right],$$

$$= 1/G_p(1 + \mathrm{j}yQ_p), \tag{7.5.1}$$

where

$$y = \frac{\omega}{\omega_{0p}} - \frac{\omega_{0p}}{\omega}. \tag{7.5.2}$$

The normalized impedance is

$$\left.\begin{aligned} \boldsymbol{Z}_p G_p &= \frac{1}{1 + \mathrm{j}yQ_p}, \\ &= \frac{1}{\sqrt{[1 + (yQ_p)^2]}} \angle -\tan^{-1}(yQ_p). \end{aligned}\right\} \tag{7.5.3}$$

This is identical with (7.2.2) if (yQ_p) replaces x, and it follows that fig. 7.2 applies to the parallel-resonant circuit if the frequency axis of the diagram is labelled (yQ_p).

The significance of the quantity y is best shown by relating it to a frequency increment above resonance. If

$$\omega = \omega_{0p} + \delta\omega,$$

then

$$y = \frac{\omega_{0p} + \delta\omega}{\omega_{0p}} - \frac{\omega_{0p}}{\omega_{0p} + \delta\omega},$$

$$= \frac{\delta\omega(2\omega_{0p} + \delta\omega)}{\omega_{0p}(\omega_{0p} + \delta\omega)},$$

$$\approx 2\frac{\delta\omega}{\omega_{0p}} \quad (\text{for } \delta\omega \ll \omega_{0p}), \tag{7.5.4}$$

$$= 2 \times \text{'fractional detuning from resonance'}.$$

In using fig. 7.2 to calculate the normalized impedance of a parallel-resonant circuit at any frequency it is convenient to use (7.5.4) to compute (yQ_p) at the given frequency. It is therefore important to know if any significant error is introduced by the approximation in (7.5.4). It turns out that the fractional error in (yQ_p) when computed from the approximate formula (7.5.4) is $1/(Q_p+1)$ at a frequency two bandwidths above resonance and $-1/(Q_p-1)$ at a frequency two bandwidths below resonance. It follows that the approximate formula is satisfactory for evaluating y for high-Q circuits at any frequency within a few bandwidths of resonance.

It is worth adding that the Bode diagram of fig. 7.2 is drawn only for positive values of x, whereas the values required for the parallel-resonant circuit are both positive (above resonance) and negative (below resonance). Since, however, the magnitude of the impedance of the parallel-resonant circuit has even symmetry about $y=0$, and the phase angle has odd symmetry, the left hand part of the diagram is redundant; it must only be remembered that at frequencies below resonance the phase angle is the negative of the corresponding value above resonance.

The bandwidth of any physical system which has a response of the form shown in fig. 7.5 is defined by

$$\text{Bandwidth} = f_2 - f_1. \qquad \text{(equation (7.4.3))}$$

The frequencies f_2 and f_1 are those at which the response is $1/\sqrt{2}$ of (3 db below) the maximum. Equation (7.5.3) shows that they are given for the parallel-resonant circuit by

$$yQ_p = \pm 1.$$

That is, by

$$\left.\begin{aligned}\left(\frac{f_2}{f_{0p}} - \frac{f_{0p}}{f_2}\right) Q_p &= +1,\\ \left(\frac{f_1}{f_{0p}} - \frac{f_{0p}}{f_1}\right) Q_p &= -1.\end{aligned}\right\} \qquad (7.5.5)$$

It follows that

$$\left.\begin{aligned}f_2^2 - f_{0p}^2 &= f_{0p} f_2/Q_p,\\ f_1^2 - f_{0p}^2 &= -f_{0p} f_1/Q_p,\end{aligned}\right\} \qquad (7.5.6)$$

and that

$$f_2 - f_1 = f_{0p}/Q_p. \qquad (7.5.7)$$

Thus the bandwidth and the Q-factor have been so defined that they satisfy the relation

$$\text{Bandwidth} = \frac{\text{Resonant frequency}}{Q \text{ factor}},$$

which emphasizes the statement made in the last section, that Q is a measure of the sharpness of the response curve.

Problem. Find the bandwidth of the circuit shown in fig. 7.4 given that $G_p = 0.1$ mS, $L_p = 1$ mH, and that the circuit resonates at 100 kHz. What is its impedance at resonance and at the half-power points? At what frequencies does the magnitude of the impedance fall to one half of the impedance at resonance?

The Q-factor of the circuit is

$$Q_p = \frac{1}{\omega_{0p} L_p G_p} = \frac{1}{(2\pi)(10^5)(10^{-3})(10^{-4})} = 15.92, \text{ or } 16 \text{ (say).}$$

The bandwidth is

$$B = \frac{f_{0p}}{Q_p} = \frac{100}{16}\,\text{kHz} = 6.3\ \text{kHz}.$$

The impedance at resonance ($y = 0$) is

$$\mathbf{Z}_{p0} = \frac{1}{G_p} = 10\angle 0^\circ\ \text{k}\Omega.$$

The impedance at the upper half power point ($y_2 = 1/Q_p$) is

$$\mathbf{Z}_{p2} = \frac{1}{\sqrt{2}G_p}\ \angle -\tan^{-1} 1 = 7.1\angle -45^\circ\ \text{k}\Omega.$$

The impedance at the lower half power point ($y_1 = -1/Q_p$) is

$$\mathbf{Z}_{p1} = \frac{1}{\sqrt{2}G_p}\ \angle -\tan^{-1}(-1) = 7.1\angle 45^\circ\ \text{k}\Omega.$$

The frequencies at which the impedance has a magnitude of $1/2G_p$ are given by

$$\frac{1}{2G_p} = \frac{1}{G_p\sqrt{[1+(yQ_p)^2]}},$$

or

$$yQ_p = \pm\sqrt{3},$$

and

$$\delta f \approx \pm\tfrac{1}{2}\sqrt{3}\cdot\frac{f_{0p}}{Q_p} = \pm 5.4\ \text{kHz},$$

i.e. at frequencies of 105.4 and 94.6 kHz, 5.4 kHz above and below the resonant frequency (100 kHz).

7.6 Q-factors of inductors and capacitors

The Q-factor described in §7.4 is that of a resonant *circuit*, and it is evaluated at the resonant frequency of the circuit. The term Q-factor is, however, applied not only to resonant *circuits* but also to inductors and capacitors, the elements which constitute the circuit. The Q-factor of an inductor or capacitor is defined for any angular frequency ω by the equation

$$Q = 2\pi . \frac{\text{Maximum energy stored in the element}}{\text{Energy dissipated by the element in one cycle}}. \qquad (7.6.1)$$

For example, the inductor shown in fig. 7.6(*a*) stores energy, which is accounted for by the ideal element L, and dissipates energy, which is

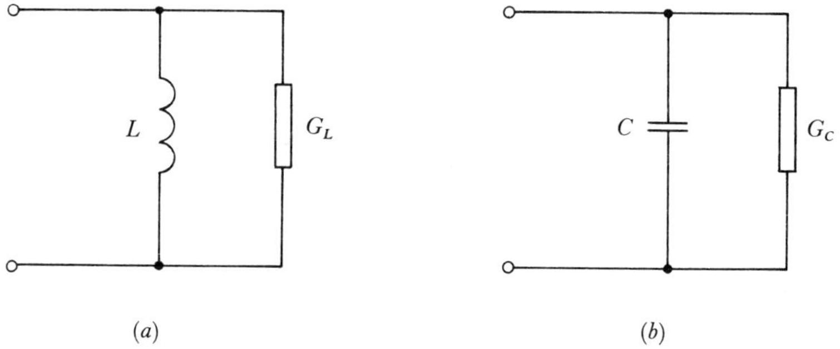

Fig. 7.6 Representation of (*a*) a lossy inductor, (*b*) a lossy capacitor.

accounted for by the ideal element G_L. (The choice of parallel elements rather than series elements to represent the inductor will be discussed in a moment). If a sinusoidal forcing voltage of maximum value V_m at an angular frequency ω is applied to this dissipative or 'lossy' inductor, the maximum energy stored in it is $\frac{1}{2}L(V_m/\omega L)^2$ and the energy dissipated in one cycle is $\frac{1}{2}G_L V_m^2(2\pi/\omega)$. The Q-factor of the inductor is therefore

$$Q_L = \frac{1}{\omega L G_L}. \qquad (7.6.2)$$

Similarly, a lossy capacitor (fig. 7.6(*b*)) has a Q-factor given by

$$Q_C = \frac{\omega C}{G_C}. \qquad (7.6.3)$$

If now the two elements are connected in parallel to form a parallel-resonant circuit and the Q-factor of this *circuit* is evaluated (at its resonant frequency f_0) the result can be expressed in the form

$$\frac{1}{Q_p} = \frac{G_C + G_L}{\omega_0 C},$$

$$= \frac{G_C}{\omega_0 C} + \omega_0 L G_L,$$

$$= \frac{1}{Q_{C0}} + \frac{1}{Q_{L0}}, \tag{7.6.4}$$

where Q_{C0} and Q_{L0} are the values of Q_C and Q_L at the resonant frequency f_0. In a parallel-resonant circuit the Q-factor of the capacitor is usually so much

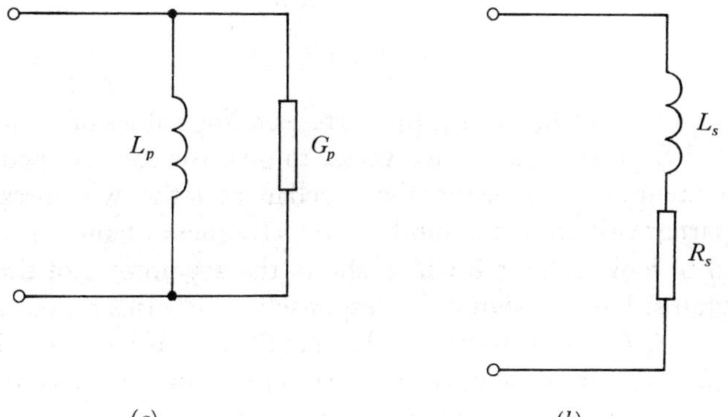

Fig. 7.7 Equivalent models of an inductor at a specific frequency.

greater than the Q-factor of the inductor that $1/Q_{C0}$ is negligible with respect to $1/Q_{L0}$, and the Q of the resonant circuit is very close to the Q of the inductor used in it. This is equivalent to ignoring G_C with respect to G_L.

We now return to the representation of the inductor by parallel elements rather than by series elements. The alternative representations, which are shown in fig. 7.7, are identical at only one frequency, but are sufficiently close to one another (for high-Q elements) over a band of frequencies to be interchangeable over this band. The parallel arrangement gives the simpler analytical expressions for the parallel-resonant circuit, and is chosen for this reason; the series arrangement is more convenient for the series-resonant circuit to be discussed later.

The conversion of the parameters of a lossy inductor from one form to the other is done by equating the real and imaginary parts of the admittances of the two forms at the centre-frequency ω_0 of the frequency band of interest. This gives

$$\left.\begin{aligned} G_p &= \frac{R_s}{R_s^2 + (\omega_0 L_s)^2}, \\ \frac{1}{\omega_0 L_p} &= \frac{\omega_0 L_s}{R_s^2 + (\omega_0 L_s)^2}, \end{aligned}\right\} \tag{7.6.5}$$

from which it can be seen that the Q of the inductor at the frequency ω_0 is

$$Q_{L0} = 1/\omega_0 L_p G_p = \omega_0 L_s/R_s. \tag{7.6.6}$$

Substitution of (7.6.6) in (7.6.5) gives

$$\left.\begin{aligned} G_p &= \frac{1}{R_s(1 + Q_{L0}^2)}, \\ L_p &= L_s(1 + 1/Q_{L0}^2). \end{aligned}\right\} \tag{7.6.7}$$

Given the values of L_s and R_s, the corresponding values of L_p and G_p can be evaluated from (7.6.7), or vice versa, to give the same impedance at a specific frequency ω_0. To see that the impedances of the two representations do not depart greatly from one another over a frequency band centred on ω_0, it is useful to look at fig. 7.8 which shows the asymptotes of the relevant Bode diagrams. Part (*a*) shows the response of the series representation in which $\omega_1 = R_s/L_s$, and part (*b*) the parallel representation in which $\omega_2 = 1/L_p G_p$. The impedances of the two representations have to be made equal at some frequency ω_0 and this requires that

$$\omega_2/\omega_0 = \omega_0/\omega_1,$$

as may be seen from (7.6.6) or (7.6.5). Thus

$$\omega_2 = \omega_0^2/\omega_1.$$

Now the two parts (*a*) and (*b*) of fig. 7.8 have the same shape (a rising magnitude of 6 db per octave and a constant phase angle of 90 degrees) in the region shown by the thick lines. That is, for

$$0.1\,\omega_2 > \omega > 10\,\omega_1,$$

or

$$0.1\,\omega_0^2/\omega_1 > \omega > 10\,\omega_1. \tag{7.6.8}$$

In other words, if the Q of the coil at the conversion frequency ω_0 is greater than 10 so that $(\omega_0/\omega_1) > 10$, (7.6.8) is satisfied in the neighbourhood of ω_0.

The larger the Q at the conversion frequency ω_0, the greater is the frequency range over which the equivalence is valid.

The equivalence may indeed be tolerable for a coil with a Q-factor less than 10. For example, the series representation of an inductor in which

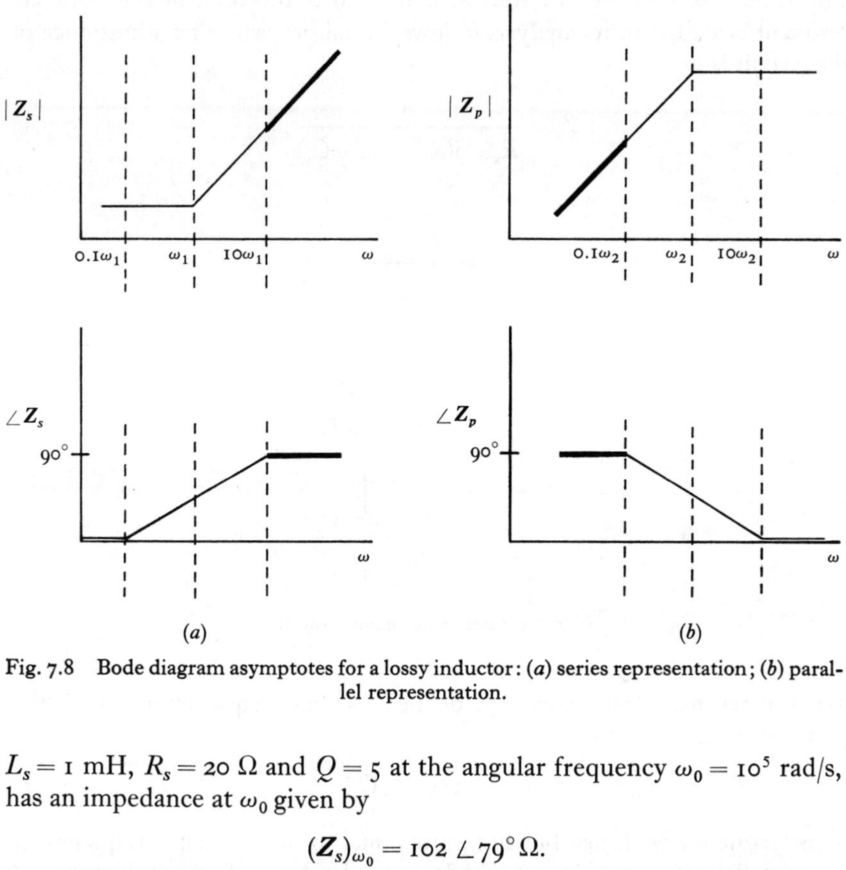

Fig. 7.8 Bode diagram asymptotes for a lossy inductor: (*a*) series representation; (*b*) parallel representation.

$L_s = 1$ mH, $R_s = 20\ \Omega$ and $Q = 5$ at the angular frequency $\omega_0 = 10^5$ rad/s, has an impedance at ω_0 given by

$$(\boldsymbol{Z}_s)_{\omega_0} = 102\ \angle 79°\ \Omega.$$

At a frequency 20 per cent higher, at $(\omega_0 + \omega_0/Q_L)$ (which is one bandwidth above ω_0 assuming the inductor to be resonated at ω_0 by a lossless capacitor), its value is

$$(\boldsymbol{Z}_s)_{(\omega_0+\omega_0/Q_L)} = 122\ \angle 81°\ \Omega.$$

The corresponding values of L_p and G_p, evaluated at ω_0, are $L_p = 1.04$ mH, $G_p = 1.92$ mS. These components, at the frequency $(\omega_0 + \omega_0/Q_L)$, have an impedance

$$(\boldsymbol{Z}_p)_{(\omega_0+\omega_0/Q_L)} = 121\ \angle 77°\ \Omega.$$

The difference is small even for a coil with a Q-factor as low as 5.

As the principle of duality suggests, similar arguments apply to the series and parallel representations of lossy capacitors.

7.7 The series-resonant circuit

The series-resonant circuit shown in fig. 7.9 is the dual of the parallel-resonant circuit, and its analysis follows a dual pattern. The admittance of the circuit is

$$\mathbf{Y}_s = \frac{1}{R_s + \mathrm{j}(\omega L_s - 1/\omega C_s)}, \tag{7.7.1}$$

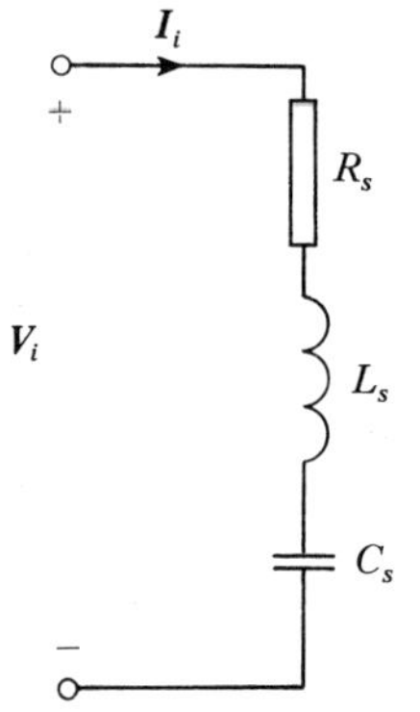

Fig. 7.9 Series-resonant circuit.

and the resonant frequency ω_{0s}, defined as that frequency at which the phase angle is zero, is given by

$$\omega_{0s} = 1/\sqrt{(L_s C_s)}. \tag{7.7.2}$$

This frequency is almost but not quite equal to the resonant frequency of the parallel-resonant circuit which uses the same lossy inductor and capacitor. For a parallel-resonant circuit, (7.4.2) gives

$$\omega_{0p} = 1/\sqrt{(L_p C_p)}. \tag{7.7.3}$$

If, for example, the capacitor is lossless and the coil is dissipative, $C_p = C_s$ but $L_p \neq L_s$ (see (7.6.7)); thus $\omega_{0p} \neq \omega_{0s}$. The difference is negligible when high-Q elements are used.

The Q-factor of the series-resonant circuit, defined by (7.4.4), is

$$Q_s = \frac{\text{Reactance}}{\text{Resistance}} = \frac{\omega_{0s} L_s}{R_s} = \frac{1}{\omega_{0s} C_s R_s} = \frac{1}{R_s}\sqrt{\frac{L_s}{C_s}}, \tag{7.7.4}$$

and if y is defined by

$$y = \frac{\omega}{\omega_{0s}} - \frac{\omega_{0s}}{\omega},$$

$$\approx 2\frac{\delta\omega}{\omega_{0s}},$$

the admittance of the series circuit becomes

$$\boldsymbol{Y}_s = \frac{1}{R_s(1 + \mathrm{j}yQ_s)}. \tag{7.7.5}$$

The Bode diagrams of fig. 7.2 are therefore applicable to the normalized admittance ($\boldsymbol{Y}_s R_s$) of the series-resonant circuit.

7.8 Cascades of synchronously-tuned and stagger-tuned circuits

Two or more circuits are said to be in *cascade* when they are connected, as in fig. 7.10, so that the response of one circuit is the stimulus of the next. The analysis is enormously simplified, and the essentials of the circuit performance made clear, if the circuits are non-interacting; that is, if the overall transfer function is the product of the transfer functions of the separate parts. This applies to fig. 7.10, and the overall transfer function $\boldsymbol{H}(\mathrm{j}\omega)$ is given by

$$\frac{\boldsymbol{V}_o}{\boldsymbol{V}_i} = \frac{\boldsymbol{V}_1 \boldsymbol{V}_o}{\boldsymbol{V}_i \boldsymbol{V}_1},$$

or

$$\boldsymbol{H}(\mathrm{j}\omega) = \boldsymbol{H}_1(\mathrm{j}\omega)\boldsymbol{H}_2(\mathrm{j}\omega), \tag{7.8.1}$$

where $\boldsymbol{H}_1(\mathrm{j}\omega)$ and $\boldsymbol{H}_2(\mathrm{j}\omega)$ are the transfer functions of the two parts separately. In this example, $\boldsymbol{H}_1$ and $\boldsymbol{H}_2$ are the transfer functions of parallel-resonant circuits which are essentially identical but whose resonant frequencies are capable of being adjusted or *tuned* by variation of L or C. The effect of a small adjustment of this kind is to shift the frequency response bodily along the frequency axis by a small amount without significantly disturbing its shape. When the resonant frequencies of the two parts (or stages) are identical, the two parts are said to be *synchronously tuned*; when the resonant frequencies are slightly offset from one another as in fig. 7.10(c), the parts are said to be *stagger-tuned*.

Consider first a cascade of two identical synchronously tuned circuits (fig. 7.10(a)) in which $L_1 = L_2 = L$ and $C_1 = C_2 = C$. The transfer function of each stage is

$$\boldsymbol{H}_1(\mathrm{j}\omega) = \boldsymbol{H}_2(\mathrm{j}\omega) = A/(1 + \mathrm{j}x), \tag{7.8.2}$$

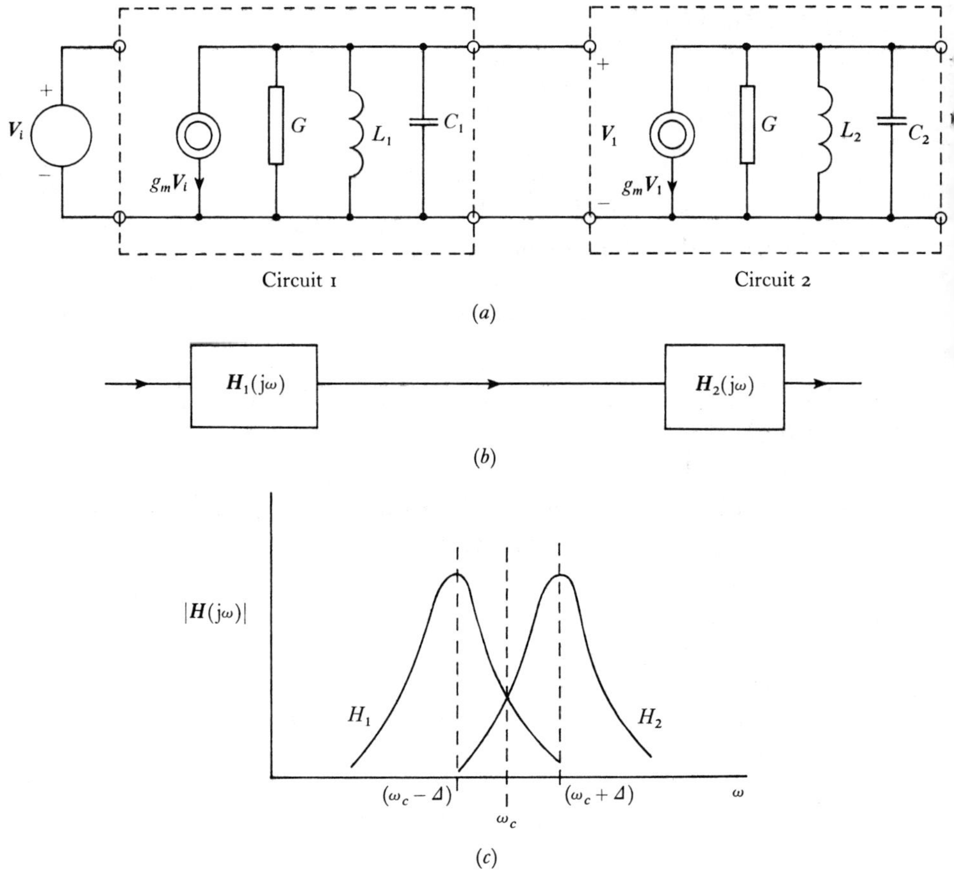

Fig. 7.10 (*a*) Cascade of two non-interacting parallel-resonant circuits. (*b*) Transfer function diagram of non-interacting stages. (*c*) Individual responses of two stagger-tuned stages.

where

$$A = -g_m/G,\ \omega_0 = 1/\surd(LC),\ Q = \omega_0 C/G, \text{ and } x = [(\omega/\omega_0) - (\omega_0/\omega)]\,Q.$$

The overall transfer function is

$$\boldsymbol{H}(j\omega) = \boldsymbol{H}_1(j\omega)\,\boldsymbol{H}_2(j\omega),$$

$$= \frac{A^2}{(1+jx)^2},$$

and its magnitude is

$$|\boldsymbol{H}(j\omega)| = \frac{A^2}{1+x^2}. \tag{7.8.3}$$

Equation (7.8.3) shows that the response falls away on either side of resonance rather more steeply than that of the single stage whose response, like that of the parallel-tuned circuit, is $A/\sqrt{(1+x^2)}$. This means that the bandwidth of the two-stage cascade is less than that of the individual stages. Its value is determined from the 3 db points. The values of x at which the denominator of (7.8.3) is equal to $\sqrt{2}$ are $\pm\sqrt{(2^{1/2}-1)}$, and since x is equal to Q multiplied by twice the fractional detuning from resonance, the bandwidth of (7.8.3) is

$$B = \frac{f_0}{Q}\sqrt{(2^{1/2}-1)}. \tag{7.8.4}$$

In passing, it is of interest to observe that the bandwidth of n identical stages in cascade is $(f_0/Q)\sqrt{(2^{1/n}-1)}$ and that the factor $\sqrt{(2^{1/n}-1)}$ has the values 0.64, 0.51, and 0.43 for n equal to 2, 3, and 4 respectively. The avoidance of this cramping effect on the overall bandwidth of synchronously tuned stages as more and more stages are cascaded is one of the reasons for staggering the tuning of the stages. For although this technique does not eliminate the cramping effect, it reduces it. A more important result of stagger-tuning, however, is that it gives to the overall transfer function a flat-topped shape that is desirable for many purposes. Indeed, it is basically this flat-topped shape that reduces the cramping effect just mentioned.

If one stage of a synchronously tuned pair of stages is detuned by a small fractional amount from the synchronous angular frequency ω_c to a lower frequency $(\omega_c - \Delta)$, the transfer function becomes

$$\boldsymbol{H}_1(\mathrm{j}\omega) = \frac{A}{1+\mathrm{j}(x_c+a)}, \tag{7.8.5}$$

where

$$x_c = \left(\frac{\omega}{\omega_c} - \frac{\omega_c}{\omega}\right)Q,$$

and

$$a \approx \frac{2\Delta Q}{\omega_c}. \tag{7.8.6}$$

If the second stage is shifted in the opposite direction by an equal amount, its transfer function becomes

$$\boldsymbol{H}_2(\mathrm{j}\omega) = \frac{A}{1+\mathrm{j}(x_c-a)}, \tag{7.8.7}$$

and the overall transfer function of the two stages is

$$\boldsymbol{H}(\mathrm{j}\omega) = \boldsymbol{H}_1(\mathrm{j}\omega)\,\boldsymbol{H}_2(\mathrm{j}\omega),$$

$$= \frac{A^2}{[1+\mathrm{j}(x_c-a)][1+\mathrm{j}(x_c+a)]}. \tag{7.8.8}$$

The normalized magnitude of (7.8.8) is sketched in fig. 7.11 for three different values of a together with the normalized response of a single stage

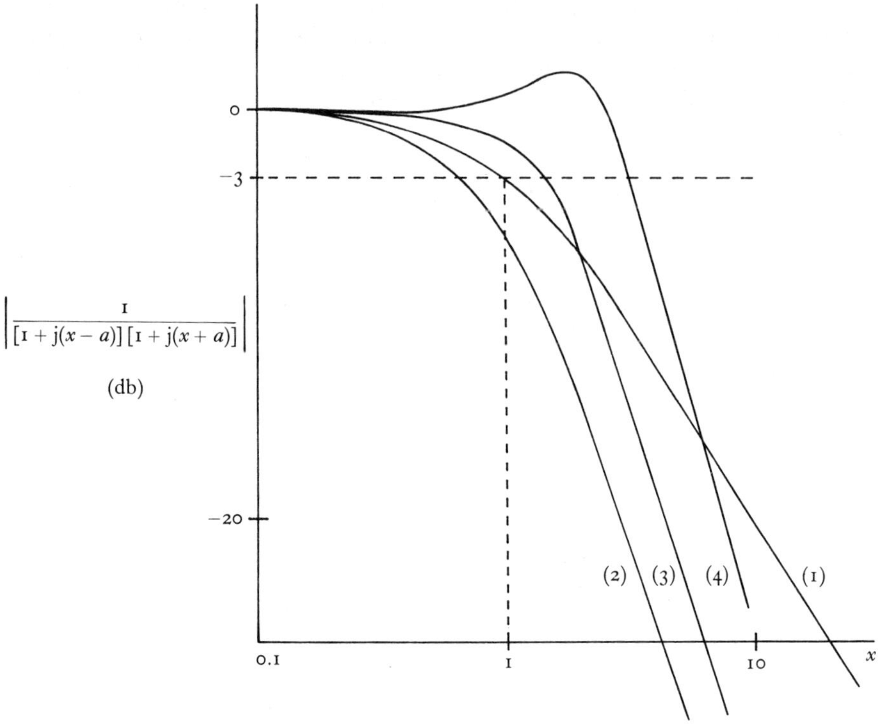

Fig. 7.11 Normalized responses of cascaded stages: (1) single parallel-resonant stage; (2) two synchronously tuned stages ($a = 0$); (3) staggered pair ($a = 1$); (4) staggered pair ($a = 2$).

for reference. The response for $a = 1$ is of special interest. When $\Delta = (\frac{1}{2})(\omega_c/Q)$, $a = 1$, and the magnitude of the transfer function is

$$|\boldsymbol{H}(j\omega)|_{a=1} = \frac{A^2}{\sqrt{\{[1+(x_c-1)^2][1+(x_c+1)^2]\}}}.$$

This reduces to

$$|\boldsymbol{H}(j\omega)|_{a=1} = \frac{A^2}{\sqrt{(x_c^4+4)}},$$

and gives a normalized response

$$\left[\frac{|\boldsymbol{H}(j\omega)|}{A^2/2}\right]_{a=1} = \frac{1}{\sqrt{[1+(x_c^4/4)]}}. \tag{7.8.9}$$

The corner frequency is given by $x_c = \surd 2$ and the bandwidth by

$$B_{a=1} = \frac{f_c}{Q} \cdot \surd 2, \tag{7.8.10}$$

which is more than twice as large as that given by (7.8.4). The asymptotic slope, like that for $a = 0$ (two synchronously tuned stages in cascade) is 12 db per octave. When a is greater than unity, there are peaks on either side of the central frequency f_c, and the un-normalized response at f_c is less than that for $a = 1$. As a increases the peaks move outwards and the response at f_c falls further.

7.9 A geometric interpretation of transfer functions

A transfer function of the form

$$\boldsymbol{H}(\mathrm{j}\omega) = K \cdot \frac{(\mathrm{j}\omega + \omega_1)(\mathrm{j}\omega + \omega_2)\ldots(\mathrm{j}\omega + \omega_i)}{(\mathrm{j}\omega + \omega_a)(\mathrm{j}\omega + \omega_b)\ldots(\mathrm{j}\omega + \omega_j)} \quad \text{(see (7.3.1))}$$

is completely represented, apart from the constant multiplier K (sometimes known as the *scale factor*), by its corner frequencies $\omega_a, \omega_b, \ldots, \omega_1, \omega_2$, etc. Each term in the numerator produces a component of the amplitude response which rises at 6 db per octave above its corner frequency, and a positive component of the phase response; each term in the denominator produces a component of the amplitude response which falls at 6 db per octave above its corner frequency, and a negative component of the phase response. A diagram which displays the corner frequencies of a transfer function thus describes it, apart from the constant multiplier K, in a convenient and compact way. Two examples will illustrate the display of corner frequencies and the geometric interpretation of the transfer function that they represent.

The transfer function $\boldsymbol{V}_i/\boldsymbol{I}_i$, the input impedance, of the RL circuit shown in fig. 7.12(*a*) is

$$\frac{\boldsymbol{V}_i}{\boldsymbol{I}_i} = \boldsymbol{H}(\mathrm{j}\omega) = K(\mathrm{j}\omega + \omega_1), \tag{7.9.1}$$

where $K = L$, and $\omega_1 = R/L$. Apart from the constant multiplier K, the magnitude and phase of $\boldsymbol{H}(\mathrm{j}\omega)$ are

$$\left.\begin{aligned} |\boldsymbol{H}(\mathrm{j}\omega)| &= r(\omega) = \surd(\omega^2 + \omega_1^2), \\ \angle \boldsymbol{H}(\mathrm{j}\omega) &= \theta(\omega) = \tan^{-1}(\omega/\omega_1). \end{aligned}\right\} \tag{7.9.2}$$

The geometric interpretation is shown in fig. 7.12(*b*). The small circle on the horizontal axis represents the corner frequency ω_1, and is drawn (to some scale or other) at this distance from the origin. Distances along the vertical axis (to the same scale) represents the angular frequencies ω. The length of the line joining the corner frequency to the point P is equal to the magnitude of the transfer function (apart from the constant multiplier K) and the angle which the line makes with the horizontal axis is equal to the phase angle of the transfer function (apart from an additional 180 degrees which will arise if the constant multiplier K is negative). As the point P

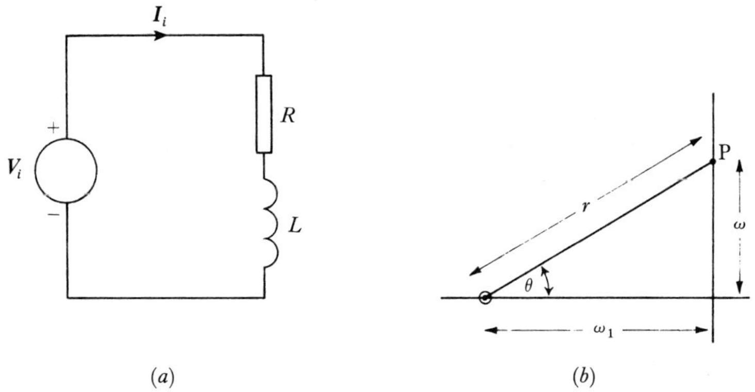

Fig. 7.12 (*a*) *RL* circuit. (*b*) Geometric representation of the transfer function of (*a*).

moves upwards from the origin (zero frequency) to infinity (infinite frequency), so the way in which r and θ vary can easily be visualized.

Another circuit is shown in fig. 7.13(*a*). Its transfer function is

$$\frac{\boldsymbol{V}_o}{\boldsymbol{V}_s} = \boldsymbol{H}(\mathrm{j}\omega) = K.\frac{(\mathrm{j}\omega + \omega_1)}{(\mathrm{j}\omega + \omega_a)(\mathrm{j}\omega + \omega_b)}, \tag{7.9.3}$$

where $K = -g_m(R_2\|R_3)/C_1 R_s$ (the symbol $\|$ means 'in parallel with'), $\omega_1 = 0$, $\omega_a = 1/C_2(R_2\|R_3)$, and $\omega_b = 1/C_1(R_1\|R_s)$. The corner frequencies of the numerator (there is only one in this example, $\omega_1 = 0$) are marked on the horizontal axis by small circles, and the corner frequencies of the denominator (ω_a and ω_b) by small crosses as shown in fig. 7.13(*b*). Apart from the constant multiplier K, the amplitude and phase angle of the transfer function are

$$\left.\begin{aligned} |\boldsymbol{H}(\mathrm{j}\omega)| &= \sqrt{\frac{(\omega^2+\omega_1^2)}{(\omega^2+\omega_a^2)(\omega^2+\omega_b^2)}} = \frac{r_1}{r_a r_b}, \\ \angle \boldsymbol{H}(\mathrm{j}\omega) &= \tan^{-1}(\omega/\omega_1) - [\tan^{-1}(\omega/\omega_a) + \tan^{-1}(\omega/\omega_b)] = \theta_1 - (\theta_a + \theta_b), \end{aligned}\right\} \tag{7.9.4}$$

where the rs and θs are the lengths and angles shown in the diagram. Once more, as P moves upwards from the origin to infinity the various lengths and angles can be visualized. Notice that lines from P which terminate on the circles contribute an amplitude component in the numerator of (7.9.4) and a positive phase angle; lines from P which terminate on the crosses contribute

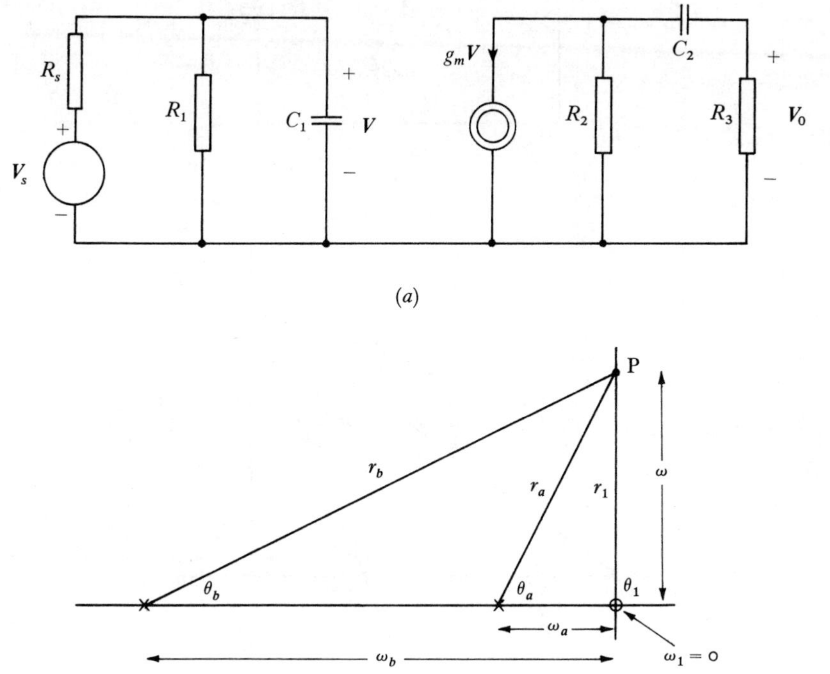

Fig. 7.13 (*a*) Amplifier circuit. (*b*) Geometric representation of the transfer function of (*a*).

an amplitude component in the denominator of (7.9.4) and a negative phase angle. The crosses and circles represent, in fact, the poles and zeros respectively of the given transfer functions, as will be seen in the next section.

7.10 Poles and zeros

The transfer functions used so far, in §7.9, have consisted of factors like $(j\omega + \omega_1)$ in which ω_1 is a real number. Many functions, however, cannot be factorized in this way unless some or all of the critical frequencies are allowed to become complex. The method of describing a transfer function by its critical frequencies ω_a, ω_b, ..., ω_1, ω_2, etc. whether they be real or complex, is known as the description by poles and zeros. It enables the

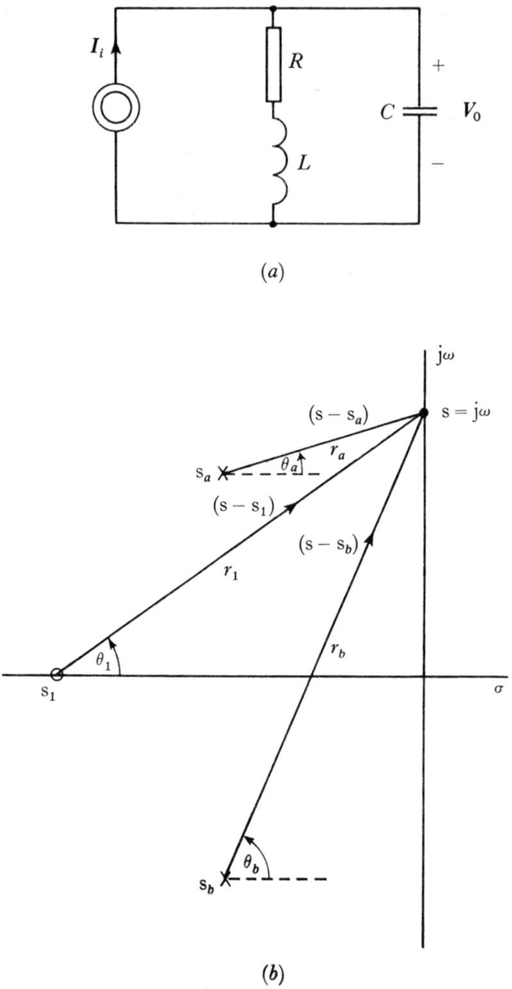

Fig. 7.14 (*a*) *RLC* circuit. (*b*) Pole–zero diagram of the transfer function (V_o/I_i).

significant properties of a transfer function to be displayed in a compact form, and gives information on the transient and steady-state behaviour of the circuit to which it applies. In this chapter emphasis is placed on the application of the pole–zero display to the steady state behaviour of circuits. Some more advanced topics are treated in chapter 12.

In developing the pole–zero concept it is helpful to replace ($j\omega$) in the transfer function by a new variable s where

$$s = \sigma + j\omega \quad (\text{see } (5.4.2)).$$

The significance of the new variable s, called the *complex frequency*, will be described in §7.11; at the moment it is best, as in chapter 5, to treat the substitution as a mathematical one, observing that the original transfer function can be recalled, if necessary, by putting σ equal to zero.

The transfer function of a linear, lumped-element circuit now takes the form

$$\boldsymbol{H}(s) = \frac{a_0 + a_1 s + a_2 s^2 + \ldots}{b_0 + b_1 s + b_2 s^2 + \ldots}, \tag{7.10.1}$$

where the coefficients a_0, b_0, a_1, b_1, etc. depend on the values of the circuit elements. For example, on writing sL for $j\omega L$, and $1/sC$ for $1/j\omega C$ in fig. 7.14(*a*), the transfer function $\boldsymbol{V}_o/\boldsymbol{I}_i$ becomes

$$\boldsymbol{H}(s) = \frac{1}{C} \cdot \frac{s + (R/L)}{s^2 + s(R/L) + (1/LC)}. \tag{7.10.2}$$

At certain values of s the transfer function will be zero; these values of s are called the zeros of the transfer function. At other values of s the transfer function will be infinite; these values of s are called the poles of the transfer function. The poles and zeros of (7.10.1) can be made explicit by factorizing the numerator and denominator polynomials as in (7.10.3):

$$\boldsymbol{H}(s) = K \cdot \frac{(s - s_1)(s - s_2)\ldots(s - s_i)}{(s - s_a)(s - s_b)\ldots(s - s_j)}. \tag{7.10.3}$$

Here the zeros of $\boldsymbol{H}(s)$ are $s_1, s_2, \ldots, s_i$, and the poles of $\boldsymbol{H}(s)$ are $s_a, s_b, \ldots, s_j$; they are the critical s-values of the network and may be either real or complex. Equation (7.10.2), for example, gives a zero at the real s-value $s_1 = -R/L$, and two poles s_a and s_b which are real if $(1/R)\sqrt{(L/C)} \leqslant \frac{1}{2}$ but complex if not.

In general the complex frequency s may have any position in the complex frequency plane as determined by its real component σ and its imaginary component ω, but for the moment we are treating s simply as a substitute for ($j\omega$), and in these circumstances s always lies on the positive imaginary axis. If we treat s, s_1, s_a, etc. as vectors drawn from the origin of the pole–zero diagram to the points s, s_1, s_a, etc. then the vectors drawn from the points s_1, s_a, etc. to the point s are $(s - s_1)$, $(s - s_a)$, etc. as shown. Suppose that they have lengths r_1, r_a, etc. and make angles θ_1, θ_a, etc. with the positive real axis. The transfer function is then given by the pair of equations

$$\left.\begin{aligned} |\boldsymbol{H}(\mathrm{s})|_{\mathrm{s}=\mathrm{j}\omega} &= \frac{\text{Product of lengths from zeros to s}}{\text{Product of lengths from poles to s}}, \\ &= \frac{r_1 . r_2 . \ldots . r_i}{r_a . r_b . \ldots . r_j}, \\ \angle \boldsymbol{H}(\mathrm{s})_{\mathrm{s}=\mathrm{j}\omega} &= (\text{Sum of angles made by zeros}) - (\text{sum of angles made by poles}), \\ &= (\theta_1 + \theta_2 + \ldots + \theta_i) - (\theta_a + \theta_b + \ldots + \theta_j). \end{aligned}\right\} \quad (7.10.4)$$

Applied to fig. 7.14 this gives

$$\left.\begin{aligned} |\boldsymbol{H}(\mathrm{s})|_{\mathrm{s}=\mathrm{j}\omega} &= \frac{r_1}{r_a r_b}, \\ \angle \boldsymbol{H}(\mathrm{s})_{\mathrm{s}=\mathrm{j}\omega} &= \theta_1 - (\theta_a + \theta_b). \end{aligned}\right\} \quad (7.10.5)$$

Further examples are given in §7.12.

7.11 Original and image worlds

The pole–zero diagram of the transfer function gives information on the steady state response of a circuit (its amplitude and phase characteristics), and this is the principal concern of this chapter; but it also gives information on the transient response of the circuit. To illuminate this, the complex frequency plane should be viewed as an image world into which we enter for the subsequent simplicity of analysis, and out of which we emerge in order to recover our time-dependent waveforms in their original form. In chapter 4, a continuous waveform such as $I_m \cos \omega t$ was represented in an image world by an expression of the form $I_m \mathrm{e}^{\mathrm{j}\omega t}$. If the idea of frequency is now generalized to include a real part σ as well as an imaginary part $\mathrm{j}\omega$, the image quantity $I_m \mathrm{e}^{\mathrm{j}\omega t}$ becomes $I_m \mathrm{e}^{\mathrm{s}t}$ and, in the image world of the complex frequency plane, represents the two original quantities

$$\left.\begin{aligned} i_1 &= I_m \mathrm{e}^{\sigma t} . \cos \omega t, \\ i_2 &= I_m \mathrm{e}^{\sigma t} . \sin \omega t. \end{aligned}\right\} \quad (7.11.1)$$

This can be confirmed by writing

$$\boldsymbol{I} = i_1 + \mathrm{j} i_2 = I_m \mathrm{e}^{\mathrm{s}t}. \quad (7.11.2)$$

The complex frequency s thus represents in the image world a wave of frequency ω expanding, constant, or decaying in accordance with the value σ, or a non-oscillatory waveform ($\omega = 0$) of increasing, constant, or decreasing amplitude as shown in fig. 7.15. This diagram shows the waveforms

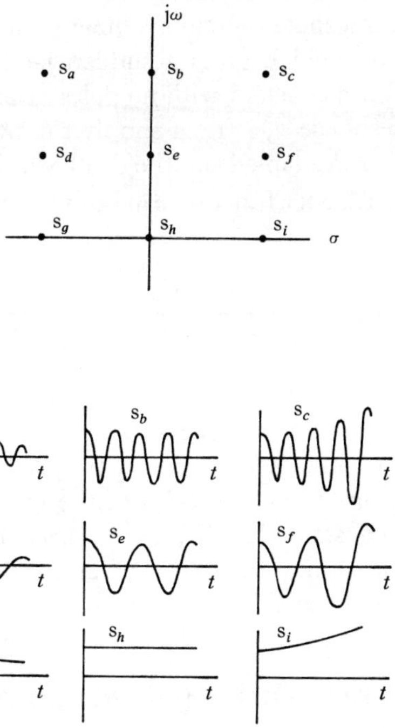

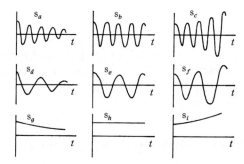

Fig. 7.15 Waveforms associated with the complex frequency plane.

associated with the upper half of the plane; the lower half is redundant in the sense that s*, the conjugate of s, is associated with the same waveform as s itself; both values are necessary for the reconstruction of the original. This is shown in (7.11.3):

$$i_1 = I_m e^{\sigma t} . \cos \omega t = \tfrac{1}{2}(I_m e^{st} + I_m e^{s^* t}). \tag{7.11.3}$$

The response of a circuit to any stimulus may be visualized with the aid of fig. 7.15. It is first of all necessary to obtain the pole–zero pattern of the response. If, for example, the response v_o to a stimulus i_i is required, it is necessary to express v_o in its image form by

$$\boldsymbol{V}_o(s) = \boldsymbol{H}(s)\boldsymbol{I}_i(s), \tag{7.11.4}$$

where $\boldsymbol{H}(s)$ is the transfer function relating $\boldsymbol{V}_o$ to $\boldsymbol{I}_i$. The pole–zero pattern of $\boldsymbol{V}_o(s)$ is the superposition of the pole–zero patterns of $\boldsymbol{H}(s)$ and $\boldsymbol{I}_i(s)$

separately, as may be seen by factorizing them both in the form of (7.10.3) and observing that the poles and zeros of the product contain all the poles and zeros of the two quantities separately. It is here assumed that a pole and a zero cancel each other out if they coincide when the two patterns are superposed; this corresponds with the cancellation of equal factors in the numerator and denominator of (7.11.4).

If the stimulus i_i is a spike function, a quantity of infinite magnitude but infinitesimal duration, which gives a sudden jar to the circuit and then leaves it alone, $\boldsymbol{I}_i(\mathrm{s})$ is a constant with no poles or zeros. (This is proved in chapter 10.) The response v_o is then simply the transient response of the circuit, and its form is available from the display of $\boldsymbol{H}(\mathrm{s})$ alone.

In many cases the transfer function can be expanded into partial fractions of a simple kind

$$\boldsymbol{H}(\mathrm{s}) = \frac{K_a}{\mathrm{s}-\mathrm{s}_a} + \frac{K_b}{\mathrm{s}-\mathrm{s}_b} + \ldots + \frac{K_j}{\mathrm{s}-\mathrm{s}_j}, \tag{7.11.5}$$

and it can be seen that each of the complex frequencies $\mathrm{s}_a, \mathrm{s}_b, \ldots, \mathrm{s}_j$ contributes a term $K_n e^{\mathrm{s}_n t}$ to the total transient response. The transient response thus consists of frequency components given by the *poles* of the transfer function in accordance with fig. 7.15. The *zeros* of the transfer function influence only the size of the constants K_a, K_b, ..., K_j; this influence may be seen by considering how the amplitudes K_n in (7.11.5) are derived from (7.10.3) during the process of expansion.

7.12 Elementary pole–zero patterns

The amplitude and phase responses associated with several elementary pole–zero patterns are described in this section.

(i) *A single real pole with or without a zero at the origin*

Figure 7.16 shows the response associated with a single pole on the negative real axis, and fig. 7.17 the effect of an additional zero at the origin. Patterns of this kind arise from circuits with only one energy-storing element.

(ii) *Two well separated real poles with a zero at the origin*

A pole–zero pattern that arises commonly in low frequency amplifier circuits and which consists of two poles on the negative real axis together with a zero at the origin is shown in fig. 7.18. Since this pattern is the superposition of the patterns of figs. 7.16 and 7.17, it could arise from a cascade of these two networks with a buffer in between to prevent interaction. Such a buffer, in the form of an ideal voltage-controlled voltage source is shown in fig. 7.18 which also gives the responses associated with this pole–zero diagram. Assuming that the poles are well separated, that is, that $R_2C_2 \ll R_1C_1$ (or vice versa), the vector from the pole at $(-1/R_2C_2)$ to the point P on the $(j\omega)$ axis lies very close to the σ-axis at angular frequencies close to zero. It is therefore almost constant over those frequencies at which the characteristics are being shaped by the other pole and zero; the

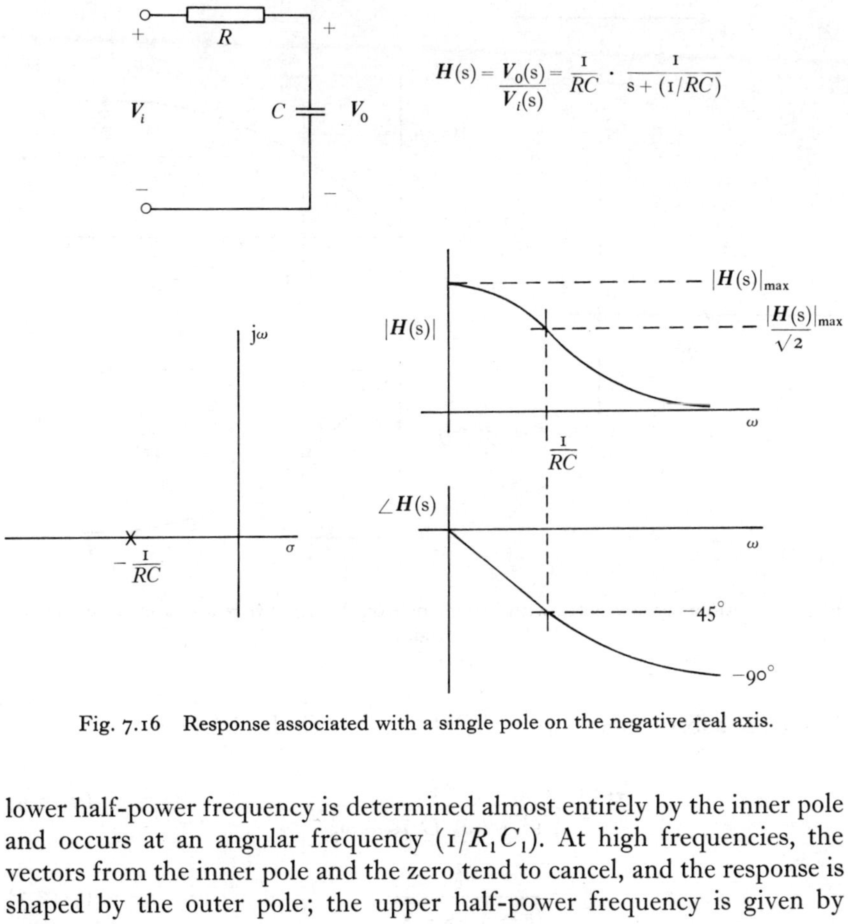

Fig. 7.16 Response associated with a single pole on the negative real axis.

lower half-power frequency is determined almost entirely by the inner pole and occurs at an angular frequency $(1/R_1C_1)$. At high frequencies, the vectors from the inner pole and the zero tend to cancel, and the response is shaped by the outer pole; the upper half-power frequency is given by $(1/R_2C_2)$.

(iii) *A pair of complex poles*

The series-resonant circuit is chosen to illustrate the next pattern which is a pair of complex poles. The transfer function of the circuit of fig. 7.19 is

$$H(s) = \frac{V_o}{V_i} = \frac{1}{LC} \cdot \frac{1}{s^2 + (R/L)s + 1/LC}. \tag{7.12.1}$$

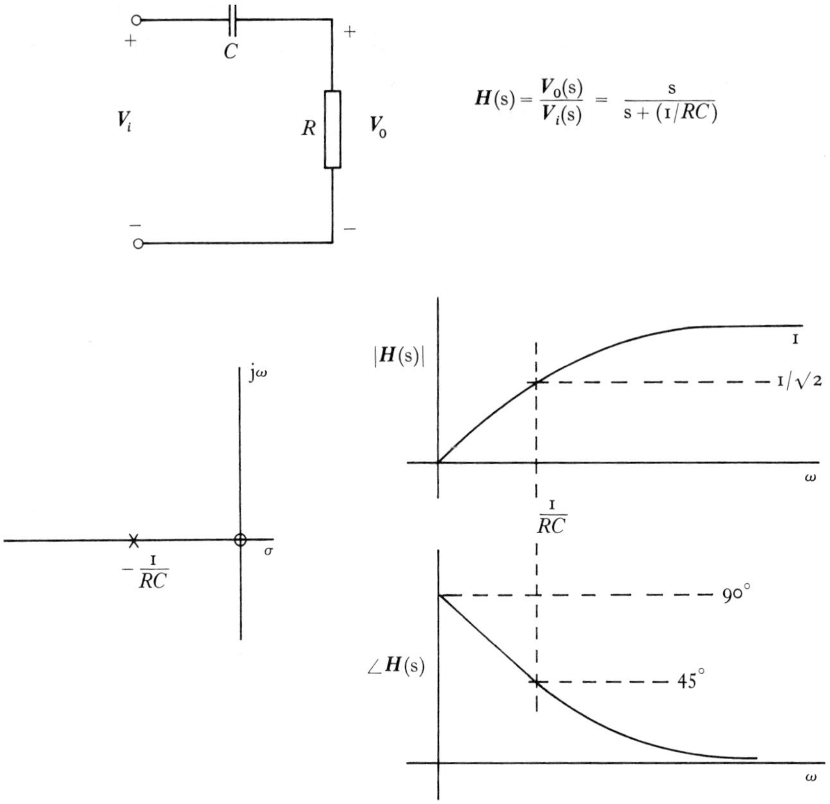

Fig. 7.17 Response associated with a single pole on the negative real axis and a zero at the origin.

Using the substitutions $\omega_0^2 = 1/LC$, $Q_s = \omega_0 L/R$, this becomes

$$H(s) = \frac{\omega_0^2}{s^2 + (\omega_0/Q_s)\,s + \omega_0^2}, \tag{7.12.2}$$

$$= \frac{\omega_0^2}{(s - s_a)(s - s_b)}, \tag{7.12.3}$$

where

$$s_a, s_b = -\frac{\omega_0}{2Q_s} \pm j\omega_0 \sqrt{\left[1 - \left(\frac{1}{2Q_s}\right)^2\right]}, \tag{7.12.4}$$

$$= -\alpha_1 \pm j\beta_1.$$

The merit in substituting ω_0 and Q_s for R, L and C is that these two parameters describe the shape of the characteristic more clearly than R, L, and

C; they are also applicable to non-electrical systems. An alternative to Q_s is sometimes used. This is called the *damping factor* ζ which is defined by

$$\zeta = \frac{\alpha_1}{\omega_0} = \frac{1}{2Q_s}, \tag{7.12.5}$$

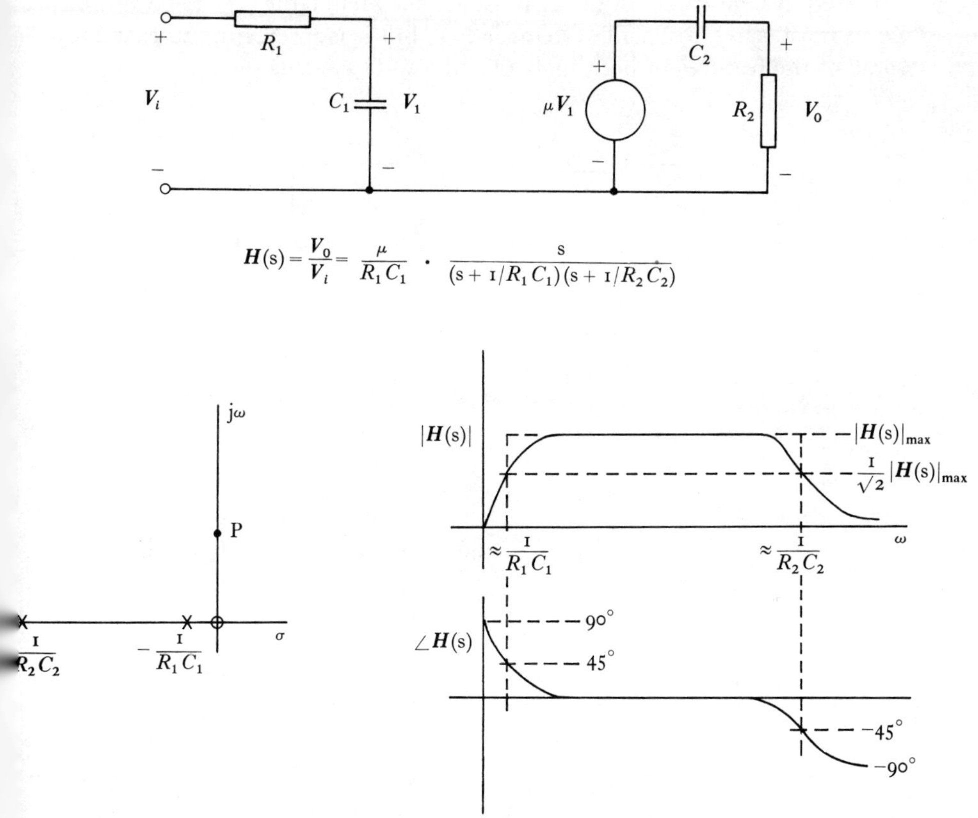

$$H(s) = \frac{V_0}{V_i} = \frac{\mu}{R_1C_1} \cdot \frac{s}{(s + 1/R_1C_1)(s + 1/R_2C_2)}$$

Fig. 7.18 Response associated with two well separated poles on the negative real axis and a zero at the origin.

and which tends to be preferred when the value of Q_s is small. Both ζ and Q_s are related to the angle ϕ shown in fig. 7.19(*b*). From (7.12.4) and the geometry of the diagram

$$\alpha_1 = \frac{\omega_0}{2Q_s} = \omega_0 \cos\phi, \tag{7.12.6}$$

and
$$\beta_1 = \omega_0 \sqrt{\left[1 - \left(\frac{1}{2Q_s}\right)^2\right]} = \omega_0 \sin\phi. \tag{7.12.7}$$

Equation (7.12.5) can thus be rewritten

$$Q_s = \frac{\omega_0}{2\alpha_1} = \frac{1}{2\zeta} = \frac{1}{2\cos\phi}. \tag{7.12.8}$$

A few equivalent values of Q_s, ζ, and ϕ are shown in table 7.1. The value of Q_s in most series-resonant systems is low, but it is interesting to examine some of the features of both high-Q and low-Q systems.

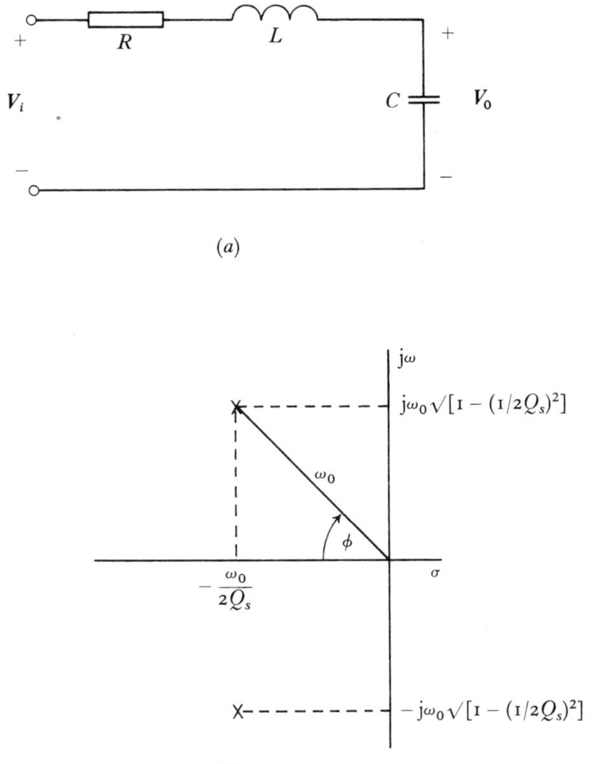

Fig. 7.19 (*a*) Series-resonant circuit. (*b*) Pole–zero pattern; a pair of complex poles.

The high-Q situation is illustrated in fig. 7.20. It is difficult, however, to make a clear diagram of high-Q poles because they are so close to the imaginary axis, and the Q of the pole-pair shown in fig. 7.20(*a*) is only 3.

It is therefore best, in the course of the discussion, to imagine the two poles to be much closer to the imaginary axis than shown. As the point P, representing the frequency, moves past the upper pole, the length r_a of the vector from this pole to P rapidly collapses to a small value and then rapidly expands again. During this collapse and expansion when P is in the neighbourhood of $j\omega_0$, the vector from the lower pole has a magnitude r_b

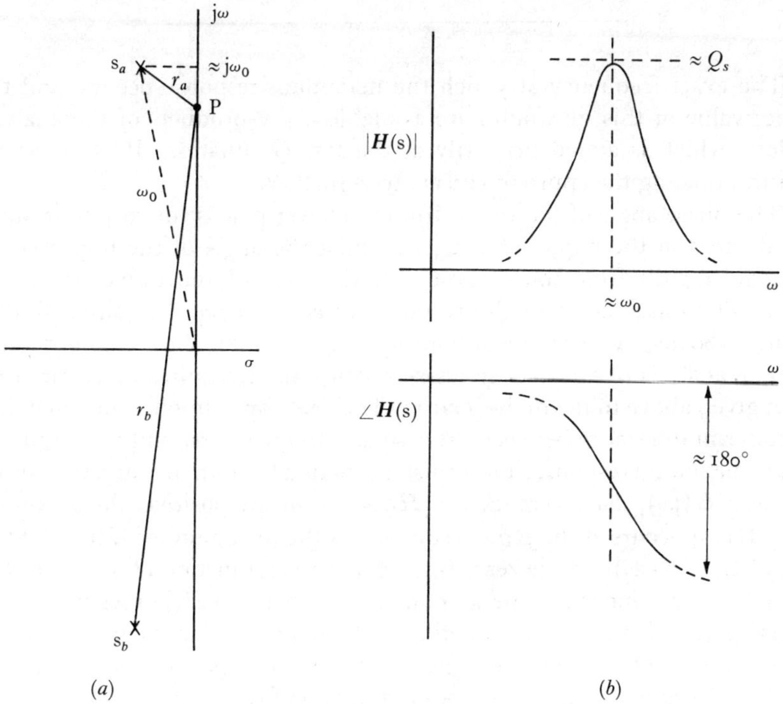

Fig. 7.20 (*a*) A complex pole-pair with a Q of 3. (*b*) Response of a high-Q pole-pair.

almost constant at about $2\omega_0$. The amplitude response of the pole-pair, the magnitude of (7.12.3), in the neighbourhood of ω_0 is thus given approximately by

$$|\boldsymbol{H}(j\omega)|_{\omega\approx\omega_0} \approx \frac{\omega_0^2}{(2\omega_0)r_a}, \tag{7.12.9}$$

and has a sharp peak at ω_0. The approximate value of the peak is found by substituting α_1, the minimum value of r_a, into (7.12.9). The result is

$$|\boldsymbol{H}(j\omega)|_{\omega=\omega_0} \approx \frac{\omega_0^2}{(2\omega_0)\alpha_1} = Q_s. \tag{7.12.10}$$

TABLE 7.1 *Some equivalent values of ϕ, ζ and Q_s*

ϕ	ζ	Q_s	ω_m/ω_0	ω_0/β_1
90°	0	∞	1	1
87°	0.05	10	>0.99	1.00
60°	0.5	1	0.71	1.15
45°	0.71	0.71	0	1.41
0°	⩾1	⩽0.5	–	–

The exact frequency at which the maximum response occurs, and the exact value of this maximum are available as by-products of the analysis below, which is aimed primarily at the low-Q situation. It is, however, worth pursuing the approximate analysis further.

The phase angle of the vector from the lower pole is roughly constant at 90 degrees in the region of $P \approx j\omega_0$, whilst the angle of the upper vector changes rapidly from about -90 to $+90$ degrees as P moves past the upper pole. The phase angle of the response thus has a rapid negative shift of almost 180 degrees in the neighbourhood of ω_0. This is shown in fig. 7.20(*b*).

The analysis of a low-Q series-resonant circuit requires more care than that given above to the high-Q circuit because, for example, there may be a significant difference between the resonant frequency ω_0 and the frequency ω_m of maximum response. The transfer function has its maximum magnitude when $d|\boldsymbol{H}(j\omega)|/d\omega$ is zero. Since $|\boldsymbol{H}(j\omega)|$ is always positive, the maximum of $|\boldsymbol{H}(j\omega)|$ occurs at the same frequency as the maximum of $|\boldsymbol{H}(j\omega)|^2$; that is, when $d|\boldsymbol{H}(j\omega)|^2/d\omega$ is zero. But $|\boldsymbol{H}(j\omega)|^2$ is a function of ω^2, as may be seen by separating the numerator and denominator of $\boldsymbol{H}(j\omega)$ separately into a real part which is an even function of ω and an imaginary part which is an odd function of ω and by recognizing that the real part is a function of ω^2 and the imaginary part is $j\omega$ times a function of ω^2.

Thus $d|\boldsymbol{H}(j\omega)|^2/d\omega$ may be written $(d|\boldsymbol{H}(j\omega)|^2/d(\omega^2)).(d(\omega^2)/d\omega)$, or $2\omega(d|\boldsymbol{H}(j\omega)|^2/d(\omega^2))$; so $\boldsymbol{H}(j\omega)$ has a maximum or minimum magnitude when either $\omega = 0$ or $d|\boldsymbol{H}(j\omega)|^2/d(\omega^2) = 0$. From (7.12.2),

$$|\boldsymbol{H}(j\omega)|^2 = \frac{\omega_0^4}{(\omega_0^2 - \omega^2)^2 + (\omega_0\omega/Q_s)^2}, \tag{7.12.11}$$

and

$$\frac{d|\boldsymbol{H}(j\omega)|^2}{d(\omega^2)} = -\frac{\omega_0^4[-2(\omega_0^2 - \omega^2) + (\omega_0/Q_s)^2]}{[(\omega_0^2 - \omega^2) + (\omega_0\omega/Q_s)^2]^2}. \tag{7.12.12}$$

This is zero at a frequency ω_m given by

$$(\omega_0/Q_s)^2 = 2(\omega_0^2 - \omega_m^2),$$

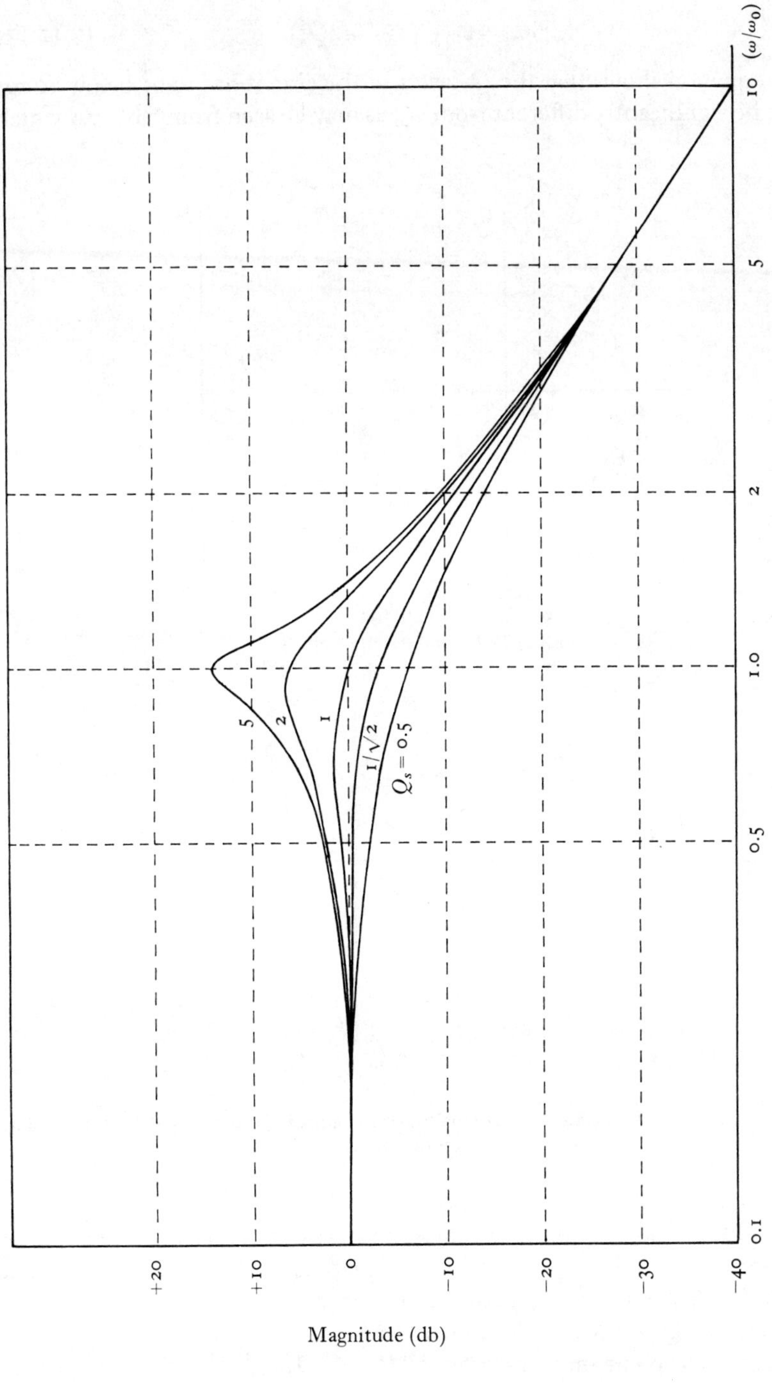

Fig. 7.21 Magnitude response of a pair of complex poles of low Q.

or

$$\omega_m = \omega_0\sqrt{(1 - 1/2Q_s^2)}. \qquad (7.12.13)$$

This equation shows that the Q-factor of the circuit has to be below 10 for ω_m to be significantly different from ω_0, as may be seen from table 7.1 which

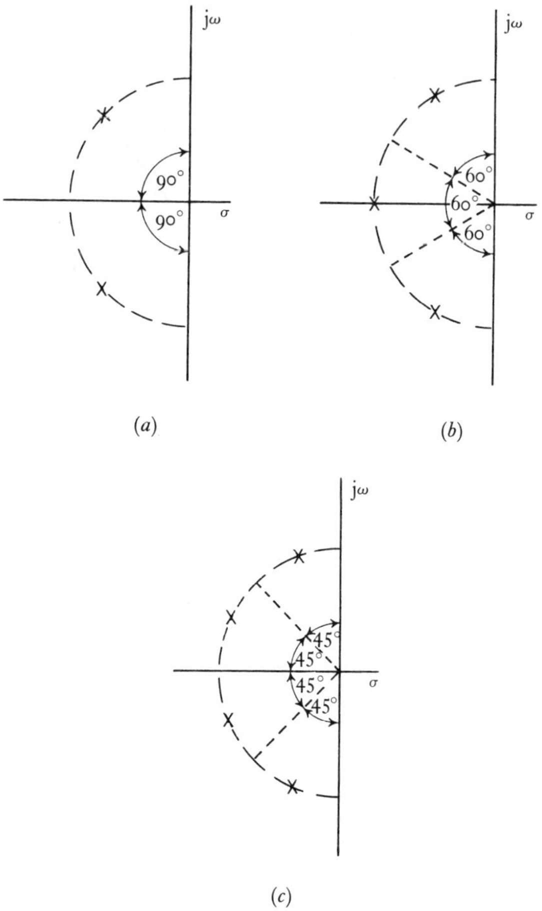

Fig. 7.22 Pole–zero patterns for maximally flat responses: (*a*) two poles, (*b*) three poles, (*c*) four poles.

gives values of (ω_m/ω_0) for four different values of Q. The magnitude of the responses at ω_0 and ω_m are found from equation (7.12.11); they are Q_s and $(\omega_0/\beta_1)Q_s$ respectively. ω_m becomes imaginary when $Q_s < 1/\sqrt{2}$.

Some low-Q responses are shown in fig. 7.21. When $Q_s > 1/\sqrt{2}$, $\omega = 0$ is a minimum point on the curve of $|\boldsymbol{H}(j\omega)|$. Interest in these responses centres on their flatness and on the frequency band over which this flatness holds. The flatness of a response is measured by the derivatives $d|\boldsymbol{H}(j\omega)|/d\omega$, $d^2|\boldsymbol{H}(j\omega)|/d\omega^2$, etc., and when as many of such derivatives as possible are zero at zero frequency in this case (or at band centre in a band-pass response, see § 7.15), the response is said to be *maximally flat*. It can be shown by double differentiation of (7.12.11) with respect to ω that a two-pole response is maximally flat when the Q-factor is $1/\sqrt{2}$, the critical value at which the type of response changes. The angle ϕ is then 45 degrees and α_1 is equal to β_1. This two-pole maximally flat response is a special case of a more general n-pole maximally flat response for which the n poles are spaced on the circumference of a semicircle as shown in fig. 7.22.

(iv) *A pair of high-Q poles with a zero at or near the origin*

The parallel resonant circuit of fig. 7.23(*a*) has for its transfer function $\boldsymbol{V}_o/\boldsymbol{I}_i$ a pole–zero pattern which consists of a pair of complex poles and a zero at the origin. The transfer function is

$$\boldsymbol{H}(s) = \frac{\boldsymbol{V}_o(s)}{\boldsymbol{I}_i(s)} = \frac{1}{C}\cdot\frac{s}{s^2+(G/C)s+1/LC}, \qquad (7.12.14)$$

$$= \frac{1}{C}\cdot\frac{s}{s^2+(\omega_0/Q_p)s+\omega_0^2}, \qquad (7.12.15)$$

$$= \frac{1}{C}\cdot\frac{s}{(s-s_a)(s-s_b)}, \qquad (7.12.16)$$

where

$$s_a, s_b = -\frac{\omega_0}{2Q_p} \pm j\omega_0\sqrt{\left[1-\left(\frac{1}{2Q_p}\right)^2\right]}, \qquad (7.12.17)$$

$$= -\alpha_1 \pm j\beta_1,$$

and it is clear that the solution for a high-Q circuit follows that of the previous example. In the neighbourhood of the frequency ω_0 the lengths r_1 and r_b of the vectors from s_1 and s_b to the point s are almost constant at about ω_0 and $2\omega_0$ respectively, and the response is almost entirely dependent on the pole s_a. An enlarged view of the area near this pole is shown in fig. 7.23(*c*). Assuming that r_1 and r_b are constant, the response is a maximum when the vector $(s-s_a)$ lies parallel to the real axis and has a length α_1. The upper and lower half-power frequencies occur when the length of this vector has increased by the factor $\sqrt{2}$. As seen from the diagram these frequencies are $(\omega_0 \pm \alpha_1)$, so the angular bandwidth is $2\alpha_1$ or ω_0/Q_p. Other features of the high-Q circuit can be similarly deduced from the diagram.

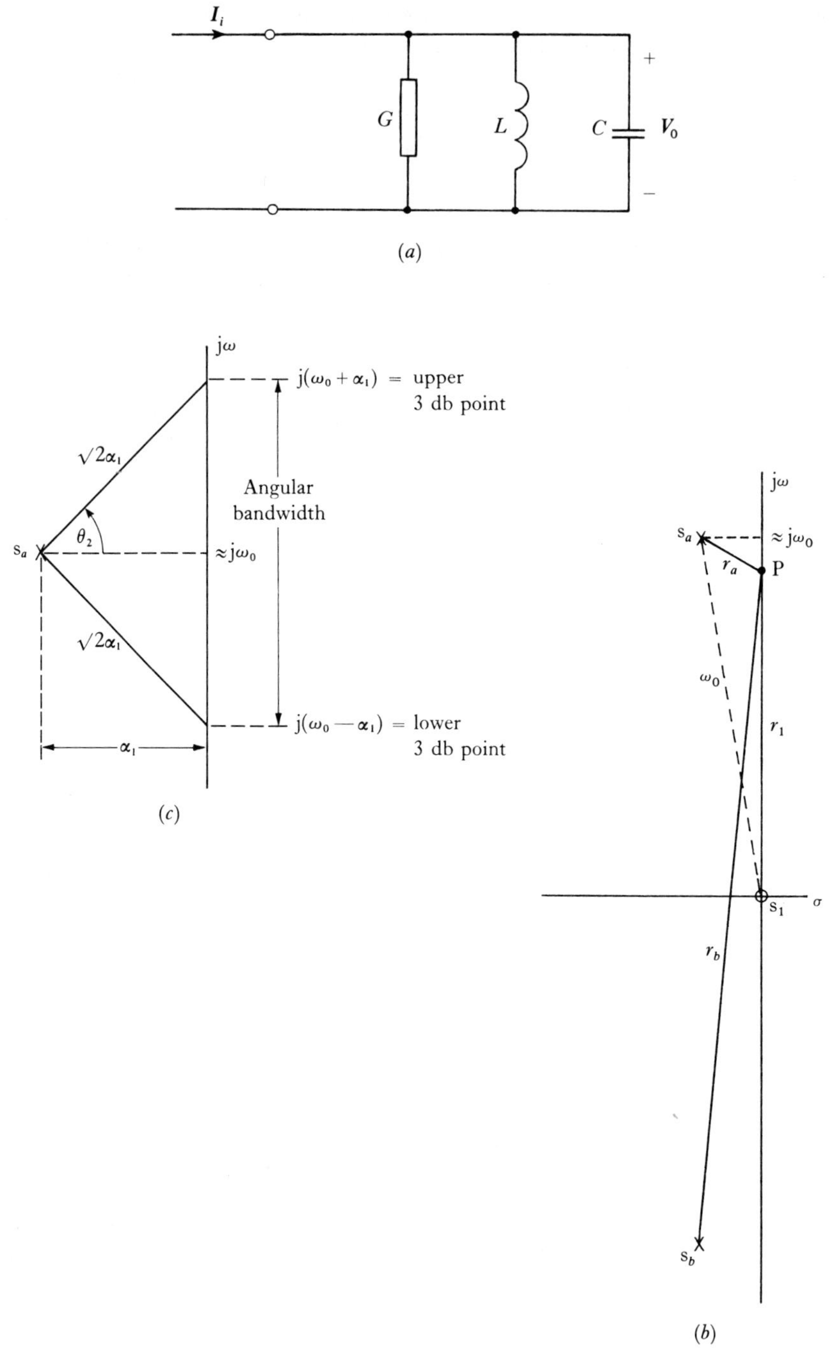

Fig. 7.23 (*a*) Parallel-resonant circuit. (*b*) Pole–zero pattern for parallel-resonant circuit. (*c*) Enlarged view of pattern near s_a.

The alternative representation of a parallel-resonant circuit (see fig. 7.24), in which the dissipative element R is placed in series with L, is better for low-Q circuits. This is because the applications of a low-Q circuit are more likely to require a wide frequency range, even including zero frequency, and the series arrangement of R and L is a much better approximation to

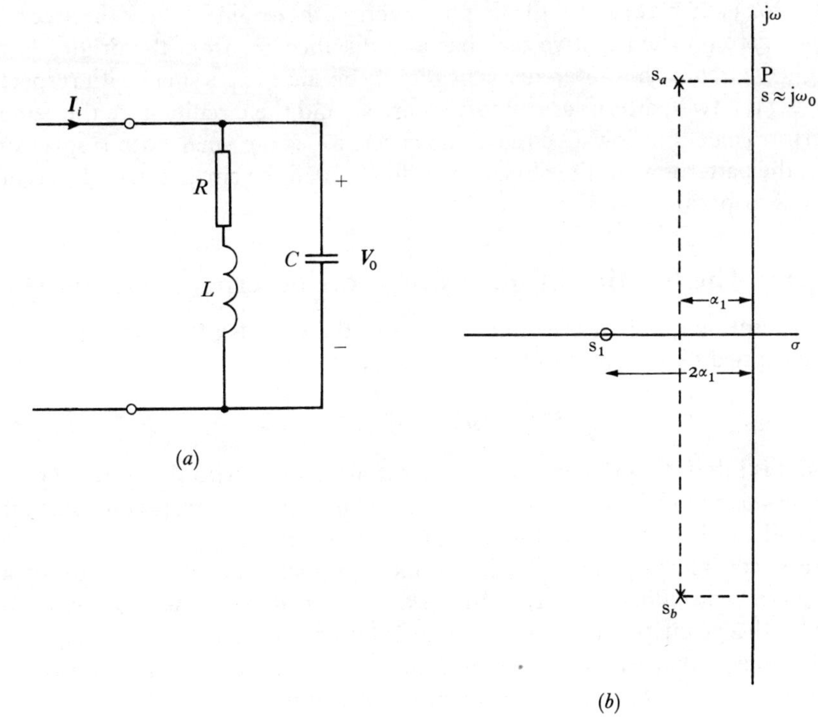

Fig. 7.24 (*a*) Alternative form of parallel-resonant circuit. (*b*) Corresponding pole–zero pattern.

reality under these conditions than is the parallel arrangement. The transfer function of this alternative representation is

$$\boldsymbol{H}(\mathrm{s}) = \frac{\boldsymbol{V}_o}{\boldsymbol{I}_i} = \frac{1}{C} \cdot \frac{\mathrm{s} + R/L}{\mathrm{s}^2 + (R/L)\mathrm{s} + 1/LC}, \tag{7.12.18}$$

$$= \frac{1}{C} \cdot \frac{\mathrm{s} + \omega_0/Q_s}{\mathrm{s}^2 + (\omega_0/Q_s)\mathrm{s} + \omega_0^2}, \tag{7.12.19}$$

$$= \frac{1}{C} \cdot \frac{(\mathrm{s} - \mathrm{s}_1)}{(\mathrm{s} - \mathrm{s}_a)(\mathrm{s} - \mathrm{s}_b)}, \tag{7.12.20}$$

where $s_1 = -\omega_0/Q_s$ and

$$\left. s_a, s_b = -\frac{\omega_0}{2Q_s} \pm j\omega_0 \sqrt{\left[1 - \left(\frac{1}{2Q_s}\right)^2\right]}. \right\} \tag{7.12.21}$$

The difference between the pole–zero patterns of the two representations (figs. 7.23 and 7.24) is that the first has a zero at the origin whereas the second has a zero on the negative real axis at a distance $2\alpha_1$ from the origin. For high-Q circuits this difference is negligible because $2\alpha_1$ is small with respect to ω_0; the two patterns are almost identical, and the circuits have the same performance. For low-Q circuits, however, $2\alpha_1$ is not small with respect to ω_0; the patterns are quite obviously different, and the circuits have different performances.

7.13 The location of quadratic roots in terms of ω_0 and Q

The location of a pair of complex poles in the complex frequency plane is determined by the roots of the quadratic equation

$$s^2 + (\omega_0/Q)s + \omega_0^2 = 0, \tag{7.13.1}$$

and it is interesting to know how the location changes with the system parameters ω_0 and Q. The roots of (7.13.1) are shown in fig. 7.25, (*a*) for constant ω_0 and varying Q, and (*b*) for constant Q and varying ω_0.

For constant ω_0 the roots, if complex, are constrained to the arc of a semicircle of radius ω_0; if real, to the negative real axis in such a way as to make the geometric mean of their magnitudes equal to the constant ω_0. This means that as one pole moves inwards (with decreasing Q) towards the origin, the other moves outwards towards infinity.

For constant Q the roots, if complex, move on a pair of equiangular radial lines from the origin. If the Q is equal to or less than 0.5, the radials become coincident on the negative real axis. The roots are then coincident for Q equal to 0.5 and separate for Q less than 0.5. Both poles move along the negative real axis towards the origin as ω_0 decreases. This movement is linear with ω_0 so that, for example, a halving of ω_0 halves the distance of each pole from the origin.

7.14 The synthesis of pole–zero patterns from elementary non-interacting circuit blocks

When two cascaded networks are isolated from one another by buffers such as ideal controlled sources so that the second does not load the first and the first acts as an ideal source for the second, the overall transfer function of the cascade is simply the product of the transfer functions of the separate parts.

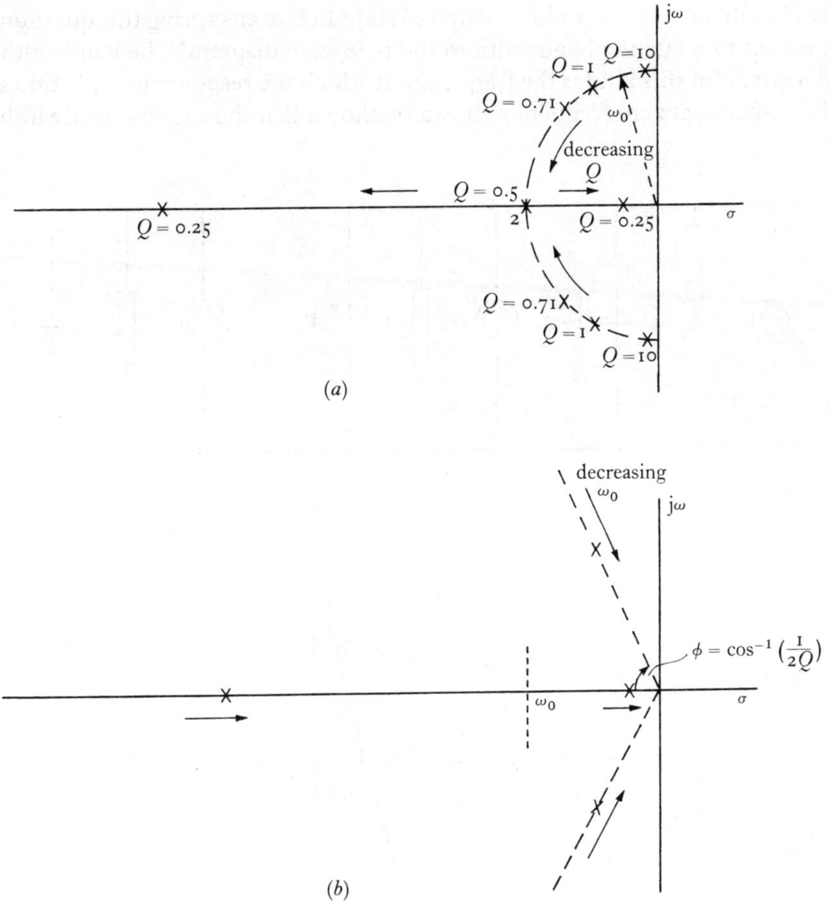

Fig. 7.25 Locus of a pair of roots: (*a*) constant ω_0, decreasing Q; (*b*) constant Q, decreasing ω_0.

As a consequence, the pole–zero pattern of a non-interacting cascade is simply the superposition of the pole–zero patterns of the separate parts, and it is a relatively simple matter to shape the overall pattern by shaping the parts separately. For example, the three-pole maximally flat array of fig. 7.26(*b*) could be synthesized by the three circuit blocks shown in fig. 7.26(*a*). The first of these, a current-fed parallel-tuned circuit with a Q-factor of unity, would produce the poles at A and C, together with an unwanted zero at B. The second and third blocks would each be proportioned to produce a single real pole at B. One of these two poles would cancel with the unwanted zero; the other would supply the requirement of the array.

Suppose that C_1, C_2, and C_3 are all equal to 10 pF. What values should be given to R_1, L_1, G_1 and G_2 to produce a maximally flat response with a bandwidth of 50×10^6 rad/s (about 8 MHz)? Before answering this question it is best to relate the bandwidth to the pole–zero diagram. The bandwidth of a circuit of this kind is the frequency at which the response is $1/\sqrt{2}$ times the response at zero frequency. It can be shown that the angular bandwidth

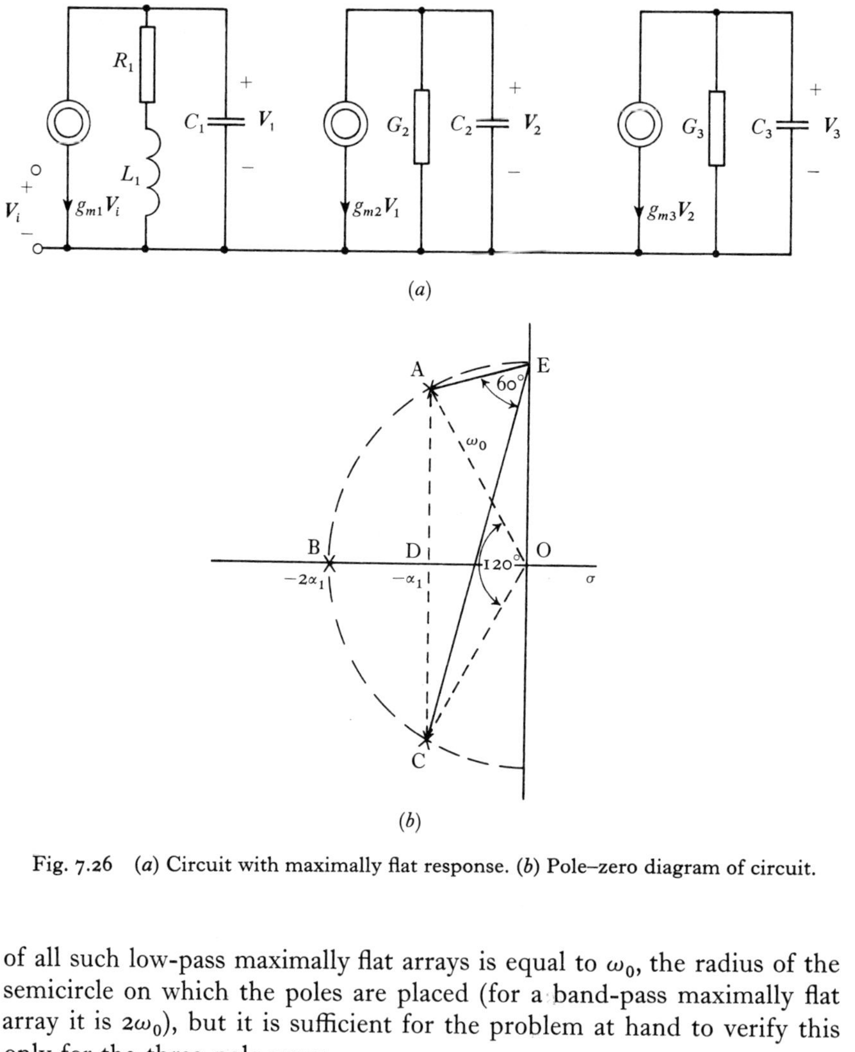

Fig. 7.26 (*a*) Circuit with maximally flat response. (*b*) Pole–zero diagram of circuit.

of all such low-pass maximally flat arrays is equal to ω_0, the radius of the semicircle on which the poles are placed (for a band-pass maximally flat array it is $2\omega_0$), but it is sufficient for the problem at hand to verify this only for the three-pole array.

The magnitude of the response at the point O (zero frequency) is given, apart from the scale factor, by the reciprocal of the product of the lengths AO, BO, and CO. Neglecting BO for the moment, we shall show that the product of AO and CO is equal to the product of AE and CE. This is true because the areas of triangles AOC and AEC are equal; they have the same base AC and the same height DO. Thus

$$\tfrac{1}{2}(\mathrm{AE})(\mathrm{CE})\sin 60^\circ = \tfrac{1}{2}(\mathrm{AO})(\mathrm{CO})\sin 120^\circ,$$

and
$$(\mathrm{AE})(\mathrm{CE}) = (\mathrm{AO})(\mathrm{CO}).$$

But
$$(\mathrm{BE}) = \sqrt{2}(\mathrm{BO}),$$

so
$$(\mathrm{AE})(\mathrm{CE})(\mathrm{BE}) = \sqrt{2}(\mathrm{AO})(\mathrm{CO})(\mathrm{BO}),$$

and the response at point E is $1/\sqrt{2}$ times the response at point O. The bandwidth is thus equal to the pole radius ω_0.

The overall transfer function of the circuit is

$$\boldsymbol{H}(\mathrm{j}\omega) = \frac{\boldsymbol{V}_3}{\boldsymbol{V}_i} = \frac{\boldsymbol{V}_1}{\boldsymbol{V}_i}\cdot\frac{\boldsymbol{V}_2}{\boldsymbol{V}_1}\cdot\frac{\boldsymbol{V}_3}{\boldsymbol{V}_2} = \boldsymbol{H}_1(\mathrm{j}\omega)\,\boldsymbol{H}_2(\mathrm{j}\omega)\,\boldsymbol{H}_3(\mathrm{j}\omega), \qquad (7.14.1)$$

where

$$\boldsymbol{H}_1(\mathrm{j}\omega) = -\frac{g_{m1}}{C_1}\cdot\left[\frac{\mathrm{s}+R_1/L_1}{\mathrm{s}^2+(R_1/L_1)\mathrm{s}+1/L_1C_1}\right] = -\frac{g_{m1}}{C_1}\frac{(\mathrm{s}-\mathrm{s}_1)}{(\mathrm{s}-\mathrm{s}_a)(\mathrm{s}-\mathrm{s}_b)}, \qquad (7.14.2)$$

and
$$\boldsymbol{H}_2(\mathrm{j}\omega) = \boldsymbol{H}_3(\mathrm{j}\omega) = -\frac{g_{m2}}{C_2}\left[\frac{1}{\mathrm{s}+G_2/C_2}\right] = -\frac{g_{m2}}{C_2}\frac{1}{(\mathrm{s}-\mathrm{s}_c)}, \qquad (7.14.3)$$

and the design of the three stages follows from these equations.

Design of stage 2

G_2 is chosen so as to produce a pole (s_c) at the point B($-\omega_0$).

Thus $\quad \mathrm{s}_c = -\omega_0 = -G_2/C_2$

and $\quad G_2 = \omega_0 C_2 = (50\times 10^6)(10\times 10^{-12})$ S

which gives $\quad R_2 = 1/G_2 = 2\,\mathrm{k}\Omega.$

Design of stage 3

G_3 is chosen so as to produce a pole at the same point B, and since C_3 equals C_2, R_3 equals R_2.

Design of stage 1

The inductance L_1 is chosen so as to place the poles s_a and s_b at the correct radius ω_0. Thus

$$L_1 = \frac{1}{\omega_0^2 C_1} = \frac{1}{(50\times 10^6)^2(10\times 10^{-12})}\,\mathrm{H} = 40\,\mu\mathrm{H}.$$

The resistance R_1 is then chosen to give the poles a Q of unity ($1/2\cos 60°$) which will place them at the points A and C. Alternatively R_1 can be chosen so as to give the poles a negative real part ($R_1/2L_1$) equal to half the angular bandwidth ω_0. Using either requirement,

$$R_1 = \omega_0 L_1 = (50 \times 10^6)(40 \times 10^{-6}) = 2\,\text{k}\Omega.$$

The zero frequency value of the transfer function is

$$\boldsymbol{H}(0) = -(g_{m1} R_1)(g_{m2} R_2)(g_{m3} R_3).$$

7.15 Low-pass to band-pass conversions

If a pole–zero pattern is shifted bodily up the jω-axis by an amount jω_c, the response is shifted bodily along the frequency axis by the amount ω_c as shown in fig. 7.27. The upper half of the new response has the same shape as the old response, and in addition there appears a lower half, symmetrical with the upper half about the point ω_c. The new response is a *band-pass* version of the original *low-pass* response; the symmetry about ω_c is a consequence of the symmetry of the original pole–zero pattern about the real axis. A glance at the two pole–zero patterns of fig. 7.27 will confirm that the lengths r_a, r_b and the angles θ_a, θ_b correspond on the two diagrams at any frequency ω_1 on the original and $(\omega_c + \omega_1)$ on the new, and that there is symmetry on the jω-axis about the point jω_c.

The realization of a shifted pole–zero pattern in order to obtain the band-pass equivalent of a low-pass response is, however, impossible because complex poles can be created only in conjugate pairs. Thus, in the creation of s_a and s_b, for example, in fig. 7.27(*b*) two other poles s_c and s_d would be simultaneously created on the other side of the real axis. In *narrow band* circuits, however, (in which the band-pass response is only a small fraction of the central frequency ω_c) this does not matter. The poles are close to the imaginary axis and their conjugates produce vectors that are almost constant over the frequency band of interest; the overall response is almost entirely that of the upper poles alone.

Consider for example the stagger-tuned circuit of fig. 7.10(*a*). The first parallel-resonant circuit, assuming a high Q-factor, which is the requirement for a narrow-band response, produces the complex poles s_a and s_b in fig. 7.28(*a*) together with one zero at the origin. The second parallel-resonant circuit produces the poles s_c and s_d together with the other zero at the origin. Near the frequency ω_c the vector from each of the poles s_b and s_d is almost constant at $2\omega_c \angle 90°$ and the vector from each of the zeros is almost constant at $\omega_c \angle 90°$. The response is thus effectively that of the poles s_a and s_c together with a factor $\frac{1}{4}$ to account for the other poles and zeros. It follows that in narrow-band situations the low-pass responses shown in fig. 7.21,

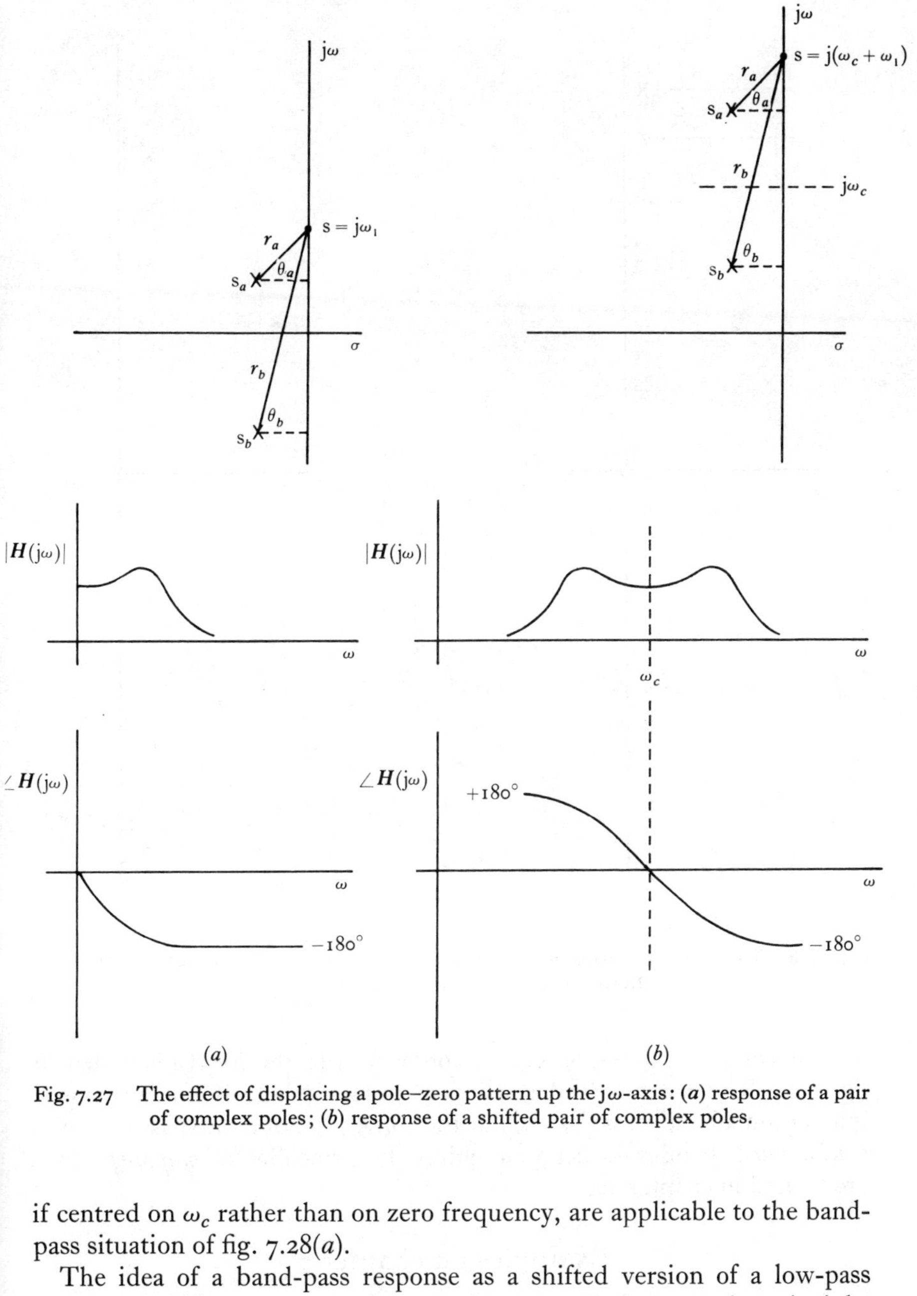

Fig. 7.27 The effect of displacing a pole–zero pattern up the $j\omega$-axis: (*a*) response of a pair of complex poles; (*b*) response of a shifted pair of complex poles.

if centred on ω_c rather than on zero frequency, are applicable to the band-pass situation of fig. 7.28(*a*).

The idea of a band-pass response as a shifted version of a low-pass response enables us to see that any low-pass design can, in principle, be translated into a narrow-band, band-pass design. For example, it becomes clear that three stages of stagger-tuned circuits would have to be arranged with their poles on the circumferences of semicircles (fig. 7.28(*b*))

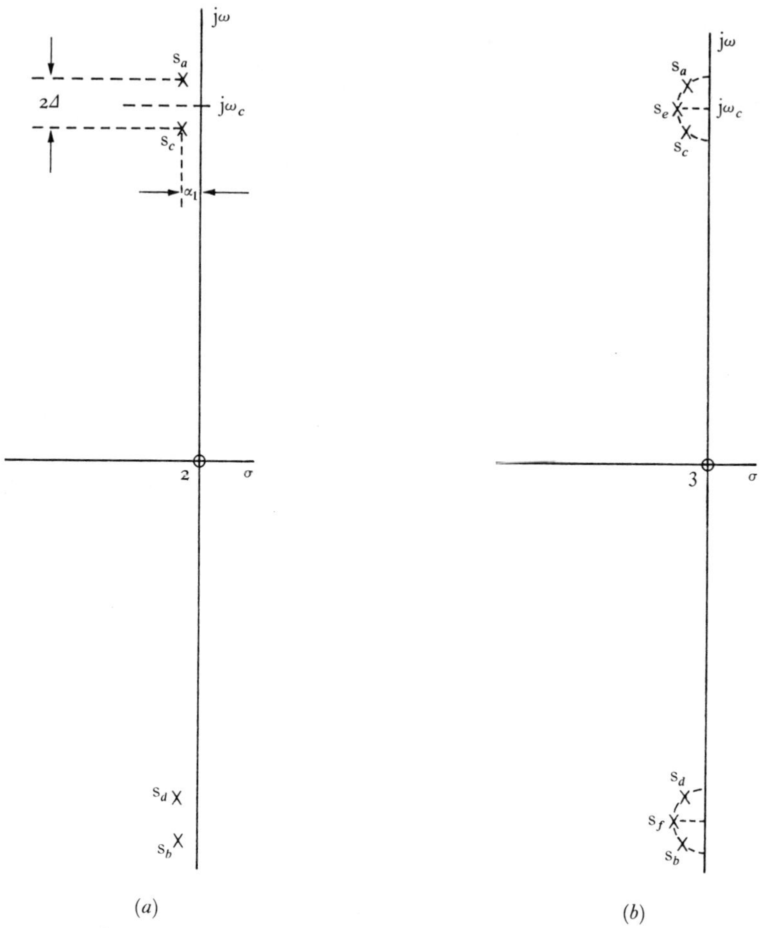

Fig. 7.28 Two pole–zero patterns for narrow-band, band-pass responses: (*a*) two-pole pattern; (*b*) three-pole maximally flat pattern.

for a maximally flat response, in accordance with the ideas illustrated in fig. 7.22. The narrow-band restriction discussed here is not, of course, inherent in a band-pass design, but the conversion of a low-pass design to a *wide-band*, band-pass design requires the principles of *mapping* which are treated in chapter 12.

Examples on chapter 7

7.1 A voltage $\boldsymbol{V}_1$ is applied to a network which consists of two resistors R_1 and R_2 in series, shunted respectively by capacitors C_1 and C_2. Obtain an expression for the voltage $\boldsymbol{V}_2$ which appears across R_2, and make a Bode plot of the magnitude and phase of $(\boldsymbol{V}_2/\boldsymbol{V}_1)$ for the case $R_2 = 10R_1$,

$C_2 = 10C_1$. Estimate the magnitude and phase of (V_2/V_1) at the angular frequency $\omega_1 = (1/C_1 R_1)$.

7.2 When a coil which has a Q-factor of 50 is shunted by a 400 pF capacitor, the resultant circuit resonates at 100 kHz. Calculate the admittance of the circuit, in the form $(G + jB)$, at the frequency of (*a*) 99 kHz, (*b*) 100 kHz, and (*c*) 101 kHz, and show by means of a sketch how B and $|\mathbf{Y}|$ vary with frequency in the neighbourhood of resonance.

7.3 An inductor L in series with a resistor R_1 is shunted by a capacitor C in series with a resistor R_2. Find the resonant frequency of the circuit and show that the input admittance at resonance has the value

$$\frac{R_1 + R_2}{R_1 R_2 + (L/C)}.$$

Assuming that $R = R_1 = R_2 \ll \sqrt{(L/C)}$, find the frequency deviation $\delta\omega_0$ from resonance at which the imaginary part of the admittance is equal to j times the admittance at resonance, and thereby show that the Q-factor of the circuit is approximately $\omega_0 L/2R$. Confirm this by adding the reciprocal Q-factors of the two arms of the circuit.

7.4 When the variable tuning capacitance C of a series resonant circuit has the value C_0 the current through the circuit is a maximum, and when it has the values C_1 (less than C_0) and C_2 (greater than C_0) the current has a value equal to $(1/\sqrt{2})$ of the maximum. Show that the Q-factor of the circuit is given by $Q = (C_2 + C_1)/(C_2 - C_1)$.

7.5 A high-Q coil, which may be represented by an ideal inductor L in series with a constant resistor R, is resonated in parallel with a lossless capacitor C. Find the ratio of the current in the coil to the supply current at the resonant frequency. If $L = 30\ \mu$H, and $Q = 100$ at the resonant frequency of 15 MHz, what power is dissipated when the circuit is driven by a 10 V r.m.s. source at the resonant frequency?

7.6 The input admittance of an electronic voltmeter is to be derived from the following measurements. A test coil, connected in series with a lossless variable tuning capacitor C, resonates at 1 MHz and has a Q-factor of 220 when C has the value of 130 pF. When the voltmeter is connected, in addition, across the capacitor, now set to 125 pF, the circuit resonates at 1 MHz and has a Q-factor of 180. What are the input capacitance and parallel input resistance of the voltmeter?

7.7 By comparing the transfer function ($\boldsymbol{V}_o/\boldsymbol{I}_i$) of the circuit shown in fig. 7.23(*a*) (using the subscript *p*) with that of fig. 7.24(*a*) (using the subscript *s*), show that the pole locations of the two circuits are the same and that the transfer functions have the same constant multiplier if $C_p = C_s$, $L_p = L_s$, and $(G_p/C_p) = (R_s/L_s)$.

7.8 Two identical parallel-tuned circuits, in which the coils have inductances of 2 mH and Q-factors of 100 each at a frequency of 120 kHz, are connected together at one end; the other ends are connected via a small coupling capacitance C_0. What values of tuning capacitance in the parallel-tuned circuits and of coupling capacitance C_0 are required to produce an open-circuit voltage response across one parallel-tuned circuit like that of the flat staggered pair (fig. 7.11(3)), when the terminals of the other are connected to an ideal current source whose frequency is centred on 120 kHz? What is the 3 db bandwidth of this circuit?

7.9 An independent current source $\boldsymbol{I}$ feeds a resistor R_1 across which is connected a capacitor C_3 in series with a second resistor R_2. Across R_1 and R_2 respectively are connected capacitors C_1 and C_2. Find (*a*) the locations of the poles and zeros of the transfer function ($\boldsymbol{V}_2/\boldsymbol{I}$) (where $\boldsymbol{V}_2$ is the voltage across R_2) for the condition $C_1 = C_2 = 0$; (*b*) the 3 db point of the circuit when $R_1 = 1$ kΩ, $R_2 = 2$ kΩ, $C_1 = C_2 = 0$, and $C_3 = 10$ μF; (*c*) the locations of the poles and zeros when C_1 and C_2 are non-zero; (*d*) the upper and lower 3 db points when $R_1 = 1$ kΩ, $R_2 = 2$ kΩ, $C_1 = 0.05$ μF, $C_2 = 0.1$ μF, and $C_3 = 10$ μF.

7.10 A source which has an open-circuit voltage $\boldsymbol{V}_1$ and an internal resistance R_1 is matched to a load R_2 by means of an inductor L and a capacitor C which are placed in series with the source. The load R_2 $(=1/G_2)$, which is connected across the capacitor, develops an output voltage $\boldsymbol{V}_2$. Show that the transfer function ($\boldsymbol{V}_2/\boldsymbol{V}_1$) has two poles

$$\mathrm{s}_a,\ \mathrm{s}_b = -\alpha \pm \mathrm{j}\beta,$$

given by the roots of the equation

$$\mathrm{s}^2 + \left(\frac{G_2}{C} + \frac{R_1}{L}\right)\mathrm{s} + \frac{(R_1 G_2 + 1)}{LC} = 0,$$

and show that if the response has a peak it occurs at the angular frequency $\omega_m = \sqrt{(\beta^2 - \alpha^2)}$ at which a circle, drawn on the pole–zero diagram with the line joining the two poles as a diameter, cuts the jω axis.

Given that $R_1 = 1$ kΩ, and $G_2 = 0.25$ mS, find the values of L and C which will: (*a*) transfer maximum power from the source to the load at an angular frequency of 10^4 rad/s; (*b*) give to the transfer function a maximally flat response with a 3 db point at the angular frequency of 10^4 rad/s.

7.11 A low-pass filter which consists of a capacitor C_1 across which is connected an inductor L in series with a second capacitor C_2 has a load G_2 connected across C_2 and is fed across C_1 from a current source $\boldsymbol{I}_1$ of internal conductance G_1. By observing that the transfer function $(\boldsymbol{V}_2/\boldsymbol{I}_1)$, (where $\boldsymbol{V}_2$ is the voltage across G_2), is maximally flat when its three poles lie on the circumference of a semicircle whose diameter lies on the imaginary axis, determine the values of L and C required to produce a maximally flat response with a half-power angular frequency of 10^5 rad/s when $C_1 = C_2 = C$ and $G_1 = G_2 = 1$ mS.

7.12 An ideal current source of value $g_m \boldsymbol{V}_s$ feeds a parallel resonant circuit G_1, L_1, C_1. The voltage $\boldsymbol{V}_1$ developed across this circuit controls a second ideal current source which has a value $g_m \boldsymbol{V}_1$ and which feeds a second parallel resonant circuit G_2, L_2, C_2. The voltage across G_2 is required to have the response of a flat-staggered tuned amplifier which has a bandwidth of 1 MHz centred on 10 MHz. Find the required values of L_1, L_2, C_1, and C_2 when $G_1 = 50\ \mu$S and $G_2 = 100\ \mu$S.

8 Twoport networks

8.1. Twoport networks

In many electrical networks interest exists not so much in the network itself as in the performance as measured at one or more pairs of terminals. These pairs of terminals, to which stimuli are applied and at which responses are measured, are called *ports*, and a network like an amplifier, for example, which accepts a stimulus at one pair of terminals and produces a response at another pair is called a *twoport network* or, simply, a *twoport*. If a network has three pairs of terminals, as in a differential amplifier where stimuli are applied separately to each of two pairs of terminals and where a response is

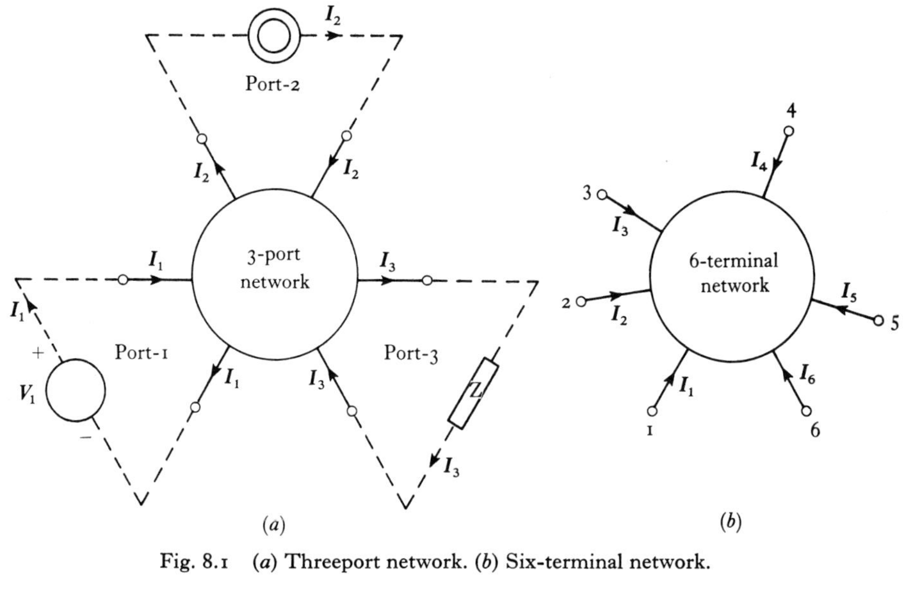

Fig. 8.1 (*a*) Threeport network. (*b*) Six-terminal network.

obtained at a third pair, the network is called a *threeport* (see fig. 8.1(*a*)); and in general a network with *n* such pairs of terminals is called an *n-port*.

A pair of terminals may legitimately be called a port only when the current entering one terminal of the pair is equal to the current leaving the other. Whether or not a pair of terminals forms a port is therefore dependent not on the network itself but on the constraints imposed by the external circuit; that is, on the method of supplying and loading the terminals. Fig. 8.1(*b*) shows a six-terminal network which may be identical with the threeport shown in fig. 8.1(*a*) but which cannot be so described unless it be known that each pair of terminals satisfies the given condition.

The reason for describing a network in terms of its ports (where this is possible) rather than its individual terminals is that fewer equations are necessary. A $2n$-terminal network normally requires $(2n-1)$ equations for its description whereas the corresponding n-port requires only n equations. In addition, each equation contains only n independent variables instead of $(2n-1)$. The importance of twoports in network theory is that these units form building blocks from which many practical systems are constructed, and a familiarity with the ways in which they may be described is of help in the synthesis or analysis of the larger system of which they form a part. This is illustrated in §12.8 of chapter 12.

In addition to a study of twoports this chapter also introduces the notation of matrix algebra which, by its compactness, is helpful in dealing with multiport and n-terminal networks and which is used in many branches of electrical engineering.

8.2 The impedance equations of a linear twoport

The twoport shown in fig. 8.2, which is assumed to have no internal independent sources, is described by the four variables $\boldsymbol{V}_1$, $\boldsymbol{V}_2$, $\boldsymbol{I}_1$, and $\boldsymbol{I}_2$, only

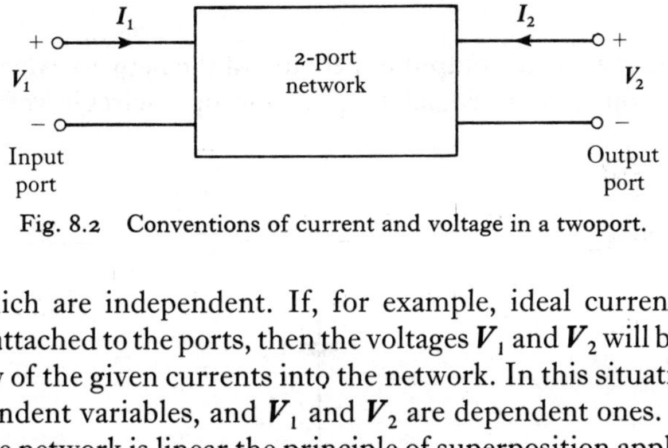

Fig. 8.2 Conventions of current and voltage in a twoport.

two of which are independent. If, for example, ideal current sources $\boldsymbol{I}_1$ and $\boldsymbol{I}_2$ are attached to the ports, then the voltages $\boldsymbol{V}_1$ and $\boldsymbol{V}_2$ will be dependent on the flow of the given currents into the network. In this situation $\boldsymbol{I}_1$ and $\boldsymbol{I}_2$ are independent variables, and $\boldsymbol{V}_1$ and $\boldsymbol{V}_2$ are dependent ones.

Since the network is linear the principle of superposition applies and each port voltage is the sum of two components, one arising from $\boldsymbol{I}_1$, the other from $\boldsymbol{I}_2$. In other words the voltages are given by equations of the form

$$\left.\begin{aligned}\boldsymbol{V}_1 &= \boldsymbol{z}_{11}\boldsymbol{I}_1 + \boldsymbol{z}_{12}\boldsymbol{I}_2,\\ \boldsymbol{V}_2 &= \boldsymbol{z}_{21}\boldsymbol{I}_1 + \boldsymbol{z}_{22}\boldsymbol{I}_2,\end{aligned}\right\} \tag{8.2.1}$$

where the $\boldsymbol{z}$ are coefficients with the dimensions of impedance. Their values depend on the internal network, and they are called the *impedance parameters* of the twoport. The physical significance of the impedance parameters may be seen by putting one or other of the two ports on open circuit. If the output port is open-circuited, $\boldsymbol{I}_2$ is zero, and (8.2.1) shows that

$$z_{11} = \left.\frac{\boldsymbol{V}_1}{\boldsymbol{I}_1}\right|_{I_2=0}, \tag{8.2.2}$$

and

$$z_{21} = \left.\frac{\boldsymbol{V}_2}{\boldsymbol{I}_1}\right|_{I_2=0}. \tag{8.2.3}$$

Thus z_{11} is the input impedance of the network when its output terminals are open-circuited, and z_{21} is the open-circuit output voltage resulting from a flow of unit current at the input.

Similarly if the input port is open-circuited, $\boldsymbol{I}_1$ is zero, and (8.2.1) become

$$z_{22} = \left.\frac{\boldsymbol{V}_2}{\boldsymbol{I}_2}\right|_{I_1=0}, \tag{8.2.4}$$

and

$$z_{12} = \left.\frac{\boldsymbol{V}_1}{\boldsymbol{I}_2}\right|_{I_1=0}, \tag{8.2.5}$$

which show that z_{22} is the output impedance of the network when the input terminals are open-circuited, and z_{12} is the open-circuit voltage which

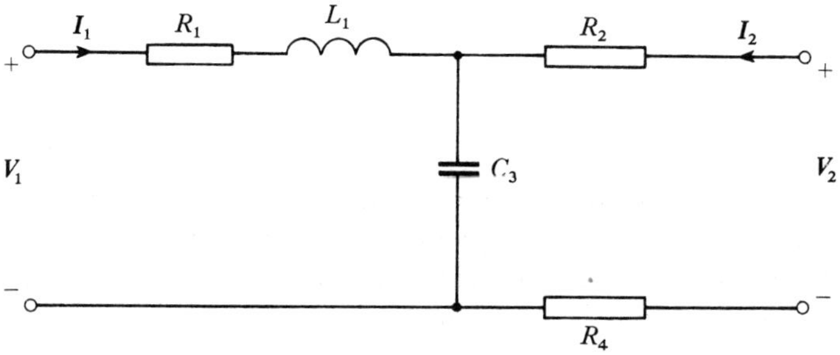

Fig. 8.3 An example of a twoport.

appears at the input terminals when unit current is applied to the output terminals.

Equations (8.2.2)–(8.2.5) may be used to calculate the impedance parameters of any twoport whose internal circuit is given, or they may be used to

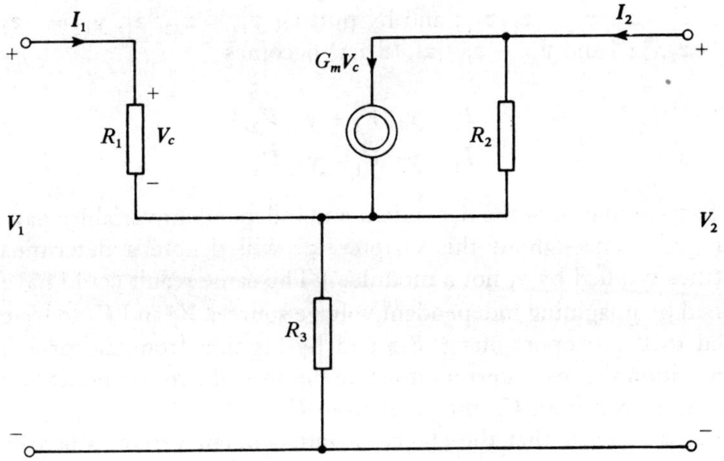

Fig. 8.4 A second example of a twoport.

obtain the parameters of an unknown network from measurements made at its open-circuited ports. The former usage, applied to fig. 8.3, gives:

$$\begin{aligned} z_{11} &= R_1 + sL_1 + 1/sC_3, \\ z_{12} &= 1/sC_3, \\ z_{21} &= 1/sC_3, \\ z_{22} &= R_2 + R_4 + 1/sC_3, \end{aligned}$$

and to fig. 8.4,

$$\begin{aligned} z_{11} &= R_1 + R_3, \\ z_{12} &= R_3, \\ z_{21} &= R_3 - G_m R_1 R_2, \\ z_{22} &= R_2 + R_3. \end{aligned}$$

Observe that in the reciprocal network of fig. 8.3, $z_{12} = z_{21}$, whereas in the non-reciprocal one of fig. 8.4 this equality does not exist.

8.3 Choice of twoport parameters

If equations (8.2.1) are solved for I_1 and I_2 in terms of V_1 and V_2, one gets

$$\left.\begin{aligned} I_1 &= \frac{z_{22}}{|z|} V_1 - \frac{z_{12}}{|z|} V_2, \\ I_2 &= -\frac{z_{21}}{|z|} V_1 + \frac{z_{11}}{|z|} V_2, \end{aligned}\right\} \tag{8.3.1}$$

where $|\boldsymbol{z}| = \boldsymbol{z}_{11}\boldsymbol{z}_{22} - \boldsymbol{z}_{12}\boldsymbol{z}_{21}$; and by putting $\boldsymbol{y}_{11} = \boldsymbol{z}_{22}/|\boldsymbol{z}|$, $\boldsymbol{y}_{12} = -\boldsymbol{z}_{12}/|\boldsymbol{z}|$, $\boldsymbol{y}_{21} = -\boldsymbol{z}_{21}/|\boldsymbol{z}|$, and $\boldsymbol{y}_{22} = \boldsymbol{z}_{11}/|\boldsymbol{z}|$, (8.3.1) becomes

$$\left.\begin{aligned} \boldsymbol{I}_1 &= \boldsymbol{y}_{11}\boldsymbol{V}_1 + \boldsymbol{y}_{12}\boldsymbol{V}_2, \\ \boldsymbol{I}_2 &= \boldsymbol{y}_{21}\boldsymbol{V}_1 + \boldsymbol{y}_{22}\boldsymbol{V}_2. \end{aligned}\right\} \tag{8.3.2}$$

This form implies that the dependent and independent variables have been exchanged. (Throughout this chapter $|\boldsymbol{z}|$ will denote a determinant of quantities typified by $\boldsymbol{z}$, not a modulus.) The same result could have been obtained by imagining independent voltage sources $\boldsymbol{V}_1$ and $\boldsymbol{V}_2$ to have been applied to the twoport of fig. 8.2 and by arguing from the principle of superposition that each current must be the sum of two components, one of which is derived from $\boldsymbol{V}_1$, the other from $\boldsymbol{V}_2$.

It becomes clear that the choice of independent variables is arbitrary; and it is immaterial, apart from convenience, which two of the four variables $\boldsymbol{V}_1, \boldsymbol{V}_2, \boldsymbol{I}_1$, and $\boldsymbol{I}_2$ are chosen. Now there are six ways of choosing two variables from a set of four, and each choice produces four parameters which characterize the network. The six choices are shown in table 8.1 and the physical

TABLE 8.1 *Six alternative sets of equations for a twoport*

Impedance parameters	$\boldsymbol{V}_1 = \boldsymbol{z}_{11}\boldsymbol{I}_1 + \boldsymbol{z}_{12}\boldsymbol{I}_2$ $\boldsymbol{V}_2 = \boldsymbol{z}_{21}\boldsymbol{I}_1 + \boldsymbol{z}_{22}\boldsymbol{I}_2$
Admittance parameters	$\boldsymbol{I}_1 = \boldsymbol{y}_{11}\boldsymbol{V}_1 + \boldsymbol{y}_{12}\boldsymbol{V}_2$ $\boldsymbol{I}_2 = \boldsymbol{y}_{21}\boldsymbol{V}_1 + \boldsymbol{y}_{22}\boldsymbol{V}_2$
Hybrid parameters	$\boldsymbol{V}_1 = \boldsymbol{h}_{11}\boldsymbol{I}_1 + \boldsymbol{h}_{12}\boldsymbol{V}_2$ $\boldsymbol{I}_2 = \boldsymbol{h}_{21}\boldsymbol{I}_1 + \boldsymbol{h}_{22}\boldsymbol{V}_2$
Inverse hybrid parameters	$\boldsymbol{I}_1 = \boldsymbol{g}_{11}\boldsymbol{V}_1 + \boldsymbol{g}_{12}\boldsymbol{I}_2$ $\boldsymbol{V}_2 = \boldsymbol{g}_{21}\boldsymbol{V}_1 + \boldsymbol{g}_{22}\boldsymbol{I}_2$
Transmission parameters	$\boldsymbol{V}_1 = \boldsymbol{a}_{11}\boldsymbol{V}_2 + \boldsymbol{a}_{12}\boldsymbol{I}_2$ $\boldsymbol{I}_1 = \boldsymbol{a}_{21}\boldsymbol{V}_2 + \boldsymbol{a}_{22}\boldsymbol{I}_2$
Inverse transmission parameters	$\boldsymbol{V}_2 = \boldsymbol{b}_{11}\boldsymbol{V}_1 + \boldsymbol{b}_{12}\boldsymbol{I}_1$ $\boldsymbol{I}_2 = \boldsymbol{b}_{21}\boldsymbol{V}_1 + \boldsymbol{b}_{22}\boldsymbol{I}_1$

significance of any one of the resulting parameters may be found by the method of §8.2.

For example,

$$\frac{1}{\boldsymbol{a}_{11}} = \left(\frac{\boldsymbol{V}_2}{\boldsymbol{V}_1}\right)_{I_2=0} = \text{open-circuit output voltage per unit input voltage,}$$

$$\frac{1}{\boldsymbol{a}_{12}} = \left(\frac{\boldsymbol{I}_2}{\boldsymbol{V}_1}\right)_{V_2=0} = \text{short-circuit output current per unit input voltage,}$$

$$\frac{1}{\boldsymbol{a}_{21}} = \left(\frac{\boldsymbol{V}_2}{\boldsymbol{I}_1}\right)_{I_2=0} = \text{open-circuit output voltage per unit input current,}$$

$$\frac{1}{\boldsymbol{a}_{22}} = \left(\frac{\boldsymbol{I}_2}{\boldsymbol{I}_1}\right)_{V_2=0} = \text{short-circuit output current per unit input current.}$$

Before going any further it is advisable to make four comments. The first is simply a restatement of the assumption implicit in (8.2.1) that the network contains no independent energy sources. If this were not so, the voltages $\boldsymbol{V}_1$ and $\boldsymbol{V}_2$ could be non-zero even when $\boldsymbol{I}_1$ and $\boldsymbol{I}_2$ are zero. The same assumption applies to all the sets of equations shown in table 8.1. The second comment refers to the assigned directions of current and voltage in fig. 8.2. In some applications it is convenient to change the assigned direction of the current $\boldsymbol{I}_2$. This is the invariable practice in the treatment of power circuits and has the effect of changing the sign of those coefficients which are dependent on $\boldsymbol{I}_2$. If the assigned direction of $\boldsymbol{I}_2$ is the opposite of that shown in fig. 8.2, and if the transmission equations in this sign convention are

$$\left.\begin{aligned} \boldsymbol{V}_1 &= \boldsymbol{A}\boldsymbol{V}_2 + \boldsymbol{B}\boldsymbol{I}_2, \\ \boldsymbol{I}_1 &= \boldsymbol{C}\boldsymbol{V}_2 + \boldsymbol{D}\boldsymbol{I}_2, \end{aligned}\right\} \tag{8.3.3}$$

then the relationship between these coefficients and the ones in table 8.1 is $\boldsymbol{A} = \boldsymbol{a}_{11}$, $\boldsymbol{B} = -\boldsymbol{a}_{12}$, $\boldsymbol{C} = \boldsymbol{a}_{21}$, and $\boldsymbol{D} = -\boldsymbol{a}_{22}$. The notation of (8.3.3) is standard in the power circuit context, and the coefficients are called the ***ABCD** parameters*. It is essential to know the sign convention adopted; throughout this chapter it is that of fig. 8.2. The third comment relates to an alternative convention for the suffixes attached to the parameters of the first four pairs of equations in table 8.1. The double suffixes 11, 12, 21, and 22 of the $\boldsymbol{z}$, $\boldsymbol{y}$, $\boldsymbol{h}$, and $\boldsymbol{g}$ parameters may usefully be replaced by the single suffixes i, r, f, and o respectively, which stand for the words *input*, *reverse*, *forward*, and *output*. This is commonly done when additional suffixes have to be used to distinguish the parameters of one network from those of another.

The fourth and final comment is more fundamental. Although the equations shown in table 8.1 are derived from all possible combinations of the four variables taken two at a time, they are not exhaustive because it is possible to set up new variables which are linear combinations of the old. Such new variables may be related to one another by linear equations whose coefficients form a new parameter set. One combination, which is chosen to utilize concepts taken from the theory of distributed circuits, and which produces coefficients known as the *scattering parameters* of the network, is useful not only in its applications to transmission lines and waveguides, but also to some lumped networks (such as ideal multiport transformers) for which some of the other parameter sets may not exist. The scattering parameters are, however, not discussed further in this book.

Even if we restrict our network description to the six pairs of equations shown in table 8.1, it is natural to ask why we need so many different ways of presenting the same information, for any one pair of equations contains the same information as any other. The answer is twofold. First, some parameter sets are more easily measured than others when their numerical values lie in a given range. Secondly, a particular problem may be solved

TABLE 8.2 *Conversion formulae for twoport parameter sets*

To \ From	[z]		[y]		[h]		[g]		[a]		[b]	
[z]	z_{11}	z_{12}	$\frac{y_{22}}{\lvert y\rvert}$	$-\frac{y_{12}}{\lvert y\rvert}$	$\frac{\lvert h\rvert}{h_{22}}$	$\frac{h_{12}}{h_{22}}$	$\frac{1}{g_{11}}$	$-\frac{g_{12}}{g_{11}}$	$\frac{a_{11}}{a_{21}}$	$-\frac{\lvert a\rvert}{a_{21}}$	$-\frac{b_{22}}{b_{21}}$	$\frac{1}{b_{21}}$
	z_{21}	z_{22}	$-\frac{y_{21}}{\lvert y\rvert}$	$\frac{y_{11}}{\lvert y\rvert}$	$-\frac{h_{21}}{h_{22}}$	$\frac{1}{h_{22}}$	$\frac{g_{21}}{g_{11}}$	$\frac{\lvert g\rvert}{g_{11}}$	$\frac{1}{a_{21}}$	$-\frac{a_{22}}{a_{21}}$	$-\frac{\lvert b\rvert}{b_{21}}$	$\frac{b_{11}}{b_{21}}$
[y]	$\frac{z_{22}}{\lvert z\rvert}$	$-\frac{z_{12}}{\lvert z\rvert}$	y_{11}	y_{12}	$\frac{1}{h_{11}}$	$-\frac{h_{12}}{h_{11}}$	$\frac{\lvert g\rvert}{g_{22}}$	$\frac{g_{12}}{g_{22}}$	$\frac{a_{22}}{a_{12}}$	$-\frac{\lvert a\rvert}{a_{12}}$	$-\frac{b_{11}}{b_{12}}$	$\frac{1}{b_{12}}$
	$-\frac{z_{21}}{\lvert z\rvert}$	$\frac{z_{11}}{\lvert z\rvert}$	y_{21}	y_{22}	$\frac{h_{21}}{h_{11}}$	$\frac{\lvert h\rvert}{h_{11}}$	$-\frac{g_{21}}{g_{22}}$	$\frac{1}{g_{22}}$	$\frac{1}{a_{12}}$	$-\frac{a_{11}}{a_{12}}$	$-\frac{\lvert b\rvert}{b_{12}}$	$\frac{b_{22}}{b_{12}}$
[h]	$\frac{\lvert z\rvert}{z_{22}}$	$\frac{z_{12}}{z_{22}}$	$\frac{1}{y_{11}}$	$-\frac{y_{12}}{y_{11}}$	h_{11}	h_{12}	$\frac{g_{22}}{\lvert g\rvert}$	$-\frac{g_{12}}{\lvert g\rvert}$	$\frac{a_{12}}{a_{22}}$	$\frac{\lvert a\rvert}{a_{22}}$	$-\frac{b_{12}}{b_{11}}$	$\frac{1}{b_{11}}$
	$-\frac{z_{21}}{z_{22}}$	$\frac{1}{z_{22}}$	$\frac{y_{21}}{y_{11}}$	$\frac{\lvert y\rvert}{y_{11}}$	h_{21}	h_{22}	$-\frac{g_{21}}{\lvert g\rvert}$	$\frac{g_{11}}{\lvert g\rvert}$	$\frac{1}{a_{22}}$	$-\frac{a_{21}}{a_{22}}$	$\frac{\lvert b\rvert}{b_{11}}$	$\frac{b_{21}}{b_{11}}$
[g]	$\frac{1}{z_{11}}$	$-\frac{z_{12}}{z_{11}}$	$\frac{\lvert y\rvert}{y_{22}}$	$\frac{y_{12}}{y_{22}}$	$\frac{h_{22}}{\lvert h\rvert}$	$-\frac{h_{12}}{\lvert h\rvert}$	g_{11}	g_{12}	$\frac{a_{21}}{a_{11}}$	$\frac{\lvert a\rvert}{a_{11}}$	$-\frac{b_{21}}{b_{22}}$	$\frac{1}{b_{22}}$
	$\frac{z_{21}}{z_{11}}$	$\frac{\lvert z\rvert}{z_{11}}$	$-\frac{y_{21}}{y_{22}}$	$\frac{1}{y_{22}}$	$-\frac{h_{21}}{\lvert h\rvert}$	$\frac{h_{11}}{\lvert h\rvert}$	g_{21}	g_{22}	$\frac{1}{a_{11}}$	$-\frac{a_{12}}{a_{11}}$	$\frac{\lvert b\rvert}{b_{22}}$	$\frac{b_{12}}{b_{22}}$
[a]	$\frac{z_{11}}{z_{21}}$	$-\frac{\lvert z\rvert}{z_{21}}$	$-\frac{y_{22}}{y_{21}}$	$\frac{1}{y_{21}}$	$-\frac{\lvert h\rvert}{h_{21}}$	$\frac{h_{11}}{h_{21}}$	$\frac{1}{g_{21}}$	$-\frac{g_{22}}{g_{21}}$	a_{11}	a_{12}	$\frac{b_{22}}{\lvert b\rvert}$	$-\frac{b_{12}}{\lvert b\rvert}$
	$\frac{1}{z_{21}}$	$-\frac{z_{22}}{z_{21}}$	$-\frac{\lvert y\rvert}{y_{21}}$	$\frac{y_{11}}{y_{21}}$	$-\frac{h_{22}}{h_{21}}$	$\frac{1}{h_{21}}$	$\frac{g_{11}}{g_{21}}$	$-\frac{\lvert g\rvert}{g_{21}}$	a_{21}	a_{22}	$-\frac{b_{21}}{\lvert b\rvert}$	$\frac{b_{11}}{\lvert b\rvert}$
[b]	$\frac{z_{22}}{z_{12}}$	$-\frac{\lvert z\rvert}{z_{12}}$	$-\frac{y_{11}}{y_{12}}$	$\frac{1}{y_{12}}$	$\frac{1}{h_{12}}$	$-\frac{h_{11}}{h_{12}}$	$-\frac{\lvert g\rvert}{g_{12}}$	$\frac{g_{22}}{g_{12}}$	$\frac{a_{22}}{\lvert a\rvert}$	$-\frac{a_{12}}{\lvert a\rvert}$	b_{11}	b_{12}
	$\frac{1}{z_{12}}$	$-\frac{z_{11}}{z_{12}}$	$-\frac{\lvert y\rvert}{y_{12}}$	$\frac{y_{22}}{y_{12}}$	$\frac{h_{22}}{h_{12}}$	$-\frac{\lvert h\rvert}{h_{12}}$	$-\frac{g_{11}}{g_{12}}$	$\frac{1}{g_{12}}$	$-\frac{a_{21}}{\lvert a\rvert}$	$\frac{a_{11}}{\lvert a\rvert}$	b_{21}	b_{22}

more easily in terms of one set than another. When there is a conflict in the choice of parameter set indicated by these two statements, the measured set is converted to the desired set by means of the formulae shown in table 8.2.

8.4 Some network models of twoports

No matter how complicated the internal structure of a linear twoport may be, its port behaviour can be described by any one of the six sets of equations listed in table 8.1 so long as it contains no independent sources. A twoport can thus be represented by a network model which has no more than four independent elements, one for each of the four parameters, and the derivation of such a model is made by rearranging the equations of a parameter set into a suitably recognizable form. Some examples follow.

1. Network models in terms of the **z**-parameters

Each term on the right hand side of the input equation

$$\boldsymbol{V}_1 = \boldsymbol{z}_{11}\boldsymbol{I}_1 + \boldsymbol{z}_{12}\boldsymbol{I}_2$$

can be represented by one ideal voltage source, and the two sources in series form the input side of a model of a twoport. The first source ($\boldsymbol{z}_{11}\boldsymbol{I}_1$), however, is proportional to the current $\boldsymbol{I}_1$ flowing through it and can therefore be represented by an impedance $\boldsymbol{z}_{11}$ (substitution theorem); this is shown on

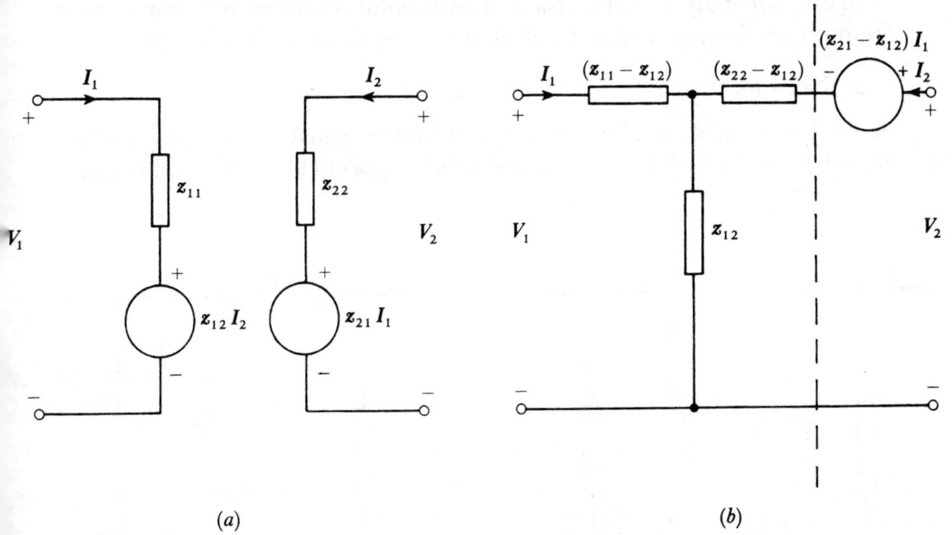

Fig. 8.5 Network models of a twoport in terms of the **z**-parameters: (*a*) **z**-parameter network; (*b*) T-network.

the left hand side of fig. 8.5(*a*). A similar argument, applied to the second equation of (8.2.1), produces the right hand part of the model; the whole diagram shows one possible circuit model of a twoport. There are several possible variants. The T-circuit shown in part (*b*), for example, is derived by putting (8.2.1) into the form

$$\left.\begin{aligned} \boldsymbol{V}_1 &= \boldsymbol{z}_{11}\boldsymbol{I}_1 + \boldsymbol{z}_{12}\boldsymbol{I}_2, \\ \boldsymbol{V}_2 &= \boldsymbol{z}_{12}\boldsymbol{I}_1 + \boldsymbol{z}_{22}\boldsymbol{I}_2 + (\boldsymbol{z}_{21} - \boldsymbol{z}_{12})\boldsymbol{I}_1. \end{aligned}\right\} \quad (8.4.1)$$

That part of (8.4.1) to the left of the dotted line is represented by the passive T shown on the left of the corresponding dotted line in the diagram (as may be verified by applying Kirchhoff's laws to it), and the remaining voltage component $(\boldsymbol{z}_{21} - \boldsymbol{z}_{12})\boldsymbol{I}_1$ is represented by the voltage source shown on the right hand side of the diagram. This is a three-terminal version of a twoport because one terminal is common to each of the two ports. If, however, complete freedom is required between the relative potentials of the two ports, an ideal transformer of unity turns ratio can be attached to either port without otherwise affecting the equivalence.

Figure 8.5(*b*) immediately raises in one's mind the question of the physical realizability of the elements. The element $(\boldsymbol{z}_{11} - \boldsymbol{z}_{12})$, for example, may turn out to have a negative real part if there are no constraints on the values of the $\boldsymbol{z}$-parameters, and this would indicate that it could not be constructed from any arrangement of passive elements. The possibility that a model may have unrealizable elements is acceptable in situations where the model is to be used only for analysis, but is crucial if an attempt is to be made to construct it from physical components. Realizability is discussed later in chapter 12.

2. Network models in terms of the $\boldsymbol{y}$-parameters

A dual argument, applied to the admittance equations (8.3.2), enables fig. 8.6 to be derived. Part (*a*) represents a twoport in terms of its admittance

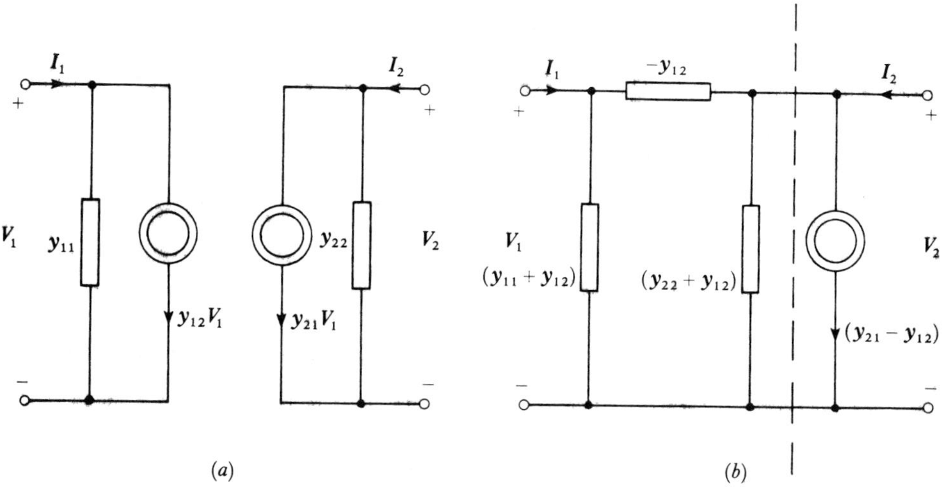

Fig. 8.6 Network models of a twoport in terms of the $\boldsymbol{y}$-parameters: (*a*) $\boldsymbol{y}$-parameter network; (*b*) Π-network.

parameters and part (*b*) represents a three-terminal variant of it. These two models are known respectively as the $\boldsymbol{y}$- and Π-network models of a twoport.

3. Network models in terms of the $\boldsymbol{h}$- and $\boldsymbol{g}$-parameters

The most commonly used models which employ the $\boldsymbol{h}$- or $\boldsymbol{g}$-parameters are shown in fig. 8.7; they are derived directly from the relevant equations of table 8.1 in the manner outlined.

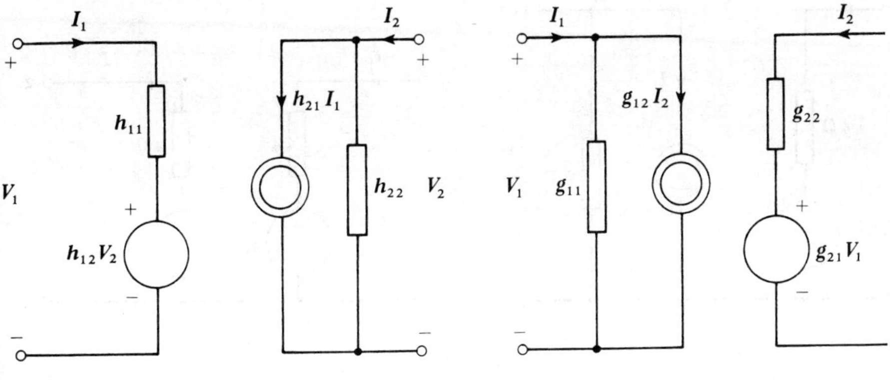

Fig. 8.7 Network models of a twoport in terms of: (*a*) the $\boldsymbol{h}$-parameters; (*b*) the $\boldsymbol{g}$-parameters.

4. Network models in terms of the $\boldsymbol{a}$- and $\boldsymbol{b}$-parameters

The transmission and inverse transmission equations of table 8.1 can be used in a similar way to derive network models in terms of the $\boldsymbol{a}$- or $\boldsymbol{b}$-parameters, but the models consist entirely of controlled sources and are not helpful.

Some of the networks described above are illustrated numerically in fig. 8.8 in which part (*a*) is assumed to be a given twoport from which are derived

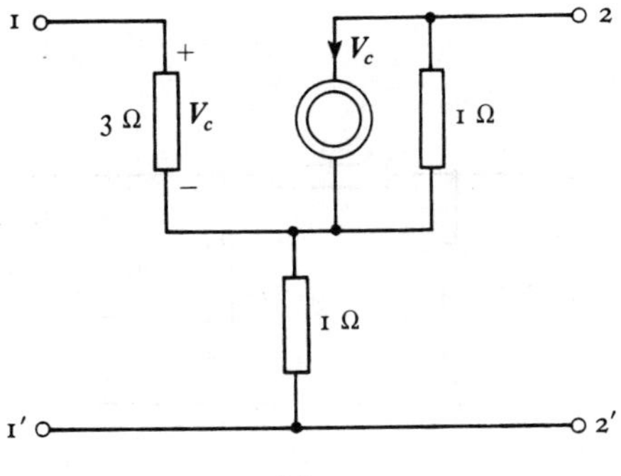

(*a*)

Fig. 8.8 A twoport for which equivalents are shown in parts (*b*) to (*f*).

the equivalent networks shown in parts (*b*) to (*f*). Illustrations of some limiting cases of the $\boldsymbol{z}$, $\boldsymbol{y}$, $\boldsymbol{h}$, and $\boldsymbol{g}$ network models are shown in fig. 8.9, which also gives, in square brackets, arrays of the appropriate parameters arranged so as to correspond in position with their place in the equations of table 8.1. Elements with the subscripts 11, 12, and 22 have been made equal to zero, a

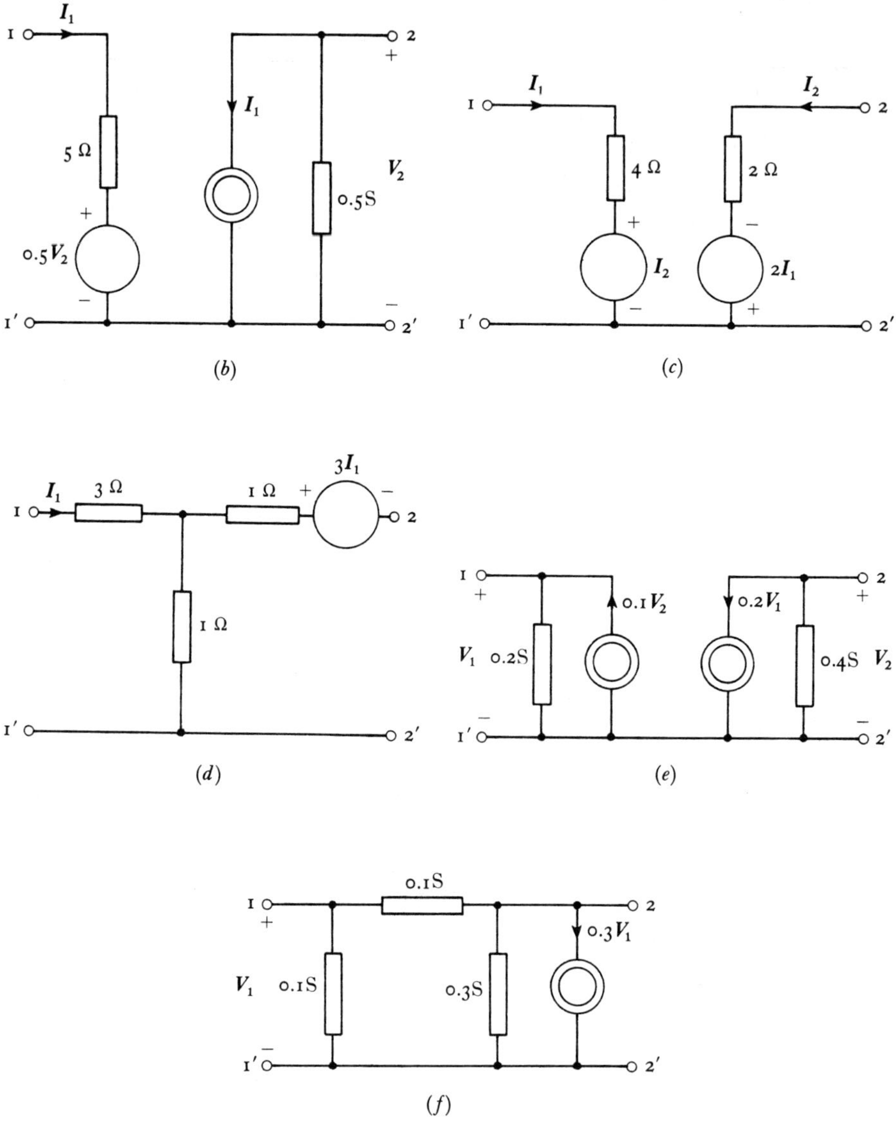

Fig. 8.8 (cont.) Equivalents of the twoport shown in part (*a*).

process which reduces each network model to that of a specific ideal amplifier (controlled source). The imperfections of a non-ideal linear amplifier may

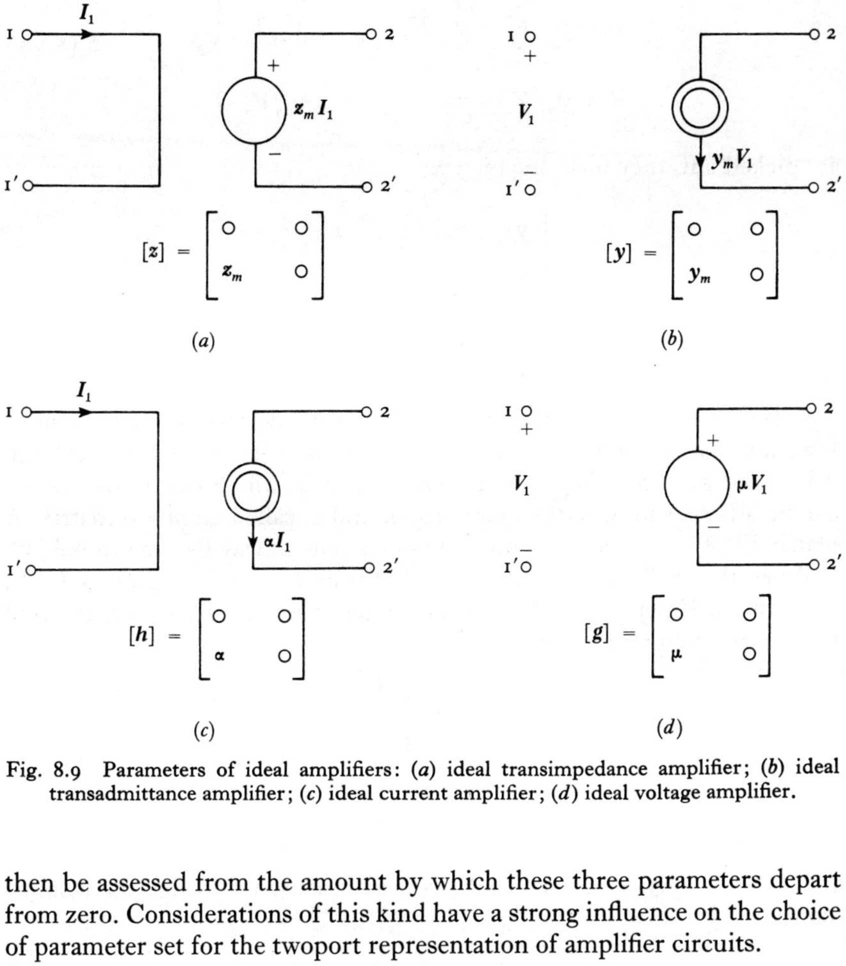

Fig. 8.9 Parameters of ideal amplifiers: (*a*) ideal transimpedance amplifier; (*b*) ideal transadmittance amplifier; (*c*) ideal current amplifier; (*d*) ideal voltage amplifier.

then be assessed from the amount by which these three parameters depart from zero. Considerations of this kind have a strong influence on the choice of parameter set for the twoport representation of amplifier circuits.

8.5 Matrix algebra

Multiport networks require for their solution the manipulation of linear simultaneous equations. Such manipulation is eased by the use of a special kind of algebra, called *matrix algebra*, devised by the English mathematician Arthur Cayley (1821–95). In multiports and in multi-terminal networks, where the number of simultaneous equations is greater than two, the algebra is of real value, but it is also of help in twoports on account of the conciseness of its notation.

A *matrix* is an array of numbers subject to certain rules of association with other matrices. If the coefficients $\boldsymbol{y}_{jk}$ in the equations

$$\left.\begin{aligned} \boldsymbol{I}_1 &= \boldsymbol{y}_{11}\boldsymbol{V}_1 + \boldsymbol{y}_{12}\boldsymbol{V}_2 + \ldots + \boldsymbol{y}_{1n}\boldsymbol{V}_n, \\ \boldsymbol{I}_2 &= \boldsymbol{y}_{21}\boldsymbol{V}_1 + \boldsymbol{y}_{22}\boldsymbol{V}_2 + \ldots + \boldsymbol{y}_{2n}\boldsymbol{V}_n, \\ &\ldots \quad \ldots \quad \ldots \quad \ldots \\ \boldsymbol{I}_n &= \boldsymbol{y}_{n1}\boldsymbol{V}_1 + \boldsymbol{y}_{n2}\boldsymbol{V}_2 + \ldots + \boldsymbol{y}_{nn}\boldsymbol{V}_n, \end{aligned}\right\} \tag{8.5.1}$$

are picked out, they form the matrix

$$\begin{bmatrix} \boldsymbol{y}_{11} & \boldsymbol{y}_{12} & \ldots & \boldsymbol{y}_{1n} \\ \boldsymbol{y}_{21} & \boldsymbol{y}_{22} & \ldots & \boldsymbol{y}_{2n} \\ \ldots & \ldots & \ldots & \ldots \\ \boldsymbol{y}_{n1} & \boldsymbol{y}_{n2} & \ldots & \boldsymbol{y}_{nn} \end{bmatrix}. \tag{8.5.2}$$

This is called a *square matrix* because it has as many rows as it has columns. If a matrix has only one row it is called a *row matrix*; if it has only one column it is called a *column matrix*; in general, a matrix which has m rows and n columns is said to have the *order* $m \times n$, and is called an $m \times n$ matrix. A matrix like (8.5.2) may be represented in a general way by the symbol $[\boldsymbol{y}]$.

When the coefficients $\boldsymbol{y}_{jk}$ are extracted from (8.5.1) the variables $\boldsymbol{I}_1$, $\boldsymbol{I}_2$, etc. and $\boldsymbol{V}_1$, $\boldsymbol{V}_2$, etc. are left. If the dependent variables $\boldsymbol{I}_1$, $\boldsymbol{I}_2$, etc. are used to form the column matrix

$$[\boldsymbol{I}] = \begin{bmatrix} \boldsymbol{I}_1 \\ \boldsymbol{I}_2 \\ \ldots \\ \boldsymbol{I}_n \end{bmatrix}, \tag{8.5.3}$$

and if the independent variables $\boldsymbol{V}_1$, $\boldsymbol{V}_2$, etc. are used to define the column matrix

$$[\boldsymbol{V}] = \begin{bmatrix} \boldsymbol{V}_1 \\ \boldsymbol{V}_2 \\ \ldots \\ \boldsymbol{V}_n \end{bmatrix}, \tag{8.5.4}$$

then (8.5.1) is regarded as defining the process of multiplying the matrices $[\boldsymbol{y}]$ and $[\boldsymbol{V}]$; this process is concisely denoted by the equation

$$[\boldsymbol{I}] = [\boldsymbol{y}][\boldsymbol{V}]. \tag{8.5.5}$$

In this context it is helpful to view $[y]$ as an operator which operates on the array of voltages $[V]$ to produce the array of currents $[I]$. This operation (of multiplication) proceeds according to rules specially devised for matrix algebra. Some of these rules will now be described.

1. *Equality*

Two matrices are equal if each element in one is equal to the corresponding element in the other. It is necessary for two matrices to have the same order $m \times n$ for them to be equal.

2. *Addition*

The sum of two matrices is obtained by adding the corresponding elements in each. For example:

$$\begin{bmatrix} (1+j2) & 3 \\ 4 & (5+j6) \end{bmatrix} + \begin{bmatrix} 4 & 3 \\ 2 & (1-j4) \end{bmatrix} = \begin{bmatrix} (5+j2) & 6 \\ 6 & (6+j2) \end{bmatrix}.$$

Addition is defined only for matrices which have the same order.

3. *Subtraction*

One matrix is subtracted from another by subtracting from each element in one the corresponding element in the other:

$$\begin{bmatrix} 4 & 3 & 0 \\ 2 & 1 & 5 \end{bmatrix} - \begin{bmatrix} 1 & 2 & 1 \\ 3 & 4 & 5 \end{bmatrix} = \begin{bmatrix} 3 & 1 & -1 \\ -1 & -3 & 0 \end{bmatrix}.$$

4. *Multiplication by a scalar*

The multiplication of a matrix by a scalar (that is, by an ordinary quantity, whether it be real, complex, or a function of the complex frequency variable s) has the effect of multiplying each element in the matrix by the scalar. Thus

$$3 \times \begin{bmatrix} 5 & 1 \\ 2 & 4 \end{bmatrix} = \begin{bmatrix} 15 & 3 \\ 6 & 12 \end{bmatrix}.$$

5. *Multiplication*

The product of two matrices $[A]$ and $[B]$ is a matrix $[C]$ whose elements are formed as follows: the element C_{jk} in the jth row and kth column of $[C]$ is equal to the sum of the terms formed by multiplying each element in the jth row of $[A]$ by the corresponding element in the kth column of $[B]$. That is,

$$C_{jk} = \sum_{t=1}^{n} A_{jt} B_{tk}. \tag{8.5.6}$$

Thus if

$$[\boldsymbol{A}] = \begin{bmatrix} \boldsymbol{A}_{11} & \boldsymbol{A}_{12} & \boldsymbol{A}_{13} \\ \boldsymbol{A}_{21} & \boldsymbol{A}_{22} & \boldsymbol{A}_{23} \\ \boldsymbol{A}_{31} & \boldsymbol{A}_{32} & \boldsymbol{A}_{32} \end{bmatrix},$$

and if

$$[\boldsymbol{B}] = \begin{bmatrix} \boldsymbol{B}_{11} & \boldsymbol{B}_{12} \\ \boldsymbol{B}_{21} & \boldsymbol{B}_{22} \\ \boldsymbol{B}_{31} & \boldsymbol{B}_{32} \end{bmatrix},$$

then

$$[\boldsymbol{A}][\boldsymbol{B}] = \begin{bmatrix} \boldsymbol{A}_{11} & \boldsymbol{A}_{12} & \boldsymbol{A}_{13} \\ \boldsymbol{A}_{21} & \boldsymbol{A}_{22} & \boldsymbol{A}_{23} \\ \boldsymbol{A}_{31} & \boldsymbol{A}_{32} & \boldsymbol{A}_{33} \end{bmatrix} \begin{bmatrix} \boldsymbol{B}_{11} & \boldsymbol{B}_{12} \\ \boldsymbol{B}_{21} & \boldsymbol{B}_{22} \\ \boldsymbol{B}_{31} & \boldsymbol{B}_{32} \end{bmatrix},$$

$$= \begin{bmatrix} (\boldsymbol{A}_{11}\boldsymbol{B}_{11} + \boldsymbol{A}_{12}\boldsymbol{B}_{21} + \boldsymbol{A}_{13}\boldsymbol{B}_{31}) & (\boldsymbol{A}_{11}\boldsymbol{B}_{12} + \boldsymbol{A}_{12}\boldsymbol{B}_{22} + \boldsymbol{A}_{13}\boldsymbol{B}_{32}) \\ (\boldsymbol{A}_{21}\boldsymbol{B}_{11} + \boldsymbol{A}_{22}\boldsymbol{B}_{21} + \boldsymbol{A}_{23}\boldsymbol{B}_{31}) & (\boldsymbol{A}_{21}\boldsymbol{B}_{12} + \boldsymbol{A}_{22}\boldsymbol{B}_{22} + \boldsymbol{A}_{23}\boldsymbol{B}_{32}) \\ (\boldsymbol{A}_{31}\boldsymbol{B}_{11} + \boldsymbol{A}_{32}\boldsymbol{B}_{21} + \boldsymbol{A}_{33}\boldsymbol{B}_{31}) & (\boldsymbol{A}_{31}\boldsymbol{B}_{12} + \boldsymbol{A}_{32}\boldsymbol{B}_{22} + \boldsymbol{A}_{33}\boldsymbol{B}_{32}) \end{bmatrix}.$$

The product $[\boldsymbol{A}][\boldsymbol{B}]$ is defined only if $[\boldsymbol{A}]$ and $[\boldsymbol{B}]$ are conformable; that is, if the number of columns in $[\boldsymbol{A}]$ is equal to the number of rows in $[\boldsymbol{B}]$. If the order of $[\boldsymbol{A}]$ is $m \times t$ and if the order of $[\boldsymbol{B}]$ is $t \times n$, then the order of the product $[\boldsymbol{A}][\boldsymbol{B}]$ is $m \times n$. It follows from the requirement of conformability that a product $[\boldsymbol{A}][\boldsymbol{B}]$ may exist while the product $[\boldsymbol{B}][\boldsymbol{A}]$ may not. Moreover, even if both products exist, in general

$$[\boldsymbol{A}][\boldsymbol{B}] \neq [\boldsymbol{B}][\boldsymbol{A}]. \tag{8.5.7}$$

This fact, illustrated below, has physical significance in twoport theory:

$$\begin{bmatrix} 1 & 2 \\ 3 & 4 \end{bmatrix} \begin{bmatrix} 0 & 3 \\ 2 & 1 \end{bmatrix} = \begin{bmatrix} 4 & 5 \\ 8 & 13 \end{bmatrix},$$

but

$$\begin{bmatrix} 0 & 3 \\ 2 & 1 \end{bmatrix} \begin{bmatrix} 1 & 2 \\ 3 & 4 \end{bmatrix} = \begin{bmatrix} 9 & 12 \\ 5 & 8 \end{bmatrix}.$$

The product $[\boldsymbol{A}][\boldsymbol{B}]$ is distinguished verbally from the product $[\boldsymbol{B}][\boldsymbol{A}]$ by saying of the first that $[\boldsymbol{A}]$ is *post-multiplied* by $[\boldsymbol{B}]$, and of the second that $[\boldsymbol{A}]$ is *pre-multiplied* by $[\boldsymbol{B}]$.

Matrix algebra obeys the associative, distributive, and commutative laws of ordinary algebra apart from the important exception discussed above. Thus

$$[A]+[B]=[B]+[A],$$

$$[A]+([B]+[C])=([A]+[B])+[C],$$

$$[A]([B][C])=([A][B])[C],$$

and

$$[A]([B]+[C])=[A][B]+[A][C].$$

6. *Division*

There is no process of division in matrix algebra. The corresponding process is to multiply by the *inverse* of the given matrix; the meaning of the word *inverse* is described in the next section.

8.6 Special types of matrices and further operations

1. *Unit Matrix*

The *unit matrix* $[I]$ is a square matrix which has elements equal to unity on its principal diagonal and equal to zero elsewhere. It is defined by

$$[I]=\begin{bmatrix} 1 & 0 & \dots & 0 \\ 0 & 1 & \dots & 0 \\ \dots & \dots & \dots & \dots \\ 0 & 0 & \dots & 1 \end{bmatrix}. \tag{8.6.1}$$

It acts like unity in ordinary algebra; when used to pre-multiply or post-multiply another matrix it leaves the matrix unchanged.

$$[I][A]=[A]=[A][I]. \tag{8.6.2}$$

In this application the unit matrix is assumed to be adjusted to conformability with the matrix it is multiplying. If, therefore, $[A]$ in (8.6.2) is not itself square, the unit matrix on the left hand side of the equation is of different order from that on the right.

2. *Transpose of a matrix*

A matrix is *transposed* by interchanging its rows and columns. If, for example, a matrix $[A]$ is given by

$$[A]=\begin{bmatrix} a_{11} & a_{12} & a_{13} \\ a_{21} & a_{22} & a_{23} \end{bmatrix},$$

its *transpose* $[\boldsymbol{A}]^{\mathrm{T}}$ is given by

$$[\boldsymbol{A}]^{\mathrm{T}} = \begin{bmatrix} a_{11} & a_{21} \\ a_{12} & a_{22} \\ a_{13} & a_{23} \end{bmatrix}.$$

A column matrix transposes into a row matrix and vice versa. The product of the column matrix $[\boldsymbol{V}]$ of (8.5.4) and the transpose of the column matrix of (8.5.3) is thus

$$[\boldsymbol{V}][\boldsymbol{I}]^{\mathrm{T}} = \boldsymbol{V}_1 \boldsymbol{I}_1 + \boldsymbol{V}_2 \boldsymbol{I}_2 + \ldots + \boldsymbol{V}_n \boldsymbol{I}_n. \qquad (8.6.3)$$

It can be shown that the transpose of the product of two matrices is equal to the product of the individual transposes taken in the reverse order. That is

$$([\boldsymbol{A}][\boldsymbol{B}])^{\mathrm{T}} = [\boldsymbol{B}]^{\mathrm{T}}[\boldsymbol{A}]^{\mathrm{T}}. \qquad (8.6.4)$$

It is a simple matter to extend this relationship to multiple products.

3. *Determinant of a matrix*

The determinant whose elements are identical with those of a square matrix $[\boldsymbol{A}]$ is called the *determinant* of $[\boldsymbol{A}]$ and is denoted in this chapter by $|[\boldsymbol{A}]|$ or $|\boldsymbol{A}|$. If $|\boldsymbol{A}| = 0$, the matrix $[\boldsymbol{A}]$ is said to be *singular*; if $|\boldsymbol{A}| \neq 0$, $[\boldsymbol{A}]$ is *non-singular*. It can be shown that the determinant of the product of two matrices is independent of the order in which the matrices are taken. Thus

$$|[\boldsymbol{A}][\boldsymbol{B}]| = |[\boldsymbol{B}][\boldsymbol{A}]|.$$

Moreover, if two matrices are square and of the same order

$$|[\boldsymbol{A}][\boldsymbol{B}]| = |[\boldsymbol{A}]|\,|[\boldsymbol{B}]|$$

and

$$|[\boldsymbol{A}]^{\mathrm{T}}| = |[\boldsymbol{A}]|.$$

The determinant of a matrix may be used in the process of obtaining the inverse of the matrix.

4. *Inverse of a matrix*

A matrix $[\boldsymbol{A}]$ has an inverse $[\boldsymbol{A}]^{-1}$ defined by

$$[\boldsymbol{A}][\boldsymbol{A}]^{-1} = [\boldsymbol{1}] = [\boldsymbol{A}]^{-1}[\boldsymbol{A}]. \qquad (8.6.5)$$

For the inverse to exist $[\boldsymbol{A}]$ must be square and non-singular.

One method of obtaining the inverse of a matrix is to replace its elements by their co-factors, transpose the result, and divide it by the determinant of the original matrix. For example, the matrix

$$[A] = \begin{bmatrix} \boldsymbol{A}_{11} & \boldsymbol{A}_{12} & \boldsymbol{A}_{13} \\ \boldsymbol{A}_{21} & \boldsymbol{A}_{22} & \boldsymbol{A}_{23} \\ \boldsymbol{A}_{31} & \boldsymbol{A}_{32} & \boldsymbol{A}_{33} \end{bmatrix},$$

has co-factors

$$\boldsymbol{a}_{11} = \begin{vmatrix} \boldsymbol{A}_{22} & \boldsymbol{A}_{23} \\ \boldsymbol{A}_{32} & \boldsymbol{A}_{33} \end{vmatrix} = \boldsymbol{A}_{22}\boldsymbol{A}_{33} - \boldsymbol{A}_{23}\boldsymbol{A}_{32},$$

$$\boldsymbol{a}_{12} = -\begin{vmatrix} \boldsymbol{A}_{21} & \boldsymbol{A}_{23} \\ \boldsymbol{A}_{31} & \boldsymbol{A}_{33} \end{vmatrix} = \boldsymbol{A}_{23}\boldsymbol{A}_{31} - \boldsymbol{A}_{21}\boldsymbol{A}_{33},$$

and so on.

Replacement of the elements $\boldsymbol{A}_{jk}$ by their co-factors $\boldsymbol{a}_{jk}$ gives

$$\begin{bmatrix} \boldsymbol{a}_{11} & \boldsymbol{a}_{12} & \boldsymbol{a}_{13} \\ \boldsymbol{a}_{21} & \boldsymbol{a}_{22} & \boldsymbol{a}_{23} \\ \boldsymbol{a}_{31} & \boldsymbol{a}_{32} & \boldsymbol{a}_{33} \end{bmatrix}.$$

If this is transposed and divided by $|A|$, the result is

$$[A]^{-1} = \frac{1}{|[A]|} \begin{bmatrix} \boldsymbol{a}_{11} & \boldsymbol{a}_{21} & \boldsymbol{a}_{31} \\ \boldsymbol{a}_{12} & \boldsymbol{a}_{22} & \boldsymbol{a}_{32} \\ \boldsymbol{a}_{13} & \boldsymbol{a}_{23} & \boldsymbol{a}_{33} \end{bmatrix}. \tag{8.6.6}$$

The application of this process to a non-singular 2×2 matrix shows that the inverse is obtained, in this special case, by interchanging the elements on the principal diagonal, changing the algebraic sign of the other two elements, and dividing by the determinant. If

$$[A] = \begin{bmatrix} \boldsymbol{A}_{11} & \boldsymbol{A}_{12} \\ \boldsymbol{A}_{21} & \boldsymbol{A}_{22} \end{bmatrix},$$

then

$$[A]^{-1} = \frac{1}{|A|} \begin{bmatrix} \boldsymbol{A}_{22} & -\boldsymbol{A}_{12} \\ -\boldsymbol{A}_{21} & \boldsymbol{A}_{11} \end{bmatrix},$$

where $|A| = A_{11}A_{22} - A_{12}A_{21}$. This is the relationship shown in table 8.2 between the y and z parameters, the g and h parameters, and the a and b parameters, and it occurs because one parameter set in each of these three pairs is the inverse of the other, as may be seen by the following example. The problem is to solve the twoport hybrid equations for the variables I_1 and V_2.

The hybrid equations are

$$\begin{aligned} V_1 &= h_{11}I_1 + h_{12}V_2, \\ I_2 &= h_{21}I_1 + h_{22}V_2. \end{aligned} \tag{8.6.7}$$

By writing

$$[W] = \begin{bmatrix} V_1 \\ I_2 \end{bmatrix}, \quad [h] = \begin{bmatrix} h_{11} & h_{12} \\ h_{21} & h_{22} \end{bmatrix}, \quad [X] = \begin{bmatrix} I_1 \\ V_2 \end{bmatrix},$$

(8.6.7) becomes

$$[W] = [h][X]. \tag{8.6.8}$$

Pre-multiplying both sides by $[h]^{-1}$ gives

$$[h]^{-1}[W] = [h]^{-1}[h][X] = [1][X] = [X],$$

or

$$[X] = [h]^{-1}[W]. \tag{8.6.9}$$

Equation (8.6.9) is the matrix solution of (8.6.8). Writing

$$[h]^{-1} = \begin{bmatrix} \dfrac{h_{22}}{|h|} & -\dfrac{h_{12}}{|h|} \\ -\dfrac{h_{21}}{|h|} & \dfrac{h_{11}}{|h|} \end{bmatrix} = \begin{bmatrix} g_{11} & g_{12} \\ g_{21} & g_{22} \end{bmatrix} = [g] \tag{8.6.10}$$

gives

$$[X] = [g][W],$$

or

$$\begin{bmatrix} I_1 \\ V_2 \end{bmatrix} = \begin{bmatrix} g_{11} & g_{12} \\ g_{21} & g_{22} \end{bmatrix} \begin{bmatrix} V_1 \\ I_2 \end{bmatrix}, \tag{8.6.11}$$

in which the g-coefficients are related to the h-coefficients by (8.6.10). The expansion of (8.6.11) gives

$$\begin{aligned} I_1 &= g_{11}V_1 + g_{12}I_2, \\ V_2 &= g_{21}V_1 + g_{22}I_2, \end{aligned} \tag{8.6.12}$$

which is the required result.

In some situations the matrix whose inverse is required is itself the product of two matrices. Evaluation of the inverse may be simplified by observing that inverse of the product of two square matrices $[\boldsymbol{A}]$ and $[\boldsymbol{B}]$ is the product of the separate inverses taken in the reverse order. Thus

$$([\boldsymbol{A}][\boldsymbol{B}])^{-1} = [\boldsymbol{B}]^{-1}[\boldsymbol{A}]^{-1}. \tag{8.6.13}$$

It is also worth noticing that if a matrix has an inverse, the transpose of the inverse is equal to the inverse of the transpose. Thus

$$([\boldsymbol{A}]^{-1})^{\mathrm{T}} = ([\boldsymbol{A}]^{\mathrm{T}})^{-1}.\ † \tag{8.6.14}$$

8.7 Interconnected twoports

Figure 8.10 shows five different ways of interconnecting two twoports. The question then arises: what are the twoport parameters of the overall networks? To answer this, consider part (*a*) which is redrawn in fig. 8.11(*a*) to display the current and voltage designations.

For twoport (*a*)

$$[\boldsymbol{V}_a] = [\boldsymbol{z}_a][\boldsymbol{I}_a], \tag{8.7.1}$$

where

$$[\boldsymbol{V}_a] = \begin{bmatrix} V_{1a} \\ V_{2a} \end{bmatrix}, \quad [\boldsymbol{z}_a] = \begin{bmatrix} z_{ia} & z_{ra} \\ z_{fa} & z_{oa} \end{bmatrix}, \quad \text{and} \quad [\boldsymbol{I}_a] = \begin{bmatrix} I_{1a} \\ I_{2a} \end{bmatrix},$$

and for twoport (*b*)

$$[\boldsymbol{V}_b] = [\boldsymbol{z}_b][\boldsymbol{I}_b], \tag{8.7.2}$$

where

$$[\boldsymbol{V}_b] = \begin{bmatrix} V_{1b} \\ V_{2b} \end{bmatrix}, \quad [\boldsymbol{z}_b] = \begin{bmatrix} z_{ib} & z_{rb} \\ z_{fb} & z_{ob} \end{bmatrix}, \quad \text{and} \quad [\boldsymbol{I}_b] = \begin{bmatrix} I_{1b} \\ I_{2b} \end{bmatrix}.$$

For the overall network let

$$[\boldsymbol{V}] = [\boldsymbol{z}][\boldsymbol{I}], \tag{8.7.3}$$

† For the derivation of these and other formulae the reader should consult a specialized text, e.g. A. von Weiss, (Transl. E. Brenner), *Matrix analysis for electrical engineers*, Van Nostrand (1964).

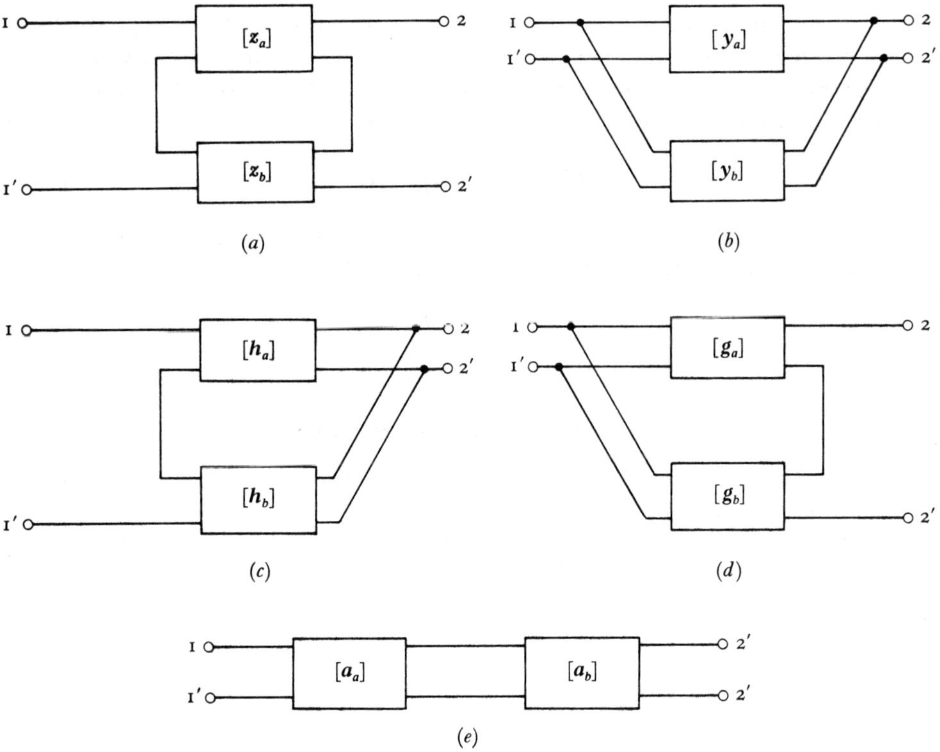

Fig. 8.10 Interconnected twoports: (*a*) series; (*b*) parallel; (*c*) series–parallel; (*d*) parallel–series; (*e*) cascade.

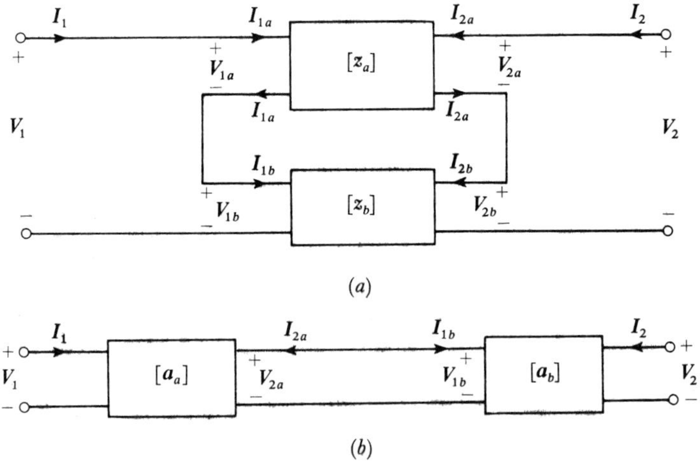

Fig. 8.11 Voltage and current designations for: (*a*) two series-connected twoports; (*b*) two cascaded twoports.

where $[\boldsymbol{V}] = \begin{bmatrix} V_1 \\ V_2 \end{bmatrix}$, $[\boldsymbol{I}] = \begin{bmatrix} I_1 \\ I_2 \end{bmatrix}$, and $[\boldsymbol{z}]$ is the overall impedance matrix which is to be found in terms of $[\boldsymbol{z}_a]$ and $[\boldsymbol{z}_b]$. Now

$$[\boldsymbol{V}] = \begin{bmatrix} V_1 \\ V_2 \end{bmatrix},$$

$$= \begin{bmatrix} (V_{1a} + V_{1b}) \\ (V_{2a} + V_{2b}) \end{bmatrix},$$

$$= \begin{bmatrix} V_{1a} \\ V_{2a} \end{bmatrix} + \begin{bmatrix} V_{1b} \\ V_{2b} \end{bmatrix},$$

$$= [\boldsymbol{V}_a] + [\boldsymbol{V}_b], \qquad (8.7.4)$$

and

$$[\boldsymbol{I}] = \begin{bmatrix} I_1 \\ I_2 \end{bmatrix} = \begin{bmatrix} I_{1a} \\ I_{2a} \end{bmatrix} = \begin{bmatrix} I_{1b} \\ I_{2b} \end{bmatrix} = [\boldsymbol{I}_a] = [\boldsymbol{I}_b], \qquad (8.7.5)$$

so the substitution of (8.7.4,5) into the equations formed by adding (8.7.1) and (8.7.2) gives

$$[\boldsymbol{V}] = ([\boldsymbol{z}_a] + [\boldsymbol{z}_b])\,[\boldsymbol{I}]. \qquad (8.7.6)$$

A comparison of this with (8.7.3) shows that

$$[\boldsymbol{z}] = [\boldsymbol{z}_a] + [\boldsymbol{z}_b]; \qquad (8.7.7)$$

that is, the impedance matrix of a network formed by the series connexion of two twoports is given by the sum of the impedance matrices of the separate twoports. A similar form of analysis shows that for parallel connected twoports,

$$[\boldsymbol{y}] = [\boldsymbol{y}_a] + [\boldsymbol{y}_b], \qquad (8.7.8)$$

for series–parallel connected twoports,

$$[\boldsymbol{h}] = [\boldsymbol{h}_a] + [\boldsymbol{h}_b], \qquad (8.7.9)$$

and for parallel–series connected twoports,

$$[\boldsymbol{g}] = [\boldsymbol{g}_a] + [\boldsymbol{g}_b]. \qquad (8.7.10)$$

For cascaded networks it is best to use the transmission parameters. Using the designations of current and voltage shown in fig. 8.11(*b*), the relevant equations are

$$\begin{aligned}\begin{bmatrix} \boldsymbol{V}_1 \\ \boldsymbol{I}_1 \end{bmatrix} &= [\boldsymbol{a}_a] \begin{bmatrix} \boldsymbol{V}_{2a} \\ \boldsymbol{I}_{2a} \end{bmatrix}, \\ &= [\boldsymbol{a}_a] \begin{bmatrix} \boldsymbol{V}_{1b} \\ -\boldsymbol{I}_{1b} \end{bmatrix}, \\ &= [\boldsymbol{a}_a] \begin{bmatrix} 1 & 0 \\ 0 & -1 \end{bmatrix} \begin{bmatrix} \boldsymbol{V}_{1b} \\ \boldsymbol{I}_{1b} \end{bmatrix}, \\ &= [\boldsymbol{a}_a] \begin{bmatrix} 1 & 0 \\ 0 & -1 \end{bmatrix} [\boldsymbol{a}_b] \begin{bmatrix} \boldsymbol{V}_2 \\ \boldsymbol{I}_2 \end{bmatrix}, \\ &= [\boldsymbol{a}] \begin{bmatrix} \boldsymbol{V}_2 \\ \boldsymbol{I}_2 \end{bmatrix},\end{aligned}$$

from which the overall transmission matrix is seen to be

$$[\boldsymbol{a}] = [\boldsymbol{a}_a] \begin{bmatrix} 1 & 0 \\ 0 & -1 \end{bmatrix} [\boldsymbol{a}_b]. \qquad (8.7.11)$$

The inverse transmission matrix $[\boldsymbol{b}]$ of two cascaded twoports, defined by the equation

$$\begin{bmatrix} \boldsymbol{V}_2 \\ \boldsymbol{I}_2 \end{bmatrix} = [\boldsymbol{b}] \begin{bmatrix} \boldsymbol{V}_1 \\ \boldsymbol{I}_1 \end{bmatrix},$$

is obtained by inverting (8.7.11):

$$\begin{aligned}[\boldsymbol{b}] &= [\boldsymbol{a}]^{-1}, \\ &= [\boldsymbol{a}_b]^{-1} \begin{bmatrix} 1 & 0 \\ 0 & -1 \end{bmatrix}^{-1} [\boldsymbol{a}_a]^{-1}, \\ &= [\boldsymbol{b}_b] \left\{ -\begin{bmatrix} -1 & 0 \\ 0 & 1 \end{bmatrix} \right\} [\boldsymbol{b}_a], \\ &= [\boldsymbol{b}_b] \begin{bmatrix} 1 & 0 \\ 0 & -1 \end{bmatrix} [\boldsymbol{b}_a]. \qquad (8.7.12)\end{aligned}$$

Equations (8.7.11,12) show that the transmission and inverse-transmission matrices of two cascaded twoports are the triple product of the two matrices of the individual twoports sandwiched by the matrix $\begin{bmatrix} 1 & 0 \\ 0 & -1 \end{bmatrix}$, which arises because of the oppositely designated directions of I_{2a} and I_{1b}. Its effect is to change the sign of the second column of its pre-multiplying matrix, and it is possible to eliminate it by designating I_2 in the opposite direction to that used in this chapter (as mentioned in §8.3). This is commonly done in such topics as transmission lines (see chapter 13) in which the cascaded connexion is paramount.

8.8 Validity tests for interconnected twoports

In the previous section it was assumed that each of the networks retained its twoport properties after being interconnected. The crucial step occurred in (8.7.5) where I_{1a} and I_{2a} were equated respectively to I_{1b} and I_{2b}. If this equality does not exist then the subsequent argument is false and the overall z-matrix is not equal to the sum of the separate z-matrices. The conditions which need to be satisfied for the previously derived results to be true are discussed below.

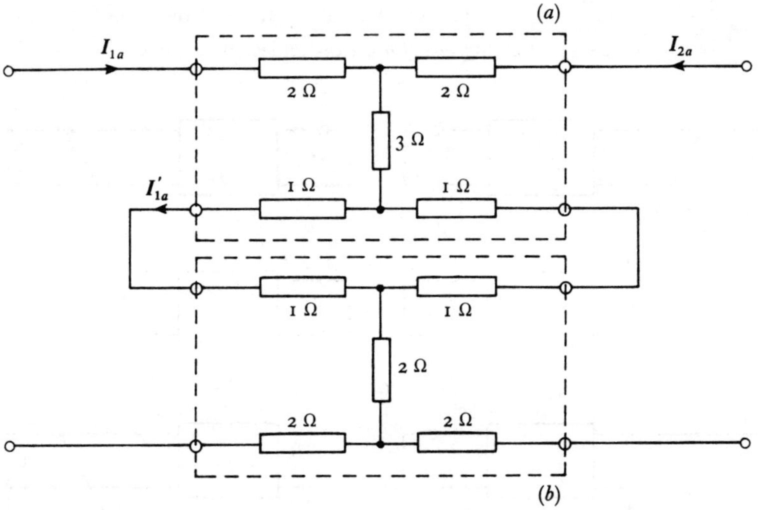

Fig. 8.12 Two series-connected networks which do not satisfy the port requirements.

Consider the series-connected twoports shown in fig. 8.12, in which the elements are purely resistive. The impedance matrices of the two separate networks are

$$[z_a] = \begin{bmatrix} 6 & 3 \\ 3 & 6 \end{bmatrix}, \quad [z_b] = \begin{bmatrix} 5 & 2 \\ 2 & 5 \end{bmatrix},$$

and the overall impedance matrix is

$$[z] = \begin{bmatrix} 10 & 6 \\ 6 & 10 \end{bmatrix},$$

which is not equal to the sum of $[z_a]$ and $[z_b]$. The difference occurs because I_{1a} is not equal to I'_{1a} when the two networks are connected together. For example, if I_{1a} is 1 A and I_{2a} is 2 A it is plain that in the given symmetrical network I'_{1a} is 1.5 A. A similar inequality occurs at the other ports.

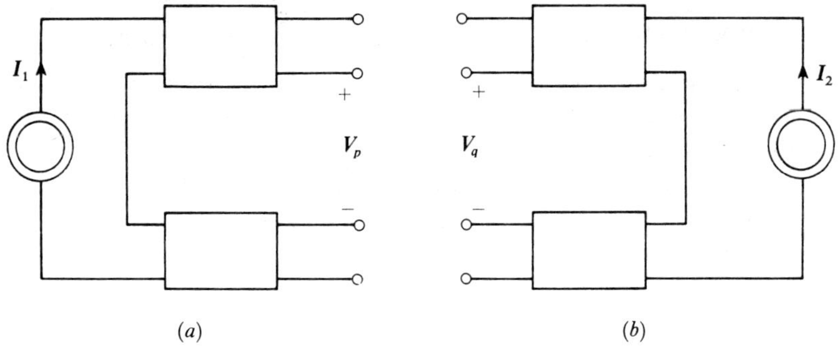

Fig. 8.13 Validity tests for series-connected networks.

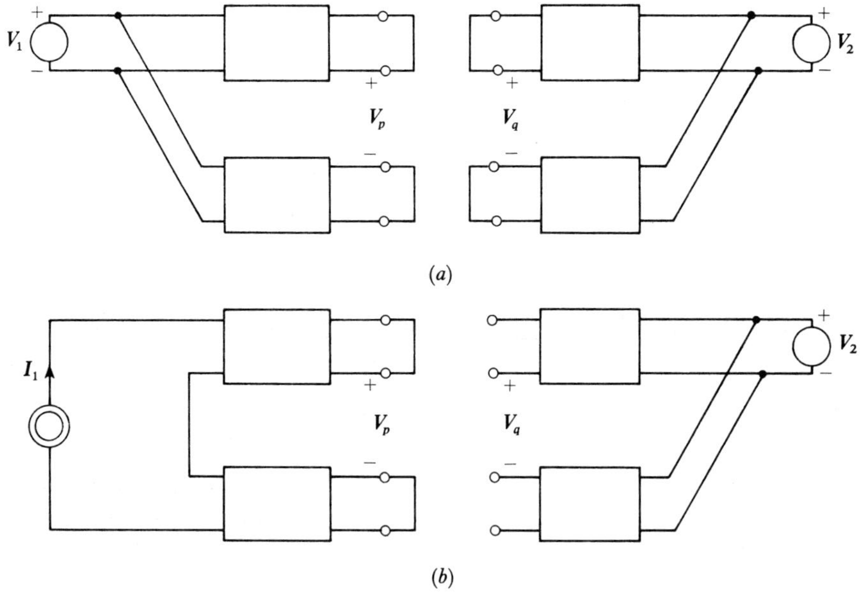

Fig. 8.14 Validity tests for: (a) parallel-connected networks; (b) series–parallel connected networks.

To test the port criteria of two series-connected networks, one first constrains each network to satisfy the input port requirements by open-circuiting its output as shown in fig. 8.13(*a*). If the voltage $\boldsymbol{V}_p$ is zero the output ports can be series-connected without disturbing the input port conditions. A similar test, made as shown in fig. 8.13(*b*), shows that only if $\boldsymbol{V}_q$ is zero will the output port requirement be maintained when the series connexion is made at the input. Both tests are required, because only then can it be shown, by superposition, that the port requirements are met for all simultaneous values of input and output currents.

Corresponding tests for the paralleling and series–paralleling of networks are shown in fig. 8.14. Part (*a*) shows the test conditions under which the voltages $\boldsymbol{V}_p$ and $\boldsymbol{V}_q$ must be zero if the port requirements are to be met for

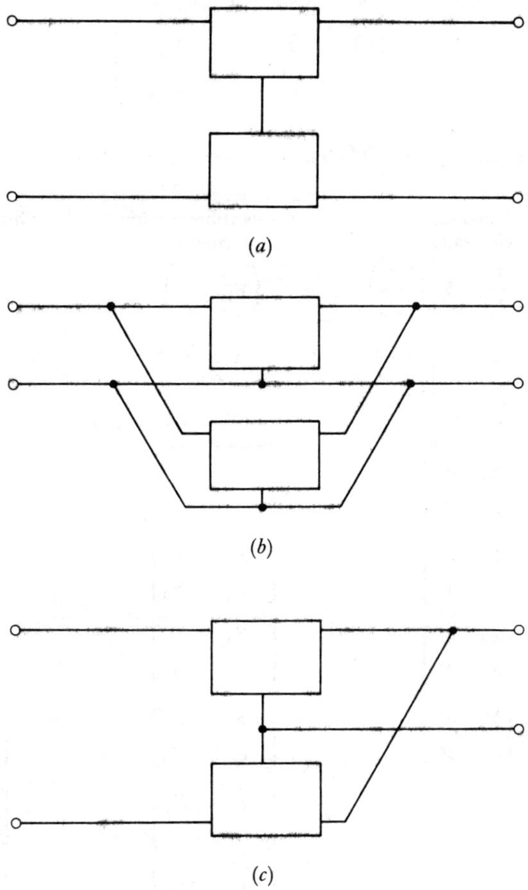

Fig. 8.15 Three-terminal network connexions which always satisfy the validity tests: (*a*) series connexion; (*b*) parallel connexion; (*c*) series–parallel connexion.

all simultaneous values of input and output voltages, and part (*b*) shows corresponding test conditions for the series–parallel connexion. Of special interest are the interconnexions shown in fig. 8.15 of three-terminal networks which always satisfy the validity conditions.

8.9 Simple applications of the interconnexion rules

The matrix of a twoport can often be built up from a knowledge of the interconnexion rules. For example, a Π-network which has shunt elements $\boldsymbol{Y}_1$ and $\boldsymbol{Y}_2$ and a series element $\boldsymbol{Y}_3$ can be viewed as the parallel connexion of the loading elements and the series transmission element of table 8.3, and has a $\boldsymbol{y}$-matrix given by

$$[\boldsymbol{y}]_{\Pi} = \begin{bmatrix} \boldsymbol{Y}_1 & 0 \\ 0 & \boldsymbol{Y}_2 \end{bmatrix} + \begin{bmatrix} \boldsymbol{Y}_3 & -\boldsymbol{Y}_3 \\ -\boldsymbol{Y}_3 & \boldsymbol{Y}_3 \end{bmatrix},$$

$$= \begin{bmatrix} (\boldsymbol{Y}_1 + \boldsymbol{Y}_3) & -\boldsymbol{Y}_3 \\ -\boldsymbol{Y}_3 & (\boldsymbol{Y}_2 + \boldsymbol{Y}_3) \end{bmatrix}. \tag{8.9.1}$$

TABLE 8.3 *Matrices of elementary twoports*

	Loading elements	Series transmission element	Shunt transmission element
	$\left(\boldsymbol{Y}_1 = \dfrac{1}{\boldsymbol{Z}_1},\ \boldsymbol{Y}_2 = \dfrac{1}{\boldsymbol{Z}_2}\right)$	$\left(\boldsymbol{Y}_3 = \dfrac{1}{\boldsymbol{Z}_3}\right)$	$\left(\boldsymbol{Z}_4 = \dfrac{1}{\boldsymbol{Y}_4}\right)$
	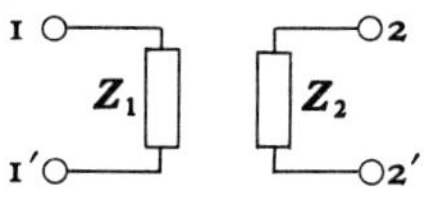	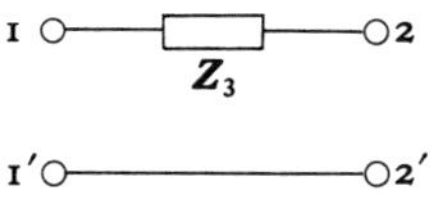	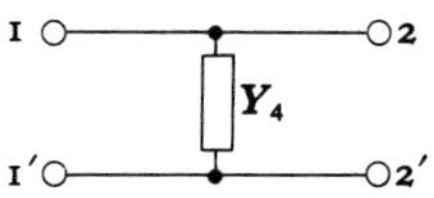
$\boldsymbol{z}$-matrix	$\begin{bmatrix} \boldsymbol{Z}_1 & 0 \\ 0 & \boldsymbol{Z}_2 \end{bmatrix}$	—	$\begin{bmatrix} \boldsymbol{Z}_4 & \boldsymbol{Z}_4 \\ \boldsymbol{Z}_4 & \boldsymbol{Z}_4 \end{bmatrix}$
$\boldsymbol{y}$-matrix	$\begin{bmatrix} \boldsymbol{Y}_1 & 0 \\ 0 & \boldsymbol{Y}_2 \end{bmatrix}$	$\begin{bmatrix} \boldsymbol{Y}_3 & -\boldsymbol{Y}_3 \\ -\boldsymbol{Y}_3 & \boldsymbol{Y}_3 \end{bmatrix}$	—
$\boldsymbol{h}$-matrix	$\begin{bmatrix} \boldsymbol{Z}_1 & 0 \\ 0 & \boldsymbol{Y}_2 \end{bmatrix}$	$\begin{bmatrix} \boldsymbol{Z}_3 & 1 \\ -1 & 0 \end{bmatrix}$	$\begin{bmatrix} 0 & 1 \\ -1 & \boldsymbol{Y}_4 \end{bmatrix}$
$\boldsymbol{g}$-matrix	$\begin{bmatrix} \boldsymbol{Y}_1 & 0 \\ 0 & \boldsymbol{Z}_2 \end{bmatrix}$	$\begin{bmatrix} 0 & -1 \\ 1 & \boldsymbol{Z}_3 \end{bmatrix}$	$\begin{bmatrix} \boldsymbol{Y}_4 & -1 \\ 1 & 0 \end{bmatrix}$
$\boldsymbol{a}$-matrix	—	$\begin{bmatrix} 1 & -\boldsymbol{Z}_3 \\ 0 & -1 \end{bmatrix}$	$\begin{bmatrix} 1 & 0 \\ \boldsymbol{Y}_4 & -1 \end{bmatrix}$
$\boldsymbol{b}$-matrix	—	$\begin{bmatrix} 1 & -\boldsymbol{Z}_3 \\ 0 & -1 \end{bmatrix}$	$\begin{bmatrix} 1 & 0 \\ \boldsymbol{Y}_4 & -1 \end{bmatrix}$

Similarly, a T-circuit with series arms $\boldsymbol{Z}_1$ and $\boldsymbol{Z}_2$ and a shunt arm $\boldsymbol{Z}_4$ can be viewed as the series connexion of the loading elements and the shunt transmission element of table 8.3, thereby having a $\boldsymbol{z}$-matrix given by

$$[\boldsymbol{z}]_T = \begin{bmatrix} \boldsymbol{Z}_1 & 0 \\ 0 & \boldsymbol{Z}_2 \end{bmatrix} + \begin{bmatrix} \boldsymbol{Z}_4 & \boldsymbol{Z}_4 \\ \boldsymbol{Z}_4 & \boldsymbol{Z}_4 \end{bmatrix},$$

$$= \begin{bmatrix} (\boldsymbol{Z}_1 + \boldsymbol{Z}_4) & \boldsymbol{Z}_4 \\ \boldsymbol{Z}_4 & (\boldsymbol{Z}_2 + \boldsymbol{Z}_4) \end{bmatrix}. \tag{8.9.2}$$

The matrix of the Π- and T-networks can also be obtained in terms of the transmission parameters by viewing each as a cascade of alternate series and shunt transmission elements. Thus the symmetrical T-circuit shown in fig. 8.16 has a transmission matrix given by

$$[\boldsymbol{a}]_T = \begin{bmatrix} 1 & -\dfrac{\boldsymbol{Z}_1}{2} \\ 0 & -1 \end{bmatrix} \begin{bmatrix} 1 & 0 \\ 0 & -1 \end{bmatrix} \begin{bmatrix} 1 & 0 \\ \boldsymbol{Y}_2 & -1 \end{bmatrix} \begin{bmatrix} 1 & 0 \\ 0 & -1 \end{bmatrix} \begin{bmatrix} 1 & -\dfrac{\boldsymbol{Z}_1}{2} \\ 0 & -1 \end{bmatrix}, \tag{8.9.3}$$

$$= \begin{bmatrix} 1 & \dfrac{\boldsymbol{Z}_1}{2} \\ 0 & 1 \end{bmatrix} \begin{bmatrix} 1 & 0 \\ \boldsymbol{Y}_2 & 1 \end{bmatrix} \begin{bmatrix} 1 & -\dfrac{\boldsymbol{Z}_1}{2} \\ 0 & -1 \end{bmatrix},$$

$$= \begin{bmatrix} \left(1 + \dfrac{\boldsymbol{Z}_1 \boldsymbol{Y}_2}{2}\right) & \dfrac{\boldsymbol{Z}_1}{2} \\ \boldsymbol{Y}_2 & 1 \end{bmatrix} \begin{bmatrix} 1 & -\dfrac{\boldsymbol{Z}_1}{2} \\ 0 & -1 \end{bmatrix},$$

$$= \begin{bmatrix} \left(1 + \dfrac{\boldsymbol{Z}_1 \boldsymbol{Y}_2}{2}\right) & -\boldsymbol{Z}_1\left(1 + \dfrac{\boldsymbol{Z}_1 \boldsymbol{Y}_2}{4}\right) \\ \boldsymbol{Y}_2 & -\left(1 + \dfrac{\boldsymbol{Z}_1 \boldsymbol{Y}_2}{2}\right) \end{bmatrix}. \tag{8.9.4}$$

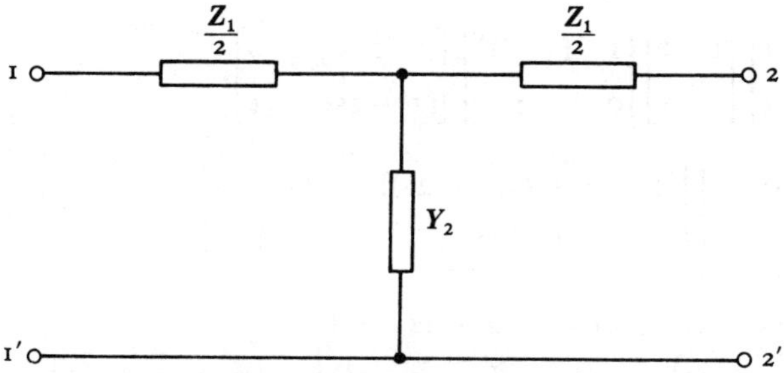

Fig. 8.16 Symmetrical T-network.

If the output variables are required as functions of the input variables, the inverse transmission matrix $[\boldsymbol{b}]$ is needed. This can be found by inverting (8.9.4), but it is simpler to go back to (8.9.3) because each term in this matrix product has an inverse that is identical with itself; so the inverse of the product is simply the product taken in the reverse order. The reverse

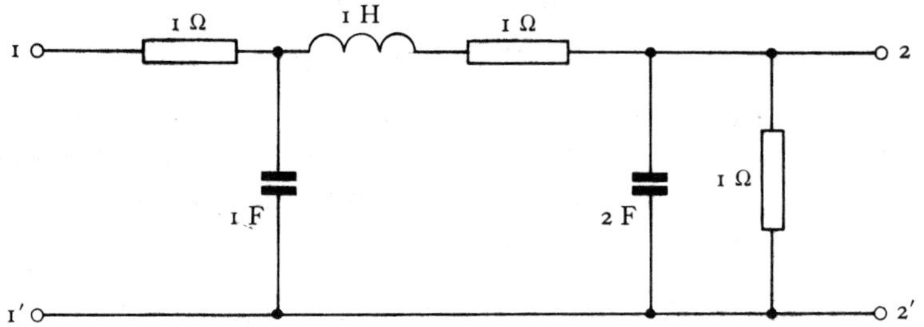

Fig. 8.17 A ladder network.

order of the product, however, is the same as the original order; so $[\boldsymbol{b}]_T$ is the same as $[\boldsymbol{a}]_T$. This is as it should be, because the circuit is symmetrical. Thus

$$\begin{bmatrix} V_2 \\ I_2 \end{bmatrix} = \begin{bmatrix} \left(1 + \dfrac{Z_1 Y_2}{2}\right) & -Z_1\left(1 + \dfrac{Z_1 Y_2}{4}\right) \\ Y_2 & -\left(1 + \dfrac{Z_1 Y_2}{2}\right) \end{bmatrix} \begin{bmatrix} V_1 \\ I_1 \end{bmatrix}. \tag{8.9.5}$$

The transmission matrix is a convenient tool for treating passive ladder-type structures. The $\boldsymbol{a}$-matrix of the network shown in fig. 8.17, for instance, is

$$[\boldsymbol{a}] = \begin{bmatrix} 1 & -1 \\ 0 & -1 \end{bmatrix} \begin{bmatrix} 1 & 0 \\ 0 & -1 \end{bmatrix} \begin{bmatrix} 1 & 0 \\ s & -1 \end{bmatrix} \begin{bmatrix} 1 & 0 \\ 0 & -1 \end{bmatrix} \begin{bmatrix} 1 & -(1+s) \\ 0 & -1 \end{bmatrix} \begin{bmatrix} 1 & 0 \\ 0 & -1 \end{bmatrix} \begin{bmatrix} 1 & 0 \\ (1+2s) & -1 \end{bmatrix}$$

$$= \begin{bmatrix} 1 & 1 \\ 0 & 1 \end{bmatrix} \begin{bmatrix} 1 & 0 \\ s & 1 \end{bmatrix} \begin{bmatrix} 1 & (1+s) \\ 0 & 1 \end{bmatrix} \begin{bmatrix} 1 & 0 \\ (1+2s) & -1 \end{bmatrix},$$

$$= \begin{bmatrix} (1+s) & 1 \\ s & 1 \end{bmatrix} \begin{bmatrix} \{1+(1+s)(1+2s)\} & -(1+s) \\ (1+2s) & -1 \end{bmatrix},$$

$$= \begin{bmatrix} (3+7s+5s^2+2s^3) & -(2+2s+s^2) \\ (1+4s+3s^2+2s^3) & -(1+s+s^2) \end{bmatrix}, \tag{8.9.6}$$

and the ratio of the open-circuit output voltage (at port-2) to an applied voltage $\boldsymbol{V}_1$ (at port-1), for example, is

$$(\boldsymbol{V}_2)_{\boldsymbol{I}_2=0} = \frac{1}{\boldsymbol{a}_{11}} . \boldsymbol{V}_1, \tag{8.9.7}$$

$$= \frac{1}{3 + 7s + 5s^2 + 2s^3} . \boldsymbol{V}_1. \tag{8.9.8}$$

8.10 Loaded twoports

When a twoport is inserted between a source and a load its input and output ports are said to be *loaded* by the source and load impedances respectively. In a particular situation both source and load may consist of several elements, but Thevenin's theorem allows them to be replaced by the equivalent models shown in fig. 8.18. In the first four parts of this diagram the derivations of

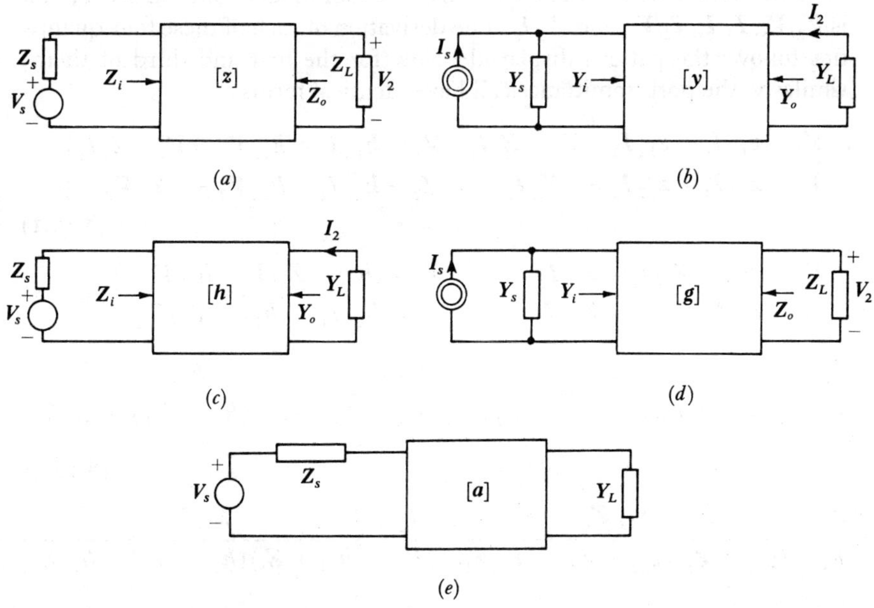

Fig. 8.18 Equivalent models of a loaded twoport: (*a*) $\boldsymbol{z}$-model; (*b*) $\boldsymbol{y}$-model; (*c*) $\boldsymbol{h}$-model; (*d*) $\boldsymbol{g}$-model; (*e*) $\boldsymbol{a}$-model.

corresponding quantities follow a single pattern. (Corresponding quantities are listed in table 8.4 in which the word *immittance*, a combination of the two words *impedance* and *admittance*, means whichever of the two words is

TABLE 8.4 *Corresponding quantities in the* $\boldsymbol{z}$, $\boldsymbol{y}$, $\boldsymbol{h}$, *and* $\boldsymbol{g}$*-parameter models*

Quantity	Model [z]	[y]	[h]	[g]
Source variable	V_s	I_s	V_s	I_s
Source immittance	Z_s	Y_s	Z_s	Y_s
Input immittance	Z_i	Y_i	Z_i	Y_i
Independent output variable	I_2	V_2	V_2	I_2
Dependent output variable	V_2	I_2	I_2	V_2
Load immittance	Z_L	Y_L	Y_L	Z_L
Output immittance	Z_o	Y_o	Y_o	Z_o

appropriate.) Consider, for example, the ratio of the dependent output variable to the source variable in each of the four models taken in order; that is V_2/V_s, I_2/I_s, I_2/V_s, and V_2/I_s. The derivation of each of these four quantities follows the pattern displayed below for the first and third of them; similarly, the port immittances all have identical forms.

$$\begin{aligned} V_1 &= z_{11}I_1 + z_{12}I_2 = V_s - Z_sI_1 & V_1 &= h_{11}I_1 + h_{12}V_2 = V_s - Z_sI_1 \\ V_2 &= z_{21}I_1 + z_{22}I_2 = -Z_LI_2 & I_2 &= h_{21}I_1 + h_{22}V_2 = -Y_LV_2 \end{aligned} \qquad (8.10.1)$$

$$\begin{aligned} V_s &= (z_{11}+Z_s)I_1 + z_{12}I_2 & V_s &= (h_{11}+Z_s)I_1 + h_{12}V_2 \\ 0 &= z_{21}I_1 + (z_{22}+Z_L)I_2 & 0 &= h_{21}I_1 + (h_{22}+Y_L)V_2 \end{aligned} \qquad (8.10.2)$$

$$\frac{I_2}{V_s} = \frac{-z_{21}}{(z_{11}+Z_s)(z_{22}+Z_L) - z_{12}z_{21}} \qquad \frac{V_2}{V_s} = \frac{-h_{21}}{(h_{11}+Z_s)(h_{22}+Y_L) - h_{12}h_{21}} \qquad (8.10.3)$$

$$\frac{V_2}{V_s} = \frac{z_{21}Z_L}{(z_{11}+Z_s)(z_{22}+Z_L) - z_{12}z_{21}} \qquad \frac{I_2}{V_s} = \frac{h_{21}Y_L}{(h_{11}+Z_s)(h_{22}+Y_L) - h_{12}h_{21}} \qquad (8.10.4)$$

$$Z_i = z_{11} - \frac{z_{12}z_{21}}{z_{22}+Z_L} \qquad Z_i = h_{11} - \frac{h_{12}h_{21}}{(h_{22}+Y_L)} \qquad (8.10.5)$$

$$Z_o = z_{22} - \frac{z_{12}z_{21}}{z_{22}+Z_s} \qquad Y_o = h_{22} - \frac{h_{12}h_{21}}{(h_{11}+Z_s)} \qquad (8.10.6)$$

It is apparent from these equations that the effect of loading a twoport is to augment the input and output self immittance terms ($\boldsymbol{z}_{11}$, $\boldsymbol{z}_{22}$; $\boldsymbol{h}_{11}$, $\boldsymbol{h}_{22}$; and so on) of the twoport matrix by the immittance of the source and of the load respectively, and that the source and load immittances may be treated like the loading elements of table 8.3 and combined with the twoport by

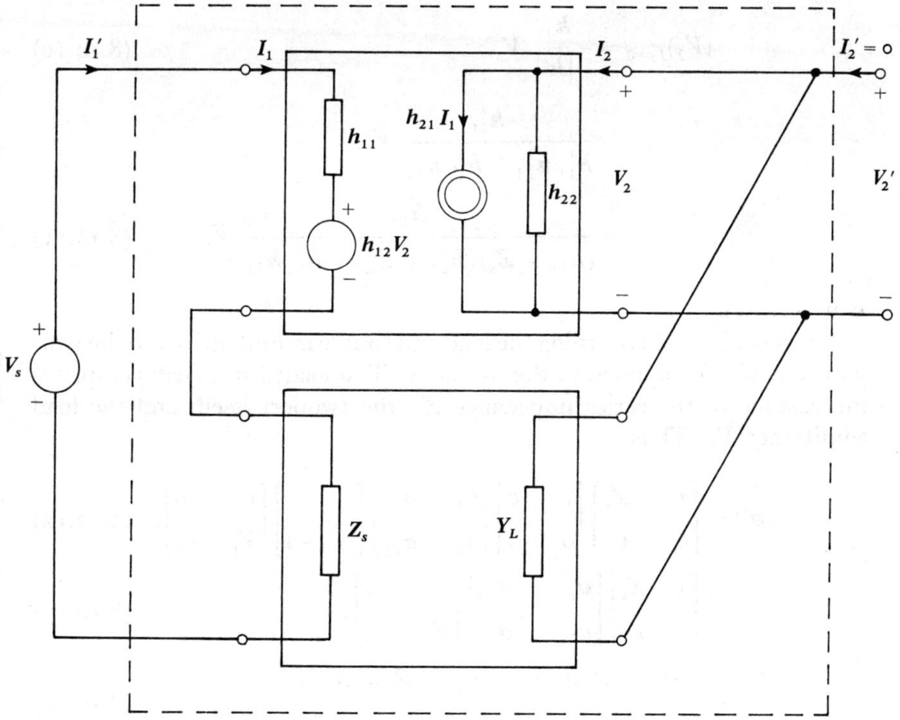

Fig. 8.19 Loading of a twoport by series–parallel connexion of loading elements.

addition. Fig. 8.19 illustrates this situation for the $\boldsymbol{h}$-model. It follows that if the $\boldsymbol{h}$-matrix of the original twoport is

$$[\boldsymbol{h}] = \begin{bmatrix} \boldsymbol{h}_{11} & \boldsymbol{h}_{12} \\ \boldsymbol{h}_{21} & \boldsymbol{h}_{22} \end{bmatrix}, \qquad (8.10.7)$$

then the matrix of the twoport loaded by $\boldsymbol{Z}_s$ and $\boldsymbol{Y}_L$ is

$$[\boldsymbol{h}'] = \begin{bmatrix} (\boldsymbol{h}_{11} + \boldsymbol{Z}_s) & \boldsymbol{h}_{12} \\ \boldsymbol{h}_{21} & (\boldsymbol{h}_{22} + \boldsymbol{Y}_L) \end{bmatrix}. \qquad (8.10.8)$$

The properties of the loaded network are then described by the equation

$$\begin{bmatrix} V_s \\ I_2' \end{bmatrix} = [h'] \begin{bmatrix} I_1' \\ V_2' \end{bmatrix}, \tag{8.10.9}$$

in which I_2' is zero because the load has been absorbed into the matrix. The output voltage, for instance, is given by

$$(V_2')_{I_2'=0} = -\frac{h_{21}'}{|h'|}.V_s, \tag{8.10.10}$$

$$= \frac{-h_{21}'}{h_{11}'h_{22}' - h_{12}'h_{21}'}.V_s,$$

$$= \frac{-h_{21}}{(h_{11}+Z_s)(h_{22}+Y_L) - h_{12}h_{21}}.V_s \tag{8.10.11}$$

as in (8.10.3).

The technique of absorbing the load and source immittances into the twoport may also be applied to the a-matrix. The loaded a-matrix comprises the cascade of the series impedance Z_s, the twoport itself, and the load admittance Y_L. Thus

$$[a'] = \begin{bmatrix} 1 & -Z_s \\ 0 & -1 \end{bmatrix} \begin{bmatrix} 1 & 0 \\ 0 & -1 \end{bmatrix} \begin{bmatrix} a_{11} & a_{12} \\ a_{21} & a_{22} \end{bmatrix} \begin{bmatrix} 1 & 0 \\ 0 & -1 \end{bmatrix} \begin{bmatrix} 1 & 0 \\ Y_L & -1 \end{bmatrix}, \tag{8.10.12}$$

$$= \begin{bmatrix} 1 & Z_s \\ 0 & 1 \end{bmatrix} \begin{bmatrix} a_{11} & -a_{12} \\ a_{21} & -a_{22} \end{bmatrix} \begin{bmatrix} 1 & 0 \\ Y_L & -1 \end{bmatrix}, \tag{8.10.13}$$

$$= \begin{bmatrix} (a_{11}+Z_s a_{21}) & -(a_{12}+Z_s a_{22}) \\ a_{21} & -a_{22} \end{bmatrix} \begin{bmatrix} 1 & 0 \\ Y_L & -1 \end{bmatrix}, \tag{8.10.14}$$

$$= \begin{bmatrix} a_{11}+Z_s a_{21} - Y_L(a_{12}+Z_s a_{22}) & (a_{12}+Z_s a_{22}) \\ (a_{21} - Y_L a_{22}) & a_{22} \end{bmatrix}, \tag{8.10.15}$$

$$= \begin{bmatrix} a_{11}' & a_{12}' \\ a_{21}' & a_{22}' \end{bmatrix}, \tag{8.10.16}$$

and the output voltage, for example, is given by

$$(V_2')_{I_2'=0} = \frac{1}{a_{11}'}.V_s, \qquad \text{(See (8.9.7))}$$

$$= \frac{1}{a_{11}+Z_s a_{21} - Y_L(a_{12}+Z_s a_{22})}.V_s.$$

Thus if an active twoport has the matrix

$$[\boldsymbol{a}] = \begin{bmatrix} -(3\times 10^{-4}) & 2 \\ -(2\times 10^{-6}) & 10^{-2} \end{bmatrix}, \tag{8.10.17}$$

and if it is loaded with a source and a load each equal to 1 kΩ, the loaded matrix is

$$[\boldsymbol{a}'] = \begin{bmatrix} 1 & 10^3 \\ 0 & 1 \end{bmatrix} \begin{bmatrix} -(3\times 10^{-4}) & -2 \\ -(2\times 10^{-6}) & -10^{-2} \end{bmatrix} \begin{bmatrix} 1 & 0 \\ 10^{-3} & -1 \end{bmatrix}. \tag{8.10.18}$$

The required term $\boldsymbol{a}'_{11}$ is obtained from the first row of the product of the first two matrices and the first column of the third matrix:

$$\begin{aligned} \boldsymbol{a}'_{11} &= (-3\times 10^{-4} - 2\times 10^{-3}) + (-2-10)\,10^{-3}, \\ &= (-2.3\times 10^{-3}) - (12\times 10^{-3}), \\ &= -14.3\times 10^{-3}, \end{aligned}$$

and

$$\begin{aligned} \frac{\boldsymbol{V}'_2}{\boldsymbol{V}_s} &= \frac{1}{\boldsymbol{a}'_{11}}, \\ &= -70. \end{aligned}$$

This is slightly less direct than the use of the $\boldsymbol{h}$-matrix. Corresponding with (8.10.17) we have for the active twoport itself

$$[\boldsymbol{h}] = \begin{bmatrix} 200 & (0.1\times 10^{-3}) \\ 100 & (0.2\times 10^{-3}) \end{bmatrix}. \tag{8.10.19}$$

The loaded matrix ($\boldsymbol{Z}_s = 1$ kΩ, $\boldsymbol{Y}_L = 1$ mS) is

$$[\boldsymbol{h}'] = \begin{bmatrix} 1200 & (0.1\times 10^{-3}) \\ 100 & (1.2\times 10^{-3}) \end{bmatrix}, \tag{8.10.20}$$

and the output voltage per unit source voltage is

$$\begin{aligned} \frac{\boldsymbol{V}'_2}{\boldsymbol{V}_s} &= \frac{-\boldsymbol{h}'_{21}}{|\boldsymbol{h}'|}, \\ &= \frac{-100}{1.44-0.01}, \\ &= -70. \end{aligned}$$

8.11 Reciprocity and symmetry in twoports

The reciprocity principle, discussed in chapter 6, states that the ratio of response to stimulus (where one is a voltage and the other a current) in a reciprocal network is independent of an exchange in the positions of the stimulus and the response. In choosing points at which to apply a stimulus and measure a response we are viewing the network as a twoport, and it is instructive to find what is needed in a twoport to satisfy the reciprocity principle. It turns out that, apart from the requirement of no independent sources in the network, there must be an internal relationship between the four parameters of any one set. By means of this relationship one of the four can be eliminated and the network can be described in terms of only three independent parameters.

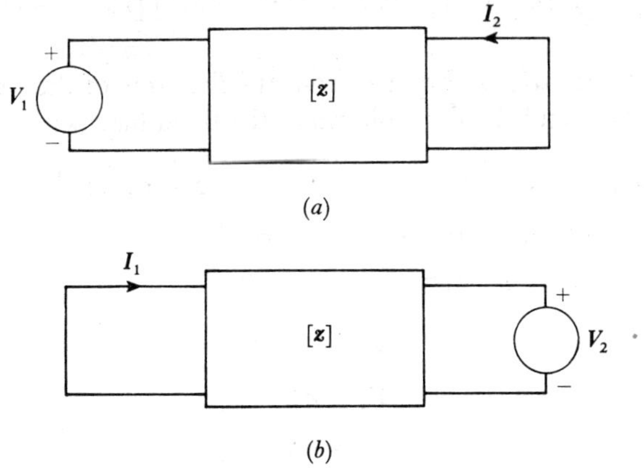

Fig. 8.20 Test conditions for reciprocity.

Figure 8.20 shows the test conditions for reciprocity in the $\boldsymbol{z}$-model of a twoport. The ratio $(\boldsymbol{I}_2/\boldsymbol{V}_1)$ in part (a) must equal the ratio $(\boldsymbol{I}_1/\boldsymbol{V}_2)$ in part (b). But from the basic equations

$$\left.\begin{aligned}\boldsymbol{V}_1 &= \boldsymbol{z}_{11}\boldsymbol{I}_1 + \boldsymbol{z}_{12}\boldsymbol{I}_2,\\ \boldsymbol{V}_2 &= \boldsymbol{z}_{21}\boldsymbol{I}_1 + \boldsymbol{z}_{22}\boldsymbol{I}_2,\end{aligned}\right\} \qquad \text{(see (8.2.1))}$$

it follows that

$$\left(\frac{\boldsymbol{I}_2}{\boldsymbol{V}_1}\right)_{V_2=0} = -\frac{\boldsymbol{z}_{21}}{|\boldsymbol{z}|}$$

and

$$\left(\frac{\boldsymbol{I}_1}{\boldsymbol{V}_2}\right)_{V_1=0} = -\frac{\boldsymbol{z}_{12}}{|\boldsymbol{z}|}$$

from which it follows that for the network to be reciprocal

$$z_{12} = z_{21}. \tag{8.11.1}$$

For the other parameter-sets, similar arguments give the results shown in table 8.5.

TABLE 8.5 *Requirements for reciprocity and symmetry in twoports*

Parameter-set	Requirement for reciprocity	Additional requirement for symmetry
$[z]$	$z_{12} = z_{21}$	$z_{11} = z_{22}$
$[y]$	$y_{12} = y_{21}$	$y_{11} = y_{22}$
$[h]$	$h_{12} = -h_{21}$	$\|h\| = 1$
$[g]$	$g_{12} = -g_{21}$	$\|g\| = 1$
$[a]$	$\|a\| = -1$	$a_{11} = -a_{22}$
$[b]$	$\|b\| = -1$	$b_{11} = -b_{22}$

A *symmetrical* twoport is one in which the two ports are electrically indistinguishable from one another; the twoport may be removed from a network of which it forms a part and reinserted with its ports interchanged without having any effect on the network as a whole. From fig. 8.5(*a*) it can be seen that, in addition to the reciprocity requirement, symmetry requires the self impedance coefficients z_{11} and z_{22} to be equal. The requirements of the other parameter-sets are shown in table 8.5. They can be derived either by recourse to the conversion formulae of table 8.2 or by a direct argument. For the **a**- and **b**-parameters this means simply equating the matrix to its inverse; for the **h**- or **g**-parameters it is necessary to equate the matrix to the original matrix altered by the exchange of elements at opposite ends of each diagonal. The reason for this is best seen by following the argument through. If

$$\begin{bmatrix} V_1 \\ I_2 \end{bmatrix} = \begin{bmatrix} h_{11} & h_{12} \\ h_{21} & h_{22} \end{bmatrix} \begin{bmatrix} I_1 \\ V_2 \end{bmatrix}, \tag{8.11.2}$$

then

$$\begin{bmatrix} I_2 \\ V_1 \end{bmatrix} = \begin{bmatrix} h_{22} & h_{21} \\ h_{12} & h_{11} \end{bmatrix} \begin{bmatrix} V_2 \\ I_1 \end{bmatrix}. \tag{8.11.3}$$

But

$$\begin{bmatrix} I_1 \\ V_2 \end{bmatrix} = \frac{1}{|h|} \begin{bmatrix} h_{22} & -h_{12} \\ -h_{21} & h_{11} \end{bmatrix} \begin{bmatrix} V_1 \\ I_2 \end{bmatrix}, \tag{8.11.4}$$

and symmetry will exist if the pre-multiplying matrices on the right hand sides of (8.11.3,4) are equal.

8.12 Multiport networks

Although this chapter is concerned primarily with twoport networks, it is interesting to look briefly at the general multiport. A linear n-port is shown

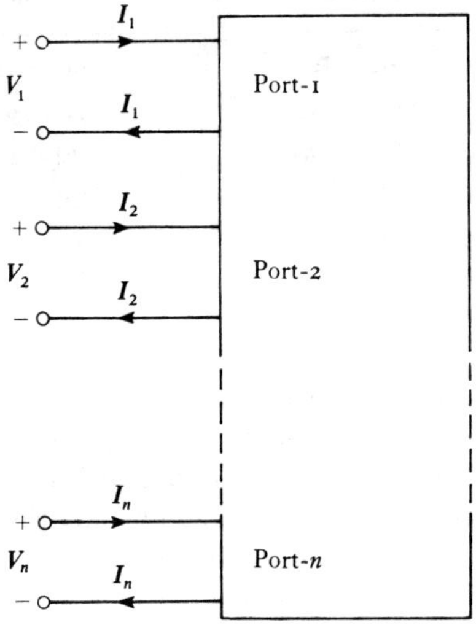

Fig. 8.21 An n-port network.

in fig. 8.21. The superposition principle tells us that the port voltages must be linearly dependent on the port currents and vice versa, so we may write

$$\left.\begin{aligned} \boldsymbol{V}_1 &= \boldsymbol{z}_{11}\boldsymbol{I}_1 + \boldsymbol{z}_{12}\boldsymbol{I}_2 + \ldots + \boldsymbol{z}_{1n}\boldsymbol{I}_n \\ \boldsymbol{V}_2 &= \boldsymbol{z}_{21}\boldsymbol{I}_1 + \boldsymbol{z}_{22}\boldsymbol{I}_2 + \ldots + \boldsymbol{z}_{2n}\boldsymbol{I}_n \\ \ldots & \quad \ldots \quad \ldots \quad \ldots \quad \ldots \\ \boldsymbol{V}_n &= \boldsymbol{z}_{n1}\boldsymbol{I}_1 + \boldsymbol{z}_{n2}\boldsymbol{I}_2 + \ldots + \boldsymbol{z}_{nn}\boldsymbol{I}_n \end{aligned}\right\}. \tag{8.12.1}$$

The physical significance of any coefficient is given by

$$\boldsymbol{z}_{ij} = \left(\frac{\boldsymbol{V}_i}{\boldsymbol{I}_j}\right)_{\boldsymbol{I}_1, \boldsymbol{I}_2, \ldots, \boldsymbol{I}_{j-1}, \boldsymbol{I}_{j+1}, \ldots, \boldsymbol{I}_n = 0} \tag{8.12.2}$$

and is the ratio of the open-circuit voltage produced at port i by a current source applied at port j when all other ports are open-circuited.

Equation (8.12.1) is conveniently written in the matrix form

$$\begin{bmatrix} V_1 \\ V_2 \\ \cdots \\ V_n \end{bmatrix} = \begin{bmatrix} z_{11} & z_{12} & \cdots & z_{1n} \\ z_{21} & z_{22} & \cdots & z_{2n} \\ \cdots & \cdots & \cdots & \cdots \\ z_{n1} & z_{n2} & \cdots & z_{nn} \end{bmatrix} \begin{bmatrix} I_1 \\ I_2 \\ \cdots \\ I_n \end{bmatrix}, \tag{8.12.3}$$

or simply

$$[V] = [z][I]. \tag{8.12.4}$$

The impedance matrix of a given network may be formed by the use of (8.12.2). This method, applied to the threeport shown in fig. 8.22, gives

$$[z] = \begin{bmatrix} \left(1 + s + \frac{1}{s}\right) & s & \frac{1}{s} \\ s & s & 0 \\ \frac{1}{s} & 0 & \left(1 + \frac{1}{s}\right) \end{bmatrix}. \tag{8.12.5}$$

Symmetry about the principal diagonal is a consequence of the reciprocal nature of this (passive) threeport and may be expressed by the equation

$$z_{ij} = z_{ji}, \tag{8.12.6}$$

which is the multiport version of the reciprocity requirement for a twoport (see (8.11.1)).

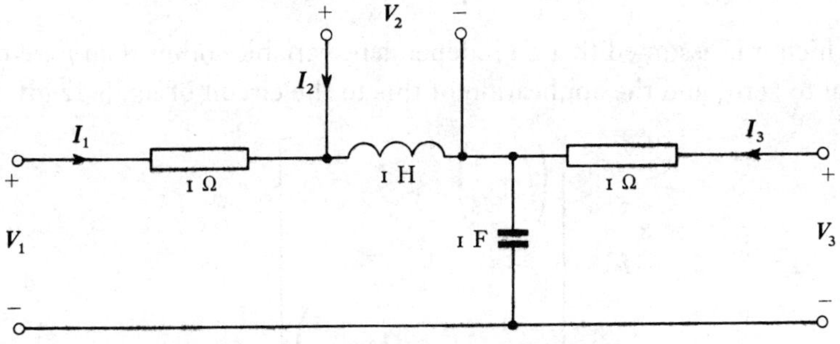

Fig. 8.22 A threeport network.

An alternative way of describing a multiport is by means of its admittance matrix. Pre-multiplying both sides of (8.12.4) by $[\boldsymbol{z}]^{-1}$ gives

$$[\boldsymbol{I}] = [\boldsymbol{y}][\boldsymbol{V}], \tag{8.12.7}$$

where

$$[\boldsymbol{y}] = [\boldsymbol{z}]^{-1}, \tag{8.12.8}$$

and

$$\boldsymbol{y}_{ij} = \left(\frac{\boldsymbol{I}_i}{\boldsymbol{V}_j}\right)_{V_1, V_2, \ldots, V_{j-1}, V_{j+1}, \ldots, V_n = 0}, \tag{8.12.9}$$

that is, $\boldsymbol{y}_{ij}$ is the short-circuit current $\boldsymbol{I}_i$ produced at port i as a result of an applied voltage $\boldsymbol{V}_j$ at port j when all other ports are short-circuited.

Descriptions corresponding with the hybrid and transmission parameters of twoports are also available to multiports. If the variables are arranged in the form

$$\begin{bmatrix} \boldsymbol{V}_1 \\ \boldsymbol{I}_2 \\ \boldsymbol{V}_3 \\ \cdots \\ \boldsymbol{I}_n \end{bmatrix} = \begin{bmatrix} \boldsymbol{h}_{11} & \boldsymbol{h}_{12} & \cdots & \boldsymbol{h}_{1n} \\ \boldsymbol{h}_{21} & \boldsymbol{h}_{22} & \cdots & \boldsymbol{h}_{2n} \\ \boldsymbol{h}_{31} & \boldsymbol{h}_{32} & \cdots & \boldsymbol{h}_{3n} \\ \cdots & \cdots & \cdots & \cdots \\ \boldsymbol{h}_{n1} & \boldsymbol{h}_{n2} & \cdots & \boldsymbol{h}_{nn} \end{bmatrix} \begin{bmatrix} \boldsymbol{I}_1 \\ \boldsymbol{V}_2 \\ \boldsymbol{I}_3 \\ \cdots \\ \boldsymbol{V}_n \end{bmatrix}, \tag{8.12.10}$$

the square matrix in the equation is called the hybrid matrix $[\boldsymbol{h}]$ of the n-port. Equation (8.12.10) is sometimes written in the mnemonic form

$$\begin{bmatrix} \boldsymbol{V}_i \\ \boldsymbol{I}_o \end{bmatrix} = [\boldsymbol{h}] \begin{bmatrix} \boldsymbol{I}_i \\ \boldsymbol{V}_o \end{bmatrix}. \tag{8.12.11}$$

The elements of the hybrid matrix are given by

$$\boldsymbol{h}_{ij} = \frac{\text{Response at port } i \text{ (o.c. voltage or s.c. current)}}{\text{Stimulus at port } j \text{ (voltage or current)}}, \tag{8.12.12}$$

in which it is assumed that all independent variables other than j are made equal to zero; and the application of this to the circuit of fig. 8.22 gives

$$[\boldsymbol{h}] = \begin{bmatrix} \left(1 + \dfrac{1}{s}\right) & 1 & \dfrac{1}{s} \\ -1 & \dfrac{1}{s} & 0 \\ \dfrac{1}{s} & 0 & \left(1 + \dfrac{1}{s}\right) \end{bmatrix}.$$

The inverse hybrid matrix $[\boldsymbol{g}]$ results from a rearrangement of the n-port variables into the form

$$\begin{bmatrix} I_1 \\ V_2 \\ I_3 \\ \cdots \\ V_n \end{bmatrix} = \begin{bmatrix} g_{11} & g_{12} & \cdots & g_{1n} \\ g_{21} & g_{22} & \cdots & g_{2n} \\ g_{31} & g_{32} & \cdots & g_{3n} \\ \cdots & \cdots & \cdots & \cdots \\ g_{n1} & g_{n2} & \cdots & g_{nn} \end{bmatrix} \begin{bmatrix} V_1 \\ I_2 \\ V_3 \\ \cdots \\ I_n \end{bmatrix}, \tag{8.12.13}$$

which is sometimes mnemonically described by

$$\begin{bmatrix} I_i \\ V_o \end{bmatrix} = [\boldsymbol{g}] \begin{bmatrix} V_i \\ I_o \end{bmatrix} \tag{8.12.14}$$

and where

$$[\boldsymbol{g}] = [\boldsymbol{h}]^{-1}. \tag{8.12.15}$$

Corresponding equations in terms of the transmission parameters arise when the variables are put in the form

$$\begin{bmatrix} V_1 \\ I_1 \\ V_3 \\ I_3 \\ \cdots \\ V_{n-1} \\ I_{n-1} \end{bmatrix} = \begin{bmatrix} a_{11} & a_{12} & \cdots & a_{1n} \\ a_{21} & a_{22} & \cdots & a_{2n} \\ a_{31} & a_{32} & \cdots & a_{3n} \\ a_{41} & a_{42} & \cdots & a_{4n} \\ \cdots & \cdots & \cdots & \cdots \\ a_{n-1,1} & a_{n-1,2} & \cdots & a_{n-1,n} \\ a_{n1} & a_{n2} & \cdots & a_{nn} \end{bmatrix} \begin{bmatrix} V_2 \\ I_2 \\ V_4 \\ I_4 \\ \cdots \\ V_n \\ I_n \end{bmatrix}, \tag{8.12.16}$$

or

$$\begin{bmatrix} V_i \\ I_i \end{bmatrix} = [\boldsymbol{a}] \begin{bmatrix} V_o \\ I_o \end{bmatrix} \tag{8.12.17}$$

and the inverse transmission parameters are given by

$$[\boldsymbol{b}] = [\boldsymbol{a}]^{-1}. \tag{8.12.18}$$

A formula for the coefficients $\boldsymbol{a}_{ij}$ similar to (8.12.2,9,12) can be found from (8.12.16) but, unlike the others, this formula is not helpful because it requires both the current and the voltage at one port or another to be zero simultaneously. This prevents the evaluation of the $\boldsymbol{a}$-parameters by short-circuiting or open-circuiting the correct ports in the manner used for evaluating the $\boldsymbol{z}$-, $\boldsymbol{y}$-, $\boldsymbol{h}$- and $\boldsymbol{g}$-parameters.

The matrices of interconnected n-ports are obtained from the formulae already derived for twoports provided that the port requirements still obtain after the interconnexions are made. The series connexion of two

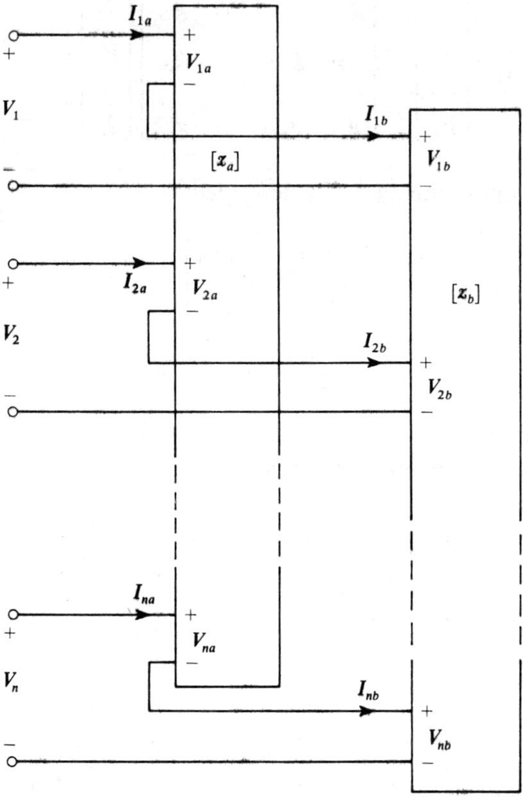

Fig. 8.23 Series connexion of two n-ports.

n-ports a and b is illustrated in fig. 8.23. If the two networks are described by the equations

$$[V_a] = [z_a][I_a], \tag{8.12.19}$$

and

$$[V_b] = [z_b][I_b], \tag{8.12.20}$$

then since

$$[V] = [V_a] + [V_b], \tag{8.12.21}$$

and

$$[I] = [I_a] = [I_b], \tag{8.12.22}$$

by virtue of the series interconnexions, it follows that

$$[V] = ([z_a] + [z_b])[I], \tag{8.12.23}$$

and the matrix of the interconnected n-ports is therefore

$$[\boldsymbol{z}] = [\boldsymbol{z}_a] + [\boldsymbol{z}_b]. \tag{8.12.24}$$

Analogous results, in terms of the $\boldsymbol{y}$-, $\boldsymbol{h}$- and $\boldsymbol{g}$-parameters, apply respectively to the parallel, series–parallel, and parallel–series interconnexions of n-ports. The sign-reversing matrix

$$\begin{bmatrix} 1 & 0 \\ 0 & -1 \end{bmatrix}, \tag{8.12.25}$$

however, used with the $\boldsymbol{a}$-matrices of cascaded twoports, now takes an order appropriate to the size of the multiport. For the cascaded fourports shown in fig. 8.24, for example, the sign-reversing matrix is

$$\begin{bmatrix} 1 & 0 & 0 & 0 \\ 0 & -1 & 0 & 0 \\ 0 & 0 & 1 & 0 \\ 0 & 0 & 0 & -1 \end{bmatrix}, \tag{8.12.26}$$

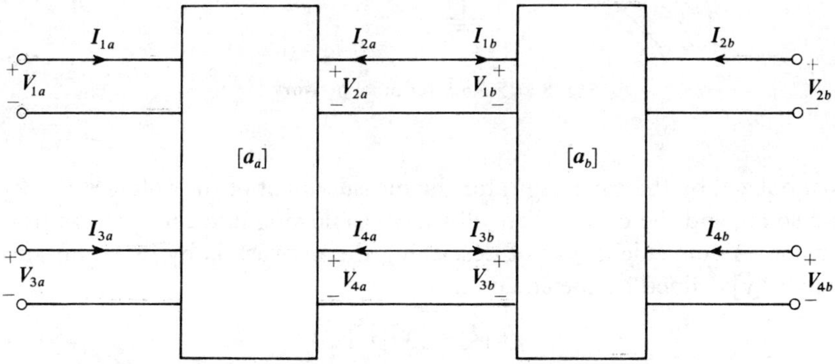

Fig. 8.24 Two cascaded fourports.

so that the output variables $\boldsymbol{V}_{2a}$, $\boldsymbol{I}_{2a}$, $\boldsymbol{V}_{4a}$ and $\boldsymbol{I}_{4a}$ of network a are related to the input variables $\boldsymbol{V}_{1b}$, $\boldsymbol{I}_{1b}$, $\boldsymbol{V}_{3b}$ and $\boldsymbol{I}_{3b}$ of network b by

$$\begin{bmatrix} \boldsymbol{V}_{2a} \\ \boldsymbol{I}_{2a} \\ \boldsymbol{V}_{4a} \\ \boldsymbol{I}_{4a} \end{bmatrix} = \begin{bmatrix} 1 & 0 & 0 & 0 \\ 0 & -1 & 0 & 0 \\ 0 & 0 & 1 & 0 \\ 0 & 0 & 0 & -1 \end{bmatrix} \begin{bmatrix} \boldsymbol{V}_{1b} \\ \boldsymbol{I}_{1b} \\ \boldsymbol{V}_{3b} \\ \boldsymbol{I}_{3b} \end{bmatrix} = \begin{bmatrix} \boldsymbol{V}_{1b} \\ -\boldsymbol{I}_{1b} \\ \boldsymbol{V}_{3b} \\ -\boldsymbol{I}_{3b} \end{bmatrix}. \tag{8.12.27}$$

8.13 The *n*-terminal network

Matrix terminology is fortunately applicable not only to multiport networks but also to networks not operated on a port basis. In the *n*-terminal network shown in fig. 8.25 the *n*th terminal is assumed to act as a reference point

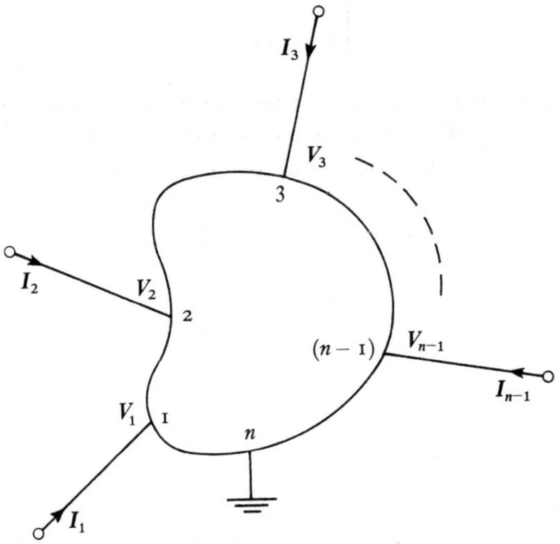

Fig. 8.25 An *n*-terminal network.

(symbolized by the earth sign) for the measurement of the voltages $\boldsymbol{V}_1$, $\boldsymbol{V}_2$, and so on, and the currents are all shown as flowing into the network from outside. A convenient way of describing the network is by its admittance matrix $[\boldsymbol{y}]$ defined by the equation

$$[\boldsymbol{I}] = [\boldsymbol{y}][\boldsymbol{V}] \tag{8.13.1}$$

in which

$$\boldsymbol{y}_{ij} = \left(\frac{\boldsymbol{I}_i}{\boldsymbol{V}_j}\right)_{V_1, V_2, \ldots, V_{j-1}, V_{j+1}, \ldots, V_{n-1}=0}. \tag{8.13.2}$$

The known $\boldsymbol{y}$-matrix of a given network can readily be modified to account for the addition of some extra two-terminal elements connected across its terminals by observing that an admittance $\boldsymbol{Y}_L$ connected between terminals p and n produces an additional flow of current $\boldsymbol{Y}_L \boldsymbol{V}_p$ in terminal p, and that this has the effect of increasing the matrix element $\boldsymbol{y}_{pp}$ to $(\boldsymbol{y}_{pp} + \boldsymbol{Y}_L)$. Similarly, a mutual admittance element $\boldsymbol{Y}_m$ connected between terminals p and q causes an increased flow of current $\boldsymbol{Y}_m(\boldsymbol{V}_p - \boldsymbol{V}_q)$ in terminal p and an increase of $\boldsymbol{Y}_m(\boldsymbol{V}_q - \boldsymbol{V}_p)$ in terminal q. To take account of this in the matrix it is only necessary to add $\boldsymbol{Y}_m$ to each of the elements $\boldsymbol{y}_{pp}$ and $\boldsymbol{y}_{qq}$, and to subtract $\boldsymbol{Y}_m$ from each of the elements $\boldsymbol{y}_{pq}$ and $\boldsymbol{y}_{qp}$. These ideas make it possible to construct the $\boldsymbol{y}$-matrix of many networks by inspection. At the

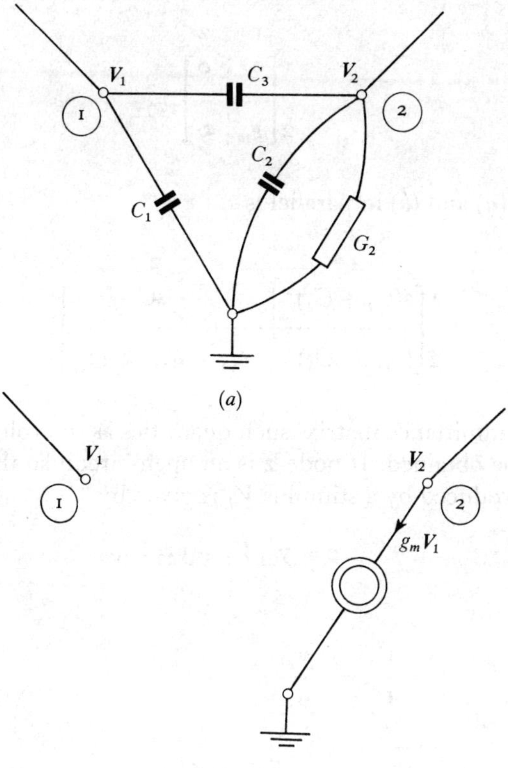

Fig. 8.26 Three-terminal networks: (*a*) parasitic capacitances of an active device together with a load; (*b*) ideal voltage-controlled current source.

start, the network is hypothetically denuded of all its elements, leaving a set of unconnected nodes 1, 2, ..., $(n-1)$, for which the $\boldsymbol{y}$-matrix is a set of zeros (a zero matrix of order $(n-1) \times (n-1)$). One by one, each element is replaced in the network and the $\boldsymbol{y}$-matrix is built up according to the process of addition described above. Applied to fig. 8.26(*a*) this technique gives

$$\begin{array}{c} \\ 1 \\ 2 \end{array}\begin{array}{c} \begin{array}{cc} \quad 1 \quad & \quad\quad 2 \quad \end{array} \\ \left[\begin{array}{c:c} s(C_1 + C_3) & -sC_3 \\ \hdashline -sC_3 & G_2 + s(C_2 + C_3) \end{array}\right]. \end{array} \tag{8.13.3}$$

Applied to more complicated circuits it produces the required result with increased labour but with no increase in conceptual difficulty. Used in conjunction with the known admittance matrix of a three-terminal element such as an ideal voltage-controlled current source (see fig. 8.26(*b*)) it enables the admittance matrix of an active circuit to be written down by inspection. The admittance matrix of fig. 8.26(*b*) is

$$\begin{array}{c} \\ 1 \\ 2 \end{array}\begin{array}{c} \begin{array}{cc} 1 & 2 \end{array} \\ \left[\begin{array}{c|c} 0 & 0 \\ \hline g_m & 0 \end{array}\right] \end{array}, \tag{8.13.4}$$

and of parts (*a*) and (*b*) in parallel is

$$\begin{array}{c} \\ 1 \\ 2 \end{array}\begin{array}{c} \begin{array}{cc} 1 & 2 \end{array} \\ \left[\begin{array}{c|c} s(C_1+C_3) & -sC_3 \\ \hline (g_m - sC_3) & G_2 + s(C_2+C_3) \end{array}\right] \end{array}. \tag{8.13.5}$$

From the admittance matrix, such quantities as the voltage amplification can readily be obtained. If node-2 is an open-circuit so that $\boldsymbol{I}_2$ is zero, the voltage $\boldsymbol{V}_2$ produced by a stimulus $\boldsymbol{V}_1$ is given by

$$0 = y_{21}\boldsymbol{V}_1 + y_{22}\boldsymbol{V}_2, \tag{8.13.6}$$

or

$$\frac{\boldsymbol{V}_2}{\boldsymbol{V}_1} = -\frac{y_{21}}{y_{22}}, \tag{8.13.7}$$

$$= -\frac{(g_m - sC_3)}{G_2 + s(C_2 + C_3)}. \tag{8.13.8}$$

8.14 The indefinite admittance matrix and its applications

It is often convenient to measure the node voltages of an n-terminal network not with respect to any one terminal of the network but with respect to a terminal isolated from the network and usually thought of as being at earth (or reference) potential. The matrix formed by using these n external voltages rather than the $(n-1)$ internal voltages is called the *indefinite matrix* or the *floating matrix* of the network. The relationship between the internal voltages $\boldsymbol{V}_1'$, $\boldsymbol{V}_2'$, and $\boldsymbol{V}_3'$ with respect to terminal-4 of the four-terminal network shown in fig. 8.27 and the external voltages V_1, V_2, V_3, and V_4 with respect to the isolated reference terminal is typically

$$V_1' = V_1 - V_4. \tag{8.14.1}$$

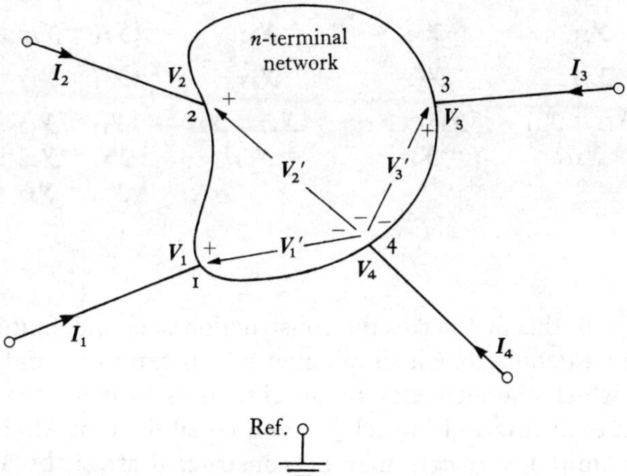

Fig. 8.27 Four-terminal network with additional isolated reference terminal.

If the admittance matrix equation of the network is

$$\begin{bmatrix} I_1 \\ I_2 \\ I_3 \end{bmatrix} = \begin{bmatrix} y_{11} & y_{12} & y_{13} \\ y_{21} & y_{22} & y_{23} \\ y_{31} & y_{32} & y_{33} \end{bmatrix} \begin{bmatrix} V_1' \\ V_2' \\ V_3' \end{bmatrix}, \tag{8.14.2}$$

then each of the three equations represented by (8.14.2) has the form

$$\begin{aligned} I_1 &= y_{11}(V_1 - V_4) + y_{12}(V_2 - V_4) + y_{13}(V_3 - V_4), \\ &= y_{11} V_1 + y_{12} V_2 + y_{13} V_3 - (y_{11} + y_{12} + y_{13}) V_4. \end{aligned} \tag{8.14.3}$$

That is, by transferring the voltage reference point from terminal-4 of the network itself to some external point, there has been established a new column in the matrix corresponding with the voltage V_4; and the elements in this column are such as to make the sum of the elements in each row equal to zero.

The final step is to introduce the current I_4 into the matrix by means of the equation

$$I_4 = -(I_1 + I_2 + I_3). \tag{8.14.4}$$

The result is the indefinite admittance matrix (8.14.5) in which it can be seen that the sum of the elements in any row or column is zero.

$$\begin{array}{c} \\ 1 \\ 2 \\ 3 \\ 4 \end{array}\begin{array}{c} \begin{array}{cccc} \quad 1 \quad & \quad 2 \quad & \quad 3 \quad & \quad 4 \quad \end{array} \\ \begin{bmatrix} y_{11} & y_{12} & y_{13} & -(y_{11}+y_{12}+y_{13}) \\ y_{21} & y_{22} & y_{23} & -(y_{21}+y_{22}+y_{23}) \\ y_{31} & y_{32} & y_{33} & -(y_{31}+y_{32}+y_{33}) \\ -(y_{11}+y_{21}+y_{31}) & -(y_{12}+y_{22}+y_{32}) & -(y_{13}+y_{23}+y_{33}) & (y_{11}+y_{12}+y_{13}+y_{21}+y_{22}+y_{23}+y_{31}+y_{32}+y_{33}) \end{bmatrix} \end{array}. \tag{8.14.5}$$

A knowledge of this fact makes the construction of an indefinite admittance matrix from a definite one a simple matter; an extra row and column are created in which the elements are so chosen as to make the sum of the elements in each row and in each column equal to zero. Alternatively the indefinite admittance matrix may be constructed afresh by applying the method of §8.13 to the circuit nodes plus the earth terminal.

Before giving some examples of the ways in which the indefinite admittance matrix may be used, it is convenient to consider the following specific features of both kinds of matrix:

1. Connecting a terminal to earth

If a terminal p of a network is connected to earth so that V_p becomes zero, the pth column of the matrix may be deleted. Furthermore, if the current I_p in the pth terminal is no longer of interest, the pth row may be deleted. (8.14.6)

2. Connecting two terminals together

If two terminals p and q are connected together so that $V_p = V_q$ and the current of the combined terminals is $(I_p + I_q)$, the effect on an admittance matrix is that of adding rows p and q and adding columns p and q. For example, if terminals-2 and -3 of the admittance matrix (8.14.7)

$$\begin{array}{c} \\ 1 \\ 2 \\ 3 \end{array}\begin{array}{c} \begin{array}{ccc} 1 & 2 & 3 \end{array} \\ \begin{bmatrix} a & b & c \\ d & e & f \\ g & h & i \end{bmatrix} \end{array}$$

are connected together, the resulting matrix is

$$\begin{array}{c} \\ 1 \\ 2,3 \end{array}\begin{array}{c} \quad 1 \qquad\qquad 2,3 \\ \begin{bmatrix} a & (b+c) \\ (d+g) & (e+f+h+i) \end{bmatrix}. \end{array} \tag{8.14.8}$$

3. *Suppression of a terminal*

If a terminal p is on open-circuit so that $\boldsymbol{I}_p$ is zero, row p and column p may be eliminated from the matrix, if desired, by a technique which will be described and whose justification will now be developed.

From row p of a set of admittance equations, assuming $\boldsymbol{I}_p$ to be zero, the following expression may be obtained for $\boldsymbol{V}_p$:

$$\boldsymbol{V}_p = -\frac{1}{\boldsymbol{y}_{pp}}(\boldsymbol{y}_{p1}\boldsymbol{V}_1 + \boldsymbol{y}_{p2}\boldsymbol{V}_2 + \ldots + \boldsymbol{y}_{p,p-1}\boldsymbol{V}_{p-1}$$

$$+ \boldsymbol{y}_{p,p+1}\boldsymbol{V}_{p+1} + \ldots + \boldsymbol{y}_{pn}\boldsymbol{V}_n). \tag{8.14.9}$$

Now the equation for any other row r is

$$\boldsymbol{I}_r = \boldsymbol{y}_{r1}\boldsymbol{V}_1 + \boldsymbol{y}_{r2}\boldsymbol{V}_2 + \ldots + \boldsymbol{y}_{rp}\boldsymbol{V}_p + \ldots + \boldsymbol{y}_{rn}\boldsymbol{V}_n. \tag{8.14.10}$$

Putting (8.14.9) into (8.14.10) gives

$$\boldsymbol{I}_r = \left(\boldsymbol{y}_{r1} - \frac{\boldsymbol{y}_{rp}\boldsymbol{y}_{p1}}{\boldsymbol{y}_{pp}}\right)\boldsymbol{V}_1 + \left(\boldsymbol{y}_{r2} - \frac{\boldsymbol{y}_{rp}\boldsymbol{y}_{p2}}{\boldsymbol{y}_{pp}}\right)\boldsymbol{V}_2 + \ldots + \left(\boldsymbol{y}_{rn} - \frac{\boldsymbol{y}_{rp}\boldsymbol{y}_{pn}}{\boldsymbol{y}_{pp}}\right)\boldsymbol{V}_n + \ldots,$$

$$= \frac{1}{\boldsymbol{y}_{pp}}\Big\{(\boldsymbol{y}_{pp}\boldsymbol{y}_{r1} - \boldsymbol{y}_{rp}\boldsymbol{y}_{p1})\boldsymbol{V}_1$$

$$+ (\boldsymbol{y}_{pp}\boldsymbol{y}_{r2} - \boldsymbol{y}_{rp}\boldsymbol{y}_{p2})\boldsymbol{V}_2 + \ldots + (\boldsymbol{y}_{pp}\boldsymbol{y}_{rn} - \boldsymbol{y}_{rp}\boldsymbol{y}_{pn})\boldsymbol{V}_n\Big\} \tag{8.14.11}$$

which contains no term in $\boldsymbol{V}_p$, so the number of columns has been reduced by one. The number of rows is reduced by one simply by omitting row p, whose information content has now been absorbed by the rest of the matrix.

Equation (8.14.11) looks more tedious to apply than it really is. If attention is given to the element $\boldsymbol{y}_{pp}$ as illustrated in fig. 8.28(*a*) it will be seen that, apart from an external factor $1/\boldsymbol{y}_{pp}$, the first term in (8.14.11) is obtained by taking the difference between the products of elements at the corners of a rectangle bounded in one corner by $\boldsymbol{y}_{pp}$ and in the opposite corner by the

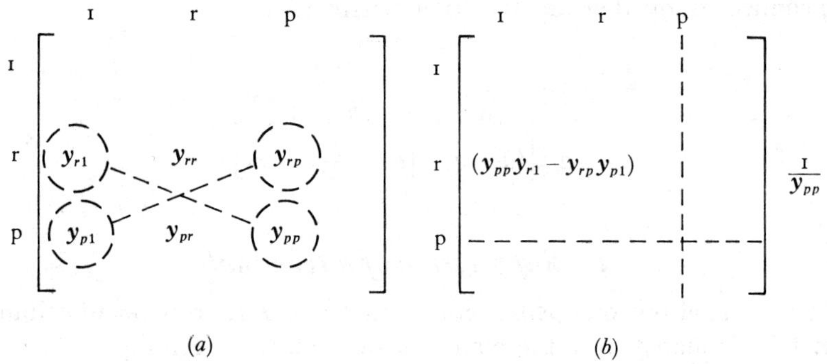

Fig. 8.28 Elimination of row p and column p from an admittance matrix when $I_p = 0$: (a) original matrix; (b) matrix with row p and column p deleted.

element which is to be replaced in the new array. For example, the elimination of terminal-2 (assuming $I_2 = 0$) in the admittance matrix

$$\begin{array}{c} \\ 1 \\ 2 \\ 3 \end{array}\begin{array}{c} \begin{array}{ccc} 1 & 2 & 3 \end{array} \\ \begin{bmatrix} a & b & c \\ d & e & f \\ g & h & i \end{bmatrix} \end{array}, \tag{8.14.12}$$

gives

$$\begin{array}{c} \\ 1 \\ 3 \end{array}\begin{array}{c} \begin{array}{cc} 1 & 3 \end{array} \\ \begin{bmatrix} (ea - bd) & (ec - bf) \\ (eg - dh) & (ei - hf) \end{bmatrix} \end{array} \cdot \frac{1}{e}, \tag{8.14.13}$$

or, if I_3 is zero instead of I_2,

$$\begin{array}{c} \\ 1 \\ 2 \end{array}\begin{array}{c} \begin{array}{cc} 1 & 2 \end{array} \\ \begin{bmatrix} (ia - gc) & (ib - hc) \\ (id - gf) & (ie - hf) \end{bmatrix} \end{array} \cdot \frac{1}{i}. \tag{8.14.14}$$

The indefinite admittance matrix is used in the first of the three following examples to produce a change in the internal reference terminal of a definite matrix. Suppose that the known $\boldsymbol{y}$-matrix of a three-terminal active device such as a transistor is, with respect to terminal-3,

$$\begin{array}{c} \\ 1 \\ 2 \end{array}\begin{array}{c} \begin{array}{cc} 1 & 2 \end{array} \\ \begin{bmatrix} (5 \times 10^{-3}) & -(0.5 \times 10^{-6}) \\ 0.5 & (0.15 \times 10^{-3}) \end{bmatrix} \end{array}, \tag{8.14.15}$$

then the indefinite admittance matrix of the device is, to three significant figures,

$$\begin{array}{c} \\ 1 \\ 2 \\ 3 \end{array}\begin{array}{c} \begin{array}{ccc} \quad 1 \quad & \quad 2 \quad & \quad 3 \quad \end{array} \\ \begin{bmatrix} (5\times 10^{-3}) & -(0.5\times 10^{-6}) & -(5\times 10^{-3}) \\ 0.5 & (0.15\times 10^{-3}) & -0.5 \\ -0.505 & -(0.150\times 10^{-3}) & 0.505 \end{bmatrix} \end{array}, \tag{8.14.16}$$

and the definite matrix with respect to either terminal-1 or terminal-2 is obtained by deleting row-1 and column-1, or row-2 and column-2 respectively (see (8.14.6) above). The definite matrix with respect to terminal-2 is

$$\begin{array}{c} \\ 1 \\ 3 \end{array}\begin{array}{c} \begin{array}{cc} \quad 1 \quad & \quad 3 \quad \end{array} \\ \begin{bmatrix} (5\times 10^{-3}) & -(5\times 10^{-3}) \\ -0.505 & 0.505 \end{bmatrix} \end{array}, \tag{8.14.17}$$

and a 200 Ω load resistor connected between terminal-3 and the new reference terminal will be absorbed into the matrix to give

$$\begin{array}{c} \\ 1 \\ 3 \end{array}\begin{array}{c} \begin{array}{cc} \quad 1 \quad & \quad 3 \quad \end{array} \\ \begin{bmatrix} (5\times 10^{-3}) & -(5\times 10^{-3}) \\ -0.505 & 0.51 \end{bmatrix} \end{array}, \tag{8.14.18}$$

from which the voltage amplification, for example, is seen to be

$$\begin{aligned} \frac{\boldsymbol{V}_3}{\boldsymbol{V}_1} &= -\frac{\text{bottom left hand element}}{\text{bottom right hand element}}, \\ &= \frac{0.505}{0.51}, \\ &= 0.99. \end{aligned}$$

A second example shows how the technique of terminal suppression may be used to find the 2 × 2 matrix of a pair of cascaded amplifiers. Fig. 8.29(*a*) gives the circuit, and parts (*b*) and (*c*) show the network decomposed into two parts whose parallel connexion restores the original. (The parallel connexion of networks (*b*) and (*c*) here means that terminal-1 in (*b*) is connected to terminal-1 in (*c*), terminal-2 in (*b*) to terminal-2 in (*c*), and so on, including the connexion of the earth line in (*b*) to that in (*c*).) The **y**-matrix of part (*b*) is

$$\begin{array}{c} \\ 1 \\ 2 \\ 3 \end{array}\begin{array}{c} \begin{array}{ccc} \quad 1 \quad & \quad 2 \quad & \quad 3 \quad \end{array} \\ \begin{bmatrix} y_{ia} & y_{ra} & 0 \\ y_{fa} & (y_{oa} + Y_1) & 0 \\ 0 & 0 & 0 \end{bmatrix} \end{array}, \tag{8.14.19}$$

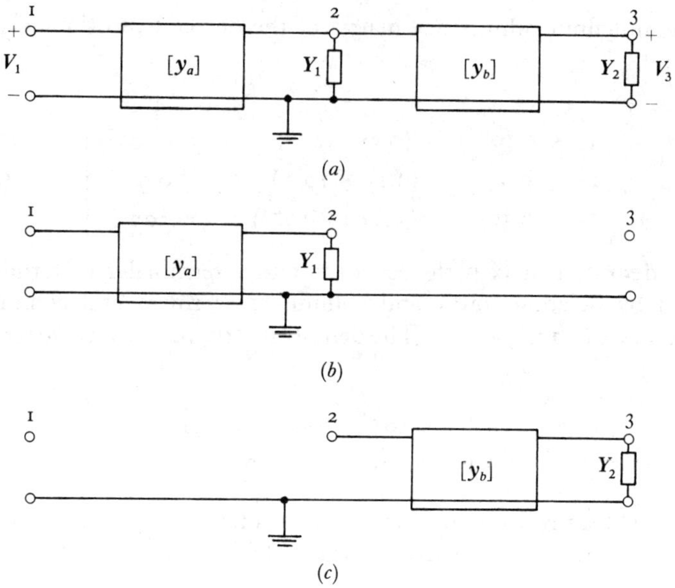

Fig. 8.29 Two amplifiers in cascade: (*a*) complete circuit; (*b*) and (*c*) decomposed circuit for parallel connexion.

and of part (*c*) is

$$\begin{array}{c} \\ 1 \\ 2 \\ 3 \end{array}\begin{array}{c} \begin{array}{ccc} 1 & 2 & 3 \end{array} \\ \left[\begin{array}{ccc} 0 & 0 & 0 \\ 0 & y_{ib} & y_{rb} \\ 0 & y_{fb} & (y_{ob}+Y_2) \end{array}\right] \end{array}. \tag{8.14.20}$$

The admittance matrix of the full circuit is the sum of (8.14.19) and (8.14.20) and is

$$\begin{array}{c} \\ 1 \\ 2 \\ 3 \end{array}\begin{array}{c} \begin{array}{ccc} 1 & 2 & 3 \end{array} \\ \left[\begin{array}{ccc} y_{ia} & y_{ra} & 0 \\ y_{fa} & (y_{ib}+y_{oa}+Y_1) & y_{rb} \\ 0 & y_{fb} & (y_{ob}+Y_2) \end{array}\right] \end{array}. \tag{8.14.21}$$

Now the current flowing into terminal-2 from outside the network is zero, and this terminal may be eliminated from the matrix if we are concerned only with the transmission between terminal-1 and terminal-3. The result is

$$\begin{array}{c} \\ 1 \\ 3 \end{array}\begin{array}{c} \begin{array}{cc} 1 & 3 \end{array} \\ \left[\begin{array}{cc} y_{ia}(y_{ib}+y_{oa}+Y_1)-y_{ra}y_{fa} & -y_{ra}y_{rb} \\ -y_{fa}y_{fb} & \begin{array}{c}(y_{ib}+y_{oa}+Y_1)(y_{ob}+Y_2) \\ -y_{rb}y_{fb}\end{array} \end{array}\right] \end{array} \cdot \frac{1}{(y_{ib}+y_{oa}+Y_1)} \tag{8.14.22}$$

and from this the open-circuit voltage amplification (V_3/V_1), for example, is

$$\frac{V_3}{V_1} = \frac{y_{fa}y_{fb}}{(y_{ib}+y_{oa}+Y_1)(y_{ob}+Y_2)-y_{rb}y_{fb}}. \tag{8.14.23}$$

The third example uses the indefinite matrix together with terminal suppression to find the composite parameters of a pair of transistors connected

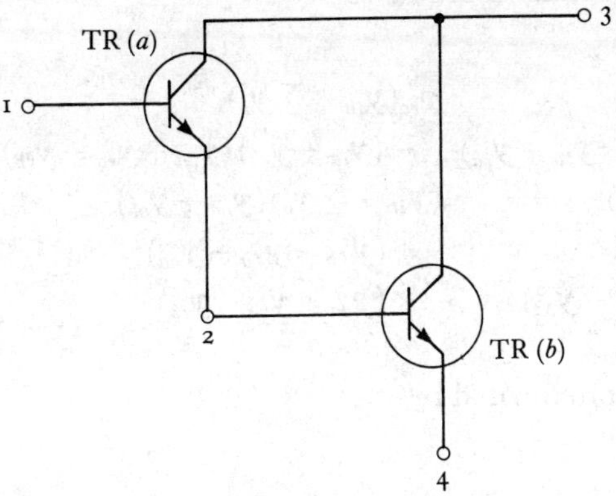

Fig. 8.30 Darlington pair of transistors.

together in the Darlington configuration shown in fig. 8.30. The **y**-matrix of transistor (a) alone with respect to terminal-2 is

$$\begin{array}{c} \\ 1 \\ 3 \end{array}\overset{\begin{array}{cc} 1 & 3 \end{array}}{\begin{bmatrix} y_{ia} & y_{ra} \\ y_{fa} & y_{oa} \end{bmatrix}}, \tag{8.14.24}$$

and its indefinite matrix with respect to terminal-4 is

$$\begin{array}{c} \\ 1 \\ 2 \\ 3 \end{array}\overset{\begin{array}{ccc} 1 & 2 & 3 \end{array}}{\begin{bmatrix} y_{ia} & -(y_{ia}+y_{ra}) & y_{ra} \\ -(y_{fa}+y_{ia}) & \sum y_a & -(y_{oa}+y_{ra}) \\ y_{fa} & -(y_{fa}+y_{oa}) & y_{oa} \end{bmatrix}}, \tag{8.14.25}$$

where
$$\sum y_a = y_{ia}+y_{ra}+y_{fa}+y_{oa}. \tag{8.14.26}$$

But the **y**-matrix of the network, with respect to terminal-4, with transistor (a) removed, is

$$\begin{array}{c} \\ 1 \\ 2 \\ 3 \end{array}\overset{\begin{array}{ccc} 1 & 2 & 3 \end{array}}{\begin{bmatrix} 0 & 0 & 0 \\ 0 & y_{ib} & y_{rb} \\ 0 & y_{fb} & y_{ob} \end{bmatrix}}, \tag{8.14.27}$$

and the $\boldsymbol{y}$-matrix of the Darlington pair is the sum of these:

$$\begin{array}{c} \\ 1 \\ 2 \\ 3 \end{array}\begin{array}{c} \begin{array}{ccc} 1 & 2 & 3 \end{array} \\ \begin{bmatrix} y_{ia} & -(y_{ia}+y_{ra}) & y_{ra} \\ -(y_{fa}+y_{ia}) & y_{ib}+\sum y_a & -(y_{oa}+y_{ra}-y_{rb}) \\ y_{fa} & (y_{fb}-y_{fa}-y_{oa}) & (y_{oa}+y_{ob}) \end{bmatrix} \end{array}. \quad (8.14.28)$$

Suppression of terminal-2 produces

$$\begin{array}{c} 1 \\ \\ 3 \end{array}\begin{bmatrix} y_{ia}(y_{ib}+\sum y_a) - (y_{fa}+y_{ia})(y_{ia}+y_{ra}) & y_{ra}(y_{ib}+\sum y_a) - (y_{ia}+y_{ra})(y_{oa}+y_{ra}-y_{rb}) \\ y_{fa}(y_{ib}+\sum y_a) + (y_{fa}+y_{ia}) \times (y_{fb}-y_{fa}-y_{oa}) & (y_{ib}+\sum y_a)(y_{oa}+y_{ob}) + (y_{fb}-y_{fa}-y_{oa}) \times (y_{oa}+y_{ra}-y_{rb}) \end{bmatrix} \cdot \frac{1}{(y_{ib}+\sum y_a)}, \quad (8.14.29)$$

which may be approximated by

$$\begin{array}{c} 1 \\ 3 \end{array}\begin{bmatrix} y_{ia}(y_{ib}+y_{oa}) & -y_{ia}y_{oa} \\ (y_{fa}y_{fb}+y_{fa}y_{ib}+y_{fb}y_{ia}) & (y_{oa}y_{fb}+y_{ob}y_{fa}) \end{bmatrix} \cdot \frac{1}{y_{fa}}, \quad (8.14.30)$$

if the magnitudes of y_{ra} and y_{rb} are zero and if those of y_{fa} and y_{fb} are large. The short-circuit current amplification, for example, is thus seen to be

$$\left(\frac{I_3}{I_1}\right)_{V_3=0} = \frac{\text{bottom left hand element of (8.14.30)}}{\text{top left hand element of (8.14.30)}},$$

$$\approx \frac{y_{fa}y_{fb}}{y_{ia}(y_{ib}+y_{oa})}. \quad (8.14.31)$$

The procedures described in this section are not limited to admittance matrices; they may also be applied to impedance matrices when the alteration to the circuit is the dual of that described above. For example, the procedure that follows (8.14.7) arises in the generalized theory of electrical machines, in which impedance matrices are used, and corresponds with the connexion in series of two coils so that $\boldsymbol{I}_p = \boldsymbol{I}_q$ and the significant voltage is $(\boldsymbol{V}_p + \boldsymbol{V}_q)$.

8.15 Matrix partitioning

It has been shown that if one of the currents in the admittance equations of a network is zero the order of the admittance matrix may be reduced by suppressing the terminal which has no external current. This process may be carried out successively where more than one terminal has zero external current, but it becomes laborious if the matrix is a large one. A better method in this instance is to use the principle of *matrix partitioning*, which leads to a relatively simple reduction procedure.

$$[A][B]=\left[\begin{array}{cc:c} \boldsymbol{A}_{11} & \boldsymbol{A}_{12} & \boldsymbol{A}_{13} \\ \boldsymbol{A}_{21} & \boldsymbol{A}_{22} & \boldsymbol{A}_{23} \\ \hdashline \boldsymbol{A}_{31} & \boldsymbol{A}_{32} & \boldsymbol{A}_{33} \\ \boldsymbol{A}_{41} & \boldsymbol{A}_{42} & \boldsymbol{A}_{43} \end{array}\right]\left[\begin{array}{c:cc} \boldsymbol{B}_{11} & \boldsymbol{B}_{12} & \boldsymbol{B}_{13} \\ \boldsymbol{B}_{21} & \boldsymbol{B}_{22} & \boldsymbol{B}_{23} \\ \hdashline \boldsymbol{B}_{31} & \boldsymbol{B}_{32} & \boldsymbol{B}_{33} \end{array}\right]. \qquad (8.15.1)$$

If matrices $[A]$ and $[B]$ in (8.15.1) are subdivided by the dotted lines into components called *submatrices*, the original matrices are said to be partitioned. In (8.15.1), the submatrices in $[A]$ are

$$\left.\begin{aligned} [\boldsymbol{\alpha}_{11}] &= \begin{bmatrix} \boldsymbol{A}_{11} & \boldsymbol{A}_{12} \\ \boldsymbol{A}_{21} & \boldsymbol{A}_{22} \end{bmatrix}; \qquad [\boldsymbol{\alpha}_{12}] = \begin{bmatrix} \boldsymbol{A}_{13} \\ \boldsymbol{A}_{23} \end{bmatrix}; \\ [\boldsymbol{\alpha}_{21}] &= \begin{bmatrix} \boldsymbol{A}_{31} & \boldsymbol{A}_{32} \\ \boldsymbol{A}_{41} & \boldsymbol{A}_{42} \end{bmatrix}; \qquad [\boldsymbol{\alpha}_{22}] = \begin{bmatrix} \boldsymbol{A}_{33} \\ \boldsymbol{A}_{43} \end{bmatrix}; \end{aligned}\right\} \qquad (8.15.2)$$

and in $[B]$ are

$$\left.\begin{aligned} [\boldsymbol{\beta}_{11}] &= \begin{bmatrix} \boldsymbol{B}_{11} \\ \boldsymbol{B}_{21} \end{bmatrix}; \qquad [\boldsymbol{\beta}_{12}] = \begin{bmatrix} \boldsymbol{B}_{12} & \boldsymbol{B}_{13} \\ \boldsymbol{B}_{22} & \boldsymbol{B}_{23} \end{bmatrix}; \\ [\boldsymbol{\beta}_{21}] &= [\boldsymbol{B}_{31}]; \qquad [\boldsymbol{\beta}_{22}] = [\boldsymbol{B}_{32} \quad \boldsymbol{B}_{33}]. \end{aligned}\right\} \qquad (8.15.3)$$

Equation (8.15.1) may then be written:

$$[A][B]=\begin{bmatrix} [\boldsymbol{\alpha}_{11}] & [\boldsymbol{\alpha}_{12}] \\ [\boldsymbol{\alpha}_{21}] & [\boldsymbol{\alpha}_{22}] \end{bmatrix}\begin{bmatrix} [\boldsymbol{\beta}_{11}] & [\boldsymbol{\beta}_{12}] \\ [\boldsymbol{\beta}_{21}] & [\boldsymbol{\beta}_{22}] \end{bmatrix}, \qquad (8.15.4)$$

and it can be shown that the product may be formed by the same rules as would apply to ordinary elements, provided that the proper order of the product terms so formed is retained. For this process to work there must be as many rows in $[\boldsymbol{\alpha}_{11}]$, for example, as columns in $[\boldsymbol{\beta}_{11}]$; that is, the submatrices must be conformable; and this requirement means that when the original matrices are partitioned, the first c columns in $[A]$ must be matched by an equal number of rows in $[B]$. Once this has been done, the rows in $[A]$ and the columns in $[B]$ may be subdivided in any desired manner without losing conformability in the submatrices. If this principle is applied to

(8.15.5), in which $\boldsymbol{I}_2$, $\boldsymbol{I}_4$, and $\boldsymbol{I}_5$ are assumed to be zero, in order to suppress terminals-2, -4 and -5:

$$\begin{bmatrix} \boldsymbol{I}_1 \\ \boldsymbol{I}_2 \\ \boldsymbol{I}_3 \\ \boldsymbol{I}_4 \\ \boldsymbol{I}_5 \end{bmatrix} = \begin{bmatrix} a_1 & a_2 & a_3 & a_4 & a_5 \\ b_1 & b_2 & b_3 & b_4 & b_5 \\ c_1 & c_2 & c_3 & c_4 & c_5 \\ d_1 & d_2 & d_3 & d_4 & d_5 \\ e_1 & e_2 & e_3 & e_4 & e_5 \end{bmatrix} \begin{bmatrix} \boldsymbol{V}_1 \\ \boldsymbol{V}_2 \\ \boldsymbol{V}_3 \\ \boldsymbol{V}_4 \\ \boldsymbol{V}_5 \end{bmatrix}, \tag{8.15.5}$$

the equations should first be rearranged as shown in (8.15.6) before partitioning.

$$\left[\begin{array}{c} \boldsymbol{I}_1 \\ \boldsymbol{I}_3 \\ \hdashline \boldsymbol{I}_2 \\ \boldsymbol{I}_4 \\ \boldsymbol{I}_5 \end{array}\right] = \left[\begin{array}{cc:ccc} a_1 & a_3 & a_2 & a_4 & a_5 \\ c_1 & c_3 & c_2 & c_4 & c_5 \\ \hdashline b_1 & b_3 & b_2 & b_4 & b_5 \\ d_1 & d_3 & d_2 & d_4 & d_5 \\ e_1 & e_3 & e_2 & e_4 & e_5 \end{array}\right] \left[\begin{array}{c} \boldsymbol{V}_1 \\ \boldsymbol{V}_3 \\ \hdashline \boldsymbol{V}_2 \\ \boldsymbol{V}_4 \\ \boldsymbol{V}_5 \end{array}\right]. \tag{8.15.6}$$

Writing

$$\left.\begin{aligned} [\boldsymbol{I}_a] &= \begin{bmatrix} \boldsymbol{I}_1 \\ \boldsymbol{I}_3 \end{bmatrix}; & [\boldsymbol{V}_a] &= \begin{bmatrix} \boldsymbol{V}_1 \\ \boldsymbol{V}_3 \end{bmatrix}; \\ [\boldsymbol{I}_b] &= \begin{bmatrix} \boldsymbol{I}_2 \\ \boldsymbol{I}_4 \\ \boldsymbol{I}_5 \end{bmatrix}; & [\boldsymbol{V}_b] &= \begin{bmatrix} \boldsymbol{V}_2 \\ \boldsymbol{V}_4 \\ \boldsymbol{V}_5 \end{bmatrix}; \\ [\boldsymbol{\alpha}_{11}] &= \begin{bmatrix} a_1 & a_3 \\ c_1 & c_3 \end{bmatrix}; & [\boldsymbol{\alpha}_{12}] &= \begin{bmatrix} a_2 & a_4 & a_5 \\ c_2 & c_4 & c_5 \end{bmatrix}; \\ [\boldsymbol{\alpha}_{21}] &= \begin{bmatrix} b_1 & b_3 \\ d_1 & d_3 \\ e_1 & e_3 \end{bmatrix}; & [\boldsymbol{\alpha}_{22}] &= \begin{bmatrix} b_2 & b_4 & b_5 \\ d_2 & d_4 & d_5 \\ e_2 & e_4 & e_5 \end{bmatrix}; \end{aligned}\right\} \tag{8.15.7}$$

(8.15.6) becomes

$$\begin{bmatrix} \boldsymbol{I}_a \\ \boldsymbol{I}_b \end{bmatrix} = \begin{bmatrix} [\boldsymbol{\alpha}_{11}] & [\boldsymbol{\alpha}_{12}] \\ [\boldsymbol{\alpha}_{21}] & [\boldsymbol{\alpha}_{22}] \end{bmatrix} \begin{bmatrix} \boldsymbol{V}_a \\ \boldsymbol{V}_b \end{bmatrix}, \tag{8.15.8}$$

and, for $[\boldsymbol{I}_b]$ equal to zero,

$$\left.\begin{aligned} [\boldsymbol{I}_a] &= [\boldsymbol{\alpha}_{11}][\boldsymbol{V}_a] + [\boldsymbol{\alpha}_{12}][\boldsymbol{V}_b] \\ [0] &= [\boldsymbol{\alpha}_{21}][\boldsymbol{V}_a] + [\boldsymbol{\alpha}_{22}][\boldsymbol{V}_b] \end{aligned}\right\}, \tag{8.15.9}$$

from which $[V_b]$ may be eliminated to give

$$[I_a] = [\alpha_{11}][V_a] + [\alpha_{12}]\{-[\alpha_{22}]^{-1}[\alpha_{21}][V_a]\},$$
$$= \{[\alpha_{11}] - [\alpha_{12}][\alpha_{22}]^{-1}[\alpha_{21}]\}[V_a], \tag{8.15.10}$$

or, in the particular example,

$$\begin{bmatrix} I_1 \\ I_3 \end{bmatrix} = \{[\alpha_{11}] - [\alpha_{12}][\alpha_{22}]^{-1}[\alpha_{21}]\}\begin{bmatrix} V_1 \\ V_3 \end{bmatrix}. \tag{8.15.11}$$

As an illustration of this process, consider the amplifier shown in fig. 8.31, which is intended to have a band-pass response without recourse to inductors. It does this by converting the two real poles of the passive filter network

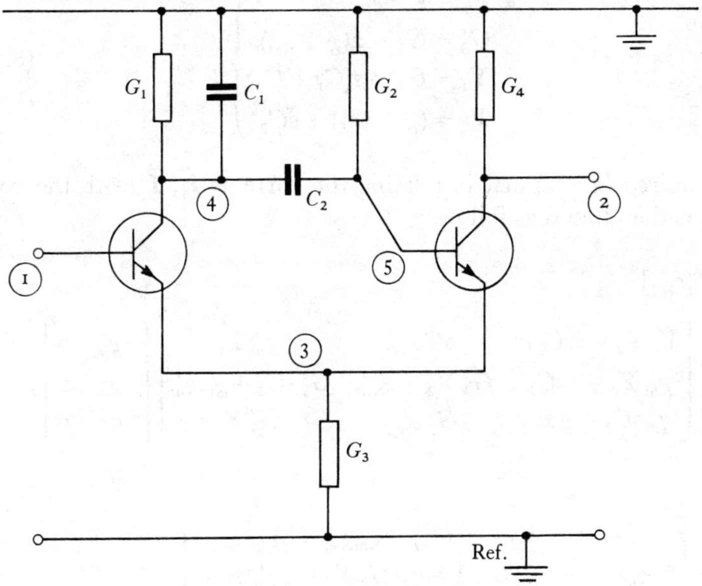

Fig. 8.31 Amplifier with a response similar to that of a tuned circuit (bias arrangements omitted).

$G_1 C_1 G_2 C_2$ into a complex pair by means of positive feedback introduced by the common emitter conductance G_3. Assuming that each of the two identical transistors has a sufficiently high cutoff frequency to be approximated by the definite admittance matrix†

$$[y] = \begin{bmatrix} g & 0 \\ g_m & 0 \end{bmatrix}, \tag{8.15.12}$$

† The validity of the low-frequency model of a transistor, to which this admittance matrix applies, is discussed in specialized texts, e.g. C. L. Searle, *et al. Elementary circuit properties of transistors*, Wiley (1964).

(in which the emitter is the reference terminal), the (5×5) admittance matrix for the whole circuit is

$$\begin{bmatrix} I_1 \\ I_2 \\ \hdashline 0 \\ 0 \\ 0 \end{bmatrix} = \left[\begin{array}{cc:ccc} g & 0 & -g & 0 & 0 \\ 0 & G_4 & -g_m & 0 & g_m \\ \hdashline -g'_m & 0 & G'_3 & 0 & -g'_m \\ g_m & 0 & -g_m & Y_1 & -sC_2 \\ 0 & 0 & -g & -sC_2 & Y_5 \end{array}\right] \begin{bmatrix} V_1 \\ V_2 \\ \hdashline V_3 \\ V_4 \\ V_5 \end{bmatrix}, \tag{8.15.13}$$

where

$$\left.\begin{aligned} g'_m &= g + g_m \\ G'_3 &= G_3 + 2(g + g_m) \\ Y_1 &= G_1 + s(C_1 + C_2) \\ Y_2 &= (g + G_2) + sC_2 \end{aligned}\right\}. \tag{8.15.14}$$

The required (2×2) matrix relating the currents I_1, I_2 with the voltages V_1, V_2 is then found as follows:

$[\alpha_{22}]^{-1}[\alpha_{21}]$

$$= \frac{1}{|\alpha_{22}|} \begin{bmatrix} Y_1 Y_2 - (sC_2)^2 & sC_2 g'_m & g'_m Y_1 \\ g_m Y_2 + gsC_2 & G'_3 Y_2 - gg'_m & G'_3 sC_2 + g_m g'_m \\ g_m sC_2 + gY_1 & G'_3 sC_2 & G'_3 Y_1 \end{bmatrix} \begin{bmatrix} -g'_m & 0 \\ g_m & 0 \\ 0 & 0 \end{bmatrix}, \tag{8.15.15}$$

$$= \frac{1}{|\alpha_{22}|} \begin{bmatrix} \{-g'_m(Y_1 Y_2 - s^2 C_2^2) + g_m g'_m sC_2\} & 0 \\ \{-g'_m(g_m Y_2 + sC_2 g) + g_m(G'_3 Y_2 - gg'_m)\} & 0 \\ \{-g'_m(g_m sC_2 + gY_1) + g_m G'_3 sC_2\} & 0 \end{bmatrix}, \tag{8.15.16}$$

$[\alpha_{12}][\alpha_{22}]^{-1}[\alpha_{21}]$

$$= \frac{1}{|\alpha_{22}|} \begin{bmatrix} -g & 0 & 0 \\ -g_m & 0 & g_m \end{bmatrix} \begin{bmatrix} \{-g'_m(Y_1 Y_2 - s^2 C_2^2) + g_m g'_m sC_2\} & 0 \\ \{-g'_m(g_m Y_2 + sC_2 g) + g_m(G'_3 Y_2 - gg'_m)\} & 0 \\ \{-g'_m(g_m sC_2 + gY_1) + g_m G'_3 sC_2\} & 0 \end{bmatrix},$$

$$= \frac{1}{|\alpha_{22}|} \begin{bmatrix} g\{g'_m(Y_1 Y_2 - s^2 C_2^2) - g_m g'_m sC_2\} & 0 \\ g_m\{g'_m(Y_1 Y_2 - s^2 C_2^2) - g_m g'_m sC_2\} & \\ \quad + g_m\{-g'_m(g_m sC_2 + gY_1) + g_m G'_3 sC_2\} & 0 \end{bmatrix}, \tag{8.15.17}$$

and

$$[\boldsymbol{\alpha}_{11}]-[\boldsymbol{\alpha}_{12}][\boldsymbol{\alpha}_{22}]^{-1}[\boldsymbol{\alpha}_{21}]$$

$$=\frac{1}{|\boldsymbol{\alpha}_{22}|}\begin{bmatrix} g\{|\boldsymbol{\alpha}_{22}|-g_m'(\boldsymbol{Y}_1\boldsymbol{Y}_2-s^2C_2^2)+g_mg_m'sC_2\} & 0 \\ g_m\{g_mg_m'sC_2-g_m'(\boldsymbol{Y}_1\boldsymbol{Y}_2-s^2C_2^2) & \\ \quad +g_m'(g_msC_2+g\boldsymbol{Y}_1)-g_mG_3'sC_2\} & G_4|\boldsymbol{\alpha}_{22}| \end{bmatrix}, \quad (8.15.18)$$

$$=\frac{1}{|\boldsymbol{\alpha}_{22}|}\begin{bmatrix}\gamma_{11} & \gamma_{12}\\ \gamma_{21} & \gamma_{22}\end{bmatrix}, \quad (8.15.19)$$

where

$$\gamma_{11}=g_m(g+g_m)C_1C_2\left(1+\frac{G_3}{g+g_m}\right)\left\{s^2+s\left[\frac{1}{C_1}\left(G_1+G_2+\frac{gG_3/g_m'}{1+(G_3/g_m')}\right)\right.\right.$$

$$\left.\left.+\frac{1}{C_2}\left(G_2+\frac{gG_3/g_m'}{1+(G_3/g_m')}\right)\right]+\frac{G_1G_2}{C_1C_2}\left[1+\frac{gG_3/g_m'G_2}{1+(G_3/g_m')}\right]\right\}, \quad (8.15.20)$$

$$\gamma_{12}=0, \quad (8.15.21)$$

$$\gamma_{21}=-g_m(g+g_m)C_1C_2\left\{s^2+s\left[\frac{G_1}{C_1}+\frac{1}{C_2}\left(G_1+G_2+\frac{g_m}{g+g_m}\cdot G_3\right)\right]+\frac{G_1G_2}{C_1C_2}\right\}, \quad (8.15.22)$$

$$\gamma_{22}=G_4|\boldsymbol{\alpha}_{22}|, \quad (8.15.23)$$

and

$$|\boldsymbol{\alpha}_{22}|=C_1C_2[G_3+2(g+g_m)]\left\{s^2+s\left[\frac{1}{C_1}\left(g+G_1+G_2-\frac{(g+g_m)^2}{G_3+2(g+g_m)}\right)\right.\right.$$

$$\left.+\frac{1}{C_2}\left(g+G_2-\frac{g(g+g_m)}{G_3+2(g+g_m)}\right)\right]$$

$$\left.+\frac{G_1G_2}{C_1C_2}\left[1+\frac{g}{G_2}\left(1-\frac{g+g_m}{G_3+2(g+g_m)}\right)\right]\right\}. \quad (8.15.24)$$

For values of g_m large with respect to both g and G_3 (a typical condition), (8.15.20) to (8.15.24) become

$$y_{11} \approx g_m^2 C_1 C_2 \left\{ s^2 + s\left[\frac{G_1+G_2}{C_1} + \frac{G_2}{C_2}\right] + \frac{G_1 G_2}{C_1 C_2}\right\}, \qquad (8.15.25)$$

$$y_{12} = 0,$$

$$y_{21} \approx -g_m^2 C_1 C_2 \left\{ s^2 + s\left[\frac{G_1}{C_1} + \frac{G_1+G_2+G_3}{C_2}\right] + \frac{G_1 G_2}{C_1 C_2}\right\}, \qquad (8.15.26)$$

$$y_{22} = G_4|\alpha_{22}|,$$

and

$$|\alpha_{22}| \approx 2g_m C_1 C_2 \left\{ s^2 + s\left[\frac{G_1+G_2+(g/2)-(g_m/2)}{C_1} + \frac{G_2+(g/2)}{C_2}\right] + \frac{G_1 G_2}{C_1 C_2}[1+(g/2G_2)]\right\}. \qquad (8.15.27)$$

The open-circuit voltage amplification, which is $-y_{21}/y_{22}$, can now be seen to have a pair of zeros due to y_{21} and a pair of poles due to y_{22}. The transfer function, therefore, differs from that of a parallel tuned circuit in that it possesses an extra zero, but the magnitude of the response can be made closely similar to that of a tuned circuit.

It will be noted that the elements y_{11}, y_{12}, etc. can be derived as determinants from the original matrix; y_{11} is the determinant of the matrix formed by deleting row-2 and column-2, y_{12} by deleting row-2 and column-1, and so on, and $|\alpha_{22}|$ by deleting the first two rows and the first two columns.

8.16 Transmission matrices of ideal controlled sources

Twoports of special kinds, like those which are used for their impedance inverting properties, often use active devices, whose essential properties are those of controlled sources. Before considering the properties of such special twoports, therefore, it is useful to look at the four types of ideal controlled sources shown, in three-terminal form, in table 8.6. In this table the **a**-parameters are used because, of the six sets discussed in this chapter, the **a**-set is the only one which exists for all four sources. The outstanding feature of these **a**-matrices is the three zero elements. The two zeros in one row of each matrix arise because one or other of the two input variables $\boldsymbol{V}_1$ and $\boldsymbol{I}_1$ is constrained to zero in the ideal source; the other zero arises because only one of the two output variables $\boldsymbol{V}_2$ and $\boldsymbol{I}_2$ is dependent on the non-zero input variable. The single non-zero element of the matrix thus appears in column-1 (2) if the source is of the voltage (current) kind, and in row-1 (2) if it is controlled by a voltage (current).

A cascade of two ideal voltage-controlled voltage sources (VVS + VVS) produces a twoport which is also an ideal VVS, and a cascade of two ideal

TABLE 8.6 *The **a**-matrices of ideal controlled sources*

Source	Three-terminal circuit	***a***-matrix
Voltage-controlled voltage source (VVS)	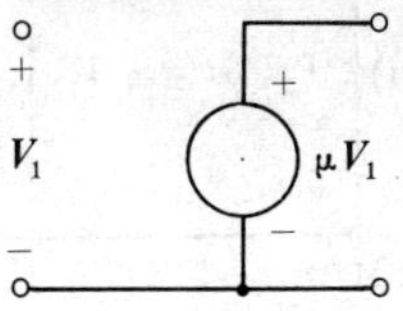	$\begin{bmatrix} 1/\mu & 0 \\ 0 & 0 \end{bmatrix}$
Voltage-controlled current source (VCS)	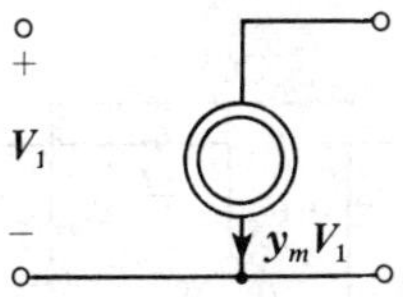	$\begin{bmatrix} 0 & 1/y_m \\ 0 & 0 \end{bmatrix}$
Current-controlled voltage source (CVS)	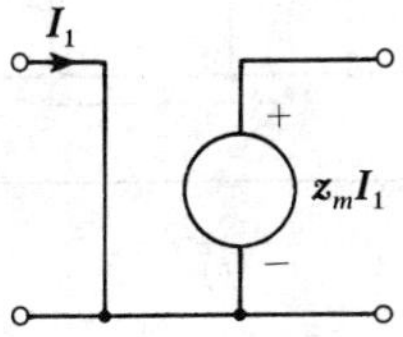	$\begin{bmatrix} 0 & 0 \\ 1/z_m & 0 \end{bmatrix}$
Current-controlled current source (CCS)	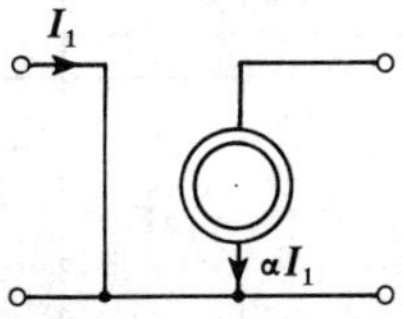	$\begin{bmatrix} 0 & 0 \\ 0 & 1/\alpha \end{bmatrix}$

current-controlled current sources (CCS + CCS) produces a twoport which is also an ideal CCS. Such arrangements as the cascade (VCS + VCS) or (CVS + CVS), however, which place an ideal current source on open-circuit or an ideal voltage source on short-circuit are best considered to be inadmissible. There remain the arrangements (VCS + CVS), which produces an ideal VVS, and (CVS + VCS), which produces an ideal CCS, which show that it is possible to construct all four kinds of sources from the VCS and the CVS alone. For this reason it is possible to view this pair as more fundamental than the other. Of more significance, however, is the physical nature of the active devices which are used in practice as controlled sources and the model to which their characteristics most nearly fit.

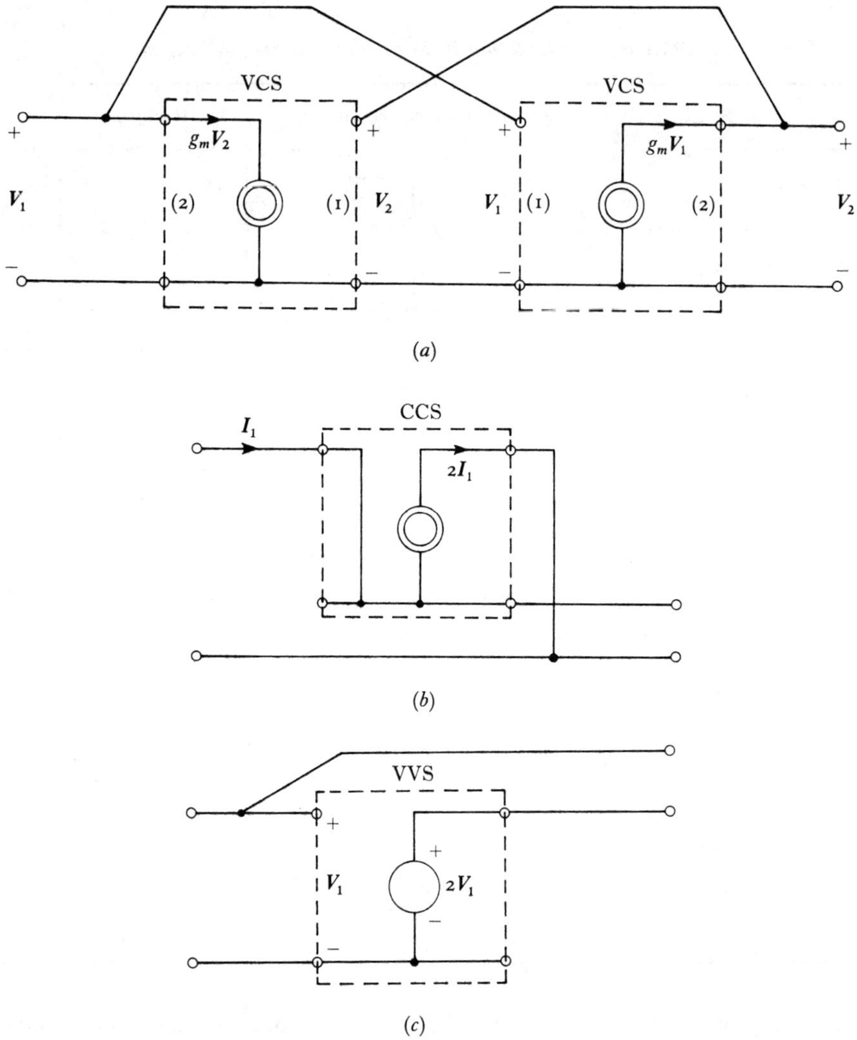

Fig. 8.32 Realization by controlled sources of: (*a*) gyrator; (*b*) INIC; (*c*) VNIC.

The important feature of controlled sources, apart from their inherent property of power amplification, is that they may be interconnected to form twoports of special kinds. For example, the parallel connexion of two voltage-controlled current sources, in which port-1 of each source is paralleled with port-2 of the other (fig. 8.32(*a*)) produces the gyrator, briefly described in chapter 6, some of whose interesting properties are discussed in the next section. Similarly, the parallel–series connexion of an ideal voltage-controlled voltage source with an oppositely directed ideal current-controlled current source, with reversed input terminals, produces a twoport equivalent to an ideal transformer, and is therefore capable in principle of operating at frequencies down to and including zero. A variety of such twoports may be formed in this way; the necessary interconnexions of the controlled sources are often found by decomposing the matrix of a wanted twoport into a number of elementary forms like those shown in table 8.6. Twoports which have impedance-transforming properties are the subject of the next section.

8.17 Impedance converters and inverters

An ideal transformer, loaded at port-2 by an impedance $\boldsymbol{Z}_L$, has an input impedance (at port-1) given by $\boldsymbol{Z}_L/k^2$ where k is a real number which is called, in this instance, the turns ratio. If it is loaded at port-1 by an impedance $\boldsymbol{Z}_s$ it has an output impedance (at port-2) given by $k^2\boldsymbol{Z}_s$. Any twoport which has impedance transforming properties of this kind is called, in general, a *positive impedance converter*. The word *positive* is used to distinguish this kind of impedance transformation from one in which there is included a negative sign so that a load impedance $\boldsymbol{Z}_L$ produces an input impedance $-\boldsymbol{Z}_L/k^2$ and a source impedance $\boldsymbol{Z}_s$ produces an output impedance $-k^2\boldsymbol{Z}_s$. A twoport with these negative impedance-transforming properties is called a *negative impedance converter*. Impedance converters have a port impedance that is simply a scaled version of the impedance attached to the distant port, possibly with a change of sign.

The significance of another type of impedance-transforming twoport, called the impedance *inverter*, is not so much that it has a port impedance which depends upon a scaled version of the impedance attached to the distant port as that the port impedance is the inverse of this scaled impedance; the form of the port impedance may therefore be quite different from the form of the impedance attached to the distant port. Thus a *positive impedance inverter* (more commonly known as a *gyrator*) which has a load $\boldsymbol{Z}_L$ attached to port-2 will have an input impedance (at port-1) given by $1/G^2\boldsymbol{Z}_L$, where G is a constant with the dimension of conductance; so that if, for example, $\boldsymbol{Z}_L$ is an inductor sL, the input impedance is that of a capacitance of value $C_{\text{IN}} = G^2L$. A *negative impedance inverter* is a twoport which has a port impedance equal to a scaled version of the negative inverse of the impedance attached to the distant port. Input and output impedances for the four varieties of twoport are shown in table 8.7. Since the transformation process is valid in terms of admittances as well as impedances, these transforming twoports are sometimes known as immittance transformers.

The transmission matrices of ideal impedance-transforming twoports have non-zero elements on only one of the two diagonals and zero elements on the other (see table 8.7). Converters have their non-zero elements on the

TABLE 8.7 *Input impedances, output impedances, and transmission matrices of impedance converters and inverters*

	Impedance converters		Impedance inverters	
	Positive (transformer) Reciprocal	Negative Anti-reciprocal	Positive (gyrator) Anti-reciprocal	Negative Reciprocal
$\boldsymbol{Z}_{\mathrm{IN}}$	$\boldsymbol{Z}_L/k^2$	$-\boldsymbol{Z}_L/k^2$	$1/G^2\boldsymbol{Z}_L$	$-1/G^2\boldsymbol{Z}_L$
$\boldsymbol{Z}_{\mathrm{OUT}}$	$k^2\boldsymbol{Z}_s$	$-k^2\boldsymbol{Z}_s$	$1/G^2\boldsymbol{Z}_s$	$-1/G^2\boldsymbol{Z}_s$
$[\boldsymbol{a}]$	$\begin{bmatrix} 1/k & 0 \\ 0 & -k \end{bmatrix}$	$\begin{bmatrix} 1/k & 0 \\ 0 & k \end{bmatrix}$	$\begin{bmatrix} 0 & -1/G \\ G & 0 \end{bmatrix}$	$\begin{bmatrix} 0 & 1/G \\ G & 0 \end{bmatrix}$
$[\boldsymbol{a}]^{-1}$	$\begin{bmatrix} k & 0 \\ 0 & -1/k \end{bmatrix}$	$\begin{bmatrix} k & 0 \\ 0 & 1/k \end{bmatrix}$	$\begin{bmatrix} 0 & 1/G \\ -G & 0 \end{bmatrix}$	$\begin{bmatrix} 0 & 1/G \\ G & 0 \end{bmatrix}$

principal diagonal, and inverters have theirs on the non-principal diagonal. In the positive impedance converter (PIC) and the positive impedance inverter (PII) the two non-zero elements have opposite signs, and in the negative impedance converter (NIC) and the negative impedance inverter (NII) the two non-zero elements have like signs; this makes the PIC and the NII reciprocal ($|\boldsymbol{a}| = -1$) and the NIC and the PII anti-reciprocal ($|\boldsymbol{a}| = +1$).

The sign of k (or G) has no effect on the impedance-transforming properties of the twoport, which depend on k^2 (or G^2), but it does affect the transmission properties. A positive value of k gives to the PIC a non-inverted output whilst a negative one gives to it an inverted output (like that of a transformer with the polarity of one winding reversed). In the NIC a positive value of k creates a twoport known as a *current* NIC, or INIC, (colloquially pronounced *eye-nick*), whilst a negative value of k creates a *voltage* NIC, or VNIC, (colloquially pronounced *vee-nick*). This terminology treats the NIC as though it were a PIC (historically the first of the two) to which negativeness has been added either by a reversal of the output current or by a reversal of the output voltage. The equations of a PIC are

$$\left.\begin{aligned} \boldsymbol{V}_1 &= \frac{1}{k}\boldsymbol{V}_2, \\ \boldsymbol{I}_1 &= -k\boldsymbol{I}_2, \end{aligned}\right\} \tag{8.17.1}$$

and if the internal operation of the twoport is now somehow changed so as to reverse the direction of the output current, the equations become those of an INIC:

$$\left.\begin{aligned} \boldsymbol{V}_1 &= \frac{1}{k}\boldsymbol{V}_2, \\ \boldsymbol{I}_1 &= k\boldsymbol{I}_2. \end{aligned}\right\} \tag{8.17.2}$$

If, however, the internal mechanism is changed so as to produce a reversal of the output voltage the equations become those of a VNIC

$$\left.\begin{aligned} \boldsymbol{V}_1 &= -\frac{1}{k}\boldsymbol{V}_2, \\ \boldsymbol{I}_1 &= -k\boldsymbol{I}_2. \end{aligned}\right\} \tag{8.17.3}$$

Equations (8.17.3) are the same as (8.17.2) if, in the latter, k is allowed to assume negative values.

In the PII (gyrator) a change in the sign of G is equivalent to an interchange in the two ports; in the NII it is equivalent to a change in the polarities of the output current and voltage without affecting the symmetry of the twoport.

Since many impedance transforming twoports employ active devices for their realization (see fig. 8.32), the parameters k (or G) in each pair of equations may not in practice always be equal. This will cause the twoports to lose their properties of reciprocity or anti-reciprocity but will not affect their impedance-transforming properties as may be seen by replacing the parameter k (G) in each pair of equations by two unequal parameters k_1 (G_1) and k_2 (G_2) such that $k_1 k_2 = k^2$ ($G_1 G_2 = G^2$).

Much interest in impedance-transforming twoports stems from the manufacture of microelectronic circuits. It has been found difficult or impossible to make microelectronic inductors except in values of only a few microhenrys, and since inductors are conventionally used in association with capacitors (as in resonant circuits) to produce frequency selective networks it has not been easy to make such networks in microelectronic form. It can be done, as illustrated in fig. 8.31, by using transistors as controlled sources to enhance the mild selectivity possessed by *RC* networks; but such *active RC networks* are more sensitive in performance to their element values, both passive and active, than are the corresponding passive *LC* networks. This undesirable feature has discouraged the use of active *RC* networks but has stimulated interest in the gyrator which, because of its passiveness, gives hope of producing an inductorless selective network which is not so sensitive to its element values.

In principle each inductor in an *LC* network can be replaced by a capacitance-loaded gyrator, but since the active elements used in practice to produce a gyrator are three-terminal ones like transistors, one side of each inductor would inevitably be tied to a fixed potential (symbolized in diagrams by an earth sign) in an arrangement like that of fig. 8.33(*a*). A floating inductor can, however, be produced as shown in part (*b*); this may be verified by multiplying the transmission matrices of the three cascaded twoports and comparing the result with the series transmission element of table 8.3:

$$\begin{bmatrix} 0 & 1/G \\ -G & 0 \end{bmatrix}\begin{bmatrix} 1 & 0 \\ 0 & -1 \end{bmatrix}\begin{bmatrix} 1 & 0 \\ sC & -1 \end{bmatrix}\begin{bmatrix} 1 & 0 \\ 0 & -1 \end{bmatrix}\begin{bmatrix} 0 & 1/G \\ -G & 0 \end{bmatrix}$$

$$= \begin{bmatrix} 0 & -1/G \\ -G & 0 \end{bmatrix}\begin{bmatrix} 1 & 0 \\ sC & 1 \end{bmatrix}\begin{bmatrix} 0 & 1/G \\ -G & 0 \end{bmatrix},$$

$$= \begin{bmatrix} sC/G & -1/G \\ -G & 0 \end{bmatrix}\begin{bmatrix} 0 & 1/G \\ -G & 0 \end{bmatrix},$$

$$= \begin{bmatrix} 1 & -sC/G^2 \\ 0 & -1 \end{bmatrix}. \qquad (8.17.4)$$

The NIC has also been used in the synthesis of *RC* active networks and it has the remarkable property of being amenable to a reduction in its own imperfections. The parameters most appropriate to the description of a NIC are the ***a***- or ***h***-parameters or their inverses, as may be seen from table 8.1. Assuming that, of these, the ***g***-parameters are appropriate for a NIC constructed from active devices, its equations take the form

$$\left.\begin{aligned} \boldsymbol{I}_1 &= 0 + k\boldsymbol{I}_2, \\ \boldsymbol{V}_2 &= k\boldsymbol{V}_1 + 0, \end{aligned}\right\} \qquad (8.17.5)$$

and the ***g***-matrix is

$$[\boldsymbol{g}] = \begin{bmatrix} \boldsymbol{g}_i & \boldsymbol{g}_r \\ \boldsymbol{g}_f & \boldsymbol{g}_o \end{bmatrix} = \begin{bmatrix} 0 & k \\ k & 0 \end{bmatrix}, \qquad (8.17.6)$$

so that $\boldsymbol{g}_i = 0$, $\boldsymbol{g}_o = 0$, $\boldsymbol{g}_r = \boldsymbol{g}_f = k$, and the port immittances are

$$\left.\begin{aligned} \boldsymbol{Y}_{\mathrm{IN}} &= \boldsymbol{g}_i - \frac{\boldsymbol{g}_r\boldsymbol{g}_f}{\boldsymbol{g}_o + \boldsymbol{Z}_L} = -k^2\,\boldsymbol{Y}_L, \\ \boldsymbol{Z}_{\mathrm{OUT}} &= \boldsymbol{g}_o - \frac{\boldsymbol{g}_r\boldsymbol{g}_f}{\boldsymbol{g}_i + \boldsymbol{Y}_s} = -k^2\,\boldsymbol{Z}_s. \end{aligned}\right\} \qquad (8.17.7)$$

Supposing now that the only departure from perfection in a practical NIC is that $\boldsymbol{g}_i$ has the non-zero value $\boldsymbol{Y}_a$, it is possible to eliminate the

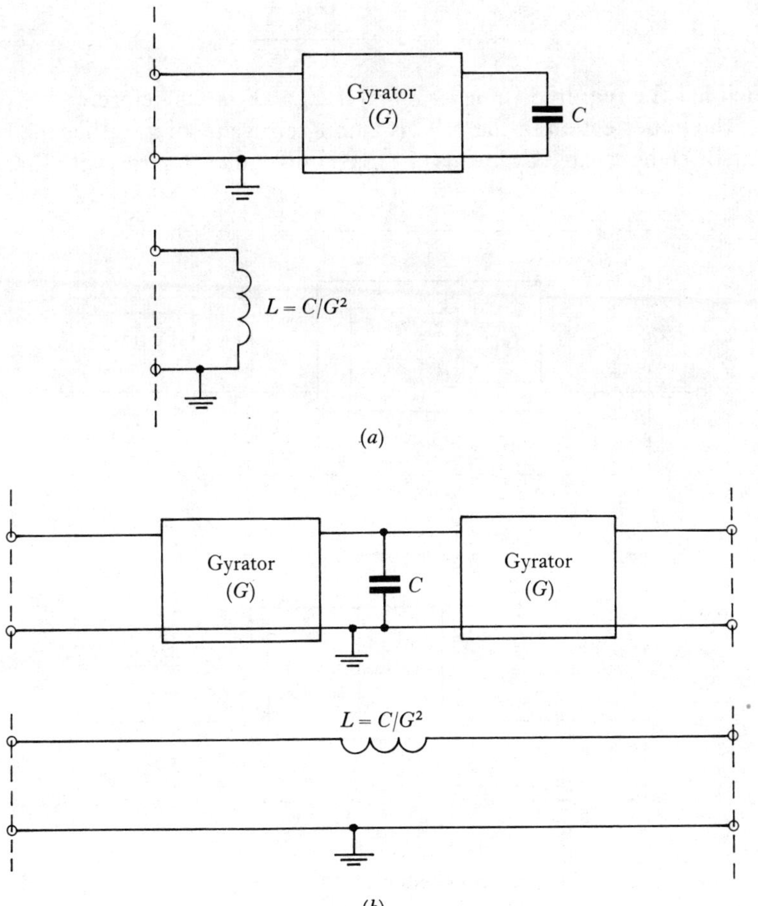

Fig. 8.33 Realization of an inductor by capacitance-loaded gyrators: (*a*) earthed inductor; (*b*) floating inductor.

imperfection by placing a fixed admittance $\boldsymbol{Y}'_a$ across the output terminals as shown in fig. 8.34(*a*). Then

$$\boldsymbol{Y}_{\text{IN}} = \boldsymbol{Y}_a - \frac{k^2}{1/(\boldsymbol{Y}'_a + \boldsymbol{Y}_L)},$$
$$= (\boldsymbol{Y}_a - k^2\,\boldsymbol{Y}'_a) - k^2\,\boldsymbol{Y}_L, \tag{8.17.8}$$

and this has the required value $(-k^2\,\boldsymbol{Y}_L)$ if $\boldsymbol{Y}'_a = \boldsymbol{Y}_a/k^2$. Similarly,

$$\boldsymbol{Z}_{\text{OUT}} = \boldsymbol{Z}'_a \left\| \left(-\frac{k^2}{\boldsymbol{Y}_a + \boldsymbol{Y}_s} \right) \right.,$$

or

$$Y_{\text{OUT}} = Y_a' - \frac{Y_a + Y_s}{k^2}, \tag{8.17.9}$$

which has the required value $(-Y_s/k^2)$ if $Y_a' = Y_a/k^2$ as before.

If the imperfection in the NIC is a non-zero value of g_o rather than of g_i it can be compensated by means of a fixed impedance in series with the input

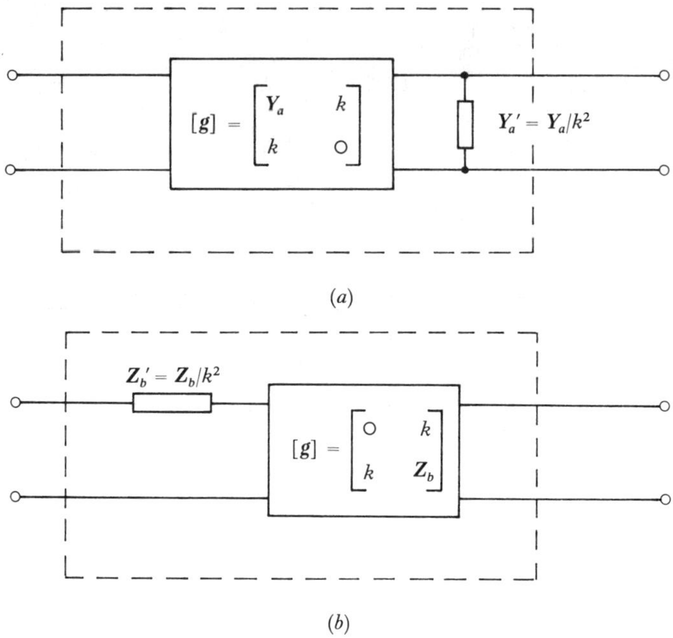

Fig. 8.34 NIC compensated for: (*a*) non-zero g_i; (*b*) non-zero g_0.

terminals as shown in fig. 8.34(*b*). It does not follow, however, that *simultaneous* compensation for non-zero values of g_i and g_o is possible. Nevertheless, a first order compensation is possible if g_i and g_o are not too large.

In many practical networks which employ active elements, the stability of the network needs to be examined by means of principles which are outlined in chapter 12; this is particularly true of negative immittance twoports and is likely to impose some limits on the kind and value of the immittances with which these twoports may successfully be terminated.

Examples on chapter 8

8.1 The series arms of a symmetrical ladder network are: an inductor of reactance X_L, a coil of series resistance R and reactance X, and a second inductor of reactance X_L, in that order. The two shunt arms consist of equal capacitors of reactance X_C, one of which is connected between the first end of the coil and a common earth lead, the second between the other end of the coil and the common earth lead. Find the steady state short-circuit transfer admittance of the network by conventional loop or nodal analysis and confirm the result by using (*a*) the cascade rules of the transmission parameters, and (*b*) the parallel rules of the admittance parameters.

8.2 (*a*) A Π-network has equal 1 μF capacitors for its shunt arms and a 2 H inductor for its series arm. Find the transmission parameters of this twoport.
(*b*) Derive an expression, in terms of the transmission parameters, for the input impedance of a twoport when a load resistor R_L is connected to its output port.
(*c*) A twoport has the transmission matrix

$$\begin{bmatrix} 0 & -R \\ G & 0 \end{bmatrix}$$

where $RG = 1$. Is the twoport (i) lossless, (ii) reciprocal, (iii) symmetrical?

8.3 A capacitor C forms the vertical arm of a T-network in which the two horizontal arms are inductors L_1 and L_2 on left and right respectively. The inductors have a mutual inductance M which is positive when the currents $\boldsymbol{I}_1$ and $\boldsymbol{I}_2$ in them are defined as flowing into the network from outside. Derive the impedance matrix of the twoport and find the frequency at which there will be no transmission through it independent of the values of any terminating impedances which may be coupled to the ports.

8.4 The hybrid parameters of a three-terminal non-reactive twoport are $\boldsymbol{h}_i = 1.6\ \text{k}\Omega$, $\boldsymbol{h}_r = 0.8 \times 10^{-3}$, $\boldsymbol{h}_f = 50$, $\boldsymbol{h}_o = 60\ \mu\text{S}$. What are the values of a pair of resistors R_1, R_2 (known as the *image resistances* of the twoport) such that when R_1 is connected to the input port the output resistance of the network is R_2, and when R_2 is connected to the output port the input resistance of the network is R_1?

8.5 The twoport of example 8.4 above is modified by putting a 1 kΩ resistor in series with the common terminal to form a new common terminal. What are the hybrid parameters of the new network? What is its output resistance when fed from a 1 kΩ source? What is the current in a 1 kΩ load across its output port when fed from a 1 kΩ source with an open-circuit voltage of 1 V?

8.6 Two identical, symmetrical transmission lines, which can be approximated by networks having parameters $\boldsymbol{A}_1, \boldsymbol{B}_1, \boldsymbol{C}_1$, are connected in cascade, and a shunt capacitor which has a susceptance B_c is connected across the junction between the networks. What are the ***ABC*** parameters of the overall system?

Each phase of a balanced, 275 kV, three-phase transmission line has parameters $\boldsymbol{A} = 0.94\angle 1°$, $\boldsymbol{B} = 105\angle 75°\ \Omega$, $\boldsymbol{C} = 0.00115\angle 90°$ S. What power can be taken from one end of the line when it is supplied at unity power factor from the other end and when the voltages at both ends are equal to 275 kV? If the receiving end voltage and power remains unchanged but the sending end voltage rises to 300 kV, what reactive power is delivered to the receiving end and what power and reactive power are supplied at the sending end?

8.7 A voltage amplifier has an open-circuit input admittance of 5 mS and a short-circuit reverse current transmittance of −0.7. On connecting a 1 V ideal voltage source to its input terminals, the output voltage is −20 V when the load resistor is 10 kΩ, and −30 V when the load resistor is 30 kΩ. Find the inverse hybrid parameters of the amplifier and the output voltage across a 100 kΩ load resistor when a source of 50 Ω internal resistance and 1 V open-circuit voltage is connected to the input terminals.

8.8 The power dissipated in a resistive load R_o when it is connected to a source of equal internal resistance is changed from P_L to P'_L on inserting a twoport between the source and the load. Show, in terms of the transmission parameters, that

$$\frac{P_L}{P'_L} = \frac{1}{4}\left|\boldsymbol{a}_{11} - \frac{\boldsymbol{a}_{12}}{R_o} + \boldsymbol{a}_{21}R_o - \boldsymbol{a}_{22}\right|^2.$$

If the twoport consists of a parallel resonant circuit LCG strung across a pair of wires directly connecting its input terminals to its output terminals, show that the insertion loss (P_L/P'_L) of the twoport is given by

$$\frac{P_L}{P'_L} = \left(1 + \frac{R_o G}{2}\right)^2 + \left[\frac{R_o C}{2}\left(\frac{\omega^2 - \omega_o^2}{\omega}\right)\right]^2$$

where $\omega_0 = 1/\sqrt{(LC)}$.

8.9 A network has nodes A, B, C, and D. Between A and B is a resistor R_2; between B and C a resistor R_3; between C and D a resistor R_6 shunted by a capacitor C_6; between B and D a capacitor C_4; between A and C a resistor R_5 in series with an inductor L_5; and between A and D a resistor R_1 in series with an ideal voltage source V_1 directed towards A.

Find the impedance matrix of the network and from this obtain the dual matrix. By inspection of this dual matrix set up the dual of the given network and confirm the result by comparing it with a graphically derived dual.

8.10 A twoport is formed from an $(N+1)$ node network by using terminals 1 and $(N+1)$ as port-1 and terminals P and $(N+1)$ as port-2. Show how the impedance parameters of the twoport may be derived in terms of the admittances of the elements of the $(N+1)$ node network, using the terminology that the admittance of the element between node P and node $(N+1)$ is $\boldsymbol{y}_P$ and the admittance of the element between node P and node Q is $\boldsymbol{y}_{PQ}$.

A four-node network has 1 Ω resistors between nodes 1 and 2, 1 and 3, 1 and 4, and 3 and 4; and 1 F capacitors either separately or in shunt with the resistors between nodes 1 and 3, 1 and 4, and 2 and 4. What are the impedance parameters, in terms of the complex frequency s, of the twoport formed by taking nodes 1 and 4 as port-1, and nodes 2 and 4 as port-2?

8.11 Two Π-networks with terminals ABC and A′B′C′ are cascaded by connecting B to A′ and C to C′; the terminals AC form the input of the cascade and B′C′ the output. The internal arrangement of the first network is as follows: between A and C is a 0.5 mS conductor in parallel with a 100 pF capacitor; between A and B is a 5 pF capacitor; and between B and C is a 1 mS conductor shunted by an ideal voltage-controlled current source (directed towards C) of value $(g_{m1} V_1)$ where g_{m1} is 100 mS and V_1 is the potential of terminal A with respect to that of C. The corresponding numerical values for the second similar network are 1 mS and 200 pF; 5 pF; and zero conductance shunted by a source of value $(g_{m2} V_2)$, directed towards C′, where g_{m2} is 200 mS and V_2 is the potential of A′ relative to C′. What is the approximate frequency at which the output voltage of the cascade, when terminated at B′C′ by a conductor of 5 mS and fed at AC from a 50 Ω source of constant open-circuit voltage and variable frequency, is 3 db down on its zero-frequency value?

8.12 Show that

(i) An ideal, three-terminal gyrator of gyration conductance G acts as a gyrator no matter which terminal is used as the common one,

(ii) The effect of connecting an admittance $\boldsymbol{Y}$ in series with one terminal of an ideal gyrator is the same as connecting an admittance $G^2/\boldsymbol{Y}$ across the other two,

(iii) The Q-factor of the floating inductor realized by shunting a capacitor across the junction of two identical ideal gyrators in cascade is equal to the Q-factor of the capacitor.

9 The harmonic analysis of alternating quantities

9.1 Harmonics

Alternating quantities were defined in chapter 3 in a manner which covered any repetitive wave shape; but it was then pointed out that a sine wave was the basic type of alternating quantity, and the theory of phasors and the 'j' method was entirely founded on the assumption of sinusoidal variations. But it is by no means possible to forget about non-sinusoidal waveforms, which play an important part in both power and communication engineering. In a.c. power circuits the non-sinusoidal waveform appears chiefly as a result of unavoidable imperfections in apparatus designed to produce or reproduce, as closely as possible, the sinusoidal shape. In communication circuits, on the other hand, trains of recurrent pulses of various shapes are used deliberately. Moreover the sinusoidal a.c. theory is a theory of a.c. variations infinitely continued; this of course is realized to all intents and purposes in power systems, but communication, by its very essence, demands the use of quantities which are varied in a controlled manner determined by the message, and this means that the sine waves employed must be started and stopped, or changed in amplitude or frequency. Thus the theory based on sinusoidal waves requires augmentation.

If the speech coil of a loudspeaker is supplied with alternating current at a frequency of 440 Hz, the note known as a′ (the A above 'middle C') will be emitted; if the apparatus is perfect, the pressure at any point near the speaker will vary sinusoidally at 440 Hz. If the same note is sounded on a violin or flute, the sounds are recognizably different from the 'pure' 440 Hz tone, each deriving its special character from the presence of 'overtones' or *harmonics*, namely oscillations at higher frequencies which are integral multiples of the *fundamental* 440 Hz frequency. This mixture of frequencies results in the pressure waveform near the instrument being non-sinusoidal.

The theory of non-sinusoidal waveforms is based upon resolving them into sinusoidal fundamental and harmonic components, and the first necessary step is to show that (subject to certain conditions) any waveform can be so resolved.

9.2 The Fourier series

The method used is due to Baron de Fourier (1768–1830), one of Napoleon's military engineers. The statement is most simply made, not in terms of functions of t having a period $T = 2\pi/\omega$, but in terms of functions of $\theta = \omega t$, having a period 2π:

Any alternating quantity $f(\theta)$, having a period 2π, can be represented by a series of the following form:

$$f(\theta) = a_0 + \sum_{n=1}^{\infty} (a_n \cos n\theta + b_n \sin n\theta). \tag{9.2.1}$$

The values of the coefficients in this *Fourier series* are obtained by integration over one cycle. Thus, integrating both sides of (9.2.1),

$$\int_0^{2\pi} f(\theta)\,d\theta = a_0 \int_0^{2\pi} d\theta + \sum_{n=1}^{\infty}\left(a_n \int_0^{2\pi} \cos n\theta\,d\theta + b_n \int_0^{2\pi} \sin n\theta\,d\theta\right).$$

The first integral on the right hand side is 2π, the others are zero; hence

$$a_0 = \frac{1}{2\pi}\int_0^{2\pi} f(\theta).d\theta. \tag{9.2.2}$$

Also, multiplying both sides by $\cos m\theta$ and integrating, we have

$$\int_0^{2\pi} f(\theta)\cos m\theta\,d\theta$$

$$= a_0 \int_0^{2\pi} \cos m\theta\,d\theta + \sum_{n=1}^{\infty}\left(a_n \int_0^{2\pi} \cos m\theta \cos n\theta\,d\theta + b_n \int_0^{2\pi} \cos m\theta \sin n\theta\,d\theta\right).$$

The integral which multiplies a_0 is zero. Of the others,

$$\int_0^{2\pi} \cos m\theta \cos n\theta\,d\theta = \frac{1}{2}\int_0^{2\pi} \{\cos (m+n)\,\theta + \cos (m-n)\,\theta\}\,d\theta.$$

When $m \neq n$, the two parts of the right hand side each integrate to zero; but $m = n$ is a special case, giving

$$\int_0^{2\pi} \cos^2 n\theta\,d\theta = \frac{1}{2}\int_0^{2\pi} (1 + \cos 2n\theta)\,d\theta = \pi.$$

Moreover

$$\int_0^{2\pi} \cos m\theta \sin n\theta\,d\theta = \frac{1}{2}\int_0^{2\pi} \{\sin (m+n)\,\theta - \sin (m-n)\,\theta\}\,d\theta,$$

which is zero for all values of m. Thus, when $f(\theta)$ is multiplied by $\cos m\theta$ and integrated, only the term containing a_m contributes to the right hand side, and

$$a_m = \frac{1}{\pi}\int_0^{2\pi} f(\theta)\cos m\theta\,d\theta,$$

or, changing the index m to n,

$$a_n = \frac{1}{\pi}\int_0^{2\pi} f(\theta)\cos n\theta \, d\theta. \tag{9.2.3}$$

In an exactly similar manner it is proved that

$$b_n = \frac{1}{\pi}\int_0^{2\pi} f(\theta)\sin n\theta \, d\theta. \tag{9.2.4}$$

This shows how the Fourier series is found for any given periodic function $f(\theta)$. It is necessary, of course, for the integrals to have finite values and for the resulting series to be convergent; the conditions for this are known as the Dirichlet conditions, but as these are satisfied by virtually all functions arising in physical problems they will not be discussed here.

The two terms containing $n\theta$ may be written

$$a_n \cos n\theta + b_n \sin n\theta = \sqrt{(a_n^2+b_n^2)}\left\{\frac{a_n}{\sqrt{(a_n^2+b_n^2)}}.\cos n\theta + \frac{b_n}{\sqrt{(a_n^2+b_n^2)}}.\sin n\theta\right\},$$

$$= c_n \cos(n\theta - \phi_n),$$

where

$$c_n = \sqrt{(a_n^2+b_n^2)}, \tag{9.2.5}$$

and

$$\phi_n = \tan^{-1}(b_n/a_n). \tag{9.2.6}$$

Thus, writing c_0 for a_0, the series becomes

$$f(\theta) = c_0 + \textstyle\sum_{n=1}^{\infty} c_n \cos(n\theta - \phi_n). \tag{9.2.7}$$

The term $c_1 \cos(\theta - \phi_1)$ is the fundamental component; $c_2 \cos(2\theta - \phi_2)$, the second harmonic; and so on. If $f(\theta)$ is an electrical quantity, the constant (c_0) term will be called the d.c. component.

We shall require the r.m.s. value of an alternating quantity expressed as a Fourier series. If F is the r.m.s. value of $f(\theta)$,

$$F^2 = \frac{1}{2\pi}\int_0^{2\pi} [f(\theta)]^2 \, d\theta,$$

$$= \frac{1}{2\pi}\int_0^{2\pi} \{c_0 + \textstyle\sum_{n=1}^{\infty} c_n \cos(n\theta - \phi_n)\}^2 \, d\theta,$$

$$= \frac{1}{2\pi}\int_0^{2\pi} \{c_0^2 + c_1^2\cos^2(\theta - \phi_1) + \ldots + 2c_0 c_1 \cos(\theta - \phi_1) + 2c_1 c_2 \cos(\theta - \phi_1)\cos(2\theta - \phi_2)\ldots\} \, d\theta.$$

The last line shows the four types of term which can occur in the squared series. The argument already used can be extended to show that

$$\int_0^{2\pi} \cos(m\theta - \phi_m)\cos(n\theta - \phi_n)\, d\theta = 0$$

except when $m = n$, when its value is π. Thus all the cross-terms with coefficients like $c_1 c_2$ vanish, and we are left with

$$F^2 = c_0^2 + \tfrac{1}{2}(c_1^2 + c_2^2 + \ldots). \tag{9.2.8}$$

But c_0 is the r.m.s. value of the d.c. component, while the r.m.s values of the a.c. components are $c_1/\sqrt{2}$, $c_2/\sqrt{2}$, etc. Therefore:

The r.m.s. value of a quantity analysed into its Fourier components is the square root of the sum of the squares of the r.m.s. values of the separate components. (9.2.9)

9.3 Examples of Fourier series analysis

Before carrying out the analysis of typical waveforms we shall note one or two general points.

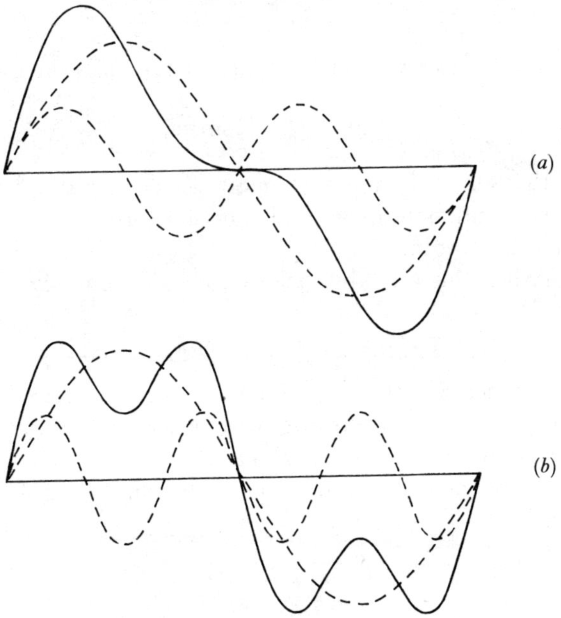

Fig. 9.1 (*a*) Waveform containing second harmonic. (*b*) Waveform containing third harmonic.

Figure 9.1 shows two non-sinusoidal waves: (*a*) is the sum of a fundamental and a second harmonic, while (*b*) contains a fundamental and a third harmonic. It is apparent that, in the second case,

$$f(\theta + \pi) = -f(\theta),$$

so that the positive and negative half-cycles are similar; this is not so in case (*a*). It is immediately concluded that:

A waveform in which the positive and negative half-cycles are similar contains odd harmonics only. (9.3.1)

For the purpose of this and other general statements, the fundamental is an 'odd harmonic', while a d.c. component is an 'even harmonic'.

A function such that $f(-\theta)=f(\theta)$ is called an *even function*; when $f(-\theta)=-f(\theta)$ it is an *odd function*. The series expansion

$$f(\theta) = a_0 + \textstyle\sum_{n=1}^{\infty} (a_n \cos n\theta + b_n \sin n\theta), \quad \text{(equation (9.2.1))}$$

contains both even functions a_0, $a_n \cos n\theta$, and odd functions $b_n \sin n\theta$; but the sum of a number of even and a number of odd functions can be neither odd nor even. This leads to the conclusions that:

The Fourier series expansion of an even function is of the form

$$f(\theta) = a_0 + \textstyle\sum_{n=1}^{\infty} a_n \cos n\theta, \tag{9.3.2}$$

and

The Fourier series expansion of an odd function is of the form

$$f(\theta) = \textstyle\sum_{n=1}^{\infty} b_n \sin n\theta. \tag{9.3.3}$$

A moment's thought will show that many waveforms can be represented either as an even function, or an odd function, or both, according to the point chosen as the origin.

These points are illustrated by the examples which follow.

1. Triangular wave (fig. 9.2(a))

This is an even function if the origin is taken at a point where the wave has a maximum, an odd function if the origin is at a zero. In either case, since the positive and negative half-waves are similar, only odd harmonics can occur. Taking the origin at a zero as shown and the maximum amplitude as unity, it follows that

$$f(\theta) = \textstyle\sum_{n \text{ odd}} b_n \sin n\theta, \tag{9.3.4}$$

where

$$\left.\begin{aligned} f(\theta) &= \frac{2\theta}{\pi} \text{ from } \theta = 0 \text{ to } \frac{\pi}{2}, \\ &= \frac{2}{\pi}(\pi-\theta) \text{ from } \theta = \frac{\pi}{2} \text{ to } \frac{3\pi}{2}, \\ &= \frac{2}{\pi}(\theta - 2\pi) \text{ from } \theta = \frac{3\pi}{2} \text{ to } 2\pi. \end{aligned}\right\} \tag{9.3.5}$$

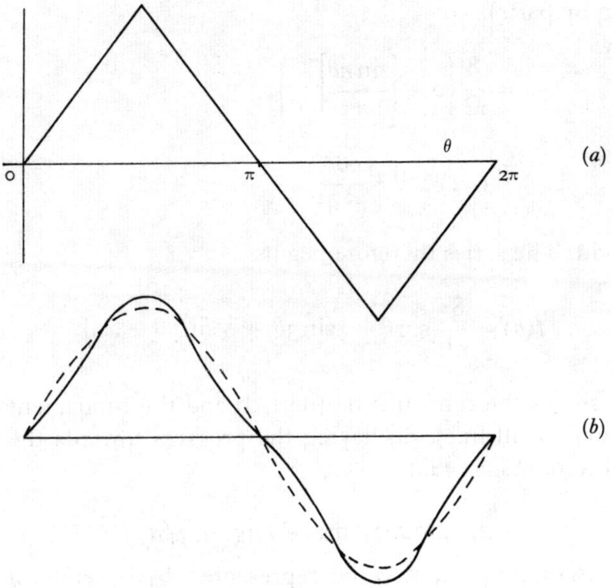

Fig. 9.2 Fourier analysis of a triangular wave: (*a*) wave to be analysed; (*b*) effect of taking one term (---) and two terms (——).

From (9.2.4)

$$b_n = \frac{1}{\pi}\int_0^{2\pi} f(\theta)\sin n\theta\,\mathrm{d}\theta;$$

but since both $f(\theta)$ and $\sin n\theta$ (n being odd) change sign when θ is increased by π, this may also be written

$$b_n = \frac{2}{\pi}\int_0^{\pi} f(\theta)\sin n\theta\,\mathrm{d}\theta. \tag{9.3.6}$$

Substituting from (9.3.5),

$$b_n = \frac{4}{\pi^2}\left\{\int_0^{\pi/2}\theta\sin n\theta\,\mathrm{d}\theta + \int_{\pi/2}^{\pi}(\pi-\theta)\sin n\theta\,\mathrm{d}\theta\right\}.$$

If θ is replaced by $\pi-\theta'$ in the second integral, it becomes identical with the first; thus

$$b_n = \frac{8}{\pi^2}\int_0^{\pi/2}\theta\sin n\theta\,\mathrm{d}\theta,$$

$$= \frac{8}{\pi^2}\left\{\left[\frac{-\theta\cos n\theta}{n}\right]_0^{\pi/2} + \frac{1}{n}\int_0^{\pi/2}\cos n\theta\,\mathrm{d}\theta\right\},$$

(integrating by parts),

$$= \frac{8}{\pi^2}\left\{0 + \left[\frac{\sin n\theta}{n^2}\right]_0^{\pi/2}\right\},$$

$$= (-1)^{(n-1)/2}\frac{8}{\pi^2 n^2}, \tag{9.3.7}$$

since n is odd. The series therefore begins

$$f(\theta) = \frac{8}{\pi^2}\{\sin\theta - \tfrac{1}{9}\sin 3\theta + \tfrac{1}{25}\sin 5\theta - \ldots\}. \tag{9.3.8}$$

Fig. 9.2(*b*) shows the fundamental (dotted) and the fundamental plus the third harmonic (full line), displaying the progress towards the triangular form as more terms are taken.

2. *Square wave (fig. 9.3(a))*

This wave, like the last, may be represented by a series of the form $\sum b_n \sin n\theta$, where n is odd. Equation (9.3.6) is operative in this problem as in the preceding one, but $f(\theta)$ now equals 1 (taken as the amplitude of the square wave) over the range $\theta = 0$ to π. Thus

$$b_n = \frac{2}{\pi}\int_0^{\pi} \sin n\theta \, d\theta,$$

$$= \frac{2}{\pi}\left[\frac{-\cos n\theta}{n}\right]_0^{\pi},$$

$$= \frac{4}{\pi n}, \tag{9.3.9}$$

since n is odd. Therefore the series is

$$f(\theta) = \frac{4}{\pi}\{\sin\theta + \tfrac{1}{3}\sin 3\theta + \tfrac{1}{5}\sin 5\theta + \ldots\}. \tag{9.3.10}$$

This is much less convergent than (9.3.8), because the wave is more unlike a sine-wave. Fig. 9.3(*b*) shows the approach to the square wave as one takes one, two and three terms. The overshoot at the corners is always found when a sudden change of amplitude occurs in a wave and the attempt is made to represent it by a finite number of terms in a Fourier series. As the number of terms is increased, the rectangular shape is approached but the overshoot is still found at points very near the corner (fig. 9.3(*c*)). This is called the *Gibbs phenomenon*.

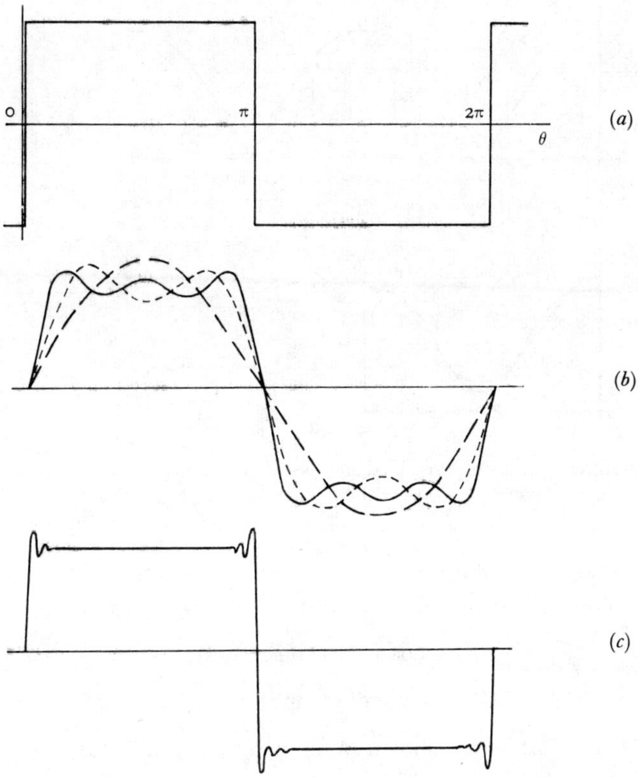

Fig. 9.3 Fourier analysis of a square wave: (*a*) wave to be analysed; (*b*) effect of taking one (–-), two (- -) and three (——) terms; (*c*) the Gibbs phenomenon.

3. *Rectified sine wave (fig. 9.4(a))*

This waveform cannot be represented as an odd function, but it is an even function if the origin is taken at a point where the wave has a maximum. The positive and negative half-waves are dissimilar; thus the Fourier expansion is

$$f(\theta) = a_0 + \sum_{n=1}^{\infty} a_n \cos n\theta, \qquad \text{(equation (9.3.2))}$$

where n takes all values. It should be noted that one cycle of the fundamental frequency corresponds to half a cycle of the wave from which, by reversing the sign of alternate half-cycles, the given wave is derived. If the given wave is described by

$$F(t) = \cos \omega t \text{ from } t = 0 \text{ to } \pi/2\omega,$$

$$F(t) = -\cos \omega t \text{ from } t = \pi/2\omega \text{ to } 3\pi/2\omega,$$

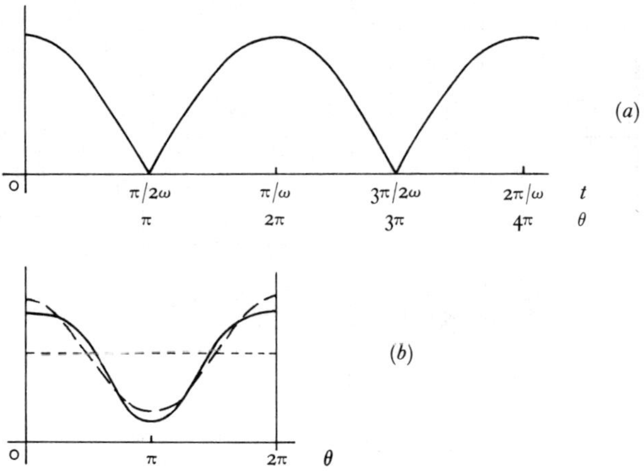

Fig. 9.4 Fourier analysis of a rectified sine wave: (a) wave to be analysed; (b) d.c. component (---) plus one (--) and two (——) terms.

etc., the expansion will therefore be in terms of $\theta = 2\omega t$.

The function of θ is thus defined as follows over one cycle:

$$f(\theta) = \cos\theta/2 \text{ from } \theta = 0 \text{ to } \pi,$$

$$f(\theta) = -\cos\theta/2 \text{ from } \theta = \pi \text{ to } 2\pi.$$

Both the t and θ scales are shown in fig. 9.4(a). From (9.2.2),

$$\begin{aligned} a_0 &= \frac{1}{2\pi}\int_0^{2\pi} f(\theta)\,d\theta, \\ &= \frac{1}{2\pi}\left\{\int_0^{\pi}\cos\tfrac{1}{2}\theta\,d\theta + \int_{\pi}^{2\pi}(-\cos\tfrac{1}{2}\theta)\,d\theta\right\}, \\ &= \frac{1}{\pi}\int_0^{\pi}\cos\tfrac{1}{2}\theta\,d\theta, \\ &= \frac{2}{\pi}. \end{aligned} \tag{9.3.11}$$

Furthermore, from (9.2.3),

$$\begin{aligned} a_n &= \frac{1}{\pi}\int_0^{2\pi} f(\theta)\cos n\theta\,d\theta, \\ &= \frac{1}{\pi}\left\{\int_0^{\pi}\cos n\theta\cos\tfrac{1}{2}\theta\,d\theta + \int_{\pi}^{2\pi}(-\cos n\theta\cos\tfrac{1}{2}\theta)\,d\theta\right\}. \end{aligned}$$

But $$\cos n\theta\cos\tfrac{1}{2}\theta = \tfrac{1}{2}\{\cos(n+\tfrac{1}{2})\theta + \cos(n-\tfrac{1}{2})\theta\};$$

and if θ is replaced in the second integral by $2\pi - \theta'$, it becomes identical with the first. Thus

$$a_n = \frac{1}{\pi}\int_0^{\pi} \{\cos(n+\tfrac{1}{2})\theta + \cos(n-\tfrac{1}{2})\theta\}\,d\theta,$$

$$= \frac{1}{\pi}\left[\frac{\sin(n+\frac{1}{2})\theta}{n+\frac{1}{2}} + \frac{\sin(n-\frac{1}{2})\theta}{n-\frac{1}{2}}\right]_0^{\pi},$$

$$= \frac{1}{\pi}\left(\frac{-1}{n+\frac{1}{2}} + \frac{1}{n-\frac{1}{2}}\right) \text{ when } n \text{ is odd,}$$

$$\text{or } \frac{1}{\pi}\left(\frac{1}{n+\frac{1}{2}} + \frac{1}{n-\frac{1}{2}}\right) \text{ when } n \text{ is even.}$$

The general expression for a_n is therefore

$$a_n = \frac{(-1)^{n+1}}{\pi} \cdot \frac{4}{4n^2-1}. \qquad (9.3.12)$$

The first few terms of the series may be written

$$f(\theta) = \frac{2}{\pi}\{1 + \tfrac{2}{3}\cos\theta - \tfrac{2}{15}\cos 2\theta + \tfrac{2}{35}\cos 3\theta - \ldots\}. \qquad (9.3.13)$$

Again the effect of taking only a few terms has been illustrated in fig. 9.4(*b*).

9.4 The response of a circuit to a non-sinusoidal stimulus

Non-linear circuit elements will produce a non-sinusoidal response to a sinusoidal stimulus; but only linear circuits are dealt with in this book. The non-sinusoidal effects with which we are concerned arise from the application of non-sinusoidal voltage or current stimuli to linear circuits. The foregoing analysis shows that these can be calculated by analysing the stimulus into its harmonic components, evaluating the response to each harmonic, and invoking the principle of superposition to obtain the total result. Some examples will illustrate this procedure.

1. A voltage source, developing a square wave of maximum value V_1 and period $2\pi/\omega$, is applied to a coil of inductance L and resistance R (fig. 9.5(*a*)). It is required to find the current.

It is shown in (9.3.10) that the output of this voltage source may be expressed in the form

$$v_1(t) = \frac{4V_1}{\pi}\sum_{n \text{ odd}} \frac{1}{n}\sin n\omega t. \qquad (9.4.1)$$

Each term produces its own response; but the image circuit for each harmonic is different, since the impedance of L is frequency-dependent. Fig.

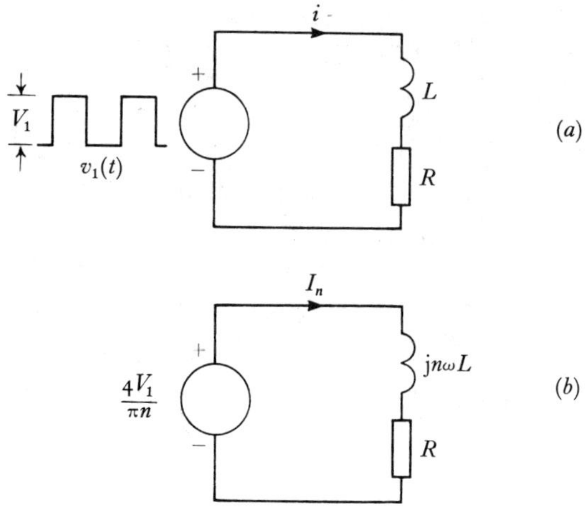

Fig. 9.5 (*a*) Coil excited by square voltage wave. (*b*) Image circuit for *n*th harmonic in (*a*).

9.5(*b*) shows the image circuit for the *n*th harmonic, the source voltage being taken as the reference phasor. The current is given by

$$\boldsymbol{I}_n = \frac{4V_1}{\pi n} \cdot \frac{1}{R + jn\omega L},$$

$$= \frac{4V_1}{\pi n} \cdot \frac{R - jn\omega L}{R^2 + n^2\omega^2 L^2}. \tag{9.4.2}$$

As $\sin n\omega t$ is the imaginary part of $e^{jn\omega t}$, the conversion of (9.4.2) into a function of t is effected by multiplying by $e^{jn\omega t}$ and taking the imaginary part of the product. Thus

$$i_n = \frac{4V_1}{\pi n} . \operatorname{Im} \frac{(R - jn\omega L)\, e^{jn\omega t}}{R^2 + n^2\omega^2 L^2},$$

$$= \frac{4V_1}{\pi n} \cdot \frac{R \sin n\omega t - n\omega L \cos n\omega t}{R^2 + n^2\omega^2 L^2},$$

$$= \frac{4V_1}{\pi n\sqrt{(R^2 + n^2\omega^2 L^2)}} . \sin(n\omega t - \phi_n), \tag{9.4.3}$$

where

$$\phi_n = \tan^{-1}(n\omega L/R). \tag{9.4.4}$$

The current is thus given by

$$i = \frac{4V_1}{\pi} \sum_{n \text{ odd}} \frac{\sin(n\omega t - \phi_n)}{n\sqrt{(R^2 + n^2\omega^2 L^2)}}. \tag{9.4.5}$$

The correctness of this may be verified by considering the extreme cases $L = 0$ and $R = 0$. When $L = 0$, $\phi_n = 0$, and (9.4.5) reduces to (9.4.1) divided by R, as it should. When $R = 0$, $\phi_n = \pi/2$, and the form taken by (9.4.5) is

$$i = \frac{4V_1}{\pi\omega L} \sum_{n \text{ odd}} \left(-\frac{1}{n^2}\cos n\omega t\right). \tag{9.4.6}$$

If, in the triangular wave expressed by the Fourier series (9.3.8), the origin is shifted by writing $\theta' - (\pi/2)$ for θ, the result is

$$f(\theta') = \frac{8}{\pi^2} \sum_{n \text{ odd}} \left(-\frac{1}{n^2}\cos n\theta'\right). \tag{9.4.7}$$

This shows that (9.4.6) represents a triangular wave having a negative maximum at $t = 0$; a moment's consideration will show why the current taken by a pure inductor from a square-wave supply must have this form.

2. A voltage source having a rectified sinusoidal waveform feeds current to a resistance R, but the ripple superimposed on the direct current component is reduced by the interposition of a smoothing circuit consisting of inductance L and capacitance C (fig. 9.6(*a*)). It is required to find the voltage across R.

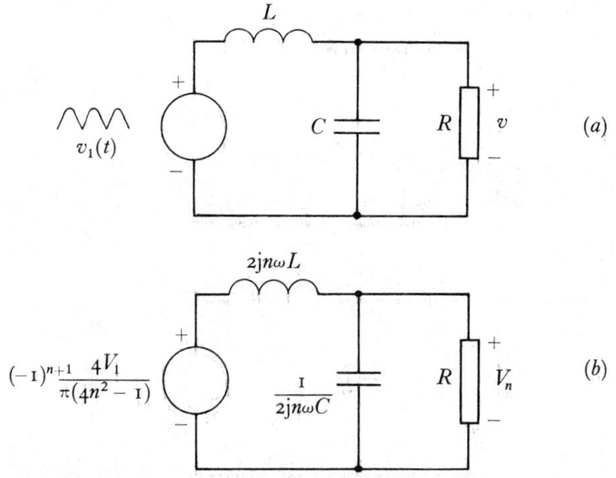

Fig. 9.6 (*a*) Circuit excited by rectified sine-wave voltage. (*b*) Image circuit for harmonics in (*a*).

(This example illustrates the operation of a rectifier, but must not be applied unthinkingly to a practical case. With some types of rectifier and under some conditions the circuit is effectively disconnected from the source during a part of each cycle.)

The Fourier series for the output of the voltage source is derived from equations (9.3.12,13) by writing $\theta = 2\omega t$:

$$v_1(t) = \frac{2V_1}{\pi}\left\{1 + \sum_{n=1}^{\infty} (-1)^{n+1} \frac{2\cos 2n\omega t}{4n^2 - 1}\right\}. \tag{9.4.8}$$

For the d.c. term, L has zero and C has infinite impedance; thus the d.c. component in the voltage across R is the same as the d.c. component in $v_1(t)$, namely $2V_1/\pi$. The a.c. terms are dealt with by image circuits of which the nth is shown in fig. 9.6(*b*). In this circuit,

$$\boldsymbol{V}_n + 2jn\omega L\left(\frac{\boldsymbol{V}_n}{R} + 2jn\omega C\boldsymbol{V}_n\right) = (-1)^{n+1}.\frac{4V_1}{\pi(4n^2 - 1)};$$

the bracketed term being the total current in L. Therefore

$$\boldsymbol{V}_n\left\{\frac{R(1 - 4n^2\omega^2 LC) + 2jn\omega L}{R}\right\} = (-1)^{n+1}\frac{4V_1}{\pi(4n^2 - 1)},$$

or

$$\boldsymbol{V}_n = (-1)^{n+1}\frac{4V_1 R}{\pi(4n^2 - 1)}.\frac{R(1 - 4n^2\omega^2 LC) - 2jn\omega L}{R^2(1 - 4n^2\omega^2 LC)^2 + 4n^2\omega^2 L^2}. \tag{9.4.9}$$

As the harmonic terms in (9.4.8) are cosines, $\boldsymbol{V}_n$ is converted into a function of t by taking the real part of $\boldsymbol{V}_n e^{2jn\omega t}$. Thus

$$v_n = (-1)^{n+1}\frac{4V_1 R}{\pi(4n^2 - 1)}.\frac{R(1 - 4n^2\omega^2 LC)\cos 2n\omega t + 2n\omega L\sin 2n\omega t}{R^2(1 - 4n^2\omega^2 LC)^2 + 4n^2\omega^2 L^2},$$

$$= (-1)^{n+1}\frac{4V_1 R\cos(2n\omega t - \phi_n)}{\pi(4n^2 - 1)\sqrt{[R^2(1 - 4n^2\omega^2 LC)^2 + 4n^2\omega^2 L^2]}}, \tag{9.4.10}$$

where
$$\phi_n = \tan^{-1}\frac{2n\omega L}{R(1 - 4n^2\omega^2 LC)}. \tag{9.4.11}$$

The voltage across R is therefore given by

$$v = \frac{2V_1}{\pi}\left\{1 + \sum_{n=1}^{\infty} (-1)^{n+1}\frac{2R\cos(2n\omega t - \phi_n)}{(4n^2 - 1)\sqrt{[R^2(1 - 4n^2\omega^2 LC)^2 + 4n^2\omega^2 L^2]}}\right\}. \tag{9.4.12}$$

The function of L and C in such a circuit would be to reduce the ripple, as represented by the sum of the harmonic terms in (9.4.12), in proportion to the d.c. component. To accomplish this it is evidently desirable to avoid values which make $1 - 4n^2\omega^2 LC = 0$ for any value of n; this, if permitted, would represent resonance at the harmonic in question. If $4\omega^2 LC > 1$, resonance is impossible for all values of n; assuming this condition satisfied, so that no harmonic can be unnaturally enhanced, we may feel justified in approximating to the ripple by taking its fundamental component only. Since the peak-to-peak variation in $\cos\theta$ is 2, (9.4.12) shows that

$$\frac{\text{peak-to-peak fundamental ripple}}{\text{d.c. component}} = \frac{4R}{3\sqrt{[R^2(4\omega^2 LC - 1)^2 + 4\omega^2 L^2]}}. \tag{9.4.13}$$

In the next chapter (§10.10), an explicit solution is obtained for the same problem; in fig. 9.7 this is plotted for particular circuit values, together

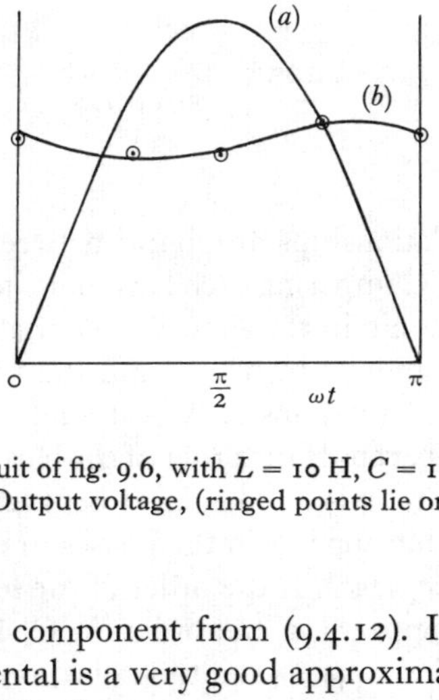

Fig. 9.7 Response of circuit of fig. 9.6, with $L = 10$ H, $C = 1\ \mu$F, $R = 1000\ \Omega$, $f = 50$ Hz. (*a*) Source voltage. (*b*) Output voltage, (ringed points lie on the fundamental of (*b*)).

with the fundamental component from (9.4.12). It will be seen that, for this case, the fundamental is a very good approximation to the total ripple, both in magnitude and in shape.

9.5 Harmonics in a.c. power systems

The voltage sources in a.c. power systems are normally rotating machines with similarity between the positive and negative poles; thus the positive and negative half-waves are similar and even harmonics do not arise. Harmonics of odd order do arise, however, for a number of reasons, and give rise to special problems in relation to the three-phase system.

Suppose that an a.c. quantity, either voltage or current, is designated in one phase of a three-phase system by the Fourier series

$$w(\omega t) = W_1 \sin(\omega t - \phi_1) + W_3 \sin(3\omega t - \phi_3) + W_5 \sin(5\omega t - \phi_5) + \dots \quad (9.5.1)$$

Assuming that the system is symmetrical, the same quantity will have its counterpart in the next phase, identical but delayed by an electrical angle of $2\pi/3$:

$$w\left(\omega t - \frac{2\pi}{3}\right) = W_1 \sin\left(\omega t - \phi_1 - \frac{2\pi}{3}\right) + W_3 \sin(3\omega t - \phi_3 - 2\pi)$$

$$+ W_5 \sin\left(5\omega t - \phi_5 - \frac{10\pi}{3}\right) + \dots \quad (9.5.2)$$

But

$$\sin(3\omega t - \phi_3 - 2\pi) = \sin(3\omega t - \phi_3), \quad (9.5.3)$$

and

$$\sin\left(5\omega t - \phi_5 - \frac{10\pi}{3}\right) = \sin\left(5\omega t - \phi_5 + \frac{2\pi}{3}\right). \quad (9.5.4)$$

Thus three different relationships are found between harmonic quantities in the successive phases. Comparing each harmonic in the second phase with the corresponding harmonic in the first, we see that the fundamental lags by $2\pi/3$, the third harmonic is in phase, and the fifth harmonic leads by $2\pi/3$. The same kinds of relations will be found to be true of all other harmonics, both odd and even. If the order of the harmonic can be expressed in the form $3m + 1$, where m is an integer, it behaves like the fundamental in that the maximum value appears in the phases in the order A, B, C – the so-called *positive phase sequence*. If the order of the harmonic is of the form $3m - 1$, the maximum appears in the order A, C, B – the *negative phase sequence*. And if the order is $3m$, a multiple of 3, the harmonic reaches its maximum at the same instant in each phase, and the relationship is called a *zero phase sequence*. Harmonics of order $3m$ are sometimes called *triplen* harmonics.

It may be added that in unsymmetrical three-phase systems, which are not discussed in this book, the fundamental, the $(3m + 1)$-type harmonics and the $(3m - 1)$-type harmonics may be found to have positive, negative and zero sequence components; but in no case will the $3m$-type harmonics have anything but zero sequence.

Consider now a star-connected three-phase voltage source in which the phase voltage contains a third harmonic as well as the fundamental. The phase voltages may be written, if the origin of t be suitably chosen,

$$\left.\begin{aligned} v_A &= V_1 \sin \omega t + V_3 \sin(3\omega t - \phi_3), \\ v_B &= V_1 \sin\left(\omega t - \frac{2\pi}{3}\right) + V_3 \sin(3\omega t - \phi_3), \\ v_C &= V_1 \sin\left(\omega t - \frac{4\pi}{3}\right) + V_3 \sin(3\omega t - \phi_3). \end{aligned}\right\} \quad (9.5.5)$$

The voltage between the terminals a, b is $v_A - v_B$; plainly this contains no third harmonic. Thus, when harmonics are present in the phase voltages of a three-phase voltage source, all those whose order is a multiple of 3 are eliminated from the line voltages. If this generator supplies a symmetrical linear load through a three-wire line, the currents will be sine waves of fundamental frequency.

What difference, then, is the third harmonic making? To see this, let the load be star-connected (Fig. 9.8) and let its neutral point be taken as the

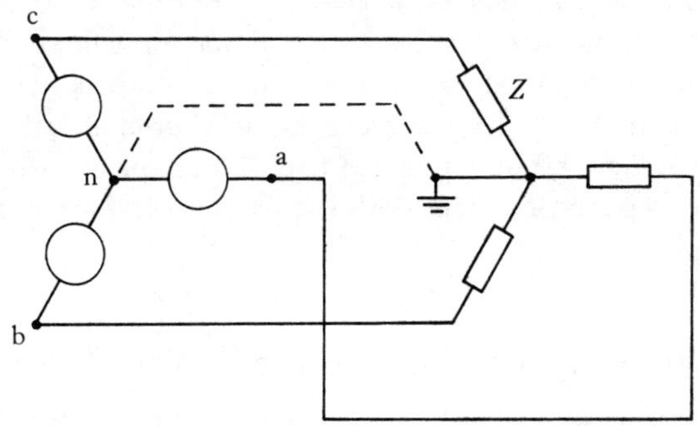

Fig. 9.8 Effect of third harmonic in a star-connected voltage source.

point of zero potential (it may be imagined to be earthed). The mean potential of the generator terminals a, b, c is then zero; but the potential differences between an, bn and cn all contain a component $V_3 \sin(3\omega t - \phi_3)$. It follows that the point n must have this potential to earth – alternating at three times the fundamental frequency.

If the neutral points of the source and load are joined by a fourth wire, we have essentially three single-phase sources each supplying its own load. The third-harmonic current will return via the neutral connexion, which carries the sum of the third-harmonic currents of the other three wires. Assuming a linear system and neglecting the impedance of the line, the neutral-wire third-harmonic current may be deduced from the voltage which appears between n and earth when this wire is removed; the peak value is $I_{N3} = 3V_3/Z_3$, where Z_3 is the impedance of each phase of the load at the third-harmonic frequency.

A three-phase voltage source may also be delta-connected. If, however, such a source contained a third harmonic in the phase voltage, there would be an e.m.f. of $3V_3 \sin(3\omega t - \phi_3)$ to circulate current round the delta; in an ideal source this current would be infinite. A delta-connected source would therefore have to be designed so as to generate little or no third-harmonic voltage. As has been stated, the aim in a power system is to use voltage sources which are as free from harmonics as possible. By employing the star connexion which eliminates all triplen harmonics, the designer is free to concentrate his effort on designing the voltage source itself so as to minimize other harmonics, particularly the fifth and seventh. Were he to employ the delta connexion he would lose this freedom.

The property of the three-wire connexion in providing no path for third-harmonic current becomes a liability in supplying certain types of non-linear load by which harmonic currents are demanded. An important case is that of an iron-cored transformer, which cannot have a sinusoidal e.m.f. unless it is allowed to draw a non-sinusoidal current, including any triplen harmonics which may be necessary. For this reason, three-phase transformers are built with at least one delta-connected winding; the triplen harmonic currents which cannot flow in the three-wire supply are able to circulate in the delta, where they are automatically induced. A floating tertiary winding, not connected to either the primary or the secondary, can serve this purpose where both the primary and secondary are star-connected.

9.6 Complex form of the Fourier series

When a periodic function $f(\theta)$ is expressed in the Fourier series

$$f(\theta) = c_0 + \sum_{n=1}^{\infty} c_n \cos(n\theta - \phi_n), \qquad \text{(equation (9.2.7))}$$

it is possible to obtain a graphical representation by drawing a series of vertical lines at intervals on a horizontal axis, the intervals representing increases of n and the vertical lines being proportional to c_n. This has been done in fig. 9.9 for the series representing a square wave (9.3.10); the diagram is called the *amplitude spectrum* of the function concerned. For time-varying functions, angular frequencies $n\omega$ may replace n as the abscissa. The amplitude spectrum is independent of the choice of origin for θ; for if, in (9.2.7) quoted above, θ is replaced by $\theta - \theta_0$, the equation takes the form

$$f(\theta) = c_0 + \sum_{n=1}^{\infty} c_n \cos(n\theta - \phi_n')$$

where $\phi'_n = \phi_n + n\theta_0$; so that the coefficients c_n are unchanged.

The sequence of numbers c_n, however, does not contain all the information embodied in the Fourier series; the phase angles ϕ_n have been ignored.

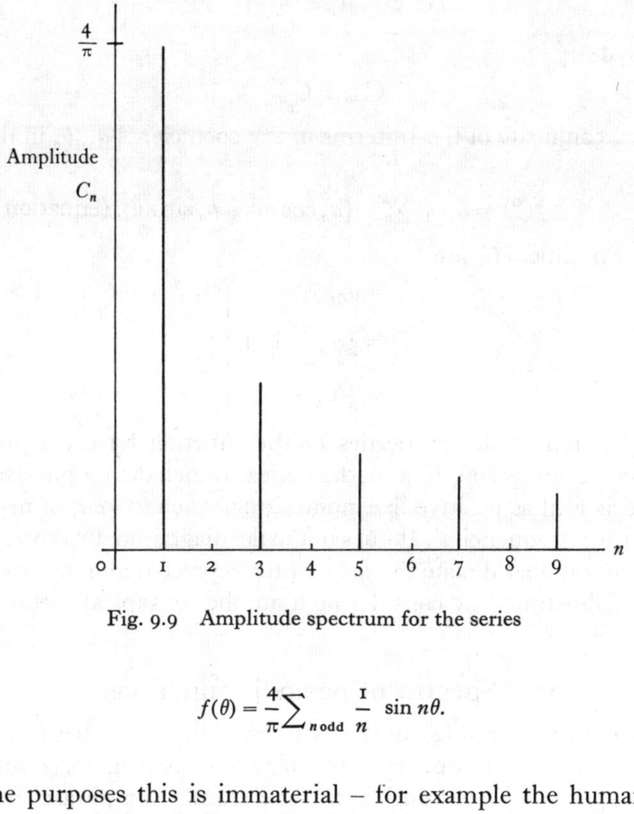

Fig. 9.9 Amplitude spectrum for the series

$$f(\theta) = \frac{4}{\pi}\sum_{n\,\mathrm{odd}} \frac{1}{n} \sin n\theta.$$

For some purposes this is immaterial – for example the human ear distinguishes sounds by their amplitude spectra but is unable to discern differences due to changes in phase. The question nevertheless arises whether it is possible to incorporate phase information as well as amplitude information in a single sequence of coefficients. This is achieved by changing to a complex form.

Using the identity $\cos x \equiv \frac{1}{2}(e^{jx} + e^{-jx})$, (9.2.7) is transformed to

$$f(\theta) = c_0 + \sum_{n=1}^{\infty} \tfrac{1}{2}c_n\{e^{jn\theta}\, e^{-j\phi_n} + e^{-jn\theta}\, e^{j\phi_n}\}.$$

The form of this series is that of a summation over negative as well as positive values of n; it may be written

$$f(\theta) = \sum_{n=-\infty}^{+\infty} C_n\, e^{jn\theta}, \tag{9.6.1}$$

where
$$C_0 = c_0 \tag{9.6.2}$$
and, if n is any positive number,
$$\left.\begin{aligned} C_n &= \tfrac{1}{2}c_n\,\mathrm{e}^{-\mathrm{j}\phi_n}, \\ C_{-n} &= \tfrac{1}{2}c_n\,\mathrm{e}^{\mathrm{j}\phi_n}. \end{aligned}\right\} \tag{9.6.3}$$
It is seen that
$$C_{-n} = C_n^*, \tag{9.6.4}$$
the complex conjugate of C_n. In terms of the coefficients a_n, b_n in the series form
$$f(\theta) = a_0 + \textstyle\sum_{n=1}^{\infty} (a_n \cos n\theta + b_n \sin n\theta), \text{ (equation (9.2.1))}$$
the complex numbers C_n are
$$\left.\begin{aligned} C_0 &= a_0, \\ C_n &= \tfrac{1}{2}(a_n - \mathrm{j}b_n), \\ C_{-n} &= \tfrac{1}{2}(a_n + \mathrm{j}b_n). \end{aligned}\right\} \tag{9.6.5}$$

The expression of the properties of the function by the sequence C_n amounts to the admission, in a mathematical rather than a physical sense, of negative as well as positive 'harmonics'; or, when $\theta = \omega t$, of negative as well as positive 'frequencies'. It does not make diagrammatic representation any easier; but it does dictate the form of the representation normally used. This will be illustrated by considering a number of typical spectra.

9.7 Spectra of periodic functions

The term *spectrum* is used for any description of the properties of a physical system on a base of frequency. One may distinguish, for example, an 'amplitude spectrum' from an 'energy spectrum'. The phrase 'frequency spectrum', commonly used, means just 'spectrum'; the word 'frequency' is redundant. From this point the symbol ω will be used as a general symbol for (angular) frequency, particular frequencies being denoted by ω with a suffix, or by other symbols.

In considering representations in terms of the coefficients C_n, we may note in the first place that, for an even function, all the C_n are real. The basic even periodic function $\cos \gamma t$ may be written
$$\cos \gamma t = \tfrac{1}{2}(\mathrm{e}^{\mathrm{j}\gamma t} + \mathrm{e}^{-\mathrm{j}\gamma t}), \tag{9.7.1}$$
so the spectrum consists of two lines of amplitude $\frac{1}{2}$ at $\omega = \pm\gamma$. By generalizing from this, it is seen that the spectrum of an even periodic function consists of pairs of equal lines for each pair of frequencies $\pm\omega$; or:

The spectrum of an even periodic function of t is a discontinuous even function of ω. (9.7.2)

If now the time-origin is changed by writing $t - t_0$ for t, we get

$$\cos\gamma(t-t_0) = (\tfrac{1}{2}\,\mathrm{e}^{-\mathrm{j}\gamma t_0})\,\mathrm{e}^{\mathrm{j}\gamma t} + (\tfrac{1}{2}\,\mathrm{e}^{\mathrm{j}\gamma t_0})\,\mathrm{e}^{-\gamma \mathrm{j}t}, \qquad (9.7.3)$$

and when $\gamma t_0 = \pi/2$, this becomes

$$\sin\gamma t = -\frac{\mathrm{j}}{2}\,\mathrm{e}^{\mathrm{j}\gamma t} + \frac{\mathrm{j}}{2}\,\mathrm{e}^{-\mathrm{j}\gamma t}. \qquad (9.7.4)$$

The transition is illustrated in fig. 9.10. The spectrum of an odd periodic function thus consists of elements all containing the factor j, so it can be

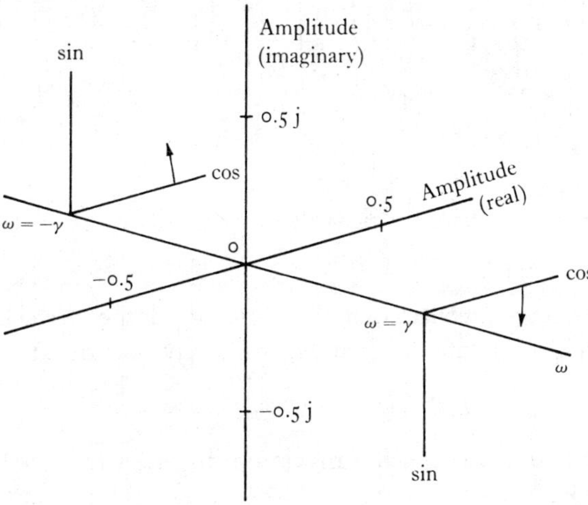

Fig. 9.10 Three-dimensional representation of the spectra of $\cos\gamma t$ and $\sin\gamma t$.

plotted on a flat sheet. When the sign of ω is changed, so is the sign of the element; in other words

The spectrum of an odd periodic function of t is a discontinuous odd function of ω. (9.7.5)

For a function which is neither even nor odd, the elements of the spectrum are seen from (9.7.3) to be complex, and the best that can be done is to plot $|C_n|$. If a function can be made even or odd by a suitable choice of origin, there is an advantage in so choosing.

Examples of typical spectra now follow.

1. Amplitude-modulated wave (fig. 9.11)

In this type of wave, widely used for communication, the signal takes the form of a variation in the amplitude of a wave, the frequency of the variation

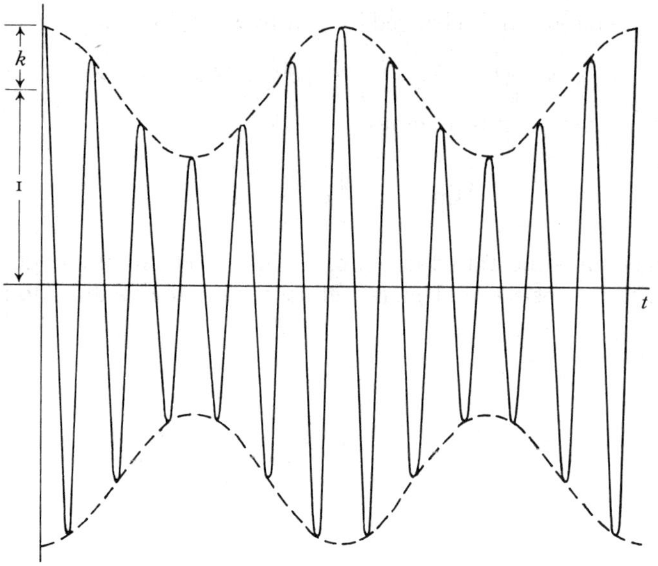

Fig. 9.11 Amplitude-modulated wave $(1 + k\cos\omega_m t)\cos\omega_0 t$.

or *modulation* being much less than the frequency of the *carrier wave*. A wave of frequency ω_0 modulated at frequency ω_m may be written

$$f(t) = (1 + k\cos\omega_m t)\cos\omega_0 t, \tag{9.7.6}$$

assuming the two waves reach a maximum together (a trivial restriction). k cannot exceed 1, and ω_m is much less than ω_0. By an elementary transformation,

$$f(t) = \cos\omega_0 t + \frac{k}{2}\{\cos(\omega_0 + \omega_m)t + \cos(\omega_0 - \omega_m)t\}. \tag{9.7.7}$$

If ω_0 and ω_m are both integral multiples of some fundamental frequency $\omega_f(\omega_0 = r\omega_f, \omega_m = s\omega_f)$, the pattern repeats itself after r cycles of the carrier and s cycles of the modulating wave. Equation (9.7.7) then becomes

$$f(t) = \cos r\omega_f t + \frac{k}{2}\{\cos(r+s)\omega_f t + \cos(r-s)\omega_f t\}, \tag{9.7.8}$$

and the three terms constitute terms $(r-s)$, r and $(r+s)$ in a Fourier series. The spectrum (fig. 9.12) comprises three lines on each side of the axis.

The restriction $\omega_0 = r\omega_f$, $\omega_m = s\omega_f$ is necessary at this stage because the chapter is solely concerned with periodic functions; but, as will appear in chapter 11, it is not fundamentally necessary. Fig. 9.12 gives correctly the spectrum for any amplitude-modulated wave of the form (9.7.6).

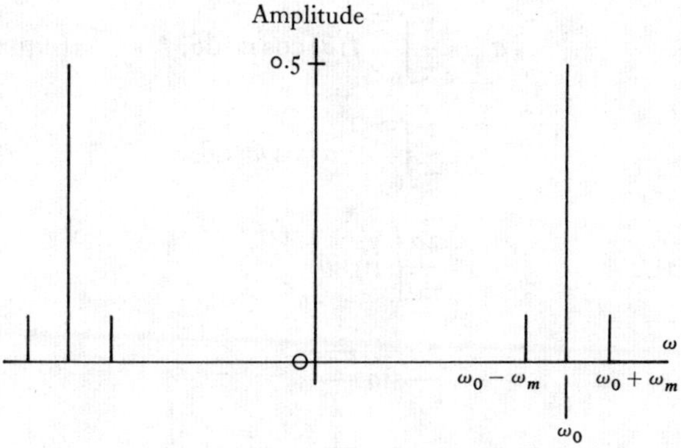

Fig. 9.12 Spectrum of the wave of fig. 9.11.

2. *Sequence of square pulses (fig. 9.13)*

The pulses to be considered will have amplitude a, duration τ and frequency ω_0, so that the time from the start of one to the start of the next is $T = 2\pi/\omega_0$. They stand on the positive side of the axis.

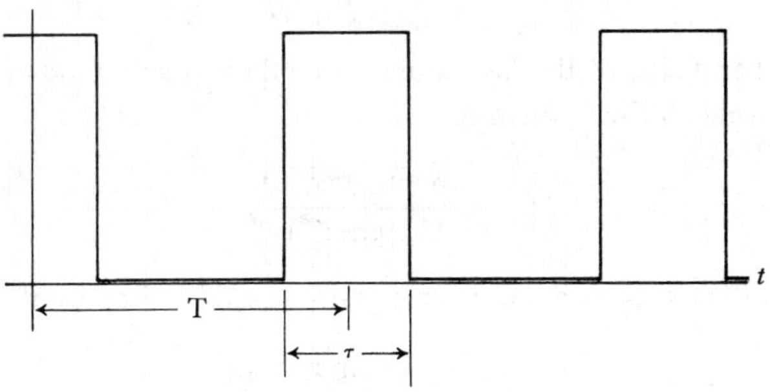

Fig. 9.13 Square pulse sequence.

With the time-origin at the centre of a pulse this is an even function. Writing θ for $\omega_0 t$,

$$f(\theta) = a_0 + \sum_{n=1}^{\infty} a_n \cos n\theta. \qquad \text{(equation (9.3.2))}$$

Here
$$a_0 = \frac{1}{2\pi}\int_0^{2\pi} f(\theta)\,d\theta, \qquad \text{(equation (9.2.2))}$$

$$= \frac{1}{\pi}\int_0^{\pi\tau/T} a\,d\theta,$$

$$= \frac{a\tau}{T}. \qquad (9.7.9)$$

Also
$$a_n = \frac{1}{\pi}\int_0^{2\pi} f(\theta)\cos n\theta\,d\theta, \qquad \text{(equation (9.2.3))}$$
$$= \frac{2}{\pi}\int_0^{\pi\tau/T} a\cos n\theta\,d\theta,$$
$$= \frac{2a}{\pi n}\Big[\sin n\theta\Big]_0^{\pi\tau/T},$$
$$= \frac{2a}{\pi n}\sin\frac{n\pi\tau}{T}. \tag{9.7.10}$$

The coefficients in the complex Fourier series are then

$$\left.\begin{aligned} C_0 &= \frac{a\tau}{T}, \\ C_n &= \frac{a}{\pi n}\sin\frac{n\pi\tau}{T}, \end{aligned}\right\} \tag{9.7.11}$$

from which
$$C_n/C_0 = \sin\frac{n\pi\tau}{T}\Big/\frac{n\pi\tau}{T}. \tag{9.7.12}$$

The right hand side of the last equation tends to 1 as n tends to 0; so one formula covers all the C_n, namely

$$C_n = \frac{a\tau}{T}\frac{\sin(n\pi\tau/T)}{(n\pi\tau/T)}. \tag{9.7.13}$$

The lines of the spectrum are thus equally spaced ordinates of the curve

$$y = \frac{a\tau}{T}\frac{\sin x}{x}. \tag{9.7.14}$$

In fig. 9.14 this curve is plotted, with $a\tau/T$ put $= 1$.

Suppose that the first zero, $x = \pi$, lies between the lines which represent the nth and $(n+1)$th harmonics. The condition for this is

$$\frac{n\pi\tau}{T} < \pi < \frac{(n+1)\pi\tau}{T}, \tag{9.7.15}$$

so that (T/τ) lies between the integers n and $(n+1)$. The greater the ratio T/τ, therefore, the greater the number of harmonics in any given range of the curve. Equation (9.7.15) may be rewritten

$$\frac{2n\pi}{T} < \frac{2\pi}{\tau} < \frac{2(n+1)\pi}{T},$$

or
$$n\omega_0 < \frac{2\pi}{\tau} < (n+1)\,\omega_0. \qquad (9.7.16)$$

This defines a frequency scale, which has been added to fig. 9.14. The scale varies with the pulse length τ.

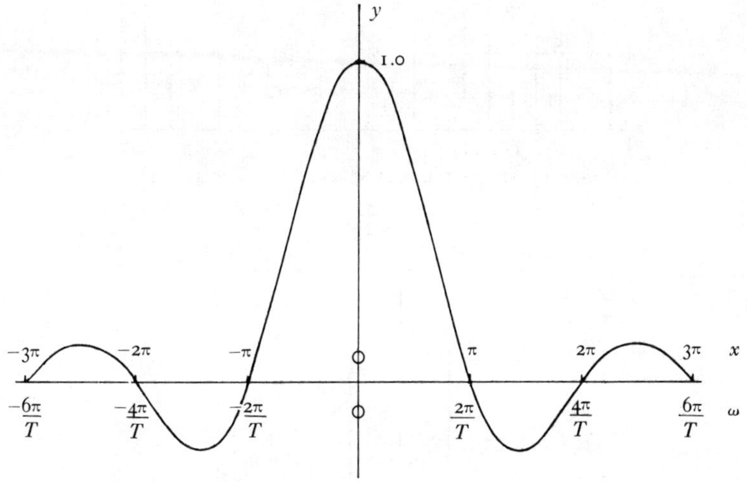

Fig. 9.14 Curve $y = \sin x/x$ with frequency scale added.

In comparing the spectra of several trains of pulses, however, it is clearer to use the same frequency scale throughout; this is accomplished by treating the period T as a constant factor, so that corresponding harmonics have the same frequency for all the pulse trains compared. It is also convenient to ascribe to each pulse an amplitude a such that $a\tau/T = 1$; this amounts to the assumption that all the pulses compared have the same area (fig. 9.15). The amplitude of the harmonic of frequency $n\omega_0$, or $2n\pi/T$, is then given by

$$y_n = \frac{\sin(n\pi\tau/T)}{(n\pi\tau/T)}. \qquad (9.7.17)$$

(It should be noted that, in terms of harmonic amplitudes as measured on a wave analyser, y_n represents the actual amplitude of the zero-frequency or d.c. component, but the half-amplitude of the a.c. components. This is because the analyser uses only positive values of n, whereas in (9.7.17) both positive and negative values are admitted.)

The pulse trains illustrated in fig. 9.15 have $\tau/T = \frac{1}{2}$, $\frac{1}{4}$ and $\frac{1}{8}$ respectively. For these cases the first few harmonics are listed in table 9.1. The three spectra are shown in fig. 9.16. The fact that their envelope is the same curve,

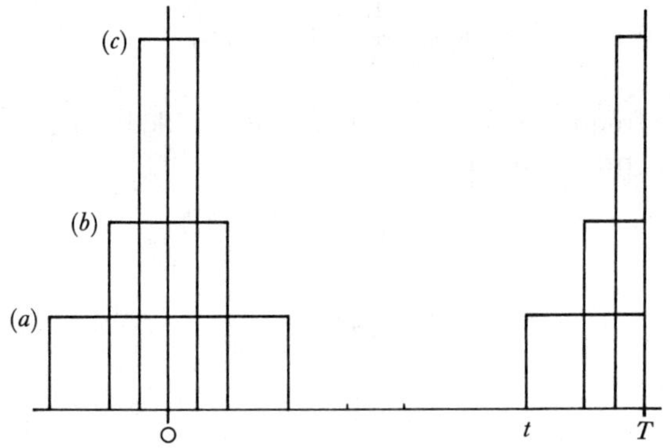

Fig. 9.15 Pulse trains of equal area ($\tau/T = \frac{1}{2}, \frac{1}{4}, \frac{1}{8}$).

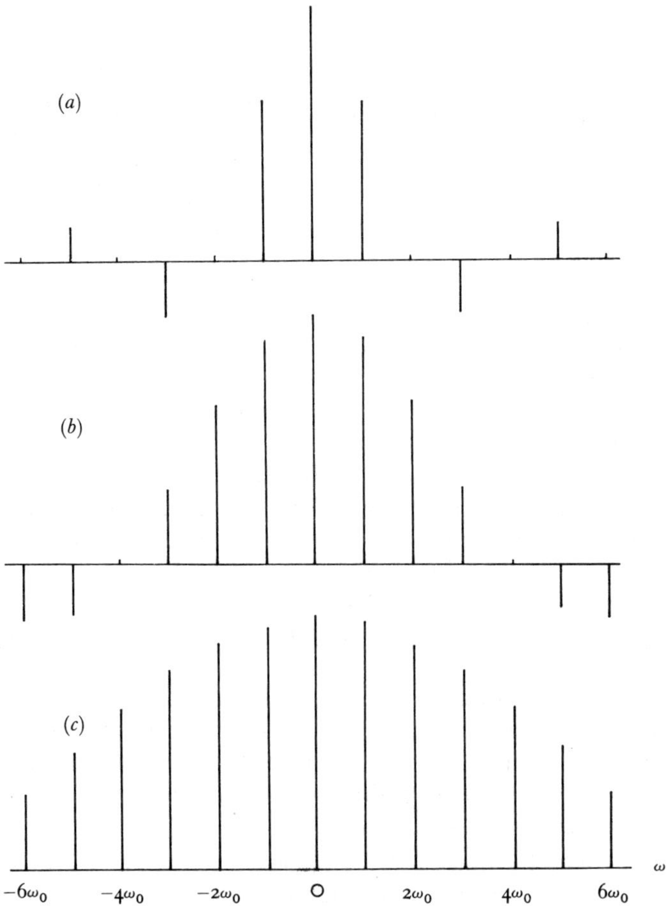

Fig. 9.16 Spectra of the pulse trains of fig. 9.15.

spreading outwards, is evident. The case $\tau/T = \frac{1}{2}$ has already been evaluated, as an odd function and without a d.c. component, in (9.3.10); the relation with the present analysis is easily seen.

TABLE 9.1

n	0	1	2	3	4	5	6
$y_n(\tau/T = \frac{1}{2})$	1	0.637	0	−0.212	0	0.127	0
$y_n(\tau/T = \frac{1}{4})$	1	0.901	0.637	0.300	0	−0.180	−0.212
$y_n(\tau/T = \frac{1}{8})$	1	0.975	0.901	0.787	0.637	0.470	0.300

A limiting case of great significance is that of very short pulses ($\tau \to 0$), the amplitude tending to infinity in such a way that $a\tau/T = 1$; such pulses will be called 'unit spikes' and are discussed in §10.3. Equation (9.7.17) shows that y_n then $= 1$ for all values of n, so that all the harmonics have the same amplitude. The tendency in this direction, as τ/T is reduced, is evident among the spectra of fig. 9.16.

The description of functions of time in terms of their spectra is yet another example of the 'image function' approach already mentioned in connexion with phasors and Laplace transforms, and will be developed as such in chapter 11.

Examples on chapter 9

9.1 An a.c. waveform is defined by

$$f(\theta) = 1 \quad (\alpha < \theta < \pi - \alpha),$$

$$f(\theta) = 0 \quad (0 \leqslant \theta < \alpha,\ \pi - \alpha < \theta \leqslant \pi),$$

where α is less than $\pi/2$. The next half-cycle is similar, but opposite in sign. Derive the Fourier series for this wave, and find the value of α which causes the third harmonic to vanish.

9.2 A periodic and unidirectional voltage consists of rectangular pulses, each of 10 μs duration, occurring at intervals of 10^{-3} s. The amplitude is 1 V. Obtain an expression giving the amplitude of each of the harmonic components which are present, and indicate the range of frequencies which should be freely transmitted by a circuit to which this voltage is applied in order that a good approximation to the rectangular shape may be retained.

9.3 Derive the simplest form of Fourier series for a half-wave rectified sine wave – that is, a wave in which the value of the function is zero over the intervals where the negative half-waves would be.

A half-wave rectifier is used in the circuit of fig. 9.6(*a*) instead of the full-wave rectifier shown. The resistor R, of 1000 Ω, draws a d.c. current of 100 mA; the capacitor C is 31.8 μF, and the supply frequency to the rectifier is 50 Hz. Assuming that there are no losses except in R, and that the rectifier functions as a voltage source with an output voltage of the form described, estimate

(i) the peak voltage supplied by the rectifier;

(ii) the value of L required to ensure that the peak-to-trough amplitude of the ripple shall be less than 10 per cent of the d.c. output.

Point out any assumptions which it may be found convenient to make.

9.4. (i) Derive the Fourier series for the a.c. waveform $f_1(\theta)$ defined as follows:

Over the range $0 < \theta < \pi/6$, $f_1(\theta)$ increases linearly from 0 to 0.5.

Over the range $\pi/6 < \theta < \pi/2$, $f_1(\theta)$ increases linearly from 0.5 to 1.0.

The half-wave is symmetrical about the ordinate $\theta = \pi/2$, and the positive and negative half-cycles are similar.

(ii) Derive the Fourier series for a wave $f_2(\theta)$ possessing similar symmetries to $f_1(\theta)$, but defined thus:

Over the range $0 < \theta < \pi/3$, $f_2(\theta)$ increases linearly from 0 to h.

Over the range $\pi/3 < \theta < \pi/2$, $f_2(\theta) = h$.

Show that the spectral components of $f_2(\theta)$ have the same relative magnitudes as those of $f_1(\theta)$, but differ in phase. What value of h will make the magnitudes equal?

9.5 Find the amplitude spectrum of a sawtooth wave which rises linearly from −0.5 to +0.5 over the range $-\pi < \theta < \pi$, then drops instantaneously to −0.5 and repeats the cycle. Deduce the spectrum of a wave which rises from 0 to 1 over the range $0 < \theta < 2\pi$, then drops to zero and repeats.

9.6 A sequence of pulses occurs at regular intervals T. Each pulse is of duration τ, and its shape is an isosceles triangle of unit vertical height. Find the amplitude spectrum of this periodic function.

9.7 Derive the Fourier series for the function $f(x)$ defined over the range from 0 to d by

$$f(x) = \frac{x}{d} - \frac{\sinh \alpha x}{\sinh \alpha d},$$

treating this as a half-cycle of a periodic function in which $f(2d - x) = -f(x)$.

9.8 A train of pulses, each of angular width 2δ (where δ is small), occurs at uniform intervals $\pi/6$ of the angle-variable θ, with one pulse at $-\delta < \theta < \delta$, the next at $(\pi/6) - \delta < \theta < (\pi/6) + \delta$, and so on. The amplitude of the pulses is modulated to the envelope $1 + k \cos \theta$; the pulses are not square, but are shaped to fit this envelope. Obtain the amplitude spectrum of this pulse train.

10 Techniques of transient analysis

10.1 Scope of the present chapter

When the Laplace transformation was introduced in chapter 5 of this book, the processes involved were essentially elementary applications of the integral calculus with a real variable t. It is true that it was stated (in §5.4) that the transform variable s was a complex number $\sigma + j\omega$, and it was pointed out that certain integrals would not converge unless σ had a suitable value; but the actual integration process in the Laplace equation

$$\mathscr{L}w(t) = \int_0^\infty w(t)\,e^{-st}\,dt, \qquad \text{(equation (5.4.9))}$$

is elementary, and so is the extension of the list of transforms by expanding functions of s in powers of 1/s and using the relation

$$\mathscr{L}^{-1}(1/s^n) = t^{n-1}/(n-1)!.$$

The full development of the method requires a knowledge of the theory of functions of a complex variable; but much development is possible within the more elementary framework, and the present chapter aims at providing this. It is broadly true to say that the techniques now to be described will suffice for analysing all networks consisting of lumped elements, failing only when distributed circuits begin to be considered in chapter 13. The first point to be attacked is the removal of the limitation imposed in §5.10, when the use of the image circuit was confined to problems in which the original circuit was initially dead.

10.2 The image of a circuit not initially dead

When the current in an inductor is changing from an initial value i_0, the image of the equation $v = L\,di/dt$ becomes

$$\boldsymbol{V} = L(s\boldsymbol{I} - i_0),$$

or

$$\boldsymbol{I} = \frac{\boldsymbol{V}}{Ls} + \frac{i_0}{s}. \qquad (10.2.1)$$

This may be interpreted as signifying that the current $\boldsymbol{I}$ contains two components: $\boldsymbol{V}/Ls$, which is the current associated with the application of voltage $\boldsymbol{V}$ to an impedance Ls; and i_0/s, which represents the output of a step function current source. A circuit in which the voltage $\boldsymbol{V}$ and total current $\boldsymbol{I}$ are related in this way is depicted in fig. 10.1(*a*); it forms a complete image for an inductor in which the initial current is not zero.

The equation for a capacitor, corresponding to (10.2.1), is

$$\boldsymbol{V} = \frac{\boldsymbol{I}}{C\mathrm{s}} + \frac{v_0}{\mathrm{s}}. \tag{10.2.2}$$

This is fully represented by a series combination of impedance $1/C\mathrm{s}$ with a step function voltage source v_0/s, as shown in fig. 10.1(*b*).

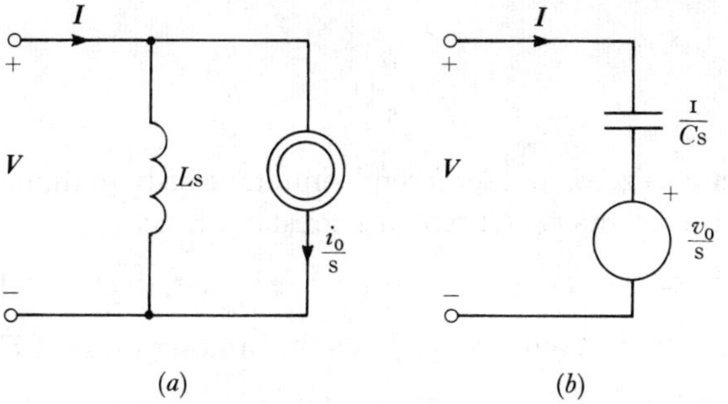

Fig. 10.1 (*a*) Step function current source representing initial current in *L*. (*b*) Step function voltage source representing initial voltage on *C*.

The equivalent circuits of fig. 10.1 are not the only possible ones. In §1.13 it was proved that a voltage source V_0 in series with a resistance R_0 was equivalent to a current source I_0 in parallel with a resistance R_0, provided that $V_0 = R_0 I_0$. Exactly the same argument may be used for the image world under either a.c. or transient conditions, to prove that a voltage source $\boldsymbol{V}_0$ in series with an impedance $\boldsymbol{Z}_0$ is equivalent to a current source $\boldsymbol{I}_0 = \boldsymbol{V}_0/\boldsymbol{Z}_0$ in parallel with an impedance $\boldsymbol{Z}_0$. When the circuits of fig. 10.1 are replaced in this way by their equivalents, the results are as shown in fig. 10.2. The inductance is now represented by an impedance $L\mathrm{s}$ in series with a voltage source Li_0, and the capacitance by an impedance $1/C\mathrm{s}$ in parallel with a current source Cv_0. These sources are of a new type, for no function of t has yet been considered which has a constant for its Laplace transform. Before the proposed circuits can be used, it is necessary for the required function to be introduced and discussed.

10.3 The spike function

A second form of basic stimulus, to be set alongside the step function as a means by which a circuit may be excited, will now be introduced. This is an idealized representation of a stimulus of very large magnitude and very short duration; in mechanical terms, a hammer-blow or impulse.

Consider the function $f(t)$ shown in fig. 10.2, which is zero when $t < 0$ and when $t > t_1$, but has the constant value h when t lies between 0 and t_1. The image of $f(t)$ is given, as usual, by

$$\boldsymbol{F}(s) = \int_0^\infty f(t)\, e^{-st}\, dt;$$

but, since $f(t)$ is only finite between 0 and t_1,

$$\boldsymbol{F}(s) = \int_0^{t_1} h\, e^{-st}\, dt,$$

$$= \frac{h}{s}(1 - e^{-st_1}). \qquad (10.3.1)$$

Now let t_1 tend to zero, and let h tend simultaneously to infinity, in such a manner that the product (ht_1) remains constant. Since

$$1 - e^{-st_1} = 1 - (1 - st_1 + (st_1)^2/2! - (st_1)^3/3! + \ldots),$$

the bracket $(1 - e^{-st_1})$ tends to st_1. Thus the limiting value of $\boldsymbol{F}(s)$ is

$$\boldsymbol{F}(s) = ht_1, \qquad (10.3.2)$$

which is a constant, and equal to the area under the curve $f(t)$.

A function which is of infinite magnitude and infinitesimal duration, but in which the product of these quantities is finite, is conveniently called a

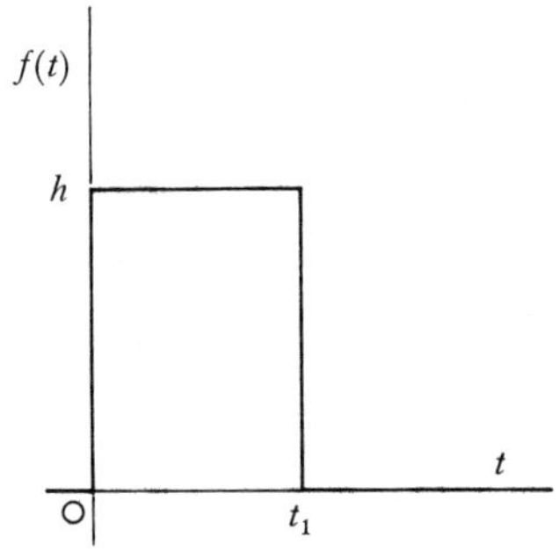

Fig. 10.2 Square pulse function.

spike function; the term *impulse function* is also used. The foregoing analysis has shown that, when the spike occurs at $t = 0$, its image is a constant, equal to its area. The standard or unit spike function, occurring at $t = 0$ and having unit area, is designated $\delta(t)$ and is also known as *Dirac's delta-function*. Its image in the s-system is equal to 1. If we regard $\delta(t)$ as the limit of a very brief square pulse occurring on the positive side of $t = 0$, as in fig. 10.2, we may formally define $\delta(t)$ by the equations

$$\left.\begin{aligned} &\delta(t) = 0 \quad (t \neq 0); \qquad \delta(t) = \infty \quad (t = 0); \\ \text{and} \qquad &\int_0^\infty \delta(t)\, \mathrm{d}t = 1. \end{aligned}\right\} \tag{10.3.3}$$

The integral just written would also equal unity if any finite positive number were substituted for infinity as the upper limit. Thus $\int_0^t \delta(\tau)\, \mathrm{d}\tau = 1$ for all positive values of t; in other words,

$$\int_0^t \delta(\tau)\, \mathrm{d}\tau = u(t). \tag{10.3.4}$$

The unit step is the integral of the unit spike. This is consistent with the proved facts that $\mathscr{L}\delta(t) = 1$, $\mathscr{L}u(t) = 1/s$; for integration in the t-world corresponds to multiplication by $1/s$ in the s-world.

10.4 Spike function sources and their application in image circuits

Generators of voltage or current pulses, singly or recurrently, are known practical devices with many applications. If the duration of the pulse is very short, it may be legitimate to idealize the pulse generator into a spike function source, in which the pulse is infinite in magnitude and infinitesimal in duration.

What is the precise meaning of a 'very short' duration in this context? There is only one possible answer: the duration of the pulse must be very short compared with any times inherent in the system to which the pulse is applied. Thus if a voltage pulse were applied to a $1\ \mu$F capacitor in series with a $1000\ \Omega$ resistor, its duration would be compared with the time $T = RC = 10^{-3}$ second, in which the capacitor, if charged from a step function voltage source, would acquire a charge equal to $(1 - e^{-1})$, or 63.3 per cent, of the ultimate full charge. This is the so-called *time-constant* of the RC circuit, and for many purposes a pulse having one-hundredth of this duration, 10 microseconds in the example cited, would be regarded as 'very short' and equivalent to a spike. With inductance L in series with resistance R, the time-constant is L/R. With an oscillatory circuit, the period of the oscillation is a natural yardstick for comparison. In general, a complicated circuit may have a number of inherent periods and time-constants; a 'very short' pulse must be very short compared with the shortest of them.

An ideal spike function current source will generate an infinite current for an infinitesimal time. During the pulse a finite charge is conveyed; if this be called Q_1, the current output of the source is to be written $Q_1\delta(t)$. A capacitor holding a charge Q_1, and suddenly discharged into a circuit of low impedance, will produce a current pulse which approaches this ideal.

In a similar manner, a spike function voltage source may be thought of as the idealization of an inductor initially carrying a current, in which the opening of a switch brings about a sudden large increase in the impedance of its circuit, so that the magnetic flux in the inductor suddenly collapses and in so doing generates a very high voltage across its terminals. The voltage output is then $\Lambda_1\,\delta(t)$, where Λ_1 is the flux linkage (that is, the mean flux per turn multiplied by the number of turns in the inductor). It is particularly to be noted that the factor multiplying $\delta(t)$ in the description of a current or voltage source is not a current or voltage, but is the integral of these quantities respectively.

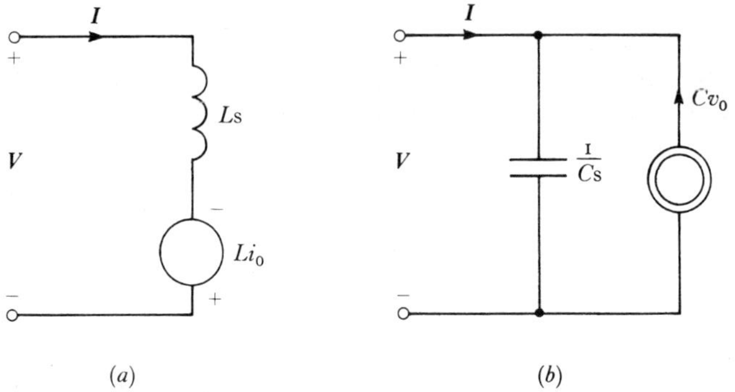

Fig. 10.3 (*a*) Spike function voltage source representing initial current in *L*. (*b*) Spike function current source representing initial voltage on *C*.

Besides being at times a useful idealization of an actual pulse generator, the spike function source makes possible, as has already been foreshadowed, a valuable development in the image-circuit method. Consider the equation

$$\boldsymbol{V} = L(\mathrm{s}\boldsymbol{I} - i_0) \tag{10.4.1}$$

which describes the performance of an inductor L in which the current at $t = 0$ is i_0. Instead of rewriting this (as in §10.2) as a relation between currents, let it be considered in its present form as a voltage equation. It signifies that the voltage $\boldsymbol{V}$ is found by subtracting from the impedance voltage $L\mathrm{s}\boldsymbol{I}$ the output of a voltage source Li_0; in other words, the inductor is represented by the image circuit shown in fig. 10.3(*a*). As the image of the source voltage is a constant, the form of the voltage is that of a spike function, and the description of it in the t-world is $Li_0\,\delta(t)$. In a similar manner, the performance of a capacitor C which is charged initially to a voltage v_0, and in which therefore

$$\boldsymbol{I} = C(\mathrm{s}\boldsymbol{V} - v_0), \tag{10.4.2}$$

is described by the image circuit of fig. 10.3(*b*).

This way of representing the initial conditions is preferable to that described in §10.2, because it has a simple physical interpretation. If fig. 10.3(*a*) is regarded as a branch of a larger network, it may be considered that, up to the instant $t = 0$, the network is completely dead. At $t = 0$, an immense pulse of voltage from the spike function voltage source instantaneously establishes a current i_0 in the inductor and in other parts of the network; then the voltage source ceases to generate, and, having zero impedance, might just as well not be there. In a similar manner, the spike function current source of fig. 10.3(*b*) charges the capacitor to voltage v_0 in a single burst and then relapses into the passive condition of an open circuit. In both diagrams it will be seen that the polarity of the source is such as will establish the initial current and voltage respectively in a positive sense.

As an example of a problem involving non-zero initial values, consider the circuit shown in fig. 10.4(*a*); this has already been partly discussed in

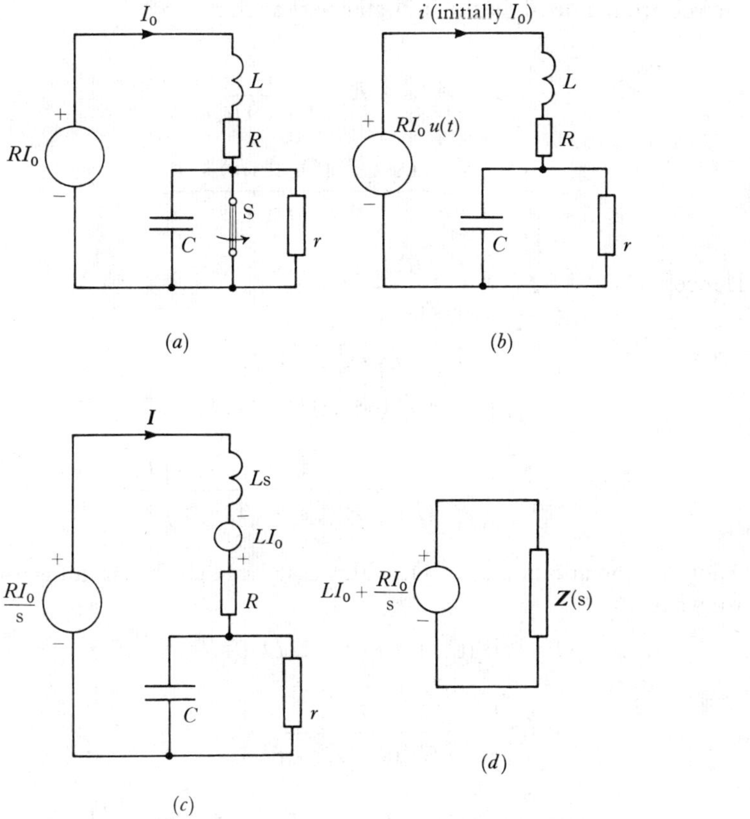

Fig. 10.4 (*a*) Circuit stimulated by opening a switch S. (*b*) Condition of circuit after opening of S. (*c*) Image circuit for (*b*). (*d*) Thevenin equivalent circuit for (*c*).

§5.10, where the solution was obtained by a superposition method. A voltage generator of e.m.f. RI_0 is passing a steady current I_0 through L, R, and the closed switch S. At $t = 0$ the switch is opened; it is required to find the subsequent current through L and R.

The state of affairs after the opening of the switch is shown in fig. 10.4(*b*), in which, as will be seen, the generator has been replaced by a step function source. We know that in fact its e.m.f. has been constant for a long time before $t = 0$; but the effect of this is summed up in the statement that $i = I_0$ when $t = 0$, and we make (as it were) a fresh start, with $i = I_0$ summarizing what has gone before, and the e.m.f. newly applied as a step. In fig. 10.4(*c*) the image circuit is drawn, with a spike function voltage source in series with the inductance; no spike function current source in parallel with the capacitance is required, since the voltage on this is zero at $t = 0$.

The image problem is best solved by using Thevenin's theorem, which is applicable in the image as well as in the real world. The equivalent circuit is shown in fig. 10.4(*d*); the source voltages are added, while the impedance viewed from a break in the inductance branch is

$$\boldsymbol{Z}(\mathrm{s}) = L\mathrm{s} + R + \frac{1}{C\mathrm{s} + 1/r},$$

$$= \frac{(L\mathrm{s} + R)(C\mathrm{s} + 1/r) + 1}{C\mathrm{s} + 1/r}. \qquad (10.4.3)$$

Hence $$\boldsymbol{I} = \frac{(L\mathrm{s} + R)I_0}{\mathrm{s}\boldsymbol{Z}(\mathrm{s})},$$

$$= \frac{(L\mathrm{s} + R)(C\mathrm{s} + 1/r)I_0}{[(L\mathrm{s} + R)(C\mathrm{s} + 1/r) + 1]\,\mathrm{s}},$$

$$= \left\{1 - \frac{1}{(L\mathrm{s} + R)(C\mathrm{s} + 1/r) + 1}\right\}\frac{I_0}{\mathrm{s}}. \qquad (10.4.4)$$

Adopting the notation of §5.11, which assumes that the circuit is oscillatory, we write

$$(L\mathrm{s} + R)(C\mathrm{s} + 1/r) + 1 \equiv LC[(\mathrm{s} + \alpha)^2 + \beta^2]. \qquad (10.4.5)$$

Then $$\boldsymbol{I} = I_0\left\{\frac{1}{\mathrm{s}} - \frac{1}{LC\mathrm{s}[(\mathrm{s} + \alpha)^2 + \beta^2]}\right\},$$

$$= I_0\left\{\frac{1}{\mathrm{s}} - \frac{1}{LC(\alpha^2 + \beta^2)}\left[\frac{1}{\mathrm{s}} - \frac{\mathrm{s} + 2\alpha}{(\mathrm{s} + \alpha)^2 + \beta^2}\right]\right\},$$

(the last line being derived by the application of the usual partial fractions method); but a scrutiny of (10.4.5) shows that $LC(\alpha^2+\beta^2) = (R/r)+1$, so

$$\boldsymbol{I} = I_0\left\{\frac{1}{s} - \frac{r}{R+r}\left[\frac{1}{s} - \frac{s+2\alpha}{(s+\alpha)^2+\beta^2}\right]\right\},$$

or

$$\boldsymbol{I} = \frac{I_0}{R+r}\left\{\frac{R}{s} + \frac{r(s+2\alpha)}{(s+\alpha)^2+\beta^2}\right\}. \tag{10.4.6}$$

The corresponding function of t is

$$i = \frac{I_0}{R+r}\left\{R + r\,e^{-\alpha t}\left(\cos\beta t + \frac{\alpha}{\beta}\sin\beta t\right)\right\}. \tag{10.4.7}$$

For the numerical values given in §5.11, the change in the current from the initial value I_0 to the final value $I_0 R/(R+r)$ is illustrated in fig. 10.5.

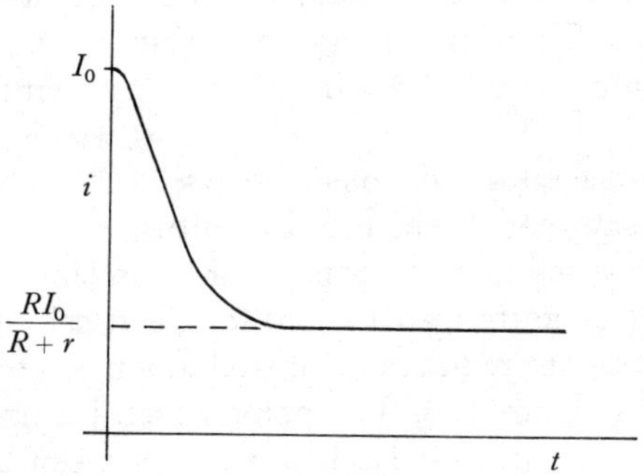

Fig. 10.5 Current drawn from voltage source in fig. 10.4(*a*).

It is of interest to compare this method with that used in §5.11. The variables there stand, not for the absolute values, but for the *changes* in the voltage and current. Equations (5.11.7), solved for $\boldsymbol{I}$, would have led to the formula

$$\boldsymbol{I} = \frac{I_0}{s[(Ls+R)(Cs+1/r)+1]} \tag{10.4.8}$$

which has to be subtracted from I_0/s, the image of the initial current I_0, in order to obtain the complete expression (10.4.4) for the image of the transient current. It would have been necessary, in using the former method, to consider carefully whether the expression (10.4.8) was to be subtracted from I_0/s, or added to it. This question is taken care of automatically in the method just described. If the simple rules for the formation of the image circuit are applied, the solution of a transient problem is scarcely any more difficult than that of a steady-state d.c. or a.c. problem.

10.5 Dimensional relationships in the Laplace transformation

It is well known that the quantities on the two sides of an equation must be of the same fundamental kind; for example, in a resistor the equation $v = Ri$ is valid because R is defined as the ratio of a voltage to a current, so that Ri is a quantity fundamentally similar to v. This underlying similarity is explored in the theory of dimensions, in which physical quantities are classified according to the change which takes place in the numerical description of them when the units of certain fundamental quantities are changed. For example, if the unit of length is changed from a centimetre to a metre, the number describing a given length (such as 537 cm) is divided by 100 (to 5.37 m). The same is true of the measure of a given velocity; therefore velocity is said to have a dimension of 1 in length. On the other hand, if the unit of time is changed from a second to a minute, a time of 90 s becomes 1.5 min, but a velocity of 3 cm/s becomes 180 cm/min; the number describing a time is divided, whereas that describing a velocity is multiplied, by 60. The dimension of a velocity is said to be -1 in time, because velocity and $(\text{time})^{-1}$ vary in proportion when the unit of time is changed.

This is not the place for a detailed discussion of units and dimensions. It is sufficient to say that mechanical quantities, such as force or energy, have dimensions expressible in terms of the fundamental dimensions of mass (M), length (L), and time (T); but for electrical quantities, the most generally valid treatment requires the introduction of a fourth fundamental dimension, such as charge (Q). The point which is important here is the investigation of the dimensional relationship between a quantity and its image.

In the equation which defines the Laplace transformation,

$$\boldsymbol{W}(\mathrm{s}) = \int_0^\infty w(t)\,\mathrm{e}^{-\mathrm{s}t}\,\mathrm{d}t, \qquad \text{(equation (5.4.5))}$$

the dimensions of the two sides must be the same. On the right hand side, the integration is essentially a summation, each term summed being of the form $w(t)\mathrm{e}^{-\mathrm{s}t}\delta t$. In this product, $w(t)$ stands for a time-varying quantity (voltage, current, etc.) which has its own appropriate dimensions; δt is a time; while the only possible way to ascribe dimensions to $\mathrm{e}^{-\mathrm{s}t}$ is to assign the dimension T^{-1} to s, so that st has zero dimensions, and therefore

$$\mathrm{e}^{-\mathrm{s}t} \equiv 1 - \mathrm{s}t + (\mathrm{s}t)^2/2! - (\mathrm{s}t)^3/3! + \dots$$

has zero dimensions. If st had any other dimensions than zero, the ascription of dimensions to $\mathrm{e}^{-\mathrm{s}t}$ would be meaningless. It follows that the dimensions of $\boldsymbol{W}(\mathrm{s})$ are those of $w(t)$ multiplied by time. The symbol $\mathscr{L}$ is dimensionally equivalent to multiplying by T; the reader may easily verify that this leads to a self-consistent system – for instance, in (5.5.3), namely

$$\mathscr{L}\left(\frac{\mathrm{d}w}{\mathrm{d}t}\right) = \mathrm{s}\boldsymbol{W} - w_0,$$

each term has the dimensions of w; and in the standard transform-pair

$$\mathscr{L}\,\mathrm{e}^{-\alpha t} = \frac{1}{\mathrm{s}+\alpha},$$

each term has dimensions T. Attention to dimensional relationships forms a valuable check upon the working.

10.6 The s-multiplied transform and the operational calculus

The statements made in the last section are equivalent to saying that $\mathrm{s}\boldsymbol{W}(\mathrm{s})$ has the same dimensions as $w(t)$. If, therefore, the image of w had been defined as $\boldsymbol{W}'$, where

$$\boldsymbol{W}'(\mathrm{s}) = \mathrm{s}\boldsymbol{W}(\mathrm{s}) = \mathrm{s}\int_0^\infty w(t)\,\mathrm{e}^{-\mathrm{s}t}\,\mathrm{d}t, \tag{10.6.1}$$

it would have been possible to state that the dimensions of the image $\boldsymbol{W}'(\mathrm{s})$ were the same as those of the original $w(t)$. $\boldsymbol{W}'(\mathrm{s})$ is the so-called s-multiplied form of the Laplace transform, which has been used by a number of writers. It has gone out of vogue, no doubt largely because the intrusion of the extra s on the right hand side of (10.6.1) appears artificial; but the simple dimensional relationship between $w(t)$ and $\boldsymbol{W}'(\mathrm{s})$ is an advantage of the s-multiplied transform, and its existence also has a historical reason which deserves mention.

Consider the very simple problem through which the Laplace transform was first introduced in this book – that of charging a capacitor C through a resistor R from a step function voltage source $V_1 u(t)$. The differential equation for the voltage v on C,

$$RC\frac{\mathrm{d}v}{\mathrm{d}t} + v = V_1\,u(t),$$

or

$$\frac{\mathrm{d}v}{\mathrm{d}t} + \alpha v = \alpha V_1\,u(t), \tag{10.6.2}$$

where $\alpha = (RC)^{-1}$, leads to the image solution

$$\boldsymbol{V} = \frac{\alpha V_1}{\mathrm{s}(\mathrm{s}+\alpha)}, \tag{10.6.3}$$

which is interpreted by the usual rules. If the s-multiplied transform had been used, the solution would have been

$$\boldsymbol{V}' = \frac{\alpha V_1}{\mathrm{s} + \alpha}, \tag{10.6.4}$$

and the rules of interpretation would have been different. Suppose now that, in the differential equation, (10.6.2), the symbol p were written instead of $\mathrm{d}/\mathrm{d}t$, to denote the operation of differentiating:

$$(\mathrm{p} + \alpha)\, v = \alpha V_1\, u(t),$$

and suppose further that it were established that, under certain conditions, the operator p could be treated like a number; then the solution would be written

$$v = \frac{\alpha V_1\, u(t)}{\mathrm{p} + \alpha}, \tag{10.6.5}$$

and again we should require rules of interpretation, so as to attach a meaning to the operator $1/(\mathrm{p} + \alpha)$ when operating upon a step function $\alpha V_1 u(t)$. This process is a simple example of the *operational calculus* employed by Heaviside for the solution of many circuit problems. A self-taught and eccentric genius, Heaviside relied greatly on intuition in the mathematically risky process of treating p like a number; the professional mathematicians criticized him severely, and he repaid them with sarcasms of a kind rarely found in scientific writing, such as 'Whether good mathematicians, when they die, go to Cambridge, I do not know.' Later Cambridge returned good for evil by playing a part, through the work of T. J. I'A. Bromwich, in justifying Heaviside's calculus in terms of orthodox mathematics, essentially by linking it with Laplace transform theory. It will be seen that (10.6.4) and (10.6.5) are practically the same; indeed, some writers use the symbol p where we have used s. Formally the two approaches are therefore closely similar, but philosophically they are very different; for in Heaviside's method the operation is conceived as being entirely carried out in the t-world – the concept of an image world disappears, and the dependent variable on the left hand side of (10.6.5) is therefore v, not $\boldsymbol{V}'$. On account of this wide difference between the two methods of approach, it appears best to reserve the symbol p for use in connexion with the operational method.

10.7 Further theorems on the Laplace transformation

(a) Differentiation with respect to a parameter

A *parameter* is a number which distinguishes one particular member in a family of curves or functions; thus, $\sin \beta t$ stands for a family of sine waves, each member of which is distinguished from all others by a particular value of the angular frequency β. In any given problem the parameter is a constant; yet differentiation with respect to it forms a valuable method of deriving new transforms from known ones.

Consider the equation

$$\mathscr{L}\,e^{-\alpha t} = \frac{1}{s+\alpha}. \qquad \text{(see (5.6.4))}$$

The Laplace transformation converts a function of t into a function of s, but the parameter α is common to both. We may view the equation as a relation between a function of t and α and a function of s and α, of the form

$$\mathscr{L}w(t,\alpha) = \boldsymbol{W}(s,\alpha).$$

This holds good no matter what the value of α may be; thus, slightly altering α to $\alpha+\delta\alpha$,

$$\mathscr{L}w(t,\alpha+\delta\alpha) = \boldsymbol{W}(s,\alpha+\delta\alpha).$$

Subtracting and dividing by $\delta\alpha$, we obtain

$$\mathscr{L}\frac{w(t,\alpha+\delta\alpha)-w(t,\alpha)}{\delta\alpha} = \frac{\boldsymbol{W}(s,\alpha+\delta\alpha)-\boldsymbol{W}(s,\alpha)}{\delta\alpha}.$$

In the limit when $\delta\alpha$ tends to 0, the two fractions become partial derivatives $\partial w/\partial\alpha$ and $\partial \boldsymbol{W}/\partial\alpha$. It has thus been proved that:

If the image of $w(t,\alpha)$ is $\boldsymbol{W}(s,\alpha)$, α being a parameter, then the image of $\partial w/\partial\alpha$ is $\partial\boldsymbol{W}/\partial\alpha$. (10.7.1)

This theorem is most useful in shortening the list of transforms which require to be memorized. Thus, when the equation

$$\mathscr{L}\,e^{-\alpha t} = \frac{1}{s+\alpha},$$

is differentiated with respect to α, we obtain

$$\mathscr{L} - t\,e^{-\alpha t} = \frac{-1}{(s+\alpha)^2},$$

or

$$\mathscr{L}t\,e^{-\alpha t} = \frac{1}{(s+\alpha)^2}, \qquad (10.7.2)$$

which is the inverse of (5.7.16), already required in the solution of a particular problem. The new transform is so easily derived that it is hardly worth while to remember it; and the same method will enable such quantities as

$$\mathscr{L}^{-1}\frac{s}{(s^2+\beta^2)^2}$$

to be evaluated when the images of $\cos\beta t$ and $\sin\beta t$ are known.

(b) *Image of a function containing the factor* $e^{-\alpha t}$

The image of the function $e^{-\alpha t} w(t)$ is given by

$$\mathscr{L}\, e^{-\alpha t}\, w(t) = \int_0^\infty \{e^{-\alpha t}\, w(t)\}\, e^{-st}\, dt,$$

$$= \int_0^\infty w(t)\, e^{-(s+\alpha)t}\, dt. \qquad (10.7.3)$$

If this equation be compared with

$$\mathscr{L} w(t) = \int_0^\infty w(t)\, e^{-st}\, dt,$$

it is seen that (10.7.3) differs from the latter only in the replacement of s by $(s+\alpha)$. Hence:

If the image of $w(t)$ is $W(s)$, then the image of $e^{-\alpha t} w(t)$ is $W(s+\alpha)$. (10.7.4)

This again reduces the need for memorizing transforms; for example, the relation

$$\mathscr{L}\, e^{-\alpha t} \cos \beta t = \frac{s+\alpha}{(s+\alpha)^2 + \beta^2} \qquad \text{(equation (5.7.12))}$$

follows at once from

$$\mathscr{L} \cos \beta t = \frac{s}{s^2 + \beta^2},$$

more quickly than by the proof outlined in §5.7.

(c) *Initial and final values of* $w(t)$

In many problems the initial value of the response $w(t)$ is specified, or can be determined by inspection. The value as t tends to infinity may also be readily discoverable when $w(t)$ tends to a steady condition at the end of the transient period. It is therefore most fortunate that these limiting values of $w(t)$ can be directly obtained from the image $W(s)$; this makes it possible to check the probable correctness of $W(s)$ before going to the trouble of interpreting it in terms of a function of t.

The required theorems are both derived from the equation

$$\mathscr{L} \frac{dw}{dt} = sW(s) - w_0, \qquad \text{(equation (5.5.3))}$$

which is

$$\int_0^\infty \frac{dw}{dt}\, e^{-st}\, dt = sW(s) - w_0. \qquad (10.7.5)$$

We shall first consider what happens when s tends to zero. The reader will recall that, when the Laplace transform was first defined in §5.4, it was pointed out that $s(=\sigma+j\omega)$ must be limited to values which make the integral meaningful, and that for most problems the condition $\sigma > 0$ sufficed. This will hardly do here; if we wish to allow s to tend to zero, we must not impose a condition more stringent than that σ is greater than *or equal to* zero; but in many cases this is good enough, and in a moment the physical meaning of this condition will become clear. Postulating, then, that the integral on the left hand side remains finite when s is put equal to zero, we obtain

$$\lim_{s\to 0}(s\boldsymbol{W}(s)) - w_0 = \int_0^\infty \frac{dw}{dt}dt,$$

$$= w(\infty) - w_0.$$

Thus is obtained the *final value theorem*:

If the image of $w(t)$ is $\boldsymbol{W}(s)$, and if $w(t)$ tends to a limit as t tends to infinity, then that limit is equal to $\lim_{s\to 0}(s\boldsymbol{W}(s))$. (10.7.6)

There is no loss of generality in conceiving s as tending to zero through real values.

Reverting to (10.7.5), consider the effect of allowing s to tend to infinity, in such a way that its real part also tends to infinity. This condition excludes, for example, a passage to $j\infty$ through purely imaginary values; we may best think of a passage to $+\infty$ through real values. The integral then tends to zero by reason of the factor e^{-st}, and we obtain

$$\lim_{s\to\infty}(s\boldsymbol{W}(s)) - w_0 = 0,$$

leading to the *initial value theorem*:

If the image of $w(t)$ is $\boldsymbol{W}(s)$, the value of $w(t)$ at $t=0$ is equal to $\lim_{s\to\infty}(s\boldsymbol{W}(s))$. (10.7.7)

A simple example is afforded by a circuit in which the closing of a switch enables a capacitor, initially charged, to share its charge with another (fig. 10.6(*a*)). The image circuit, shown in fig. 10.6(*b*), consists of a spike function function current source feeding two parallel paths whose total impedance is given by

$$\frac{1}{\boldsymbol{Z}(s)} = C_1 s + \frac{1}{R+(1/C_2 s)},$$

or $$\boldsymbol{Z}(s) = \frac{RC_2 s + 1}{RC_1 C_2 s^2 + (C_1+C_2)s}.$$

The voltage across the capacitor C_1 (in image form) is then

$$\boldsymbol{V}_1(\mathrm{s}) = \frac{(RC_2\,\mathrm{s}+1)C_1 v_0}{RC_1C_2\,\mathrm{s}^2+(C_1+C_2)\,\mathrm{s}}. \tag{10.7.8}$$

From this,

$$\lim_{\mathrm{s}\to\infty}(\mathrm{s}\boldsymbol{V}_1(\mathrm{s})) = v_0, \tag{10.7.9}$$

and

$$\lim_{\mathrm{s}\to 0}(\mathrm{s}\boldsymbol{V}_1(\mathrm{s})) = \frac{C_1}{C_1+C_2}.v_0. \tag{10.7.10}$$

It may be seen by inspection that the right hand sides of these equations are indeed equal to the initial and final values of the voltage on the capacitor C_1 in fig. 10.6(*a*).

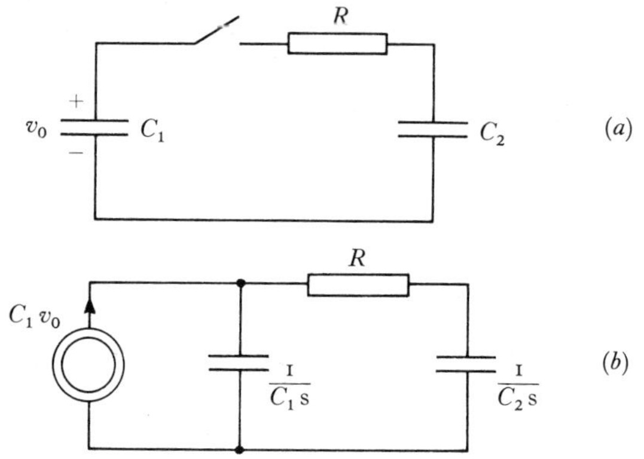

Fig. 10.6 (*a*) Sharing of charge between two capacitances. (*b*) Image circuit for (*a*).

In an oscillatory circuit, the quantities can only settle down to a final constant value if damping is taken into account; otherwise their oscillations will persist. The point may be illustrated by comparing the damped sine wave, $e^{-\alpha t}\sin\beta t$, with its undamped counterpart $\sin\beta t$. The images are $\beta/\{(\mathrm{s}+\alpha)^2+\beta^2\}$ and $\beta/(\mathrm{s}^2+\beta^2)$ respectively; in both cases $\lim_{\mathrm{s}\to 0}(\mathrm{s}\boldsymbol{W}(\mathrm{s}))=0$, but only in the damped case is this limit equal to the ultimate value of the corresponding function of t. This is the reason for the proviso in (10.7.6) – 'if $w(t)$ tends to a limit as t tends to infinity'. Harking back to the condition imposed when this equation was derived, we may say that

$$\int_0^\infty (e^{-\alpha t}\sin\beta t)\,e^{-\mathrm{s}t}\,\mathrm{d}t \quad \text{and} \quad \int_0^\infty (e^{-\alpha t}\cos\beta t)\,e^{-\mathrm{s}t}\,\mathrm{d}t$$

are meaningful for values of s such that σ is *greater than or equal to* zero, whereas

$$\int_0^\infty \sin\beta t\,e^{-\mathrm{s}t}\,\mathrm{d}t \quad \text{and} \quad \int_0^\infty \cos\beta t\,e^{-\mathrm{s}t}\,\mathrm{d}t$$

are only meaningful if σ is *greater than* zero; thus, in the latter instance a condition which was pointed out as essential is not fulfilled.

10.8 The superposition of delayed stimuli

In §5.11 there was a brief mention of the possibility of simulating a square pulse by superimposing positive and negative step functions, the latter being delayed (fig. 10.7(*a*)). This method will now be developed.

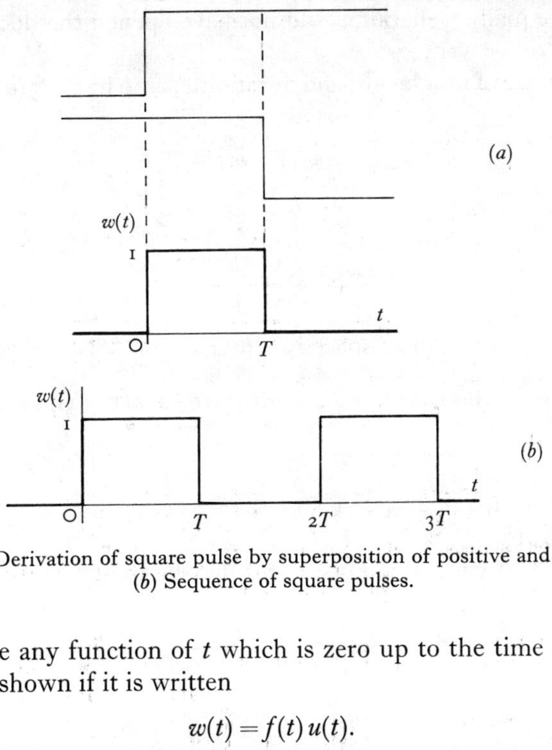

Fig. 10.7 (*a*) Derivation of square pulse by superposition of positive and negative steps. (*b*) Sequence of square pulses.

Let $w(t)$ be any function of t which is zero up to the time $t=0$; this is most clearly shown if it is written

$$w(t) = f(t)\,u(t).$$

The same function delayed by a time T is $f(t-T)\,u(t-T)$; and

$$\mathscr{L}f(t-T)\,u(t-T) = \int_0^\infty f(t-T)\,u(t-T)\,e^{-st}\,dt,$$

$$= \int_T^\infty f(t-T)\,u(t-T)\,e^{-st}\,dt,$$

since the integrand is zero for $t < T$. Substituting t' for $t-T$, we obtain

$$\mathscr{L}f(t-T)\,u(t-T) = \int_0^\infty f(t')\,u(t')\,e^{-s(t'+T)}\,dt'$$

$$= e^{-sT}\int_0^\infty f(t')\,u(t')\,e^{-st'}\,dt'.$$

The integral is merely $\mathscr{L}f(t)u(t)$, or $\boldsymbol{F}(s)$. The following theorem has therefore been proved:

If the image of $f(t)u(t)$ is $\boldsymbol{F}(s)$, then the image of $f(t-T)u(t-T)$ (the same function delayed by a time T) is $e^{-Ts}\boldsymbol{F}(s)$. (10.8.1)

This theorem is called the *shifting theorem*. Some examples of its use will now be given; but it must be stated at once that no *analytical* use can be made of the shifting formula at this stage – the factor e^{-Ts} will simply be a label used in the image circuit when a time delay is found in the original circuit. An arbitrarily invented notation would have served the purpose of this chapter equally well, but would not have opened the door, as the label e^{-Ts} does, to other writings.

A square pulse of unit height and duration T, as in fig. 10.7(*a*), is described by

$$w(t) = u(t) - u(t-T).$$

Therefore its image is

$$\boldsymbol{W}(s) = \frac{1 - e^{-Ts}}{s}. \qquad (10.8.2)$$

A succession of such pulses spaced at intervals of T (fig. 10.7(*b*)) is

$$w(t) = u(t) - u(t-T) + u(t-2T) - u(t-3T) + \dots$$

so the image is

$$\boldsymbol{W}(s) = \frac{1}{s}(1 - e^{-Ts} + e^{-2Ts} - e^{-3Ts} + \dots). \qquad (10.8.3)$$

For an infinite succession of pulses, this may be written

$$\boldsymbol{W}(s) = \frac{1}{s(1 + e^{-Ts})}, \qquad (10.8.4)$$

but to interpret such an expression it must be re-expanded in the form of a series.

The addition of two sine waves, one starting at $t = 0$ and the other delayed by half a period, gives a single half-wave (fig. 10.8(*a*)). This is

$$w(t) = \sin \beta t \, u(t) + \sin(\beta t - \pi)\, u\left(t - \frac{\pi}{\beta}\right),$$

so

$$\boldsymbol{W}(s) = \frac{\beta}{s^2 + \beta^2}(1 + e^{-\pi s/\beta}). \qquad (10.8.5)$$

The output of a half-wave rectifier consists of a succession of such half-waves displaced by times $2\pi/\beta$, $4\pi/\beta$, etc. (fig. 10.8(*b*)). The images of the

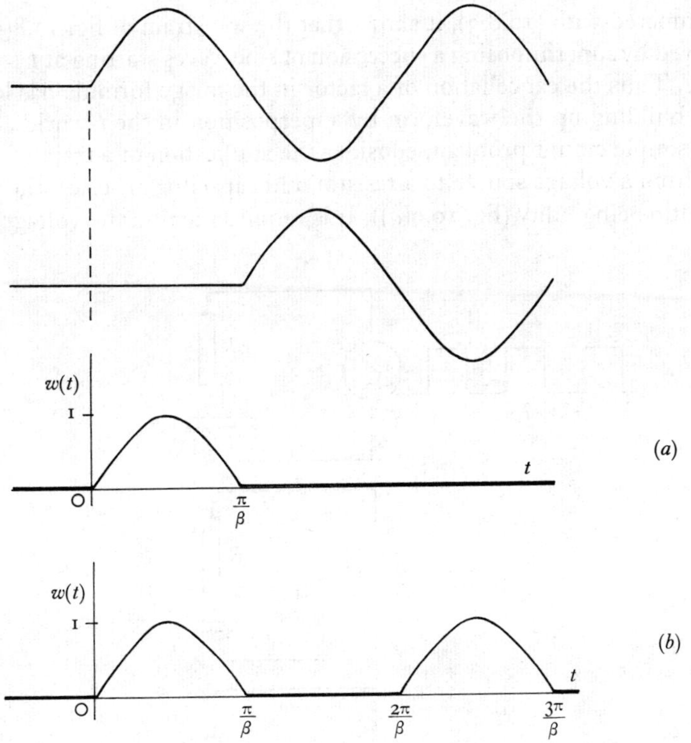

Fig. 10.8 (*a*) Derivation of half sine wave by superposition of sine waves suddenly applied. (*b*) Half-wave rectified sine function.

second, third, ... half-waves are obtained by multiplying the image of the first half-wave by $e^{-2\pi s/\beta}$, $e^{-4\pi s/\beta}$, ... ; so, for the whole sequence,

$$\boldsymbol{W}(s) = \frac{\beta}{s^2+\beta^2}(1+e^{-\pi s/\beta})(1+e^{-2\pi s/\beta}+e^{-4\pi s/\beta}+\dots), \quad (10.8.6)$$

and when the sequence is infinite,

$$\boldsymbol{W}(s) = \frac{\beta}{s^2+\beta^2}\cdot\frac{1+e^{-\pi s/\beta}}{1-e^{-2\pi s/\beta}},$$

$$= \frac{\beta}{s^2+\beta^2}\cdot\frac{1}{1-e^{-\pi s/\beta}}. \quad (10.8.7)$$

If (10.8.7) is re-expanded in the form

$$\boldsymbol{W}(s) = \frac{\beta}{s^2+\beta^2}(1+e^{-\pi s/\beta}+e^{-2\pi s/\beta}+\dots), \quad (10.8.8)$$

and compared with (10.8.5), it shows that the wavetrain of fig. 10.8(*b*) may be derived by superimposing a succession of sine waves starting at $t = 0, \pi/\beta, 2\pi/\beta, \ldots$ Thus the cancellation of a factor in the image formula has led to a way of building up the waveform by superposition in the *t*-world.

As a simple circuit problem, consider the application of a train of square pulses from a voltage source to a resistor and capacitor in series, the mark–space ratio being unity (fig. 10.9(*a*)). It is required to find the voltage across

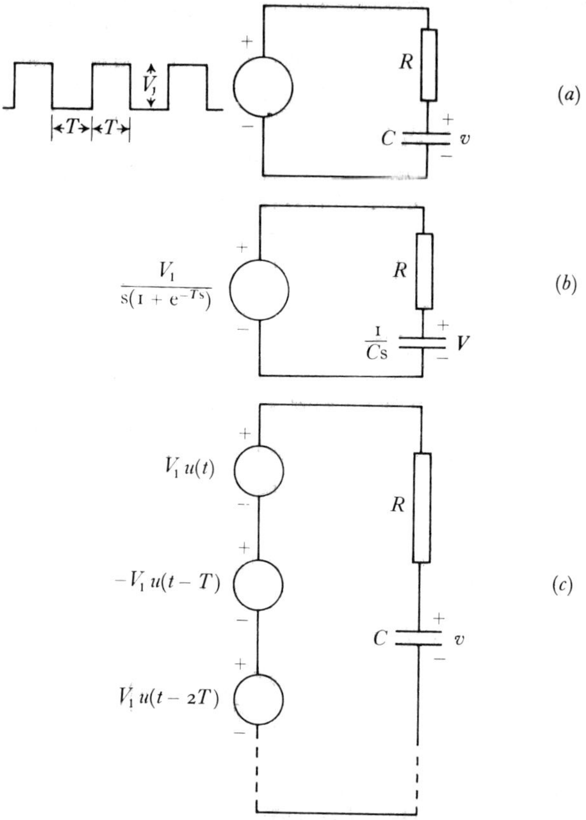

Fig. 10.9 (*a*) Stimulation of *RC* circuit by a train of square pulses of voltage. (*b*) Image circuit for (*a*). (*c*) Circuit (*a*) with pulse source replaced by a number of step function sources.

the capacitor from $t = 0$ onwards, assuming that it was initially uncharged.

The source voltage is of a form already discussed, and (from (10.8.4)) is represented by an image source

$$\boldsymbol{V}_1(s) = \frac{V_1}{s(1 + e^{-Ts})}, \tag{10.8.9}$$

as in fig. 10.9(*b*). The transfer function is

$$\frac{1/Cs}{R+1/Cs}$$

or $\alpha/(s+\alpha)$, where $\alpha=(RC)^{-1}$; thus the output is given by

$$V=\frac{V_1\alpha}{s(s+\alpha)(1+e^{-Ts})},$$

$$=V_1\left(\frac{1}{s}-\frac{1}{s+\alpha}\right)(1-e^{-Ts}+e^{-2Ts}-\ldots). \tag{10.8.10}$$

The corresponding function of t may be most clearly set down thus:

$$\begin{aligned}
v &= V_1(1-e^{-\alpha t}) && \text{starting from } t=0\\
&-V_1(1-e^{-\alpha(t-T)}) && \text{starting from } t=T\\
&+V_1(1-e^{-\alpha(t-2T)}) && \text{starting from } t=2T\\
&-\ldots
\end{aligned} \tag{10.8.11}$$

One may conceive this as due to a number of step function voltage sources in series, alternately positive and negative, and switched on at intervals T (fig. 10.9(*c*)). When the terms of (10.8.11) are successively added, the following results are obtained:

1st pulse, $t=0$ to T: $v=V_1(1-e^{-\alpha t})$

1st space, $t=T$ to $2T$: $v=V_1(-e^{-\alpha t}+e^{-\alpha(t-T)})$

2nd pulse, $t=2T$ to $3T$: $v=V_1(1-e^{-\alpha t}+e^{-\alpha(t-T)}-e^{-\alpha(t-2T)})$

2nd space, $t=3T$ to $4T$: $v=V_1(-e^{-\alpha t}+e^{-\alpha(t-T)}-e^{-\alpha(t-2T)}+e^{-\alpha(t-3T)})$

nth pulse, $t=(2n-2)T$ to $(2n-1)T$: $v=V_1(1-e^{-\alpha t}+e^{-\alpha(t-T)}-\ldots-e^{-\alpha(t-(2n-2)T)})$

nth space, $t=(2n-1)T$ to $2nT$: $v=V_1(-e^{-\alpha t}+e^{-\alpha(t-T)}-\ldots+e^{-\alpha(t-(2n-1)T)})$. (10.8.12)

The series expression for the nth pulse may be written

$$v=V_1\{1-e^{-\alpha t}(1-e^{\alpha T}+e^{2\alpha T}-\ldots+e^{(2n-2)\alpha T})\},$$

$$=V_1\left\{1-\left(\frac{e^{(2n-1)\alpha T}+1}{e^{\alpha T}+1}\right)e^{-\alpha t}\right\}, \tag{10.8.13}$$

and for the nth space,

$$v=V_1\,e^{-\alpha t}(-1+e^{\alpha T}-e^{2\alpha T}+\ldots+e^{(2n-1)\alpha T}),$$

$$=V_1\left(\frac{e^{2n\alpha T}-1}{e^{\alpha T}+1}\right)e^{-\alpha t}. \tag{10.8.14}$$

If the origin of t is moved to the beginning of the nth pulse, so that t is replaced by $t+(2n-2)T$, these expressions are changed to a more convenient form:

$$n\text{th pulse, } t = 0 \text{ to } T\text{: } v = V_1\left\{1 - \left(\frac{1+e^{-(2n-1)\alpha T}}{1+e^{-\alpha T}}\right)e^{-\alpha t}\right\}, \quad (10.8.15)$$

$$n\text{th space, } t = T \text{ to } 2T\text{: } v = V_1\left(\frac{1-e^{-2n\alpha T}}{1+e^{-\alpha T}}\right)e^{-\alpha(t-T)}. \quad (10.8.16)$$

The form of this voltage curve is indicated in fig. 10.10, in which the progress towards a quasi-steady state may be seen.

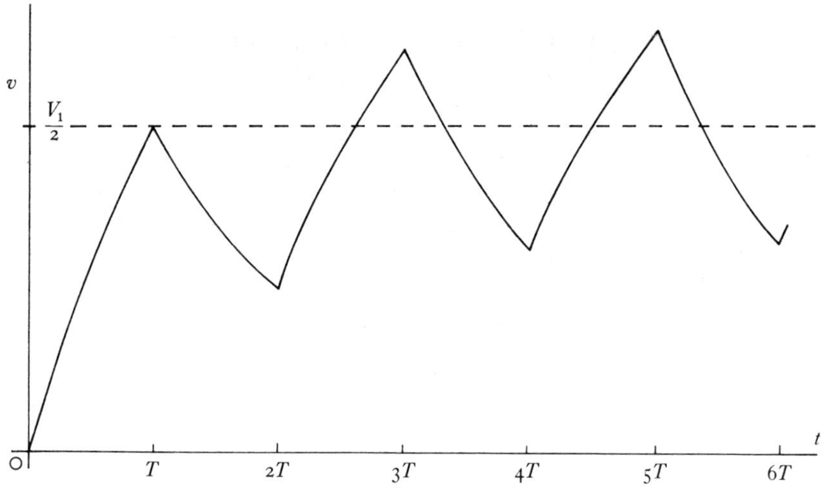

Fig. 10.10 Voltage across capacitance in fig. 10.9 ($\alpha T = 0.693$).

10.9 Convolution

It has already been remarked in §10.1 that when a stimulus $f(t) \equiv \mathscr{L}^{-1}\boldsymbol{F}(s)$ is applied to a circuit having the transfer function $\boldsymbol{H}(s)$, the image of the response is $\boldsymbol{F}(s)\boldsymbol{H}(s)$. The inverse transform of $\boldsymbol{H}(s)$, namely $h(t)$, is a quantity with a simple physical meaning; it is the response of the circuit, assumed initially dead, to a unit spike function stimulus. We are therefore led to enquire whether the response $\mathscr{L}^{-1}\boldsymbol{F}(s)\boldsymbol{H}(s)$, a function of t, can be expressed in terms of the easily intelligible functions $f(t)$ and $h(t)$.

Let the stimulus $f(t)$ be divided by ordinates into a sequence of pulses of very short duration, and let attention be concentrated on a particular pulse between ordinates τ and $\tau+\delta\tau$ (fig. 10.11). By the principle of superposition which applies to all linear systems, the response to $f(t)$ at the instant t_1 is the sum of the responses to all the pulses which have occurred before that instant. Now if $\delta\tau$ is much smaller than the shortest time-constant or period inherent in $\boldsymbol{H}(s)$, the response to the pulse $(\tau, \tau+\delta\tau)$ will be indistinguishable from the response to a spike occurring at the same instant and having the same area; namely, to $f(\tau)\delta\tau.\delta(t-\tau)$. It is known that the response to

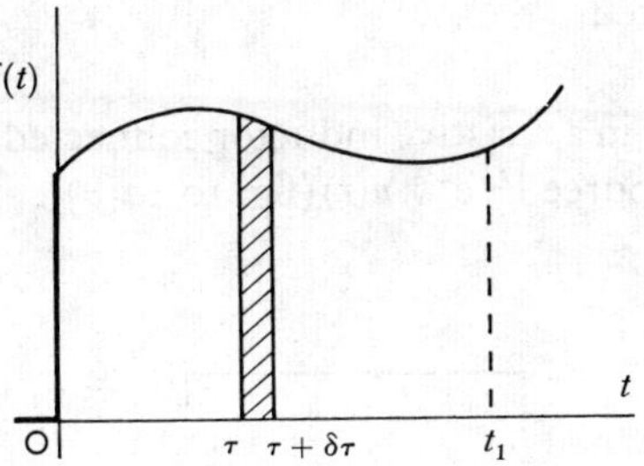

Fig. 10.11 Resolution of a stimulus into a sequence of spike function stimuli.

$\delta(t)$ is $h(t) \equiv \mathscr{L}^{-1}\boldsymbol{H}(s)$; thus the response to $f(\tau)\delta\tau.\delta(t-\tau)$ is $f(\tau)\delta\tau.h(t-\tau)$, or, at the instant t_1, $f(\tau)\delta\tau.h(t_1-\tau)$. Adding the effects of all spikes occurring prior to $t = t_1$, we obtain

$$\text{Total response} = \int_0^{t_1} f(\tau)\,h(t_1-\tau)\,d\tau.$$

This integral is a function of t_1, which may take any positive value. Thus t_1 may be replaced by t, and we conclude that

$$\mathscr{L}^{-1}\boldsymbol{F}(s)\,\boldsymbol{H}(s) = \int_0^t f(\tau)\,h(t-\tau)\,d\tau. \tag{10.9.1}$$

An alternative form, obtained by substituting τ for $t-\tau$, is

$$\mathscr{L}^{-1}\boldsymbol{F}(s)\,\boldsymbol{H}(s) = \int_0^t f(t-\tau)\,h(\tau)\,d\tau. \tag{10.9.2}$$

The integrals in these two equations are called *convolution integrals*, and the process of combining two functions of t in this way is called *convolution*. It is not necessary for $f(t)$ and $h(t)$ to have the physical interpretations laid down at the beginning of this section, and the convolution theorem may be expressed in the following more general terms:

If $\boldsymbol{W}_1(s)$, $\boldsymbol{W}_2(s)$ are the images of $w_1(t)$, $w_2(t)$ respectively, the product $\boldsymbol{W}_1(s)\,\boldsymbol{W}_2(s)$ is the image of the convolution of $w_1(t)$ and $w_2(t)$; that is, of

$$\left.\begin{aligned} &\int_0^t w_1(\tau)\,w_2(t-\tau)\,d\tau, \\ \text{or of}\quad &\int_0^t w_1(t-\tau)\,w_2(\tau)\,d\tau. \end{aligned}\right\} \tag{10.9.3}$$

In brief, multiplication in the s-world corresponds with convolution in the t-world.

10.10 Applications of convolution

The use of the convolution integral will now be illustrated, starting with a simple problem and going on to others by which its elegance and power are more fully demonstrated.

1. To find the current in a resistive inductor connected across an exponentially decaying voltage source $V_1 e^{-\lambda t} u(t)$ (fig. 10.12(*a*)).

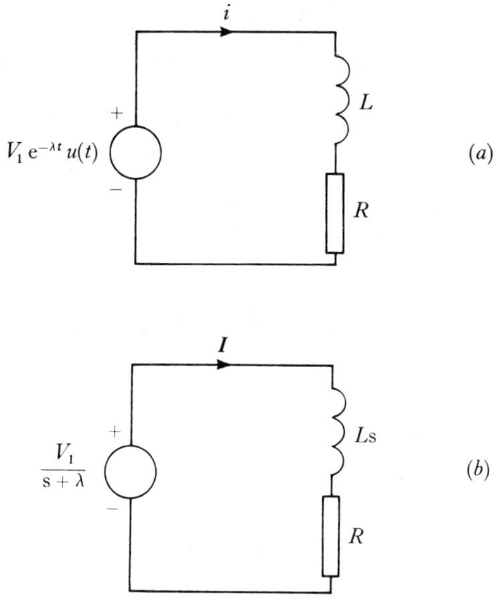

Fig. 10.12 (*a*) *RL* circuit subjected to exponentially decaying voltage. (*b*) Image circuit for (*a*).

The image circuit is shown in fig. 10.12(*b*); the transfer function is $1/(Ls+R)$, which is conveniently written

$$\boldsymbol{H}(s) = \frac{1}{L(s+\alpha)}, \tag{10.10.1}$$

where $\alpha = R/L$. Hence

$$h(t) = \frac{1}{L} e^{-\alpha t}. \tag{10.10.2}$$

This is the response of the circuit to a spike function voltage of unit value. The response to the actual stimulus $V_1 e^{-\lambda t} u(t)$ is therefore, from (10.9.1),

$$\mathscr{L}^{-1}\frac{V_1 \boldsymbol{H}(s)}{s+\lambda} = \int_0^t (V_1 e^{-\lambda\tau})\left(\frac{e^{-\alpha(t-\tau)}}{L}\right)d\tau,$$

$$= \frac{V_1}{L} e^{-\alpha t}\int_0^t e^{(\alpha-\lambda)\tau}\,d\tau,$$

$$= \frac{V_1}{L(\alpha-\lambda)} e^{-\alpha t}[e^{(\alpha-\lambda)t} - 1],$$

$$= \frac{V_1}{L(\alpha-\lambda)}(e^{-\lambda t} - e^{-\alpha t}). \qquad (10.10.3)$$

This result can also be obtained by more elementary methods.

2. The circuit of fig. 10.13(*a*) is energized from a voltage source which generates a single pulse of sinusoidal shape. The circuit is non-oscillatory

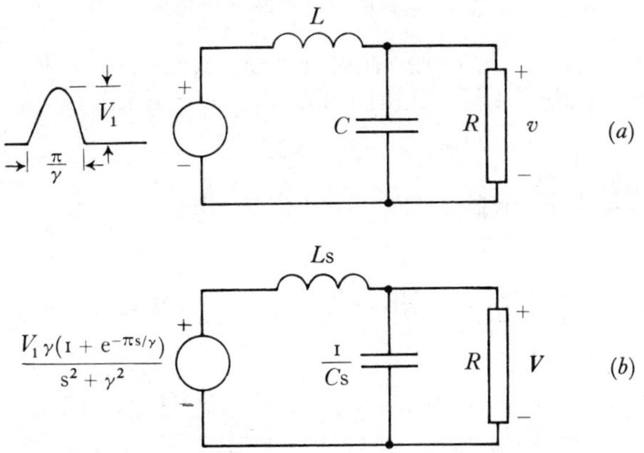

Fig. 10.13 (*a*) *RLC* circuit stimulated by half sine wave voltage. (*b*) Image circuit for (*a*).

and initially dead. It is required to find the voltage across *R* and *C*, (*a*) during the pulse, and (*b*) after the pulse has ended.

It has been shown in (10.8.5) that the image of the single half-wave takes the form

$$\frac{V_1\gamma(1+e^{-\pi s/\gamma})}{s^2+\gamma^2}.$$

In §5.8 the response of a circuit to a general stimulus $f(t)$ was discussed, but the method was limited to cases in which $\boldsymbol{F}(s)$ was a ratio of two polynomials. This is not true here; but it will be found that the convolution method is not subject to any such restriction.

The image circuit is shown in fig. 10.13(*b*); the transfer function is

$$\frac{R/(RCs+1)}{R/(RCs+1)+Ls}, \quad \text{or} \quad \frac{R}{LRCs^2+Ls+R}.$$

The circuit being non-oscillatory, this is written as

$$\boldsymbol{H}(s) = \frac{\alpha\beta}{(s+\alpha)(s+\beta)}, \qquad (10.10.4)$$

where $\alpha+\beta = (RC)^{-1}$, $\alpha\beta = (LC)^{-1}$. This may be rewritten in the form

$$\boldsymbol{H}(s) = \frac{\alpha\beta}{\alpha-\beta}\left(\frac{1}{s+\beta} - \frac{1}{s+\alpha}\right),$$

so that

$$h(t) = \frac{\alpha\beta}{\alpha-\beta}(e^{-\beta t} - e^{-\alpha t}). \qquad (10.10.5)$$

(*a*) *While the pulse is still operative*, we may write $f(\tau) = V_1 \sin\gamma\tau$ in the convolution integral; the switching off of the source still lies in the future. Thus

$$v = \frac{\alpha\beta V_1}{\alpha-\beta}\int_0^t \sin\gamma\tau(e^{-\beta(t-\tau)} - e^{-\alpha(t-\tau)})\,d\tau,$$

$$= \frac{\alpha\beta V_1}{\alpha-\beta}\left\{e^{-\beta t}\int_0^t e^{\beta\tau}\sin\gamma\tau\,d\tau - e^{-\alpha t}\int_0^t e^{\alpha\tau}\sin\gamma\tau\,d\tau\right\}. \qquad (10.10.6)$$

Also $\int_0^t e^{\beta\tau}\sin\gamma\tau\,d\tau$ is the imaginary part of $\int_0^t e^{(\beta+j\gamma)\tau}\,d\tau$;

thus it is found that

$$\int_0^t e^{\beta\tau}\sin\gamma\tau\,d\tau = \frac{\beta e^{\beta t}\sin\gamma t + \gamma(1 - e^{\beta t}\cos\gamma t)}{\beta^2+\gamma^2}. \qquad (10.10.7)$$

Substituting in (10.10.6),

$$v = \frac{\alpha\beta V_1}{\alpha-\beta}\left\{\frac{\beta\sin\gamma t - \gamma\cos\gamma t + \gamma e^{-\beta t}}{\beta^2+\gamma^2} - \text{similar terms in } \alpha\right\}. \qquad (10.10.8)$$

This holds good from $t = 0$ to π/γ.

(*b*) *After the cessation of the pulse*, the range of integration from 0 to t is to be divided into two parts, 0 to π/γ and π/γ to t. During the latter part the stimulus $f(\tau)$ is zero; only the range from 0 to π/γ contributes to the integral, and the contribution is obtained by writing π/γ instead of t as the upper limit of the integrals in (10.10.6). From (10.10.7),

$$\int_0^{\pi/\gamma} e^{\beta\tau}\sin\gamma\tau\,d\tau = \frac{\gamma(1+e^{\pi\beta/\gamma})}{\beta^2+\gamma^2}, \qquad (10.10.9)$$

so (10.10.6) yields

$$v = \frac{\alpha\beta V_1}{\alpha-\beta}\left\{\frac{\gamma(1+e^{\pi\beta/\gamma})}{\beta^2+\gamma^2}e^{-\beta t} - \text{a similar term in } \alpha\right\}. \qquad (10.10.10)$$

This holds good for all values of $t \geqslant \pi/\gamma$. The value of v at $t=\pi/\gamma$ is necessarily equal to that given by (10.10.8).

3. The same circuit is energized by a succession of half sine waves with no intervals between them. It is required to determine the response during the $(n+1)$th pulse.

The solution to this problem may easily be deduced from the last by a superposition process. Let t be measured from the beginning of the $(n+1)$th pulse; the contribution of that pulse to the response is then given by (10.10.8). The aftermath of pulses n, $(n-1)$, $(n-2)$, ..., is obtained from (10.10.10), by replacing t by the time which has elapsed since the starts of those pulses, namely $t+\pi/\gamma$, $t+2\pi/\gamma$, etc. The total effect of the first n pulses, at a time t after the end of the nth one, is thus seen to be

$$v = \frac{\alpha\beta V_1}{\alpha-\beta}\left\{\frac{\gamma(1+e^{\pi\beta/\gamma})}{\beta^2+\gamma^2}\left[e^{-\pi\beta/\gamma}+e^{-2\pi\beta/\gamma}+\ldots+e^{-n\pi\beta/\gamma}\right]e^{-\beta t}\right.$$
$$\left. - \text{a similar term in } \alpha\right\},$$

$$= \frac{\alpha\beta V_1}{\alpha-\beta}\left\{\frac{\gamma(1+e^{-\pi\beta/\gamma})(1-e^{-n\pi\beta/\gamma})}{(\beta^2+\gamma^2)(1-e^{-\pi\beta/\gamma})}e^{-\beta t} - \text{a similar term in } \alpha\right\}.$$
$$(10.10.11)$$

Adding the effect of the still operative pulse from (10.10.8), we obtain the following total response:

$$v = \frac{\alpha\beta V_1}{\alpha-\beta}\left\{\frac{1}{\beta^2+\gamma^2}\left[\beta\sin\gamma t - \gamma\cos\gamma t + \gamma\left(1+\frac{(1+e^{-\pi\beta/\gamma})(1-e^{-n\pi\beta/\gamma})}{(1-e^{-\pi\beta/\gamma})}\right)e^{-\beta t}\right]\right.$$
$$\left. - \text{a similar term in } \alpha\right\}. \qquad (10.10.12)$$

The steady-state response was plotted, for particular circuit values, in fig. 9.7 (curve (*b*)).

This method of evaluating the response of a system to a train of pulses by superimposing the effects of pulses running backwards into time may be contrasted with the treatment by means of the shifting theorem in §10.8; as the shifting theorem treats of time-delays, the successive pulses in that case run forwards into time. This method just introduced is more elegant, for the pulses are taken in diminishing order of importance so far as their effect at the present instant is concerned. It is tempting to try the same approach in the earlier treatment, starting with the last pulse and regarding the earlier ones as being subject to negative time-delays. It is not legitimate, however, to use the shifting theorem in this way; (10.8.1), as its mode of derivation shows, applies to positive values of T only.

4. A capacitor of 20 μF is charged through a resistor of 50000 Ω from a voltage source $v_1(t)$ which is approximately specified by the following numerical values:

t (seconds)	0	0.1	0.2	0.3	0.4	0.5	0.6	0.7	0.8	0.9	1.0
$v_1(t)$ (volts)	0	60	115	145	75	100	135	170	200	195	130

It is required to obtain approximately the voltage on the capacitor at 0.2, 0.4, 0.6, 0.8 and 1.0 s.

The voltage transfer function $\boldsymbol{H}(s) = \alpha/(s+\alpha)$, where $\alpha = (RC)^{-1} = 1\ \mathrm{s}^{-1}$. Thus $h(t) = e^{-t}$, and the output voltage at time t is given by the convolution integral:

$$v = \int_0^t v_1(\tau)\, e^{-(t-\tau)}\, d\tau,$$

$$= e^{-t} \int_0^t v_1(\tau)\, e^{\tau}\, d\tau. \qquad (10.10.13)$$

The values of the integrand at the stated time-points are as follows:

τ	0	0.1	0.2	0.3	0.4	0.5	0.6	0.7	0.8	0.9	1.0
$v_1(\tau)\, e^{\tau}$	0	66	140	196	112	165	246	342	445	480	353

The best available numerical approximation for the integrals to alternate points in the τ-scale is that given by Simpson's Rule, which assumes that the integrand follows, from $\tau = 0$ to 0.2, a curve of the second degree having the correct ordinates at 0, 0.1 and 0.2; from $\tau = 0.2$ to 0.4, another curve similarly defined, and so on. The data do not permit any closer approximation. The rule states that the mean height of the curve over the range 0 to 0.2 (for instance) is given by

$$\tfrac{1}{6}\{v_1(0)\, e^0 + 4v_1(0.1)\, e^{0.1} + v_1(0.2)\, e^{0.2}\}.$$

Thus the following approximations to the integrals are obtained, and (by multiplying by e^{-t}) the approximations to v:

t	0	0.2	0.4	0.6	0.8	1.0
$\int_0^t v_1(\tau)\, e^{\tau} d\tau$ (approx)	0	13.5	48.0	81.9	150.6	241.2
v (approx)	0	11	32	45	68	89

The response and the stimulus are plotted in fig. 10.14.

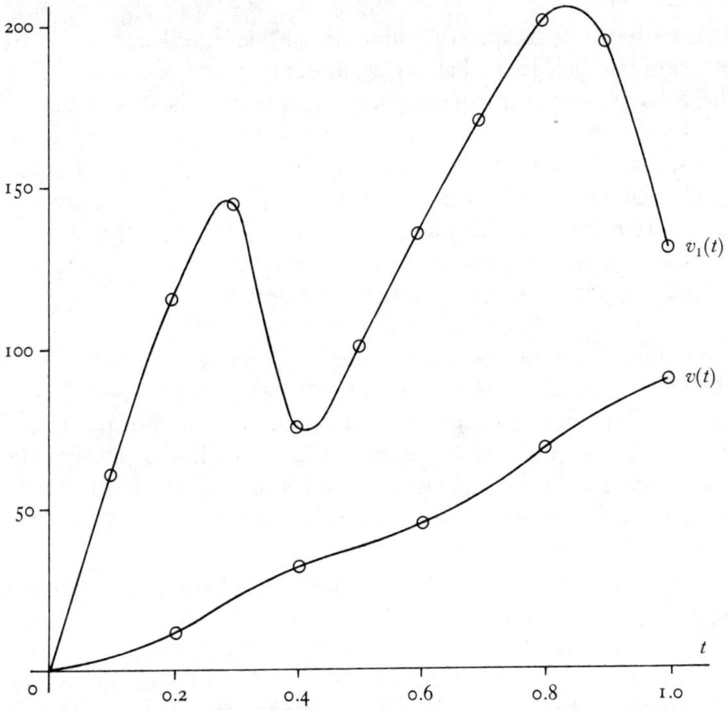

Fig. 10.14 Response $v(t)$ of an *RC* circuit subjected to a stimulus $v_1(t)$ which is numerically defined.

This kind of calculation must be performed with discretion. The fine detail of the curve of applied voltage, specified only at intervals of 0.1 s, remains unknown; but as the time-constant of the circuit is 1 s, its response is not sensitive to minor variations occurring during the intervals. The greater the ratio between the time-constant and the duration of each interval, the more probable does it become that the figures calculated are approximately right.

Examples on chapter 10

10.1 Find the functions of t corresponding to the following functions of s:

$$(a)\ \frac{s^2+5}{s^2+2s+5};\qquad (b)\ \frac{(s+a)^2}{(s+b)^2};\qquad (c)\ \frac{2\beta(s+\alpha)}{[(s+\alpha)^2+\beta^2]^2};$$

$$(d)\ \frac{2s(s^2-3\beta^2)}{(s^2+\beta^2)^3};\qquad (e)\ \frac{1}{(s^2+1)(1-e^{-\pi s})}.$$

10.2 A battery, a dissipative inductor and an ideal capacitor are all connected in parallel. If the battery is suddenly disconnected at time $t = 0$, find the voltage across the capacitor assuming that R^2 is greater than $4L/C$.

10.3 A sinusoidal voltage $V_1 \sin \omega t$ is applied to a resistor R and capacitors C_1, C_2, all in series. The capacitor C_1 is initially short-circuited by a closed switch S. After the network has reached a steady state, the switch S is opened at an instant when the applied voltage is zero and increasing. Obtain an expression for the current as a function of time.

10.4 An ideal current source of output I_0 feeds a resistor R with a capacitor C in parallel. With the circuit in a steady state, a switch is closed connecting a coil of inductance L and resistance r across the generator. Determine the subsequent voltage across C as a function of s, verifying that this is consistent with the known initial and final values; and hence derive the voltage as a function of t, assuming it is oscillatory.

10.5 A pure inductor of 20 mH is connected in parallel with a capacitor of 0.04 μF; in series with the combination is a second pure inductor, of 160 mH. To the terminals of the whole network is applied a square pulse of voltage of amplitude 10^5 V and duration 10^{-6} s. Making any reasonable approximation, calculate the subsequent voltage across the 20 mH inductor as a function of time.

10.6 A 1 Ω resistor is in parallel with a $\frac{1}{2}$ F capacitor, and the combination is in series with a $\frac{4}{3}$ H inductor. In parallel with this network is a $\frac{3}{2}$ F capacitor with terminals A, B. Find the voltage across the resistor when a spike function current is injected at AB, of magnitude such as to transfer 1 C of charge.

10.7 A voltage $V_1(1-e^{-\alpha t})u(t)$ is applied to a resistor R and capacitor C in series, the capacitor being initially uncharged. Find the voltage on the capacitor, given that $RC = \alpha^{-1}$.

10.8 Two perfect inductors L are connected in series; there is mutual inductance M between them, and in parallel with one of them is a resistor R. A current source connected across the outer terminals A, B of the inductors injects a current $\lambda t u(t)$. Find the voltage across AB.

10.9 Two twoport networks comprise symmetrical T circuits as follows:

(i) Capacitance 0.5 μF between A, M and between M, C; resistance 1 MΩ between M and the connected terminals BD.

(ii) Resistance 2 MΩ between A′, M′ and between M′, C′; capacitance 1 μF between M′ and the connected terminals B′D′.

The two networks are made into a single twoport by connecting AA′, BB′, CC′, DD′. Find the voltage transfer function for an input across AB and an output across CD (on open-circuit). If the input is a sine wave $V_1 \sin \omega t u(t)$, show that the oscillation present in the input is completely eliminated from the output for one particular value of ω, and find what that value is.

10.10 An inductor L and capacitor C are connected in parallel across the terminals of a current source which generates a square pulse of current of amplitude I_1 and duration T (less than $\pi\sqrt{(LC)}$). Obtain an expression for the voltage across the circuit after the cessation of the pulse. Derive also an expression for the voltage which would follow a pulse of very short duration conveying the same total charge, showing that the ratio of the amplitude of the oscillations in the two cases is $(\omega_1 T/2)/\sin(\omega_1 T/2)$, where $\omega_1 = (LC)^{-1/2}$.

10.11 Use the convolution integral to find the inverse transform of $1/s^2(s+\alpha)$.

10.12 A 5 mH inductor and a 0.01 μF capacitor are in parallel, and a 40 mH inductor is in series with the combination. Find the voltage response across the capacitor for a unit spike voltage input across the whole circuit, and apply the convolution integral to deduce the response to a double-exponential voltage pulse $V_1(e^{-\alpha t} - e^{-\beta t})$, where $V_1 = 6150$, $\alpha = \frac{1}{5}10^6$, $\beta = \frac{2}{3}10^6$.

10.13 A 0.1 μF capacitor is in parallel with a 10^5 Ω resistor across terminals AB. Also across AB is a 10^3 Ω resistor in series with an ideal diode which conducts when A is positive relatively to B. Terminals AB are fed from an ideal voltage source in series with another 10^5 Ω resistor; the output of the voltage source is a pulse of amplitude 200 V and duration 1 ms. Initially the capacitor is charged so that A is at -10 V with respect to B; the voltage pulse is then applied. Find the subsequent voltage across the capacitor.

11 Fourier transforms

11.1 Extension of the Fourier series concept

It was shown in chapter 9 that a periodic function $f_1(\theta)$ may be expressed as a complex Fourier series

$$f_1(\theta) = \sum_{n=-\infty}^{\infty} C_n e^{jn\theta}, \qquad \text{(equation (9.6.1))}$$

where C_n was related to the coefficients of the cosine and sine terms in the real form of the Fourier series, the relations being

$$C_0 = a_0, \quad C_n = \tfrac{1}{2}(a_n - jb_n), \quad C_{-n} = \tfrac{1}{2}(a_n + jb_n). \qquad \text{(equation (9.6.5))}$$

Furthermore, from (9.2.2,3,4),

$$a_0 = \frac{1}{2\pi}\int_0^{2\pi} f_1(\theta)\, d\theta,$$

$$a_n = \frac{1}{\pi}\int_0^{2\pi} f_1(\theta) \cos n\theta \, d\theta,$$

$$b_n = \frac{1}{\pi}\int_0^{2\pi} f_1(\theta) \sin n\theta \, d\theta.$$

When these are substituted into the expressions for C_0, C_n, C_{-n}, it is found that all three are covered by the single formula

$$C_n = \frac{1}{2\pi}\int_0^{2\pi} f_1(\theta)\, e^{-jn\theta}\, d\theta. \qquad (11.1.1)$$

The range of integration need not be 0 to 2π, however; all that is necessary is to integrate over one cycle of the (periodic) integrand, and for the present purpose we choose

$$C_n = \frac{1}{2\pi}\int_{-\pi}^{\pi} f_1(\theta)\, e^{-jn\theta}\, d\theta. \qquad (11.1.2)$$

A further slight change will put this equation into a form which directs attention at the frequency of the harmonic rather than at its order. Let the fundamental component of $f(\theta)$ have period T; then $\theta = 2\pi t/T$, and

$$n\theta = \frac{2\pi n t}{T} = \omega t,$$

where ω is the angular frequency of a typical harmonic. The equation thus becomes

$$C(\omega) = \frac{1}{T}\int_{-T/2}^{T/2} f(t)\,e^{-j\omega t}\,dt, \tag{11.1.3}$$

where $f(t)$ stands for $f_1\ (2\pi t/T)$. It is here understood that $C(\omega)$ is only defined for values of ω which are integral multiples of $2\pi/T$.

Suppose now that $f(t)$ is some *non-periodic* function of t. Given a fundamental period T, it is still possible to go through the motions of calculating $C(\omega)$ from (11.1.3); but what does the calculation mean? A moment's thought shows that in so doing one is taking account of the values of $f(t)$ over the range $-(T/2) \leqslant t \leqslant (T/2)$, and assuming that they recur cyclically outside that range (fig. 11.1). Thus the analysis represents $f(t)$ within the

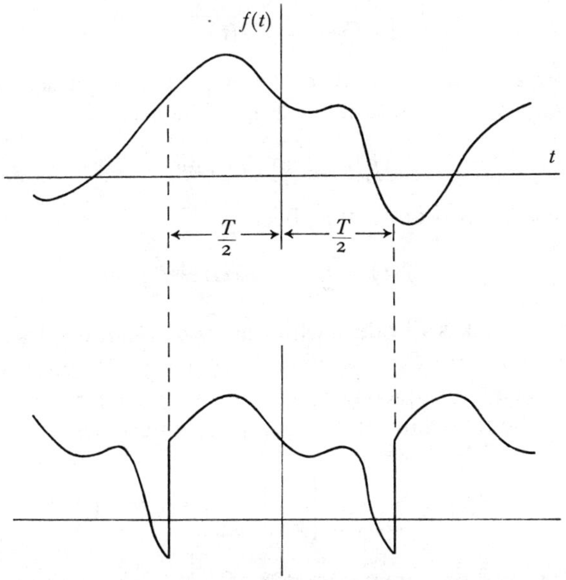

Fig. 11.1 Development of a periodic function from a non-periodic function.

limited range from $-T/2$ to $T/2$; the frequency interval between successive harmonics is $2\pi/T$. If T is increased, $f(t)$ is represented over a wider range, and the frequency interval between harmonics is smaller. Therefore if we choose any small frequency range ω to $\omega+\delta\omega$, it is possible to select a T so large that many harmonics occur in the range, however small $\delta\omega$ may be; the number of harmonics in $\delta\omega$ is the next integer below $\delta\omega \div (2\pi/T)$. More and more exactly as T is increased, we may say that the range $\delta\omega$

contains $T\delta\omega/2\pi$ harmonics, each having the amplitude $C(\omega)$ given by (11.1.3); so that the total amplitude of all the harmonics in $\delta\omega$ is

$$C(\omega)\frac{T\delta\omega}{2\pi} = \frac{\delta\omega}{2\pi}\int_{-T/2}^{T/2} f(t)\,\mathrm{e}^{-\mathrm{j}\omega t}\,\mathrm{d}t. \tag{11.1.4}$$

When T is made to tend to infinity, each individual harmonic becomes infinitely small; but their spacing becomes infinitely close, and the total amplitude of the harmonics in the range ω to $\omega+\delta\omega$ tends to the limit of $(1/2\pi)\boldsymbol{F}(\omega)\delta\omega$, where

$$\boldsymbol{F}(\omega) = \int_{-\infty}^{\infty} f(t)\,\mathrm{e}^{-\mathrm{j}\omega t}\,\mathrm{d}t. \tag{11.1.5}$$

$\boldsymbol{F}(\omega)$ is therefore defined here as the harmonic content per unit interval of frequency, not of angular frequency. It is called the *Fourier transform* of $f(t)$ – a statement which is briefly expressed as

$$\boldsymbol{F}(\omega) = \mathscr{F} f(t). \tag{11.1.6}$$

We can now go back and convert the original Fourier series into an expression which gives $f(t)$ in terms of $\boldsymbol{F}(\omega)$. The series

$$f_1(\theta) = \textstyle\sum_{-\infty}^{\infty} C_n\,\mathrm{e}^{\mathrm{j}n\theta}, \qquad \text{(equation (9.6.1))}$$

when rewritten in terms of t and ω, becomes

$$f(t) = \sum_{\omega=-\infty}^{\infty} C(\omega)\,\mathrm{e}^{\mathrm{j}\omega t}. \tag{11.1.7}$$

In this equation, ω takes all values which are integral multiples of the fundamental value $2\pi/T$, and $C(\omega)$ is given by (11.1.3). When T is so greatly increased that a small frequency range $\delta\omega$ contains a large number of harmonics having the total amplitude (11.1.4), $f(t)$ becomes

$$f(t) = \sum_{\omega=-\infty}^{\infty}\left\{\frac{\delta\omega}{2\pi}\int_{-T/2}^{T/2} f(t)\,\mathrm{e}^{-\mathrm{j}\omega t}\,\mathrm{d}t\right\}\mathrm{e}^{\mathrm{j}\omega t}$$

and as T is increased the integral tends to $\boldsymbol{F}(\omega)$. In the limit the summation becomes an integral, namely

$$f(t) = \frac{1}{2\pi}\int_{-\infty}^{\infty} \boldsymbol{F}(\omega)\,\mathrm{e}^{\mathrm{j}\omega t}\,\mathrm{d}\omega, \tag{11.1.8}$$

$f(t)$ is the *inverse Fourier transform* of $\boldsymbol{F}(\omega)$; the two functions $f(t)$ and $\boldsymbol{F}(\omega)$ form a *Fourier pair*. The resemblance between (11.1.5) and (11.1.8) is close; the symmetry is only marred by the factor $1/2\pi$, and some writers achieve perfect symmetry by so defining $\boldsymbol{F}(\omega)$ that a factor $1/\sqrt{(2\pi)}$ occurs in each equation. The reason for not doing this will, however, soon become apparent.

11.2 Nature of the Fourier transformation

The Fourier transform is an image function, in the sense already used several times in this book; the properties of $f(t)$, a real function of time, are translated into properties of $\boldsymbol{F}(\omega)$, a complex function of frequency. In view of the extensive application of the method in the analysis of communication systems, it is convenient to call $f(t)$ the *signal* possessing the spectrum $\boldsymbol{F}(\omega)$. $\boldsymbol{F}(\omega)$ has certain characteristics analogous to ones already proved for the Fourier series, namely

$$\text{If } f(t) \text{ is an even function of } t,\ \boldsymbol{F}(\omega) \text{ is real,} \tag{11.2.1}$$

$$\text{If } f(t) \text{ is an odd function of } t,\ \boldsymbol{F}(\omega) \text{ is imaginary.} \tag{11.2.2}$$

These correspond with (9.3.2) and (9.3.3) respectively. Furthermore, by the definition (11.1.5), $\boldsymbol{F}(\omega)$ is expressible as a function of $\mathrm{j}\omega$; thus changing the sign of ω is equivalent to changing the sign of j, or

$$\boldsymbol{F}(-\omega) = \boldsymbol{F}^*(\omega), \tag{11.2.3}$$

where $\boldsymbol{F}^*(\omega)$ is the complex conjugate of $\boldsymbol{F}(\omega)$.

It is impossible to overlook the close relationship between the Fourier transformation

$$\boldsymbol{F}(\omega) = \int_{-\infty}^{\infty} f(t)\,\mathrm{e}^{-\mathrm{j}\omega t}\,\mathrm{d}t, \qquad \text{(equation (11.1.5))}$$

and the Laplace transformation

$$\boldsymbol{F}(\mathrm{s}) = \int_0^{\infty} f(t)\,\mathrm{e}^{-\mathrm{s}t}\,\mathrm{d}t. \qquad \text{(see (5.4.9))}$$

In this form the Laplace transformation is both less and more general than the Fourier transformation; less general because it takes no account of the value of $f(t)$ for negative values of t, so that it is only applicable to problems in which $f(t) = 0$ when $t < 0$; more general, because the image variable s is a complex number $\sigma + \mathrm{j}\omega$, and the convergence of the integral is assured by a suitable choice of σ. Thus, for the unit step function $u(t)$,

$$\mathscr{L}u(t) = \int_0^{\infty} \mathrm{e}^{-\mathrm{s}t}\,\mathrm{d}t,$$

$$= \left[\frac{-\mathrm{e}^{-\mathrm{s}t}}{\mathrm{s}}\right]_0^{\infty},$$

which has the finite value $1/s$ provided $\sigma > 0$; but

$$\mathscr{F}u(t) = \int_0^\infty e^{-j\omega t}\,dt,$$

$$= \left[\frac{-e^{-j\omega t}}{j\omega}\right]_0^\infty, \quad \text{or} \lim_{T\to\infty} \left[\frac{j\sin\omega t - \cos\omega t}{j\omega}\right]_0^T,$$

which tends to no definite value since $\cos\omega t$ and $\sin\omega t$ are oscillatory. Hence problems of convergence – the question whether the image function can be defined at all – arise much more acutely in the Fourier case. For the Fourier transform integral to exist it is necessary for $f(t)$ to tend to zero as t tends either to $+\infty$ or $-\infty$; and the Fourier transform of $u(t)$ must be defined by a limiting process from a function which satisfies this condition, such as $e^{-\alpha t}u(t)$. Thus

$$\mathscr{F}\,e^{-\alpha t}u(t) = \int_0^\infty e^{-(\alpha+j\omega)t}\,dt,$$

$$= \frac{1}{\alpha + j\omega}, \tag{11.2.4}$$

provided $\alpha > 0$; and, allowing α to tend to zero from the positive side and writing '$\alpha \to +0$' for this process,

$$\mathscr{F}u(t) = \lim_{\alpha\to+0} \frac{1}{\alpha + j\omega},$$

$$= \frac{1}{j\omega}. \tag{11.2.5}$$

In general, as will be seen, the Fourier transform of a function which is zero for negative values of t, *provided it exists*, is obtained by writing $j\omega$ for s in the Laplace transform. This is the justification for placing a multiplier $1/2\pi$ in (11.1.8) and no multiplier in (11.1.5), rather than straining after symmetry by placing $1/\sqrt{(2\pi)}$ in each. Both the Laplace transform (as here used) and the Fourier transform are special cases of the *two-sided Laplace transform*, defined by

$$\mathscr{L}_2 f(t) = \int_{-\infty}^{\infty} f(t)\,e^{-st}\,dt. \tag{11.2.6}$$

The inverse formula (11.1.8), giving $f(t)$ when its Fourier transform $\boldsymbol{F}(\omega)$ is known, raises further problems. If applied to $\boldsymbol{F}(\omega) = 1/(\alpha + j\omega)$, for example, it gives

$$f(t) = \frac{1}{2\pi}\int_{-\infty}^{\infty} \frac{e^{j\omega t}}{\alpha + j\omega}\,d\omega.$$

It can be shown that the right hand side equals zero for $t < 0$ and $e^{-\alpha t}$ for $t > 0$; but the method of doing this is that known as *contour integration*, and lies beyond the scope of this book. Thus the Fourier transformation can speedily lead to integrals requiring advanced methods for their evaluation. (A similar integral formula exists for carrying out the inverse Laplace transformation, but in elementary applications it is not necessary to have recourse to it; for these the expansion of $\boldsymbol{F}(s)$ in powers of $1/s$, as described in §5.6, suffices.)

Notwithstanding the mathematical difficulties, the Fourier transform occupies so central a place in circuit analysis as to demand at least an introductory treatment. A major reason for this is that the most convenient way of measuring and specifying the performance of a system is on a base of frequency. In the last chapter the action of circuits was described by the transfer function $\boldsymbol{H}(s)$, which was the image of $h(t)$, the response to a unit spike function stimulus. If one were requiring to measure the transfer function of an unknown network or of an electromechanical system, it would thus be possible to apply a stimulus in the form of a very short pulse, measure the response and deduce $\boldsymbol{H}(s)$. But one would be much more likely to apply continuous sinusoidal stimuli over a wide range of frequencies, measuring the relation between response and stimulus both in magnitude and phase; the Fourier transform would then make it possible in principle to deduce the response to any other form of stimulus. Fourier transform analysis is indeed the ultimate generalization of the phasor method, and it is worth while to devote a few words to the relation between them.

11.3 Phasors and the Fourier transform

The circuit of fig. 11.2(*a*) has already been used in §4.6 in explaining the use of complex numbers to represent phasors. The source voltage $V\cos(\omega t + \alpha)$ is the stimulus, the current i is regarded as the response. From the present standpoint we regard the stimulus as containing the two frequencies $\pm\omega$; and instead of designating $V\cos(\omega t + \alpha)$ as the real part of $\boldsymbol{V}e^{j\omega t}$, where $\boldsymbol{V} = Ve^{j\alpha}$, we write

$$V\cos(\omega t + \alpha) = \boldsymbol{V}(\omega)\,e^{j\omega t} + \boldsymbol{V}(-\omega)\,e^{-j\omega t}, \tag{11.3.1}$$

where

$$\boldsymbol{V}(\omega) = \tfrac{1}{2}V e^{j\alpha}, \qquad \boldsymbol{V}(-\omega) = \tfrac{1}{2}V e^{-j\alpha}. \tag{11.3.2}$$

The real function of time is thus replaced by the sum (not merely the real part) of two complex functions which may be represented by phasors rotating in opposite directions (fig. 11.2(*b*)). The coefficients $\boldsymbol{V}(\omega)$ and $\boldsymbol{V}(-\omega)$ may be regarded as the voltage stimuli in image circuits as shown in fig. 11.2(*c*) and (*d*); the responses are

$$\boldsymbol{I}(\omega) = \boldsymbol{Y}(\omega)\,\boldsymbol{V}(\omega), \qquad \boldsymbol{I}(-\omega) = \boldsymbol{Y}(-\omega)\,\boldsymbol{V}(-\omega), \tag{11.3.3}$$

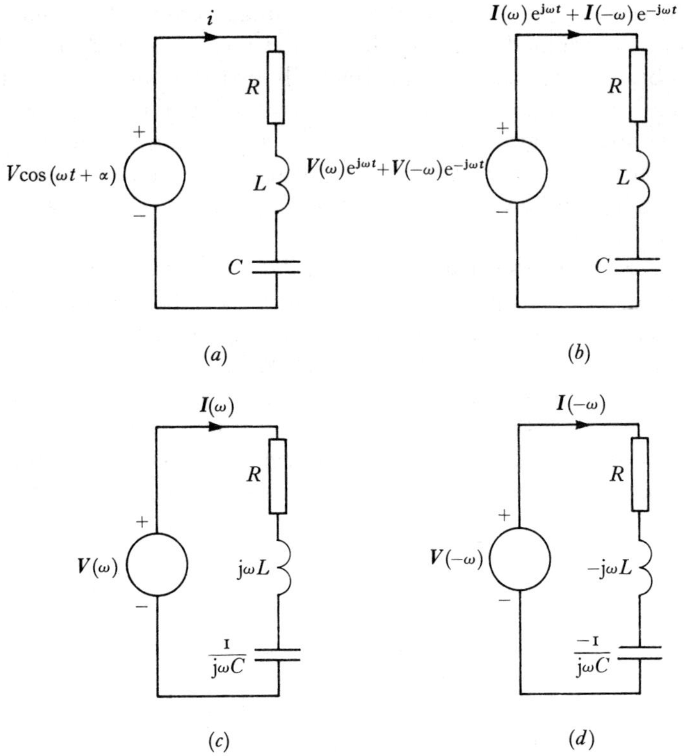

Fig. 11.2 (*a*) Original circuit. (*b*) Intermediate circuit. (*c*) and (*d*) Image circuits, positive and negative frequency.

where

$$\left.\begin{aligned} \boldsymbol{Y}(\omega) &= \frac{1}{R+j\omega L+1/j\omega C}, \\ \boldsymbol{Y}(-\omega) &= \frac{1}{R-j\omega L-1/j\omega C}. \end{aligned}\right\} \tag{11.3.4}$$

The final stage in the solution is to re-form the function of t by multiplying each component by its exponential:

$$i(t) = \boldsymbol{I}(\omega)\,e^{j\omega t} + \boldsymbol{I}(-\omega)\,e^{-j\omega t}. \tag{11.3.5}$$

In fig. 11.3 this procedure is illustrated by means of counter-rotating phasors. The lower half of this diagram is a mirror image of the upper half and adds little; nor, for that matter, does the second image circuit, fig. 11.2(*d*).

If the source had contained components at other frequencies, each could have been similarly resolved into a pair of counter-rotating phasors with the appropriate rotational speed. With a non-repetitive stimulus $v(t)$ there

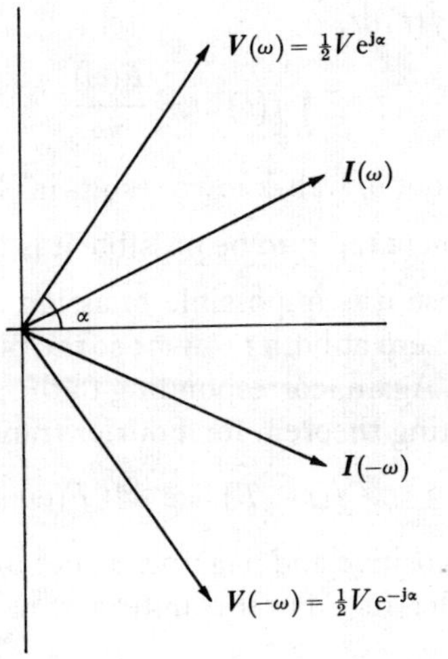

Fig. 11.3 Stationary phasor diagram derived from counter-rotating phasors.

would have been a continuous spectrum of frequencies; the formation of the Fourier transform,

$$\boldsymbol{V}(\omega) = \int_{-\infty}^{\infty} V(t)\, e^{-j\omega t}\, dt, \tag{11.3.6}$$

then amounts to the removal of the factor $e^{j\omega t}$ from each component, by multiplying it by $e^{-j\omega t}$. The component stimulus $\boldsymbol{V}(\omega)\,\delta\omega$ has a response $\boldsymbol{I}(\omega)\,\delta\omega$, where

$$\boldsymbol{I}(\omega) = \boldsymbol{Y}(\omega)\,\boldsymbol{V}(\omega), \tag{11.3.7}$$

and the final stage puts back the exponential factors through the inverse transform:

$$i(t) = \frac{1}{2\pi}\int_{-\infty}^{\infty} \boldsymbol{I}(\omega)\, e^{j\omega t}\, d\omega. \tag{11.3.8}$$

Despite the slight difference of approach in the use of counter-rotating phasors, the transfer function $\boldsymbol{Y}(\omega)$ is identical with the admittance of elementary phasor theory.

11.4 Properties of the Fourier transformation

Many theorems developed for the Laplace transformation have counterparts in the Fourier transformation, subject always to the proviso that the transforms exist. Thus, for the derivative $f'(t)$,

$$\mathscr{F} f'(t) = j\omega \boldsymbol{F}(\omega), \tag{11.4.1}$$

and for the integral $\int f(t)\,\mathrm{d}t$,

$$\mathscr{F}\int f(t)\,\mathrm{d}t = \frac{\boldsymbol{F}(\omega)}{\mathrm{j}\omega}. \tag{11.4.2}$$

These are generalizations of well-known results in elementary phasor theory. For the transformation (11.4.2) to be possible it is necessary that $\int_{-\infty}^{\infty} f(t)\,\mathrm{d}t$ shall be zero, otherwise it is impossible to assign a constant of integration which makes $\int f(t)\,\mathrm{d}t$ zero at both $t = \pm\infty$; so the positive and negative areas in $f(t)$ must be equal. Again, corresponding to the shifting theorem (10.8.1) there is the time-shifting theorem for Fourier transforms:

$$\mathscr{F} f(t-T) = \mathrm{e}^{-\mathrm{j}\omega T}\,\boldsymbol{F}(\omega). \tag{11.4.3}$$

This is true for both positive and negative values of T. Finally the convolution theorem (10.9.3) has its Fourier counterpart, which may be expressed in the forms

$$\left.\begin{aligned}\mathscr{F}^{-1}\,\boldsymbol{F}_1(\omega)\,\boldsymbol{F}_2(\omega) &= \int_{-\infty}^{\infty} f_1(\tau) f_2(t-\tau)\,\mathrm{d}\tau,\\ &= \int_{-\infty}^{\infty} f_1(t-\tau) f_2(\tau)\,\mathrm{d}\tau.\end{aligned}\right\} \tag{11.4.4}$$

A further set of properties may be deduced from the near-symmetry of the direct and inverse Fourier transformations. Thus, alongside the time-shifting theorem (11.4.3) we may lay a frequency-shifting theorem known as the *modulation theorem*:

$$\mathscr{F}^{-1}\,\boldsymbol{F}(\omega-\Omega) = \mathrm{e}^{\mathrm{j}\Omega t} f(t). \tag{11.4.5}$$

The complex function of t seems a fundamental innovation, but it is not; it leads to transforms for the real functions $\cos\Omega t\,.f(t)$ and $\sin\Omega t\,.f(t)$.

If the Fourier transformation is applied to $f(t/T)$, a function which is dimensionless in time, we obtain

$$\mathscr{F} f\left(\frac{t}{T}\right) = \int_{-\infty}^{\infty} f\left(\frac{t}{T}\right)\mathrm{e}^{-\mathrm{j}\omega t}\,\mathrm{d}t.$$

Let $t/T = t_1$, $\omega T = \omega_1$; then the integral becomes $T\int_{-\infty}^{\infty} f(t_1)\,\mathrm{e}^{-\mathrm{j}\omega_1 t_1}\,\mathrm{d}t_1$, or $T\boldsymbol{F}(\omega_1)$. In other words, if $\mathscr{F} f(t) = \boldsymbol{F}(\omega)$,

$$\mathscr{F} f\left(\frac{t}{T}\right) = T\boldsymbol{F}(\omega T). \tag{11.4.6}$$

This formula amounts to a change in the unit of time. It may be written

$$\boldsymbol{F}(\omega T) = \frac{1}{T}\int_{-\infty}^{\infty} f\left(\frac{t}{T}\right)\mathrm{e}^{-\mathrm{j}\omega t}\,\mathrm{d}t,$$

and the conjugate $\boldsymbol{F}^*(\omega T)$ is obtained by changing the sign of j:

$$\boldsymbol{F}^*(\omega T) = \frac{1}{T}\int_{-\infty}^{\infty} f\left(\frac{t}{T}\right) \mathrm{e}^{\mathrm{j}\omega t}\, dt.$$

If in this equation we interchange the dimensionless quantities t/T and ωT, we get

$$\boldsymbol{F}^*\left(\frac{t}{T}\right) = T\int_{-\infty}^{\infty} f(\omega T)\, \mathrm{e}^{\mathrm{j}\omega t}\, d\omega.$$

When this equation is compared with the inverse transform equation, (11.1.8), it is seen to mean that

$$\boldsymbol{F}^*\left(\frac{t}{T}\right) = \mathscr{F}^{-1}\, 2\pi T f(\omega T),$$

or, replacing T by $T/2\pi$,

$$\boldsymbol{F}^*\left(\frac{2\pi t}{T}\right) = \mathscr{F}^{-1}\, T f\left(\frac{\omega T}{2\pi}\right).$$

The notation $\boldsymbol{F}$ has hitherto always implied a function of ω. Now that the roles are being interchanged it is preferable to use a non-committal notation, and write the *duality theorem* just proved in this form:

$$\text{If} \qquad \mathscr{F} f\left(\frac{t}{T}\right) = T\cdot g(\omega T), \quad \text{then} \quad \mathscr{F} g^*\left(\frac{2\pi t}{T}\right) = T\cdot f\left(\frac{\omega T}{2\pi}\right). \qquad (11.4.7)$$

Each Fourier transform is thus seen to imply another, on account of the near-symmetry of the direct and inverse equations. However, for our purpose the functions of time must be real; and, as has been seen, a real transform is obtained when the original function is an even one. In practice, therefore, (11.4.7) is useful for deriving a second transform when the transform of an *even* function of t has been obtained. An example is given in the next section.

It is useful to set down explicitly the common features applying to the 'duality' of circuits and the 'duality' of Fourier transforms. Systems A, B are said to be duals of each other if the following statements are true:

(i) To each element of A there corresponds a single element of B, and vice versa.

(ii) If elements of type 1 in A correspond with elements of type 2 in B, then elements of type 2 in A correspond with elements of type 1 in B.

(iii) The relations between corresponding groups of elements in the two systems are expressed by mathematical equations of the same essential form. It will be found that this description covers both cases.

11.5 Examples of Fourier transforms

(a) Rectangular pulse signal

The function defined by

$$\left.\begin{aligned} f(t) &= 1 \quad (-T \leqslant 0 \leqslant T),\\ f(t) &= 0 \text{ outside these limits,} \end{aligned}\right\} \tag{11.5.1}$$

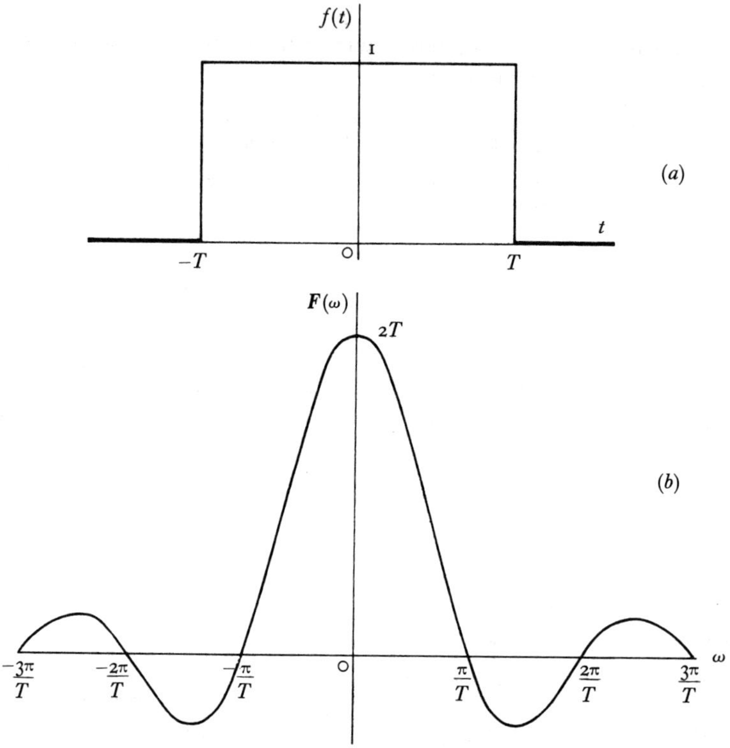

Fig. 11.4 (*a*) Rectangular pulse. (*b*) Its spectrum.

is a rectangular pulse (fig. 11.4(*a*)). Its Fourier transform is

$$\boldsymbol{F}(\omega) = \int_{-T}^{T} \mathrm{e}^{-\mathrm{j}\omega t}\,\mathrm{d}t,$$

$$= -\frac{1}{\mathrm{j}\omega}\Big[\mathrm{e}^{-\mathrm{j}\omega t}\Big]_{-T}^{T},$$

$$= 2T\left(\frac{\sin \omega T}{\omega T}\right). \tag{11.5.2}$$

This spectrum is illustrated in fig. 11.4(*b*). The connexion with the analysis given in §9.7, where the same curve was obtained as the envelope of the ordinates which formed the line spectrum of a train of pulses, is evident.

(*b*) *Signal having a rectangular spectrum*

The question asked here is what function of t is represented by a spectrum which is uniform but confined to the range $-\Omega \leqslant \omega \leqslant \Omega$ (fig. 11.5(*a*))? The

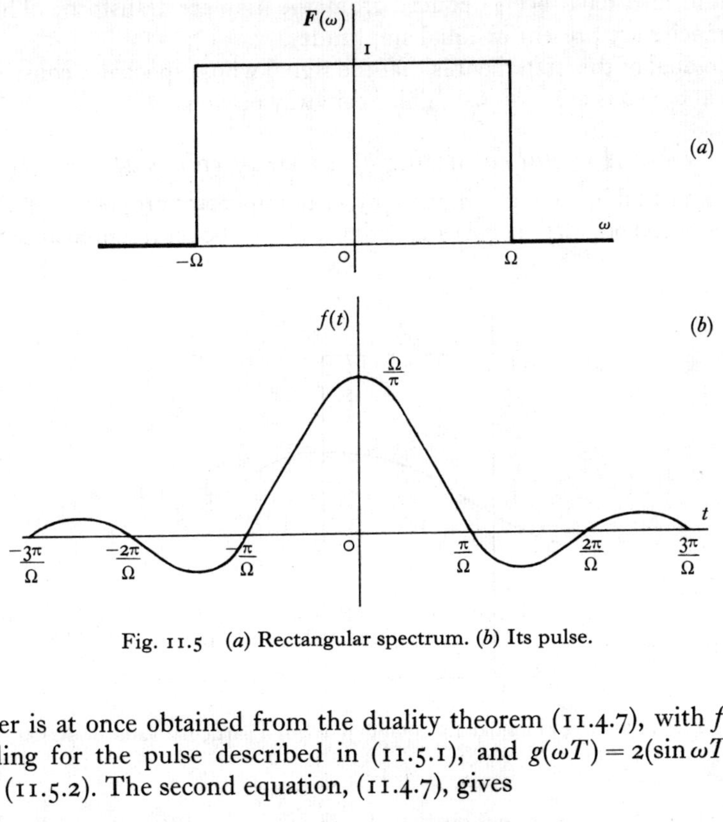

Fig. 11.5 (*a*) Rectangular spectrum. (*b*) Its pulse.

answer is at once obtained from the duality theorem (11.4.7), with $f(t/T)$ standing for the pulse described in (11.5.1), and $g(\omega T) = 2(\sin \omega T/\omega T)$ from (11.5.2). The second equation, (11.4.7), gives

$$\mathscr{F}^{-1} f\left(\frac{\omega T}{2\pi}\right) = \frac{2}{T}\frac{\sin(2\pi t/T)}{(2\pi t/T)}.$$

Writing Ω for $2\pi/T$, we obtain

$$\mathscr{F}^{-1} f\left(\frac{\omega}{\Omega}\right) = \frac{\Omega}{\pi}\left(\frac{\sin \Omega t}{\Omega t}\right). \tag{11.5.3}$$

Thus, as compared with fig. 11.4, fig. 11.5 shows the signal and its spectrum interchanged.

(c) Spike function signal

If the pulse described in (11.5.1) has a height $2T$ instead of unity, so that it has unit area, its transform becomes $\sin\omega T/\omega T$. As T tends to 0, the pulse tends to the unit spike function $\delta(t)$; therefore

$$\mathscr{F}\,\delta(t) = \lim_{T\to 0}\frac{\sin\omega T}{\omega T} = 1, \tag{11.5.4}$$

as could also have been deduced from the Laplace transform. Thus all frequencies are present in equal amplitude.

The dual of this statement is that the signal whose spectrum consists of a spike at $\omega = 0$ is a d.c. signal. This is anyway obvious.

(d) A signal consisting of a pair of spikes at $t = \pm t_1$

This is a suitable point at which to prove an important property of the spike function. Let $\delta(t - t_1)$, a spike occurring at $t = t_1$, be represented as the limit

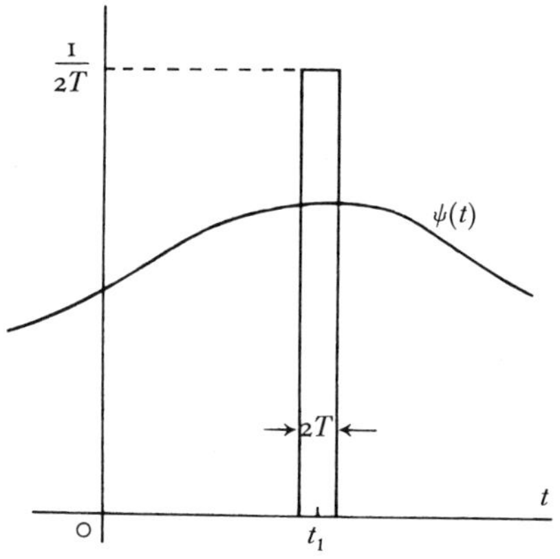

Fig. 11.6 Use of a spike function to pick out a particular value of $\psi(t)$.

of a rectangular pulse of height $1/2T$ lasting from $t_1 - T$ to $t_1 + T$ (fig. 11.6). If this function is called $\delta_T(t - t_1)$, so that

$$\delta(t - t_1) = \lim_{T\to 0}\delta_T(t - t_1), \tag{11.5.5}$$

and if $\psi(t)$ is any function of t, the product $\delta_T(t - t_1)\psi(t)$ approximates closely to $(1/2T)\psi(t_1)$ when T is small. Thus

$$\int_{-\infty}^{\infty}\delta_T(t - t_1)\,\psi(t)\,dt = \frac{1}{2T}\psi(t_1)\times 2T \text{ (nearly)};$$

and in the limit,

$$\int_{-\infty}^{\infty} \delta(t-t_1)\,\psi(t)\,\mathrm{d}t = \psi(t_1). \qquad (11.5.6)$$

This is true for positive or negative values of t_1, and for functions $\psi(t)$ which are complex as well as real.

A pair of spikes at $\mp t_1$, each of magnitude $\frac{1}{2}$, is represented by

$$f(t) = \tfrac{1}{2}\delta(t+t_1) + \tfrac{1}{2}\delta(t-t_1). \qquad (11.5.7)$$

The Fourier transform is

$$\int_{-\infty}^{\infty} \tfrac{1}{2}\{\delta(t+t_1)+\delta(t-t_1)\}\,\mathrm{e}^{-\mathrm{j}\omega t}\,\mathrm{d}t,$$

or

$$\boldsymbol{F}(\omega) = \tfrac{1}{2}(\mathrm{e}^{\mathrm{j}\omega t_1} + \mathrm{e}^{-\mathrm{j}\omega t_1}),$$

$$= \cos \omega t_1. \qquad (11.5.8)$$

Thus the spectrum of a symmetrical spike-pair is a cosine wave. The dual, namely that the spectrum of $\cos \omega_1 t$ consists of a pair of lines at frequencies $\pm\omega_1$, is already known (see (9.7.1)).

A signal consisting of a sum of cosines, like the amplitude-modulated wave

$$f(t) = \cos \omega_0 t + \frac{k}{2}\{\cos(\omega_0+\omega_m)t + \cos(\omega_0-\omega_m)t\}, \qquad \text{(equation (9.7.7))}$$

has a spectrum consisting of spikes at the various frequencies. This verifies the statement made but not proved in §9.7, that the spectrum of an amplitude-modulated wave contains a line at the carrier frequency ω_0 flanked by lines at frequencies $\omega_0 \pm \omega_m$, *whether or not* ω_0 *and* ω_m *are both integral multiples of the same fundamental frequency*. The frequencies $\omega_0 \pm \omega_m$ are known as *sideband* frequencies.

11.6 Applications of Fourier transform analysis

Just as in Laplace transform analysis, the characteristics of the network to which the signal $\boldsymbol{F}(\omega)$ is applied are defined by the transfer function $\boldsymbol{H}(\omega)$; the response of the network is then given by the inverse Fourier transform,

$$\mathscr{F}^{-1}\,\boldsymbol{F}(\omega)\,\boldsymbol{H}(\omega) = \frac{1}{2\pi}\int_{-\infty}^{\infty} \boldsymbol{F}(\omega)\,\boldsymbol{H}(\omega)\,\mathrm{e}^{\mathrm{j}\omega t}\,\mathrm{d}\omega\,. \qquad (11.6.1)$$

It may be noted to begin with that the property of the spike function, of containing all frequencies in equal amplitude, is implicit in this equation. A spike function stimulus disturbs a circuit for an infinitesimal time and then lets it alone; thus the nature of the response at subsequent instants must be determined by the circuit only. This means that, for such a stimulus, the form of the function of t defined by the integral (11.6.1) must depend upon $\boldsymbol{H}(\omega)$ only; this can only be the case if $\boldsymbol{F}(\omega)$ is a constant.

In general $\boldsymbol{F}(\omega)$ and $\boldsymbol{H}(\omega)$ are complex functions of ω, so that a pair of diagrams is required to describe either of them graphically. (The examples considered in §11.5 were all even functions of t, so that the spectra were real

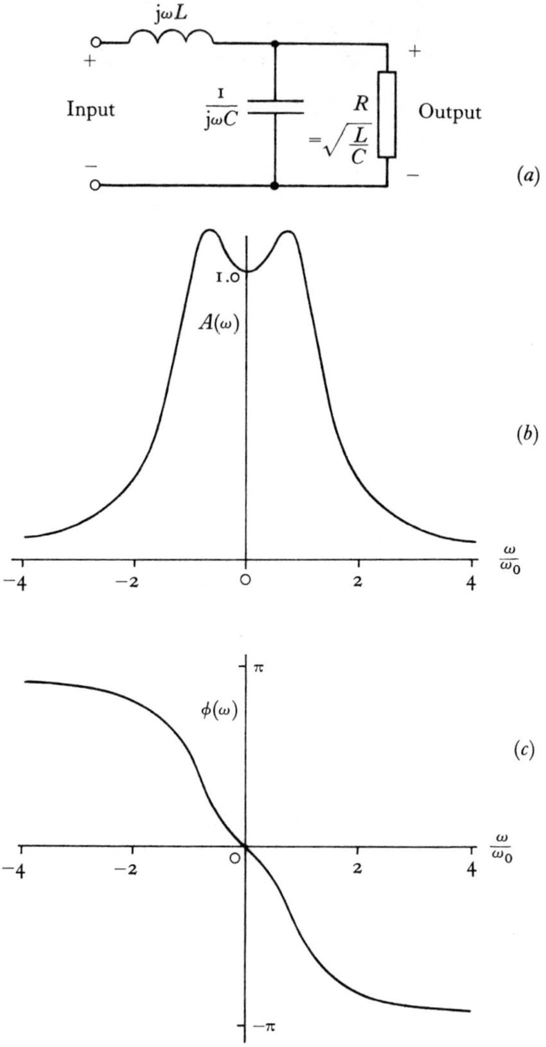

Fig. 11.7 (*a*) Simple low-pass filter. (*b*) Amplitude characteristic of (*a*). (*c*) Phase characteristic of (*a*).

and only one diagram was needed.) The diagrams might show the real and imaginary parts; but for the transfer function it is particularly informative to use the polar form

$$\boldsymbol{H}(\omega) = A(\omega)\,\mathrm{e}^{\mathrm{j}\phi(\omega)}, \tag{11.6.2}$$

so that $A(\omega)$ is the ratio of the amplitudes of the input and output for a periodic signal at frequency ω, and $\phi(\omega)$ is the phase difference. Fig. 11.7 shows $A(\omega)$ (*the amplitude characteristic*) and $\phi(\omega)$ (*the phase characteristic*) for a simple circuit which acts as a low-pass filter, since the amplitude ratio equals or exceeds 1 at low frequencies but falls towards zero at high frequencies. The fact that $A(\omega)$ is an even function and $\phi(\omega)$ an odd function of ω is always true, being a consequence of the fact that $\boldsymbol{H}(\omega)$ is a function of $\mathrm{j}\omega$; thus the simultaneous reversal of the signs of j and ω leaves $\boldsymbol{H}(\omega)$ unchanged, and

$$A(\omega)\,\mathrm{e}^{\mathrm{j}\phi(\omega)} = A(-\omega)\,\mathrm{e}^{-\mathrm{j}\phi(-\omega)},$$

whence
$$A(-\omega) = A(\omega), \qquad \phi(-\omega) = -\phi(\omega). \tag{11.6.3}$$

There are many problems in which the transfer function, to a base of frequency, is of much more interest than the calculation of the response as a function of time. The applied stimulus often consists of a mixture of sine waves which do not even add up to a repetitive pattern unless their frequencies are commensurate; the transfer function tells us what happens to each component, and this is precisely what we wish to know. When the response to a stimulus suddenly applied at $t=0$ is required as a function of time, it will be found that the Laplace transform method (as set forth in chapters 5 and 10) is much simpler than the use of the inverse Fourier transform expressed in the integral (11.6.1).

This will be illustrated by applying both methods to the circuit of fig. 11.7, when excited by a voltage pulse of magnitude V_1 and duration T, starting at $t=0$. The Laplace transfer function of the circuit is found to be

$$\boldsymbol{H}(\mathrm{s}) = \frac{\omega_0^2}{\mathrm{s}^2+\omega_0\,\mathrm{s}+\omega_0^2}, \tag{11.6.4}$$

where
$$\omega_0 = (LC)^{-1/2}. \tag{11.6.5}$$

From $t=0$ to T, the response $v(t)$ is the same as the response to a step function, of which the image is V_1/s. Thus

$$\begin{aligned} v(t) &= \mathscr{L}^{-1}\frac{\omega_0^2 V_1}{\mathrm{s}(\mathrm{s}^2+\omega_0\,\mathrm{s}+\omega_0^2)},\\ &= V_1\,\mathscr{L}^{-1}\left\{\frac{1}{\mathrm{s}}-\frac{\mathrm{s}+\omega_0}{\mathrm{s}^2+\omega_0\,\mathrm{s}+\omega_0^2}\right\},\\ &= V_1\,\mathscr{L}^{-1}\left\{\frac{1}{\mathrm{s}}-\frac{\mathrm{s}+\omega_0}{(\mathrm{s}+\omega_0/2)^2+(\omega_0\sqrt{3}/2)^2}\right\},\\ &= V_1\,u(t)\left\{1-\mathrm{e}^{-\omega_0 t/2}\left(\cos\frac{\sqrt{3}\omega_0 t}{2}+\frac{1}{\sqrt{3}}\sin\frac{\sqrt{3}\omega_0 t}{2}\right)\right\}. \end{aligned} \tag{11.6.6}$$

From $t = T$ onwards the effect of the negative step must be superimposed; this is

$$v'(t) = -V_1 u(t-T)\left\{1 - e^{-\omega_0(t-T)/2}\left(\cos\frac{\sqrt{3}\omega_0(t-T)}{2} + \frac{1}{\sqrt{3}}\sin\frac{\sqrt{3}\omega_0(t-T)}{2}\right)\right\}. \qquad (11.6.7)$$

The numerical computation for equation (11.6.6) also serves for (11.6.7); the stimulus and response are shown in fig. 11.8.

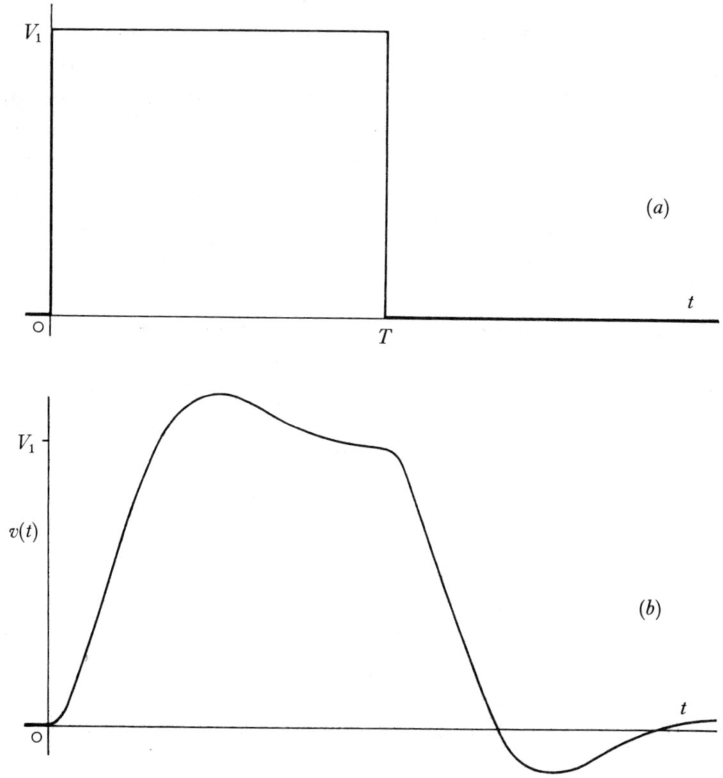

Fig. 11.8 Stimulus (*a*) and response (*b*), in the circuit of fig. 11.7.

To perform the same calculation with the Fourier transform, we write $j\omega$ for s in (11.6.4) to obtain the transfer function:

$$\boldsymbol{H}(\omega) = \frac{\omega_0^2}{(\omega_0^2 - \omega^2) + j\omega_0\omega}. \qquad (11.6.8)$$

Also the image of the stimulus is

$$\boldsymbol{V}_1(\omega) = \int_0^T V_1 \,\mathrm{e}^{-\mathrm{j}\omega t}\,\mathrm{d}t,$$
$$= \frac{V_1}{\mathrm{j}\omega}(1 - \mathrm{e}^{-\mathrm{j}\omega T}). \qquad (11.6.9)$$

Then

$$v(t) = \mathscr{F}^{-1}\,\boldsymbol{V}_1(\omega)\,\boldsymbol{H}(\omega)$$
$$= \frac{V_1}{2\pi}\int_{-\infty}^{\infty} \frac{\omega_0^2(1 - \mathrm{e}^{-\mathrm{j}\omega T})\,\mathrm{e}^{\mathrm{j}\omega t}}{\mathrm{j}\omega\{(\omega_0^2 - \omega^2) + \mathrm{j}\omega_0\,\omega\}}\,\mathrm{d}\omega. \qquad (11.6.10)$$

It would require some recondite mathematics to prove that this integral was equal to the function defined by equations (11.6.6,7).

Suppose now that the same circuit is excited by a stimulus given by

$$v_1(t) = V_1 \cos(\omega_1 t + \phi_1 \sin\omega_m t), \qquad (11.6.11)$$

where ϕ_1 is small compared with π. This is a cosine wave with a small periodic variation applied to its phase; it is said to be *phase-modulated*. Since both $\omega_1 t$ and $\phi_1 \sin\omega_m t$ are odd functions, $v_1(t)$ is an even function and its Fourier transform will be real. It is found by writing

$$v_1(t) = V_1\{\cos\omega_1 t \cos(\phi_1 \sin\omega_m t) - \sin\omega_1 t \sin(\phi_1 \sin\omega_m t)\}$$
$$= V_1\left\{\cos\omega_1 t\left(1 - \frac{\phi_1^2}{2}\sin^2\omega_m t + \ldots\right) - \sin\omega_1 t\left(\phi_1 \sin\omega_m t - \frac{\phi_1^3}{6}\sin^3\omega_m t + \ldots\right)\right\}. \qquad (11.6.12)$$

Retaining only the terms up to order ϕ_1, we get

$$v_1(t) = V_1\{\cos\omega_1 t - \phi_1 \sin\omega_1 t \sin\omega_m t\},$$
$$= V_1\{\cos\omega_1 t + \tfrac{1}{2}\phi_1 \cos(\omega_1 + \omega_m)t - \tfrac{1}{2}\phi_1\cos(\omega_1 - \omega_m)t\}. \qquad (11.6.13)$$

The retention of more terms in (11.6.12) would introduce further sideband frequencies $\omega_1 \pm 2\omega_m$, $\omega_1 \pm 3\omega_m$, etc. The Fourier transform is

$$\boldsymbol{V}_1(\omega) = \pi V_1\{\delta(\omega - \omega_1) + \delta(\omega + \omega_1) + \tfrac{1}{2}\phi_1[\delta(\omega - \omega_1 - \omega_m) + \delta(\omega + \omega_1 + \omega_m) - \delta(\omega - \omega_1 + \omega_m) - \delta(\omega + \omega_1 - \omega_m)]\}. \qquad (11.6.14)$$

The signal and the spectrum are both shown in fig. 11.9, in which the length of each line in the spectrum is proportional to the area of the spike function which it represents.

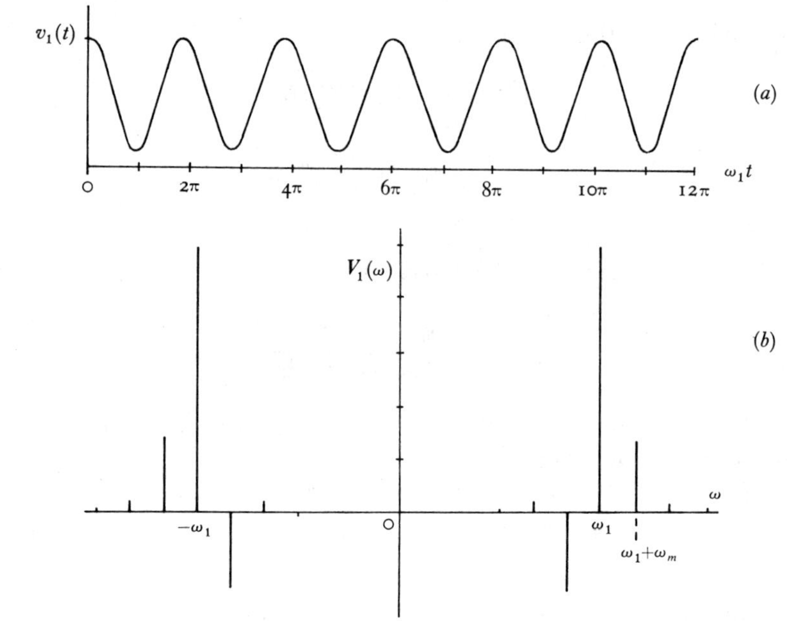

Fig. 11.9 A phase-modulated wave (*a*), with its spectrum (*b*).

The response of the circuit to each frequency component is now found by the application of (11.6.1), $\boldsymbol{H}(\omega)$ being given by (11.6.8). For example, the response to the sidebands of frequency $\pm\omega_s$, where

$$\omega_s = \omega_1 + \omega_m \tag{11.6.15}$$

is given by

$$v(t) = \frac{1}{2\pi}\int_{-\infty}^{\infty} \frac{\pi}{2} V_1 \phi_1 [\delta(\omega-\omega_s)+\delta(\omega+\omega_s)] \frac{\omega_0^2 \, e^{j\omega t} \, d\omega}{(\omega_0^2-\omega^2)+j\omega_0\omega}.$$

It was shown in equation (11.5.6) that

$$\int_{-\infty}^{\infty} \delta(\omega-\omega_s)\,\boldsymbol{F}(\omega)\,d\omega = \boldsymbol{F}(\omega_s). \tag{11.6.16}$$

Hence

$$v(t) = \frac{V_1\phi_1\omega_0^2}{4}\left\{\frac{e^{j\omega_s t}}{(\omega_0^2-\omega_s^2)+j\omega_0\omega_s} + \frac{e^{-j\omega_s t}}{(\omega_0^2-\omega_s^2)-j\omega_0\omega_s}\right\},$$

$$= \frac{V_1\phi_1\omega_0^2}{4[(\omega_0^2-\omega_s^2)^2+\omega_0^2\omega_s^2]}\cdot\{(\omega_0^2-\omega_s^2)(e^{j\omega_s t}+e^{-j\omega_s t})$$

$$-j\omega_0\omega_s(e^{j\omega_s t}-e^{-j\omega_s t})\},$$

$$= \frac{V_1\phi_1\omega_0^2\{(\omega_0^2-\omega_s^2)\cos\omega_s t+\omega_0\omega_s\sin\omega_s t\}}{2\{(\omega_0^2-\omega_s^2)^2+\omega_0^2\omega_s^2\}}. \tag{11.6.17}$$

Thus the magnitude and phase of each component in the response is found, assuming that the signal has been operative for an infinite time. In such a problem the Fourier transform comes into its own, since the Laplace transform is geared to the calculation of the transient effects immediately after the signal has been switched on, and these are irrelevant here.

Another type of problem in which the Fourier transform is required is in determining the response of a circuit for which the transfer function is known as a function of ω, while the component elements of the circuit are not known. Thus, by adding more elements to the simple low-pass filter of fig. 11.7, the cutoff in the amplitude characteristic can be made more abrupt, and the characteristic tends towards the ideal described by

$$\begin{aligned} A(\omega) &= 1 \quad \text{when} \quad |\omega| \leqslant \omega_c, \\ A(\omega) &= 0 \quad \text{when} \quad |\omega| > \omega_c, \end{aligned} \tag{11.6.18}$$

ω_c being the cutoff frequency (fig. 11.10). We shall investigate the response of a filter possessing such a characteristic to a square pulse suddenly applied,

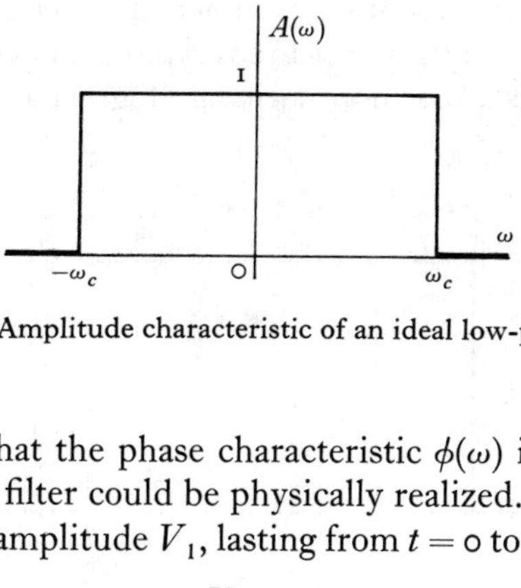

Fig. 11.10 Amplitude characteristic of an ideal low-pass filter.

on the assumption that the phase characteristic $\phi(\omega)$ is zero, and without knowing how such a filter could be physically realized.

A voltage pulse of amplitude V_1, lasting from $t = 0$ to T, has the spectrum

$$\boldsymbol{V}_1(\omega) = \frac{V_1}{\mathrm{j}\omega}(1 - \mathrm{e}^{-\mathrm{j}\omega T}). \qquad \text{(equation (11.6.9))}$$

Thus the response is

$$v(t) = \frac{1}{2\pi}\int_{-\infty}^{\infty} \boldsymbol{V}_1(\omega)\,\boldsymbol{H}(\omega)\,\mathrm{e}^{\mathrm{j}\omega t}\,\mathrm{d}\omega,$$

$$= \frac{1}{2\pi}\int_{-\omega_c}^{\omega_c} \boldsymbol{V}_1(\omega)\,\mathrm{e}^{\mathrm{j}\omega t}\,\mathrm{d}\omega,$$

$$= \frac{V_1}{2\pi}\int_{-\omega_c}^{\omega_c} \frac{\mathrm{e}^{\mathrm{j}\omega t} - \mathrm{e}^{\mathrm{j}\omega(t-T)}}{\mathrm{j}\omega}\,.\,\mathrm{d}\omega. \tag{11.6.19}$$

Also

$$\int_{-\omega_c}^{\omega_c} \frac{e^{j\omega t}}{j\omega} . d\omega = \int_{-\omega_c}^{\omega_c} \frac{\cos \omega t + j \sin \omega t}{j\omega} . d\omega,$$

and the cosine contributes nothing because $\cos \omega t/\omega$ is an odd function and the limits are symmetrical with respect to zero. Therefore

$$\int_{-\omega_c}^{\omega_c} \frac{e^{j\omega t}}{j\omega} . d\omega = \int_{-\omega_c}^{\omega_c} \frac{\sin \omega t}{\omega} . d\omega. \tag{11.6.20}$$

The function

$$\operatorname{Si} x = \int_0^x \frac{\sin \xi}{\xi} d\xi \tag{11.6.21}$$

is tabulated,† and can be shown to equal $\pm\pi/2$ at $x = \pm\infty$. It follows from equations (11.6.20,21) that (11.6.19) may be written

$$v(t) = \frac{V_1}{\pi} \{\operatorname{Si} \omega_c t - \operatorname{Si} \omega_c(t - T)\}. \tag{11.6.22}$$

The stimulus and the response are shown in fig. 11.11; the resemblance to fig. 11.8, which shows the response to a very simple low-pass filter, is evident, but the difference is even more striking. Fig. 11.11 shows the response

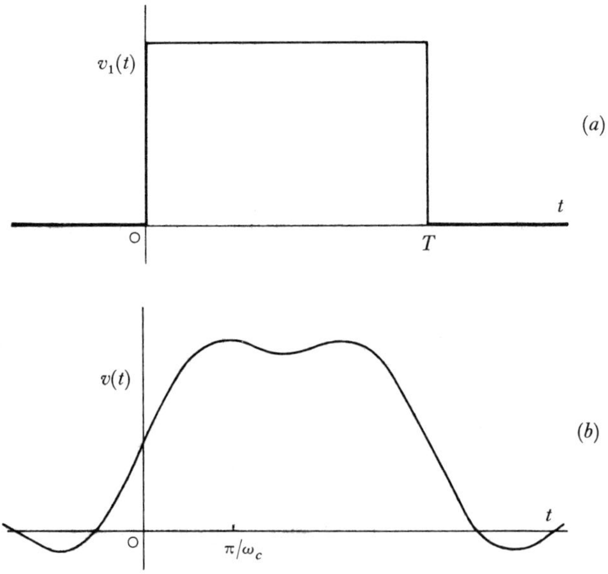

Fig. 11.11 Stimulus (*a*) and response (*b*), for an ideal low-pass filter.

† See, for example, J. B. Dale, *Five-figure tables of mathematical functions*, Arnold.

starting before the stimulus – the filter is credited with the gift of prophecy. It follows that a filter with the characteristics postulated cannot be realized in practice.

The filter is brought a step nearer to what is practicable by assuming a non-zero phase characteristic. Thus, if

$$\begin{aligned} \boldsymbol{H}(\omega) &= \mathrm{e}^{-\mathrm{j}\pi\omega/\omega_c}, \quad \text{when} \quad |\omega| \leqslant \omega_c, \\ &= 0, \qquad\qquad \text{when} \quad |\omega| > \omega_c, \end{aligned} \tag{11.6.23}$$

there is a uniform change of phase from $+\pi$ to $-\pi$ over the frequency band $-\omega_c$ to $+\omega_c$. The response is now

$$v(t) = \frac{V_1}{2\pi}\int_{-\omega_c}^{\omega_c} \frac{\mathrm{e}^{\mathrm{j}\omega(t-\pi/\omega_c)} - \mathrm{e}^{\mathrm{j}\omega(t-T-\pi/\omega_c)}}{\mathrm{j}\omega}.\,\mathrm{d}\omega$$

$$= \frac{V_1}{\pi}\{\mathrm{Si}\,(\omega_c t - \pi) - \mathrm{Si}\,(\omega_c(t-T) - \pi)\}. \tag{11.6.24}$$

The whole response curve is thus shifted to the right, so that the negative maximum which appears at $t = -\pi/\omega_c$ in fig. 11.11(*b*) now occurs at $t = 0$; but the anticipatory transient, though reduced, is not eliminated.

Examples on chapter 11

11.1 Find the Fourier transform of a single isosceles triangular pulse of height 1 and base extending from $t = -T/2$ to $T/2$.

11.2 Find the Fourier transform of a half cosine wave of height 1 with its base extending from $t = -T/2$ to $T/2$.

11.3 Find the Fourier transform of the function $\mathrm{e}^{-\alpha t}u(t)$, and express the result in polar form. Find also the transform of the symmetrical pulse $\mathrm{e}^{-\alpha|t|}$ obtained by adding to the above function its reflexion in the axis $t = 0$.

11.4 Find the Fourier transform of the Gaussian function $\mathrm{e}^{-t^2/2T^2}$, proving that the time and frequency representations are essentially the same.

$$\left[\int_{-\infty}^{\infty} \mathrm{e}^{-x^2}\cos\lambda x\,\mathrm{d}x = \sqrt{\pi}\,\mathrm{e}^{-\lambda^2/4}.\right]$$

11.5 Prove that the function which consists of a positive unit spike at $t = -T/2$ and a negative unit spike at $t = T/2$ has the Fourier transform

$2j\sin\omega T/2$. Hence obtain the transforms of the following, each derived from the integral of its predecessor:

(i) A rectangular pulse of unit area, defined by

$$f(t) = 1/T \text{ when } |t| \leqslant T/2 \qquad \text{and} \qquad f(t) = 0 \text{ when } |t| > T/2.$$

(ii) A pair of saw-teeth of unit area, defined by

$$f(t) = -8t/T^2 \text{ when } |t| \leqslant T/2 \qquad \text{and} \qquad f(t) = 0 \text{ when } |t| > T/2.$$

(iii) A parabolic pulse of unit area, defined by

$$f(t) = 3(T^2 - 4t^2)/2T^3 \text{ when } |t| \leqslant T/2$$

and

$$f(t) = 0 \text{ when } |t| > T/2.$$

11.6 Apply the method of example 11.5 to find the Fourier transform of a trapezoidal pulse which rises linearly from 0 to 1 over the range $-T/2 \leqslant t \leqslant -T/6$, remains at 1 for $-T/6 \leqslant t \leqslant T/6$, and falls linearly to zero over $T/6 \leqslant t \leqslant T/2$.

11.7 A filter has the transfer function $A(\omega)e^{j\phi(\omega)}$, where $A(\omega) = A$ (a constant) and $\phi(\omega) = -\omega T$ over the range $|\omega| \leqslant \omega_c$, while $A(\omega) = 0$ for $|\omega| > \omega_c$. Prove that if the spectrum of a signal $f(t)$ contains only frequencies less than ω_c, it will be transmitted without distortion.

11.8 A filter has the same phase characteristic as that considered in example 11.7, and is excited by a signal of the same type; but the amplitude characteristic is given by $A(\omega) = A[1 + a\cos(\pi\omega/\omega_c)]$ for $|\omega| \leqslant \omega_c$, where $a < 1$, and $A(\omega) = 0$ for $|\omega| > \omega_c$. Prove that the output now consists of the sum of a signal similar to the input and a pair of smaller echoes, also similar to the input but displaced in time from the main signal.

11.9 The Fourier transform of an amplitude-modulated carrier wave consists of two equal rectangular pulses, each of width $2\omega_1$, centred at $\omega = \pm\omega_0$. Determine the waveform of the modulation signal assuming that the carrier is cosinusoidal.

11.10 A low-pass filter consists of a Π network with an inductor of $\frac{2}{3}$ mH in the series branch; the branch in parallel with the input is a capacitor of 0.003 μF, that in parallel with the output is a capacitor of 0.001 μF. The load is a resistor of 500 Ω. Find the transfer function $A(\omega)e^{j\phi(\omega)}$ between currents in the input and in the load, and obtain an integral for the response to a spike function current. Evaluate this by solving the Laplace transform of the network.

11.11 A band-pass filter consists of a Π network, in which the series branch contains an inductor of 2.222 mH in series with a capacitor of 0.2812 μF; in parallel with the input is an inductor of 2.5 mH and also a capacitor of 0.25 μF; in parallel with the output is an inductor of 7.5 mH and also a capacitor of 0.0833 μF. The load is a resistor of 100 Ω. Find the transfer function between currents in the input and output, showing that the amplitude ratio is a maximum when $\omega = 4 \times 10^4$. Obtain in integral form the response to a rectangular current pulse of height 1 A and duration 10^{-3} s.

12 The complex plane in circuit analysis and synthesis

12.1 Impedance and admittance loci

In chapter 7 the complex frequency $\sigma + j\omega$ was introduced, not as a quantity of physical significance, but as a conceptual device which made possible the representation of the characteristics of a circuit by plotting its poles and zeros on the s-plane. It was noted that a variation in the frequency from zero to infinity is represented by a passage along the imaginary axis from the origin to infinity; if, following the idea discussed in chapter 11, negative frequencies are permitted, the negative half of the imaginary axis is included in the frequency locus.

In the present chapter the significant circuit characteristic – its impedance, admittance or transfer function – will itself be represented on a complex plane, and its properties will be explored in terms of related loci in this plane and in the s-plane. This approach is quite a general one, leading to far-reaching conclusions on such questions as the assessment of the stability of circuits and the realizability of prescribed characteristics with practical circuit elements. Two simple examples will serve as an introduction.

Consider first an inductor L and resistor R in series. The impedance at any frequency ω is represented by the complex number

$$\boldsymbol{Z} \equiv Z_R + jZ_I = R + j\omega L. \qquad (12.1.1)$$

In fig. 12.1 the line PQ is the locus of the tip of the radius which represents $\boldsymbol{Z}$ on an Argand diagram; a frequency scale, in terms of some standard frequency ω_0, has been drawn upon it. The dotted extension PQ′ relates to negative frequencies. The line is called the *impedance locus* of the circuit. Each point on it corresponds to a particular frequency, and therefore to a particular point on the imaginary axis in the s-plane.

Suppose now that complex frequencies are permitted. The impedance becomes

$$\boldsymbol{Z} = R + Ls, \qquad (12.1.2)$$

and if expressed in real and imaginary parts,

$$Z_R = R + L\sigma, \quad Z_I = L\omega. \qquad (12.1.3)$$

The relation between $\boldsymbol{Z}$ and s is expressed graphically in fig. 12.2, which shows the $\boldsymbol{Z}$-plane and the s-plane side by side. To each point such as P in

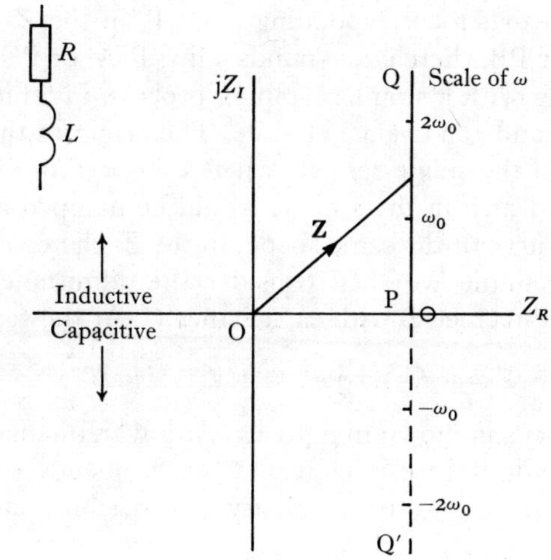

Fig. 12.1 Impedance locus of series *RL* circuit.

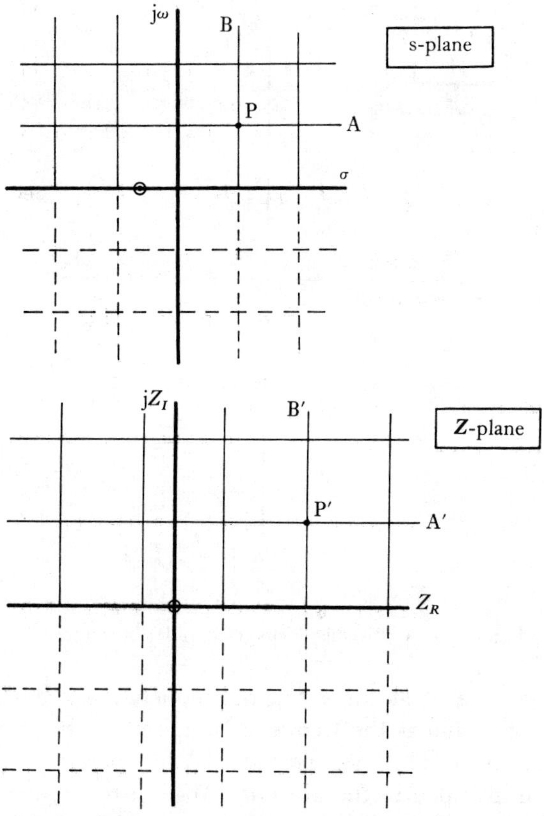

Fig. 12.2 Related s and ***Z*** planes for series *RL* circuit.

the s-plane there exists a corresponding point P′ in the $\boldsymbol{Z}$-plane. To each line, such as PA or PB, there corresponds a line P′A′ or P′B′. It is evident that we are dealing with a simple mapping process, in which R denotes a shift of the origin and L a change of scale. The origin in the $\boldsymbol{Z}$-plane is of course the 'map' of the single zero shown at $s = -R/L$ in the s-plane. Any continuous curve drawn in the s-plane would be mapped as a continuous curve (actually a curve of the same shape) in the $\boldsymbol{Z}$-plane.

As a second example we shall consider the admittance of a parallel combination of inductance L with capacitance C. At the real frequency ω,

$$\boldsymbol{Y} = j\omega C + 1/j\omega L = j(\omega C - 1/\omega L). \qquad (12.1.4)$$

The admittance locus is shown in fig. 12.3. At low frequencies the circuit is inductive and its admittance is large. As the frequency is increased, the point on the locus moves up the imaginary axis, reaching the origin (where

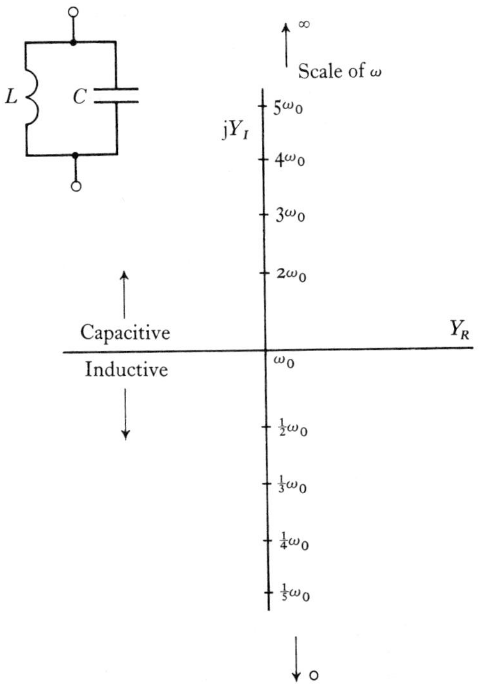

Fig. 12.3 Admittance locus for parallel LC circuit.

the admittance is zero) at the resonant frequency ω_0. The circuit now becomes capacitive, and as the frequency is raised further the locus point moves towards $+\infty$ on the imaginary axis. A few points of the frequency scale have been included in the diagram. At negative frequencies the sign of $\boldsymbol{Y}_I$ would be reversed and the same locus retraced.

When complex frequencies are permitted,

$$\boldsymbol{Y} = C(\sigma + \mathrm{j}\omega) + \frac{1}{L(\sigma + \mathrm{j}\omega)},$$

from which

$$\boldsymbol{Y}_R = C\sigma\left(1 + \frac{\omega_0^2}{\sigma^2 + \omega^2}\right), \quad \boldsymbol{Y}_I = C\omega\left(1 - \frac{\omega_0^2}{\sigma^2 + \omega^2}\right). \qquad (12.1.5)$$

Each point in the s-plane is denoted by co-ordinates (σ, ω), and there is a related point in the $\boldsymbol{Y}$-plane given by (12.1.5); we are therefore again dealing with a mapping process, but the map, which is illustrated in fig. 12.4, is

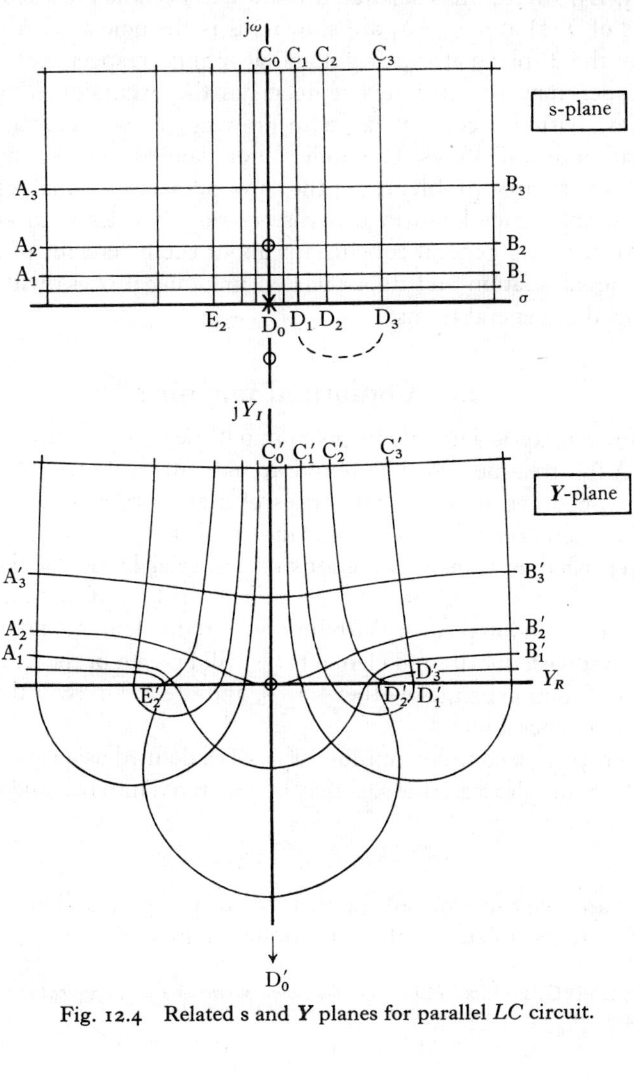

Fig. 12.4 Related s and $\boldsymbol{Y}$ planes for parallel LC circuit.

fairly complicated. To begin with, the whole $\boldsymbol{Y}$-plane is required for mapping only half of the s-plane; negative values for ω would demand a second sheet. Secondly the pattern has been fantastically distorted, as though the rectilinear diagram in the s-plane had been drawn on a rubber sheet which was then pulled quite out of shape. The straight lines A_1B_1, A_2B_2, A_3B_3, corresponding to three values of ω, become curved into $A'_1B'_1$, $A'_2B'_2$, $A'_3B'_3$. The axis C_0D_0 ($\sigma = 0$) remains straight, but becomes the line $C'_0D'_0$ extending from $+\infty$ to $-\infty$. The straight lines C_1D_1, C_2D_2, C_3D_3, corresponding to three values of σ, become curved into $C'_1D'_1$, $C'_2D'_2$, $C'_3D'_3$; and the points D'_1, D'_3 coincide, so that $C'_1D'_1D'_3C'_3$ looks like a single continuous curve. The distortion is least at the top of the diagram, and greatest near the points D'_2, E'_2, where lines separated elsewhere become crowded together. The zeros of $\boldsymbol{Y}(s)$ at $s = \pm j\omega_0$ are shown, as is the pole at $s = 0$; these are mapped in the $\boldsymbol{Y}$-plane at the origin and at infinity respectively.

The reader may be pardoned if he views the excursion into complex frequencies, with its consequent map-drawing, as a somewhat wanton multiplication of difficulties. It is indeed not claimed that the map throws light on this particular problem; but the concept of a functional relationship between complex members (in this case s and $\boldsymbol{Y}(s)$) leads to a powerful method for drawing general conclusions about the behaviour of networks. The functional relation and its associated mapping process will therefore be developed in general terms.

12.2 Conformal mapping

The required analysis will only be given in outline with a restricted purpose in mind. A full treatment will be found in mathematical texts.†

We are concerned with the properties of a network as expressed by its transfer function $\boldsymbol{H}(s)$; the impedance and admittance of a oneport network may be regarded as transfer functions of a special kind. All the transfer functions so far discussed have been rational algebraic functions; that is, ratios of two polynomials in s. With lumped circuit elements no other form can arise; but with distributed elements (as will be seen in the next chapter) the transfer function can take other forms, and the theory will therefore be developed in general terms.

Consider, then, a complex number $w = w(s)$, defined as a function of the independent complex variable s. w may be separated into real and imaginary parts:

$$w(s) = u(\sigma, \omega) + jv(\sigma, \omega), \tag{12.2.1}$$

where u, v are real functions of the two real variables σ, ω. They cannot be just any functions; related as they are to the original function w(s), they are

† For example, E. G. Phillips, *Functions of a complex variable with applications*, Oliver and Boyd (1945).

functions of a special kind known as *conjugate functions*. Equations (12.1.3) and (12.1.5) give two examples.

The derivative of w(s) at the point s is defined, as with functions of a real variable, by the equation

$$w'(s) = \lim_{\delta s \to 0} \frac{w(s+\delta s) - w(s)}{\delta s}. \tag{12.2.2}$$

A function is said to be *differentiable* when this limit exists, and if the function is one-valued as well as differentiable it is said to be *regular*. Differentiability is a fairly stringent condition in the complex plane, since the point representing $(s+\delta s)$ may lie in any direction with respect to s (fig. 12.5),

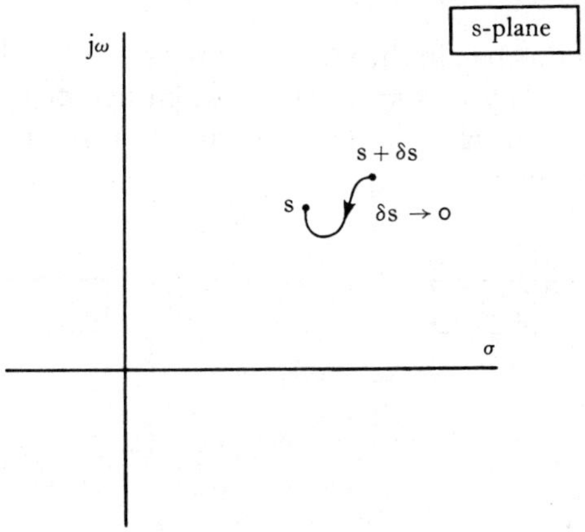

Fig. 12.5 Differentiation in the complex plane.

and the limit can only be said to exist if it is independent of the path by which δs approaches zero. A large number of functions, including all the transfer functions of linear systems, are differentiable except at isolated points; these points are known as *singularities*; and are in fact the poles of the transfer functions. Since transfer functions are also one-valued (in elementary cases palpably so, in other cases capable of being defined so that they are so), they are regular functions except at their singularities.

When the real and imaginary parts of a regular function w(s) are plotted, we get a map of the s-plane, in the sense already described. The w-plane may have to be used more than once to map the s-plane completely – that is, a single point in the w-plane may be the map of more than one point in the s-plane; but the reverse is never the case – each point in the s-plane has a unique map, because a regular function is one-valued.

Let P, Q be points close together in the s-plane, denoted by s, $s+\delta s$, where

$$\delta s = \delta l \, e^{j\alpha}, \tag{12.2.3}$$

(see fig. 12.6). The maps P′, Q′ are the points w(s), w(s + δs), and

$$\delta w = w(s+\delta s) - w(s),$$
$$= w'(s)\,\delta s, \qquad \text{to the first order.}$$

Let

$$w'(s) = M e^{j\gamma}, \tag{12.2.4}$$

where M is the modulus and γ the argument or phase angle of w′(s). Then, if

$$\delta w = \delta\lambda \, e^{j\beta}, \tag{12.2.5}$$

we obtain

$$\delta\lambda \, e^{j\beta} = M e^{j\gamma} \, \delta l \, e^{j\alpha},$$

or

$$\delta\lambda = M\,\delta l, \qquad \beta = \alpha + \gamma + 2n\pi. \tag{12.2.6}$$

M gives the scale of the map in the vicinity of the point P′; all displacements from P are multiplied by this same factor when mapped. γ is an angular rotation of PQ, and this again is the same for all displacements; the term $2n\pi$ does not affect the map.

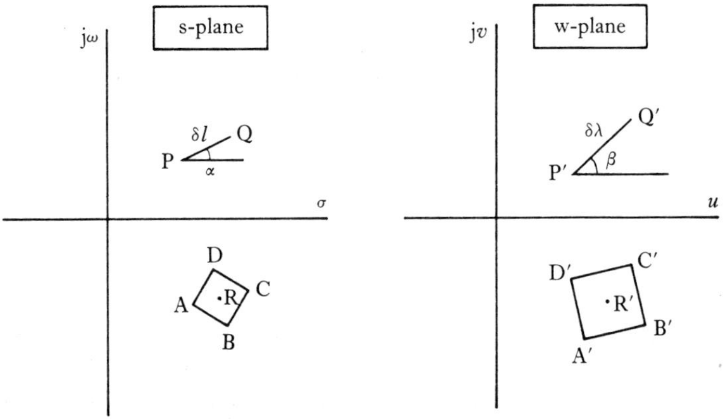

Fig. 12.6 Conformal transformation of a short line PQ, and of a small square ABCD.

Consider now the mapping of a small geometrical figure such as the square ABCD surrounding the point R. The four sides will not be subjected in mapping to exactly the same magnification M or rotation γ; but, the smaller the square, the more nearly do these quantities approach the values appropriate to the point R, and the more nearly does the map A′B′C′D′ become a square, M times the linear size of ABCD and rotated through an angle γ. It can at once be concluded that small geometrical figures retain their shape in the mapping, which is therefore described as *conformal mapping*. The relation between points in the s and w planes, defined by $w = w(s)$, is called a *conformal transformation*.

The conformal property breaks down at points where M is zero or infinite – that is, at the zeros and poles of $w'(s)$. Examples of this may be seen in fig. 12.4, which illustrated the transformation

$$\boldsymbol{Y}(s) = Cs + \frac{1}{Ls}. \tag{12.2.7}$$

Here

$$\boldsymbol{Y}'(s) = C - \frac{1}{Ls^2}, \tag{12.2.8}$$

for it is easily shown that the ordinary rules of differentiation with respect to a real variable apply also to a complex variable. The zeros of $\boldsymbol{Y}'(s)$ are at $s = \pm\omega_0$, where $\omega_0 = (LC)^{-1/2}$; these are at the points D_2, E_2 in fig. 12.4, and the distortion of the map by the crowding of the lines near D'_2, E'_2 has already been pointed out. $\boldsymbol{Y}'(s)$ also has a pole of order 2 at $s = 0$, the point denoted by D_0; in the map the corresponding point D'_0 is at infinity, so that the distortion is associated with a magnification which rapidly increases as D_0 is approached. The zeros and poles of $w'(s)$ are said to be *critical points* in the transformation. Although the mapping of the earth's surface on a plane is not a transformation of the type discussed, it illustrates the conformal property in that small areas are relatively free from distortion; moreover in Mercator's projection the poles of the earth are critical points, and the distortion of regions near the poles is very great.

12.3 The mapping of contours

The idea of a continuous plane curve is an intuitive one, although mathematically it requires careful definition which will not be attempted here. It is sufficient to think of it as the trajectory of a moving particle which never reverses its direction; if, in addition, the trajectory is free from loops, it is described as a *contour*. A closed contour such as C (fig. 12.7) divides the plane into two regions, the 'outside' and the 'inside'; these may be distinguished by noting that the angular position θ_i of the radius vector P_iQ from an inside point P_i changes by 2π when Q makes a journey round the contour, whereas the angle θ_o of the radius vector from an outside point does not change.

To the contour C in the s-plane there corresponds a mapped curve Γ in the related w-plane. Γ will be a continuous curve, if 'continuity' be so defined as to include excursions through infinity; but it need not be free from loops, since the definition of $w(s)$ as a regular function does not prevent two or more points in the s-plane from being mapped at the same point in the w-plane. A closed curve with a single loop (as illustrated) divides the plane into an 'outside' and two 'insides', characterized by radius vectors which make rotations of 0, 2π and 4π respectively. The conformal property of the transformation implies that points to the left or right of the direction

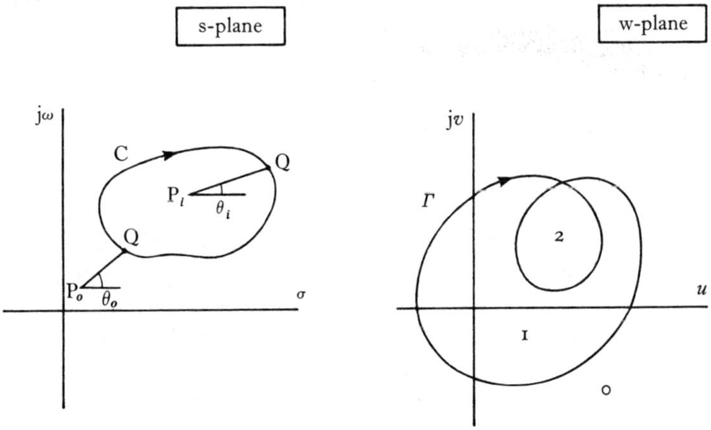

Fig. 12.7 A contour in the s-plane and its map in the w-plane.

in which C is described are mapped as points to the left or right of Γ; but, as we shall see, this does not mean that the 'outside' of C necessarily becomes the 'outside' of Γ.

In order to discover more about the relation between C and Γ, it is necessary to make an assumption about the transformation function w(s); namely that it can be expressed as a rational algebraic function

$$w(s) = \frac{K(s-s_1)(s-s_2)\ldots(s-s_i)^m\ldots}{(s-s_a)(s-s_b)\ldots(s-s_j)^n\ldots}. \qquad (12.3.1)$$

In other words, w(s) is a quotient of factors describing its zeros (from which zeros of order higher than 1 are not excluded) divided by factors describing its poles, and multiplied by a constant K. All transfer functions of lumped circuits automatically take this form, but its application is wider; for example, a certain transmission line problem discussed in §13.11 leads to the transfer function

$$\boldsymbol{H}(s) = s\,\frac{\cosh(sx/u)}{\cosh(sd/u)}, \qquad (12.3.2)$$

and both numerator and denominator may be reduced to an infinite product of linear factors – in a form differing slightly from (12.3.1) – by the use of the identity

$$\cosh z \equiv \left[1+\left(\frac{2z}{\pi}\right)^2\right]\left[1+\left(\frac{2z}{3\pi}\right)^2\right]\left[1+\left(\frac{2z}{5\pi}\right)^2\right]\ldots \text{ ad infinitum.} \qquad (12.3.3)$$

For reasons which will appear, we shall assume that the contour C in the s-plane is described in the clockwise direction, so that the angle between the radius vector from an inside point and a fixed line diminishes by 2π for each

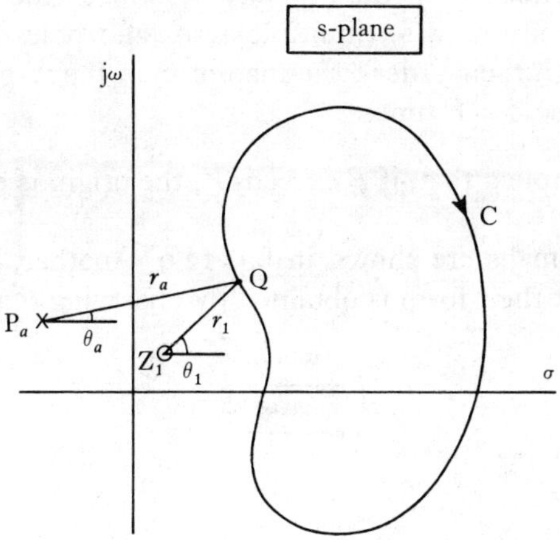

Fig. 12.8 Contour with pole and zero.

completed circuit of C. Let the point Z_1 denote the zero $s = s_1$, and P_a the pole $s = s_a$ (fig. 12.8); then, with Q representing s, the diagram shows that

$$(s - s_1) = r_1 e^{j\ \ 1}, \qquad (s - s_a) = r_a e^{j\theta_a}, \tag{12.3.4}$$

where $r_1 e^{j\theta_1}$, $r_a e^{j\theta_a}$ denote the radius vectors Z_1Q, P_aQ respectively. Extending this process, we rewrite equation (12.3.1) in the form

$$w(s) = K\frac{r_1 r_2 \ldots r_i^m \ldots}{r_a r_b \ldots r_j^n \ldots} . e^{j(\theta_1 + \theta_2 + \ldots + m\theta_i + \ldots - \theta_a - \theta_b - \ldots - n\theta_j - \ldots)}. \tag{12.3.5}$$

When Q makes a circuit of C, the factor containing the radii comes back to the same value; so does the exponential factor, so far as poles and zeros outside C are concerned. But if the zero Z_1 is inside C, the term $e^{j\theta_1}$ becomes $e^{j(\theta_1 - 2\pi)}$; if Z_i is inside C, $e^{jm\theta_i}$ becomes $e^{j(m\theta_i - 2m\pi)}$; if P_a is inside C, $e^{-j\theta_a}$ becomes $e^{-j(\theta_a - 2\pi)}$; and so on. Thus the argument or phase angle of w(s) diminishes by 2π for each zero within C, *m*-fold zeros being counted *m* times, and increases by 2π for each pole within C, *n*-fold poles being counted *n* times.

This statement is immediately reflected in the position and mode of description of the related curve Γ. Since, in the w-plane, w(s) is represented by the radius vector from the origin, a change (by a multiple of 2π) in the phase angle of w(s) denotes that the origin is inside Γ. An increase of 2π, 4π, 6π, ... means that the origin is encircled in the counterclockwise sense by the closed curve Γ, once, twice, thrice, ...; a diminution means that it is encircled in the clockwise sense. The whole conclusion may be summarized as follows:

> If a closed contour C in the s-plane, described clockwise, encircles Z zeros and P poles of w(s) (multiple zeros and poles being counted in accordance with their order), the map of C in the w-plane encircles the origin clockwise $Z-P$ times. (12.3.6)

It is of course implied that, if P exceeds Z, the origin is encircled counter-clockwise.

Some typical maps are shown in fig. 12.9. Another, more geometrical way of looking at their form is obtained by observing that all the zeros are

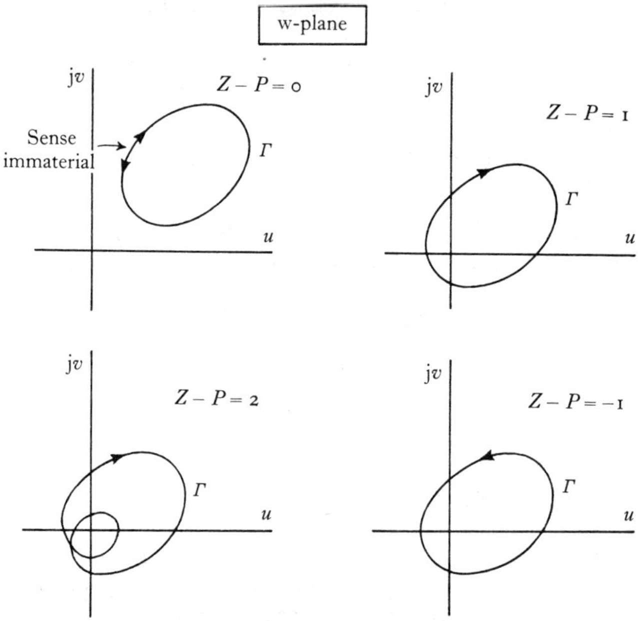

Fig. 12.9 Maps of contours C described clockwise in the s-plane and encircling Z zeros and P poles.

mapped at the origin and all the poles at infinity (which is regarded in complex variable theory as a single 'point at infinity'). Suppose C encircles one zero but no pole; then when C is described clockwise the zero lies on the right hand side of the path and any poles must lie on the left. The conformal transformation preserves these relationships, so the origin must lie to the right of Γ and the point at infinity to the left, as in the curve labelled '$Z-P=1$'. Similarly, if C encircles one pole but no zeros, the point at infinity must lie to the right of Γ and the origin to the left, as in the curve labelled '$Z - P = -1$'. It is not easy, however, to pursue this argument into the more complicated cases.

12.4 Stability

A system is said to be *stable* when the application of a stimulus which is not sustained causes a response which dies away to zero. It is *unstable* when the response is sustained or increases. For example, a sphere perched on the apex of another sphere is unstable – touch it, and it rolls off; whereas a sphere lying at the bottom of a spherical bowl is stable. The ideal instance of a non-sustained stimulus is a spike, the response to which is of a form dictated entirely by the transfer function of the system; for the image of a spike stimulus is a constant, which affects the magnitude of the response but not its form. Thus stability or instability is a property inherent in the system, and the information as to whether the system is stable is contained in its transfer function.

For a transfer function of the form

$$\boldsymbol{H}(\mathrm{s}) = \mathrm{K}\frac{(\mathrm{s}-\mathrm{s}_1)(\mathrm{s}-\mathrm{s}_2)\ldots(\mathrm{s}-\mathrm{s}_i)^m\ldots}{(\mathrm{s}-\mathrm{s}_a)(\mathrm{s}-\mathrm{s}_b)\ldots(\mathrm{s}-\mathrm{s}_j)^n\ldots}, \qquad (12.4.1)$$

the significant point is easily discerned. The response of the system to a unit spike stimulus is evaluated by resolving $\boldsymbol{H}(\mathrm{s})$ into partial fractions, and these will include terms corresponding to the poles of $\boldsymbol{H}(\mathrm{s})$:

$$\boldsymbol{H}(\mathrm{s}) = \frac{\mathrm{A}}{\mathrm{s}-\mathrm{s}_a} + \frac{\mathrm{B}}{\mathrm{s}-\mathrm{s}_b} + \ldots + \frac{\mathrm{J}_1}{\mathrm{s}-\mathrm{s}_j} + \frac{\mathrm{J}_2}{(\mathrm{s}-\mathrm{s}_j)^2} + \frac{\mathrm{J}_3}{(\mathrm{s}-\mathrm{s}_j)^3} + \ldots. \qquad (12.4.2)$$

The corresponding function of t is

$$h(t) = \mathrm{A}e^{\mathrm{s}_a t} + \mathrm{B}\,e^{\mathrm{s}_b t} + \ldots + \left(\mathrm{J}_1 + \mathrm{J}_2 t + \frac{\mathrm{J}_3 t^2}{2} + \ldots\right)e^{\mathrm{s}_j t} + \ldots. \qquad (12.4.3)$$

For this to tend to zero as t tends to infinity, it is necessary for the real parts of s_a, s_b, ..., s_j, ... to be all negative; this introduces into each term an '$e^{-\alpha t}$ factor' which causes it to tend to zero even when it also includes an increasing factor such as $(\mathrm{J}_1 + \mathrm{J}_1 t + \ldots)$.† If a term has an index in which the real part is positive, it tends to infinity; if the real part is zero, it oscillates with a constant amplitude, which by definition is a form of instability. Therefore

The necessary and sufficient condition for a system to be stable is that all the poles of its transfer function shall have negative real parts. (12.4.4)

† It can be shown that, when $\alpha > 0$, $\lim t^n e^{-\alpha t} = 0$ for all values of n. See, for example, J. M. Hyslop, *Real variable*, p.97, Oliver and Boyd (1960).

The application of this in circuit theory will now be considered. The only passive networks in which a transient stimulus brings about a sustained response are lossless networks; these have their poles on the imaginary axis, and come formally within the category of instability. In practice all networks are dissipative, and in the absence of controlled energy sources the response to a transient stimulus must die away; the poles then lie to the left of the imaginary axis. Actual circuits are therefore necessarily stable unless controlled energy sources are present. If one or more such sources are connected in such a way as to enhance the response, they may provide sufficient power to counterbalance the losses and leave a margin; the response then continually increases, and the network is unstable. In many active networks, such as control networks, instability is highly undesirable; instead of coming to the required value, the response oscillates about it with an amplitude which does not diminish, a phenomenon sometimes called 'hunting'. In other cases, such as oscillator circuits, a sustained response is precisely what is required. Either way, the location of the poles of an active network is of paramount importance.

Figure 12.10 shows a simple oscillator circuit which will serve to illustrate the stability problem. The stimulus is provided by the spike function

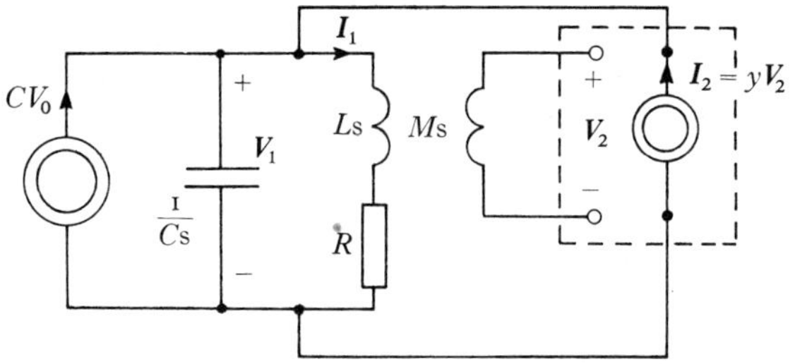

Fig. 12.10 Image circuit for a simple oscillator.

current source labelled CV_0, which charges the capacitor to an initial voltage V_0. The subsequent changes of current in the LC circuit lead to a voltage $\boldsymbol{V}_2$ across a secondary winding; this is regarded as the 'output' or response, but it is also made to inject further current into the LC circuit through the voltage-controlled current source labelled $\boldsymbol{I}_2 = y\boldsymbol{V}_2$. This, as will be shown, can enable the response to be sustained.

The circuit equations are:

$$Cs\boldsymbol{V}_1 + \boldsymbol{I}_1 = CV_0 + \boldsymbol{I}_2, \tag{12.4.5}$$

$$\boldsymbol{V}_1 = (Ls + R)\boldsymbol{I}_1, \tag{12.4.6}$$

$$\boldsymbol{V}_2 = Ms\boldsymbol{I}_1, \tag{12.4.7}$$

$$\boldsymbol{I}_2 = y\boldsymbol{V}_2. \tag{12.4.8}$$

From (12.4.5,6),

$$(LCs^2 + RCs + 1)\boldsymbol{I}_1 = C\boldsymbol{V}_0 + \boldsymbol{I}_2,$$
$$= C\boldsymbol{V}_0 + yMs\boldsymbol{I}_1,$$

using (12.4.7,8). Thus

$$[LCs^2 + (RC - yM)s + 1]\boldsymbol{I}_1 = C\boldsymbol{V}_0,$$

which gives $\boldsymbol{I}_1$; whence, from (12.4.7),

$$\boldsymbol{V}_2 = \left(\frac{Ms}{LCs^2 + (RC - yM)s + 1}\right)C\boldsymbol{V}_0. \qquad (12.4.9)$$

The bracketed quantity is the transfer function; the effect of the controlled current source, operating in opposition to the resistance R, is immediately apparent.

The poles of the transfer function are the roots of the equation

$$LCs^2 + (RC - yM)s + 1 = 0. \qquad (12.4.10)$$

These are $$s_1, s_2 = \frac{-(RC - yM) \pm \surd[(RC - yM)^2 - 4LC]}{2LC}. \qquad (12.4.11)$$

Two cases must be considered. If $(RC - yM)^2 < 4LC$, the roots are complex, with $(yM - RC)/2LC$ as their real part. The circuit is therefore unstable if $yM - RC$ is positive or zero. If $(RC - yM)^2 > 4LC$ the roots are real, and the square root term is a number smaller than $|RC - yM|$, which may be written $k|RC - yM|$, where $k < 1$. Thus:

$$\text{if } yM - RC > 0, \qquad s_1, s_2 = \frac{(yM - RC)(1 \pm k)}{2LC},$$

so both s_1 and s_2 are positive;

$$\text{if } yM - RC < 0, \qquad s_1, s_2 = \frac{(RC - yM)(-1 \pm k)}{2LC},$$

so both s_1 and s_2 are negative.

Whether the roots are real or complex, then, the condition for stability is

$$yM - RC < 0. \qquad (12.4.12)$$

If the circuit is intended as an oscillator, it must satisfy two conditions: to be oscillatory, and to be unstable. These conditions require

$$\left.\begin{aligned}(yM - RC)^2 &< 4LC,\\ yM - RC &\geqslant 0,\end{aligned}\right\} \qquad (12.4.13)$$

which may be combined into

$$0 \leqslant yM - RC < 2\sqrt{(LC)}. \tag{12.4.14}$$

Unless $yM - RC = 0$, the oscillation is of ever-increasing amplitude; but in practice the controlled current source would only have a linear characteristic over a limited range of the control voltage, and the non-linearity would limit the maximum amplitude of the response.

This example is as simple a one as could be found, yet it has required for its elucidation a careful discussion of all the possible cases. This would be impracticable in a more complicated problem, and it is plain that a more general approach must be worked out.

12.5 The transfer locus and its relation to the stability problem

The general stability criterion (12.4.4) may be restated in geometrical terms as follows: The necessary and sufficient condition for a system to be stable is that all the poles of its transfer function shall lie to the left of the imaginary axis in the s-plane. If then we describe a contour C as shown in fig. 12.11,

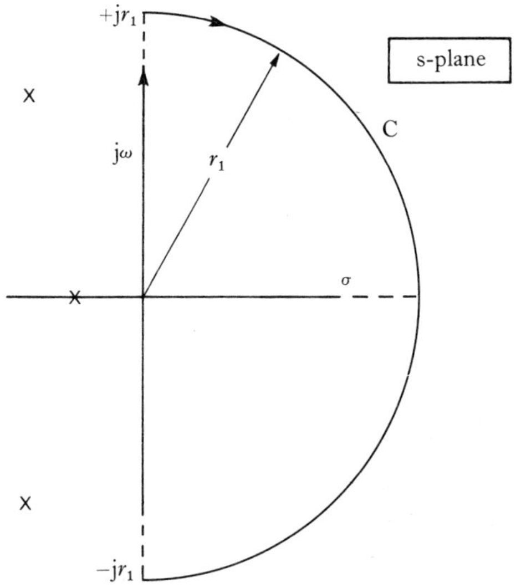

Fig. 12.11 Contour for stability discrimination; $r_1 \to \infty$.

consisting of the imaginary axis from $-jr_1$ to $+jr_1$ and a semicircle of radius r_1, where r_1 tends to infinity, stability requires that there shall be no poles either on or within C. Thus not only must poles on the imaginary axis be excluded, but also poles at infinity. $\boldsymbol{H}$(s) is said to have a pole at infinity if s is put equal to $1/s'$ and $\boldsymbol{H}(1/s')$ has a pole at the origin; this means that, when $\boldsymbol{H}(s)$ is expressed (as in (12.4.1)) as a ratio of two polynomials, the degree of the numerator must not be higher than that of the denominator. C will be called the *stability discrimination contour*.

Accepting this restriction, we proceed to discuss the map of the contour C in the $\boldsymbol{H}(s)$-plane. When $\boldsymbol{H}(s)$ is the impedance or admittance of a oneport network, this is none other than the impedance or admittance locus discussed in §12.1; more generally we can describe it as a *transfer function locus or transfer locus*, regarding the impedance and admittance of a oneport as transfer functions of a special kind. The transfer locus is simply a plot of the real and imaginary parts of the transfer function as co-ordinates in a cartesian graph, at all frequencies from zero to infinity; alternatively, polar co-ordinates may be used to plot the modulus and phase angle of the transfer function. It is indeed remarkable that this locus, which corresponds with a practical measurement under a.c. conditions, can be used to predict the performance of the system under quite other conditions; that it can be so used is due to the stringent requirements under which functions of a complex variable operate. Assuming that $\boldsymbol{H}(s)$ is a regular function within and on the contour C, and making a certain assumption about its value as s tends to infinity, it can indeed be shown that the value of $\boldsymbol{H}(s)$ at every point on the right hand half of the s-plane is determined when the values at all points of the imaginary axis are known.

The transfer locus, then, is a plot of $\boldsymbol{H}(j\omega)$. Changing the sign of ω has the same effect as changing the sign of j; thus the locus is symmetrical, the part corresponding to negative values of ω being the reflexion in the real axis of the part corresponding to positive ω. When the contour C is described in the clockwise sense, so that positive frequencies are covered from zero to infinity, the theorem numbered (12.3.6) shows that the transfer locus encircles the origin clockwise $Z-P$ times; Z and P being the number of zeros and poles of $\boldsymbol{H}(s)$ within C. Stability requires $P=0$; we therefore reach the following stability criterion:

> The transfer locus of a stable system, described in the sense $\omega=-\infty$ to $\omega=+\infty$, encircles the origin clockwise Z times ,where Z is the number of zeros of the transfer function in the right hand half of the s-plane. If the origin is encircled fewer than Z times in the clockwise sense, the system is unstable. (12.5.1)

It is understood, of course, that the description includes non-encirclement and counterclockwise encirclement. In the common case in which the transfer function has no zeros in the right hand half of the s-plane, the transfer locus does not encircle the origin at all when the system is stable, and encircles it counterclockwise when the system is unstable (fig. 12.12).

It will sometimes be found in practice that the transfer locus passes through the origin; this means that $\boldsymbol{H}(s)$ has a zero on the imaginary axis in the s-plane. The ambiguity thus caused is removed by modifying the

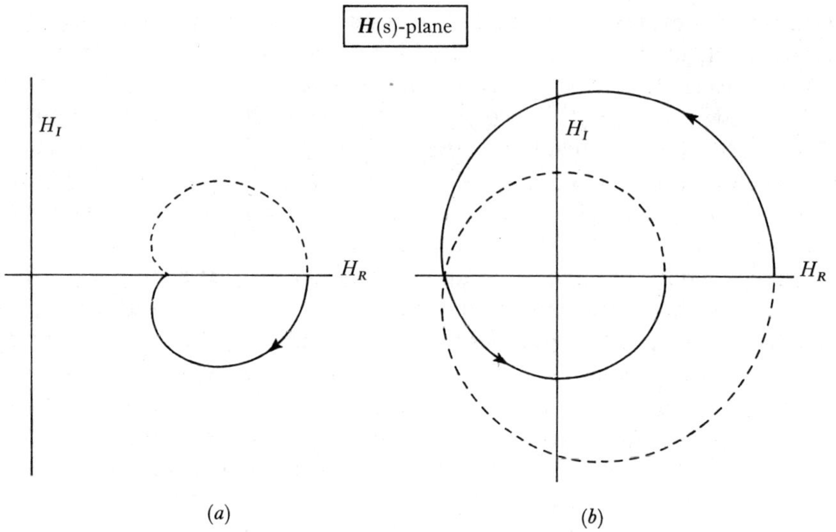

Fig. 12.12 Typical transfer loci: (a) for a stable system; (b) for an unstable system.

contour C in the manner shown in fig. 12.13; the zeros (necessarily occurring either at the origin or in conjugate pairs) are by-passed on the right by small semicircular detours. This puts corresponding detours into the transfer locus at the origin, to the right of the direction in which the locus is described; on account of the conformal property these detours will also be semicircles

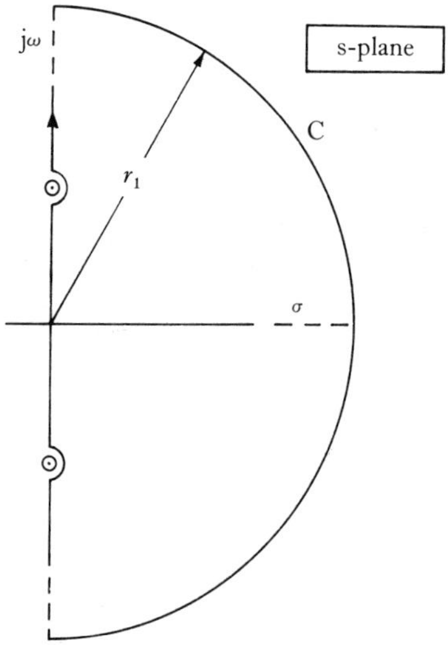

Fig. 12.13 Modification of contour to by-pass zeros on the imaginary axis.

unless the point is also a zero of $\boldsymbol{H}'(\mathrm{s})$, when the conformal property breaks down. An instance occurs when we seek to plot the locus of the transfer function in (12.4.9), namely

$$\boldsymbol{H}(\mathrm{s}) = \frac{M\mathrm{s}}{LC\mathrm{s}^2 + (RC - yM)\,\mathrm{s} + 1}. \tag{12.5.2}$$

We may simplify this, without losing the essential form, by writing $(LC/M)\boldsymbol{H}(\mathrm{s}) = \mathrm{w}(\mathrm{s})$, $(RC - yM)/LC = a$, $1/LC = b$; then

$$\mathrm{w}(\mathrm{s}) = \frac{\mathrm{s}}{\mathrm{s}^2 + a\mathrm{s} + b} \tag{12.5.3}$$

from which, writing $\mathrm{s} = \mathrm{j}\omega$, the co-ordinates in the transfer locus are given by

$$u + \mathrm{j}v = \frac{\mathrm{j}\omega}{(b - \omega^2) + \mathrm{j}a\omega},$$

$$= \frac{a\omega^2 + \mathrm{j}\omega(b - \omega^2)}{(b - \omega^2)^2 + a^2\omega^2}. \tag{12.5.4}$$

Plotting, for particular values of a and b, leads one to suspect a circular locus; this is verified by observing that

$$u^2 + v^2 = \frac{a^2\omega^4 + \omega^2(b - \omega^2)^2}{[(b - \omega^2)^2 + a^2\omega^2]^2},$$

$$= \frac{\omega^2}{(b - \omega^2)^2 + a^2\omega^2},$$

$$= \frac{u}{a}. \tag{12.5.5}$$

Fig. 12.14 shows the position of the locus for the two cases, (*a*) $RC - yM$ positive, and (*b*) $RC - yM$ negative. The locus is described twice, once for negative and once for positive values of ω. It passes through the origin when ω passes through zero and infinity.

If the origin in the s-plane is by-passed by a small semicircle as shown in fig. 12.15, we may write

$$\mathrm{s} = r_0\,\mathrm{e}^{\mathrm{j}\theta}, \tag{12.5.6}$$

on the semicircle, where r_0 is constant and θ changes from $-\pi/2$ to $+\pi/2$. When this is substituted into (12.5.2) and $\boldsymbol{H}(\mathrm{s})$ is expanded in powers of r_0, powers above the first being neglected, it is found that

$$\boldsymbol{H}(\mathrm{s}) = Mr_0\,\mathrm{e}^{\mathrm{j}\theta}, \tag{12.5.7}$$

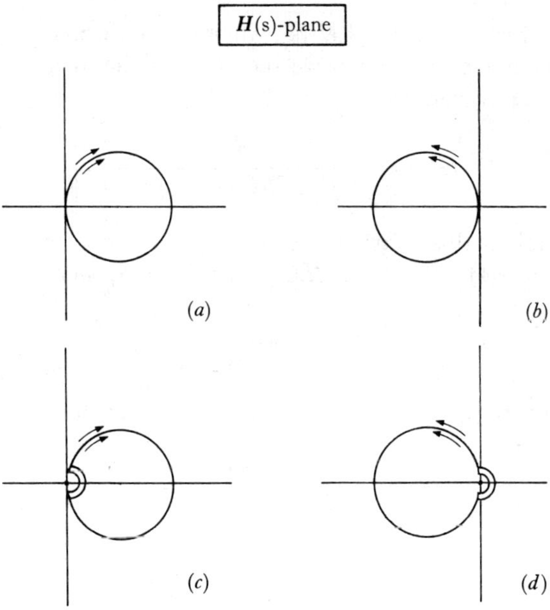

Fig. 12.14 Transfer loci for the circuit of fig. 12.10: (a) $RC - yM > 0$; (b) $RC - yM < 0$; (c, d) as (a, b), but modified by detours to the right at the origin.

so that the locus point by-passes the origin on the right by means of a circle of small radius Mr_0. At infinity the contour in the s-plane follows, as in fig. 12.13, a semicircle of large radius r_1; thus, in this region,

$$s = r_1 e^{j\theta}, \tag{12.5.8}$$

where θ changes from $+\pi/2$ to $-\pi/2$. Substitution into (12.5.2) and retention of the predominant terms gives

$$\boldsymbol{H}(s) = \frac{M}{LCr_1} e^{-j\theta}, \tag{12.5.9}$$

so that the locus point by-passes the origin, again on the right, on a circle of small radius (M/LCr_1). The transfer loci therefore take the forms shown in fig. 12.14(c) and (d); these bear out the statement that the detours in the

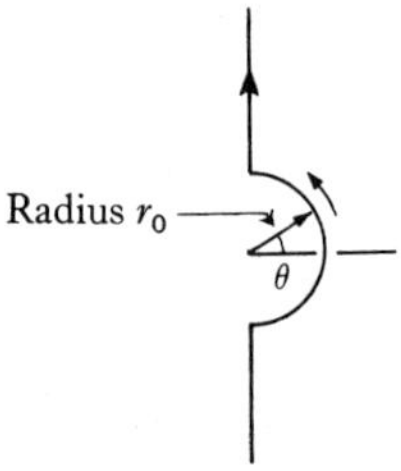

Fig. 12.15 Enlarged view of semicircular detour.

$\boldsymbol{H}$(s)-plane, as in the s-plane, are semicircles to the right of the path being described. Since the contour in the s-plane, as modified, encircles no zeros of $\boldsymbol{H}$(s), it can at once be concluded from the transfer locus of fig. 12.14(*c*) that $\boldsymbol{H}$(s) has no poles in the right hand half of the s-plane in this case, whereas fig. 12.14(*d*) indicates the presence of two poles there, since the origin is encircled twice in the counterclockwise sense.

In this problem, of course, the location of the poles is easily determined directly, since the denominator of $\boldsymbol{H}$(s) is only of the second degree. This is no longer feasible in general terms when the degree of the denominator is three or more, and the transfer locus becomes proportionately more useful as a means of assessing stability; but it may justly be objected that we have avoided determining the precise location of each pole of $\boldsymbol{H}$(s), only at the expense of having to determine the location of each zero. In many problems, however, this is easier; an example to be given in the next section will illustrate this. In such cases the transfer locus approach comes into its own – as it also does, of course, when the data take the form of a tabulation of the system's response to sinusoidal stimuli over a range of frequencies.

It may be added that the method used above for investigating the behaviour of $\boldsymbol{H}$(s) near a zero is valid even when the zero happens to be a critical point in the conformal transformation. For example, suppose that $\boldsymbol{H}$(s) approximates to $K\mathrm{s}^2$ near $\mathrm{s} = 0$, so that $\boldsymbol{H}'$(s) as well as $\boldsymbol{H}$(s) has a zero at the origin. The substitution $\mathrm{s} = r_0 \mathrm{e}^{\mathrm{j}\theta}$ then gives

$$\boldsymbol{H}(\mathrm{s}) = K r_0^2 \mathrm{e}^{2\mathrm{j}\theta}. \tag{12.5.10}$$

When the point in the s-plane moves round a semicircle of radius r_0 from $\theta = -\pi/2$ to $+\pi/2$, its map in the $\boldsymbol{H}$(s)-plane moves on a circle of radius Kr_0^2 from $\theta = -\pi$ to $+\pi$. Fig. 12.16 illustrates this; the transformation is not

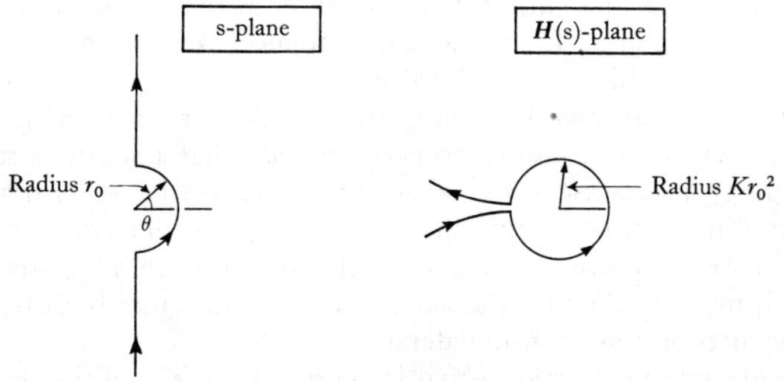

Fig. 12.16 Map of a semicircular detour round a zero which is a critical point in the transformation, namely the origin in the transformation $\boldsymbol{H}(\mathrm{s}) = K\mathrm{s}^2$.

conformal at this point, so a semicircle is not mapped as a semicircle; but the locus with its circular protuberance could still be used in a discussion of stability.

12.6 Nyquist's stability criterion for circuits with negative feedback

The subject of this section is an approach to the stability problem, related to but differing from the one just described. It is limited in its application, but, where applicable, very convenient. For understanding its range of applicability a brief preliminary discussion is necessary.

The transfer function of a twoport network was defined in §5.10 as the ratio, in the image world, of a named variable (voltage or current) on the output side to a named variable on the input side. This named variable will now be called the *principal variable*, and its mate (current or voltage) the *subsidiary variable*. The word *load* will be used to denote any network connected to the output terminals which makes the subsidiary variable different from zero; by transference it is sometimes used to signify the subsidiary variable itself. On this view, a load need not dissipate energy; for example, a short-circuit constitutes a load when the principal variable is the voltage. The definition applies to a load on any pair of output terminals including those of a oneport source. In general the transfer function of a twoport network is changed by a change in the load; but in some cases the alteration is small, and the transfer function may then be said to be nearly *load-independent* with respect to the load change in question. Thus, the transfer function of a controlled voltage source at a given frequency can be so described when the output impedance of the source is very small compared with the impedance of a load connected across the output terminals.

The transfer function of an ideal controlled source is load-independent with respect to any load; and such a source possesses a further noteworthy property in being *unilateral*. This means that a stimulus at the input produces a response at the output, but a stimulus at the output produces no response at the input. Like load-independence, the property of being unilateral is one which, in practical networks, may be approached very closely but not, in the absolute sense, attained. The term *non-unilateral* will be used to denote the contrary property, namely that a stimulus at either port produces a response at the other. The more restrictive term *bilateral* will be used to describe a non-unilateral network in which equal stimuli at the two ports produce responses which are equal in magnitude (not necessarily in phase). With one esoteric exception which may be disregarded, all passive networks are non-unilateral.†

A considerable number of active twoport networks take the form of a more or less unilateral network which is modified by the connexion of a separate network designed to have a transfer function in the reverse direction. This is the principle of *feedback*, and is illustrated in fig. 12.17(*a*). The

† A gyrator is a passive bilateral element in the sense defined above, yet a network containing a gyrator and other passive elements can be unilateral.

lower box, having a known transfer function in the sense shown, contains the main energy-producing controlled sources and will be called the *amplifier network*. The upper box, having the transfer function in the sense shown, is the *feedback network*; it may contain controlled sources or may be entirely passive. Across the output terminals is a load $\boldsymbol{Z}_L(s)$. When the circuit is opened at XX, the upper box is out of action and the overall transfer function is that of the lower box only; but when this circuit is closed, the total stimulus applied to the amplifier network is modified by a component derived through the feedback network from the output, and the overall transfer function is changed.

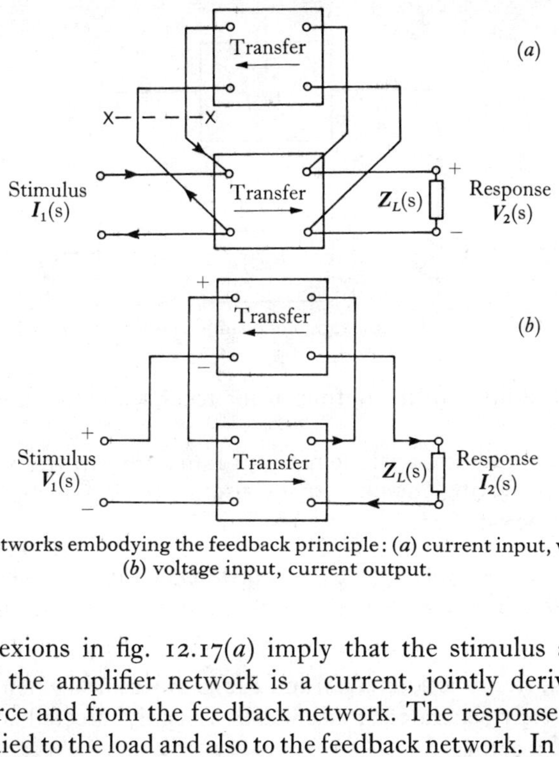

Fig. 12.17 Networks embodying the feedback principle: (*a*) current input, voltage output; (*b*) voltage input, current output.

The connexions in fig. 12.17(*a*) imply that the stimulus at the input terminals of the amplifier network is a current, jointly derived from an external source and from the feedback network. The response is a voltage, which is applied to the load and also to the feedback network. In other words, current and voltage are the principal variables on the input and output sides respectively. Of the other three possibilities (current/current, voltage/voltage, voltage/current), the last-named is illustrated in fig. 12.17(*b*).

All four cases are covered by the block diagram of fig. 12.18. The networks are shown with transfer functions $\boldsymbol{H}_0(s)$, $\boldsymbol{F}(s)$; the box labelled $\boldsymbol{H}_0(s)$ incorporates, not only the amplifier itself, but also the external load $\boldsymbol{Z}_L(s)$ and, if the source is non-ideal, the output impedance or admittance of the source. The quartered circle represents a *summing point*, where the quantity denoted by the outgoing arrow is obtained by summing the incoming quantities with the signs shown; while the dot on the right denotes a *pick-off point*, where both the outgoing quantities are identical and equal to the incoming quantity. The upper arm in fig. 12.18 is called a *feedback loop*, the term 'loop' referring to the single-line block diagrams rather than to the circuits which they represent; and the disconnexion of it from the summing point is called 'opening the feedback loop', even when (as in fig. 12.17(*b*)) this 'opening' would mean short-circuiting the output terminals of the feedback network. The feedback is described as positive when the signal fed back reinforces the effect of the original stimulus, negative when it opposes it. The polarity of the principal variable in the feedback network is normally and most naturally so chosen that positive feedback is denoted

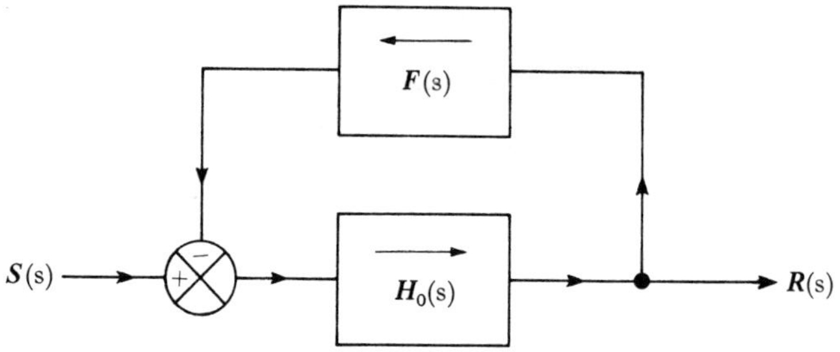

Fig. 12.18 Generalized block diagram of a simple network with negative feedback.

by algebraic addition of the output of the feedback network to the original stimulus.

For this resolution of a whole network into a forward-transferring and a backward-transferring member to be useful, the following general conditions must be satisfied.

1. It must be possible to neglect transference through the separate networks in the directions opposite to those indicated by the arrows. (12.6.1)

2. The transfer functions $\boldsymbol{H}_0(\mathrm{s})$ and $\boldsymbol{F}(\mathrm{s})$, as measured when the networks are completely separated, must not be sensibly affected when they are connected together. (12.6.2)

These conditions would be satisfied, for example, if each network were unilateral, and had a transfer function which was load-independent with respect to the load imposed by the other. This question will be more fully discussed in §12.8.

From the separate transfer functions $\boldsymbol{H}_0(\mathrm{s})$, $\boldsymbol{F}(\mathrm{s})$, we can deduce the overall transfer function of the system when the feedback loop is closed. The negative feedback case (fig. 12.18) is treated as standard. Denoting the stimulus and response by the non-committal forms $\boldsymbol{S}(\mathrm{s})$, $\boldsymbol{R}(\mathrm{s})$, we see that the input to the feedback network is $\boldsymbol{R}(s)$, and the output therefrom $\boldsymbol{F}(s)\boldsymbol{R}(s)$. The total input to the amplifier network is thus $\boldsymbol{S}(s)-\boldsymbol{F}(s)\boldsymbol{R}(s)$, so that

$$\boldsymbol{R}(s)=\boldsymbol{H}_0(s)\{\boldsymbol{S}(s)-\boldsymbol{F}(s)\,\boldsymbol{R}(s)\},$$

or

$$\boldsymbol{R}(s)=\frac{\boldsymbol{H}_0(s)}{1+\boldsymbol{H}_0(s)\,\boldsymbol{F}(s)}\boldsymbol{S}(s).$$

Thus the overall transfer function with the feedback loop closed and with negative feedback is given by

$$\boldsymbol{H}(s)=\frac{\boldsymbol{H}_0(s)}{1+\boldsymbol{H}_0(s)\,\boldsymbol{F}(s)}. \tag{12.6.3}$$

In lumped circuits $\boldsymbol{H}_0(s)$ and $\boldsymbol{F}(s)$ are ratios of polynomials in s, and the discussion will proceed on this assumption.

Let

$$\boldsymbol{H}_0(s)=\frac{A(s)}{B(s)}, \qquad \boldsymbol{F}(s)=\frac{C(s)}{D(s)}, \tag{12.6.4}$$

where A(s), etc., are polynomials. Equation (12.6.3) then gives $\boldsymbol{H}(s)$ in the same form, namely

$$\boldsymbol{H}(s)=\frac{A(s)\,D(s)}{B(s)\,D(s)+A(s)\,C(s)}. \tag{12.6.5}$$

The poles of $\boldsymbol{H}(s)$ are therefore the roots of the equation

$$B(s)\,D(s)+A(s)\,C(s)=0, \tag{12.6.6}$$

the so-called *characteristic equation* of the whole network. It is not necessary to admit the possibility of any root s_1 being such that $B(s_1)D(s_1)$ and $A(s_1)C(s_1)$ are separately zero; this would mean that the numerator and denominator of $\boldsymbol{H}_0(s)\,\boldsymbol{F}(s)$ had a common factor, and any such factor would have been cancelled. Thus the poles of $\boldsymbol{H}(s)$ are also roots of the equation

$$1+\frac{A(s)\,C(s)}{B(s)\,D(s)}=0,$$

that is, of

$$1+\boldsymbol{H}_0(s)\,\boldsymbol{F}(s)=0. \tag{12.6.7}$$

The presence or absence of zeros of $1+\boldsymbol{H}_0(s)\,\boldsymbol{F}(s)$ in the right hand half of the s-plane may be ascertained by mapping the stability discrimination contour of fig. 12.11 by means of the transformation $w(s)=1+\boldsymbol{H}_0(s)\,\boldsymbol{F}(s)$. According to the theorem (12.3.6), the mapped curve will encircle the origin clockwise $(Z-P)$ times, where Z, P are the numbers of zeros and poles of $1+\boldsymbol{H}_0(s)\,\boldsymbol{F}(s)$ in the region in question.

In this way it becomes possible to examine the stability of a system through the *loop transfer function* $\boldsymbol{H}_0(s)\boldsymbol{F}(s)$, instead of through the overall transfer function $\boldsymbol{H}(s)$. (The term *open-loop transfer function* is also used, but is less apt; it suggests $\boldsymbol{H}_0(s)$, the transfer function between the input and output terminals when the feedback loop is opened, whereas $\boldsymbol{H}_0(s)\boldsymbol{F}(s)$ is the transfer function right round the loop.) The map of the stability discrimination contour – that is, the locus obtained by plotting $\boldsymbol{H}_0(j\omega)\boldsymbol{F}(j\omega)$ – is the *loop transfer locus*. The locus $1+\boldsymbol{H}_0(j\omega)\boldsymbol{F}(j\omega)$, discussed above, is simply the loop transfer locus moved unit distance to the right; therefore the relation between the locus of $1+\boldsymbol{H}_0(j\omega)\boldsymbol{F}(j\omega)$ and the origin – the crucial question – is the same as the relation between the loop transfer locus and the point $(-1, 0)$.

The stability criterion associated with the name of H. Nyquist of the Bell Telephone Company, and first published in 1932, is based on examining the relation between the loop transfer locus and the $(-1, 0)$ point. Stability has been shown to require the absence of zeros of $1+\boldsymbol{H}_0(s)\boldsymbol{F}(s)$ from the right hand half of the s-plane; the locus of $1+\boldsymbol{H}_0(j\omega)\boldsymbol{F}(j\omega)$ encircles the origin clockwise $(Z-P)$ times, so the locus of $\boldsymbol{H}_0(j\omega)\boldsymbol{F}(j\omega)$ encircles the $(-1, 0)$ point clockwise $(Z-P)$ times. Thus we derive the following theorem:

> A system containing negative feedback has a loop transfer function with P poles in the right hand half of the s-plane. If the loop transfer locus, described in the sense $\omega=-\infty$ to $+\infty$, encircles the $(-1, 0)$ point counterclockwise P times, the system is stable. If it passes through the $(-1, 0)$ point, encircles it counterclockwise fewer than P times or encircles it clockwise, the system is unstable. (12.6.8)

This is the general statement of Nyquist's criterion of stability; the plot of the loop transfer locus is called the *Nyquist diagram*. The commonest application of Nyquist's criterion is to systems in which the loop transfer function possesses no poles in the right hand half of the s-plane; the criterion then takes a simpler form:

> A system containing negative feedback has a loop transfer function having no poles in the right hand half of the s-plane. If the loop transfer locus does not encircle the $(-1, 0)$ point, the system is stable. If (described in the sense $\omega=-\infty$ to ∞) it passes through the $(-1, 0)$ point or encircles it clockwise, the system is unstable. (12.6.9)

In essence Nyquist's criterion does not differ from the criterion (12.5.1), in which the overall transfer locus is examined. What was said in §12.5 about by-passing any zeros of $\boldsymbol{H}(s)$ on the imaginary axis by semicircular detours applies equally here to the by-passing of poles of $\boldsymbol{H}_0(s)\boldsymbol{F}(s)$; at such points the loop transfer function passes through infinity, and the detour method enables us to discover how the curve in the Nyquist diagram is to be closed. The poles of the loop transfer function have to be located in Nyquist's method, but this is often easier than locating the zeros.

For an example of stable and unstable Nyquist diagrams the reader is referred forward to fig. 12.23 in the next section.

12.7 A typical stability problem with negative feedback

An example will now be given of the application of the Nyquist criterion to a problem which could not be resolved by elementary means such as the solution of a quadratic equation. The problem is typical of many, but one way in which it can arise will be briefly described.

Every rotating dynamo–electric generator generates its e.m.f. by electromagnetic induction, and in most types the magnetic flux is brought into being by the passage of current through a field winding. Thus the machine is at heart a current-controlled voltage source. If magnetic saturation can be neglected the source has a linear characteristic, and the relation (in the image world) between the field current $\boldsymbol{I}_f$ and the armature e.m.f. $\boldsymbol{V}_a$ takes the form

$$\boldsymbol{V}_a = \boldsymbol{z}(\mathrm{s})\,\boldsymbol{I}_f, \tag{12.7.1}$$

when the machine is driven at a constant speed. In the d.c. machine, $\boldsymbol{z}(\mathrm{s})$ becomes a constant, z.

Indissolubly associated with such a machine are the resistance and inductance of its windings. In a d.c. generator as normally operated there is no mutual inductance between the field and armature circuits; the machine is therefore described by the equivalent circuit shown in fig. 12.19(*a*), in

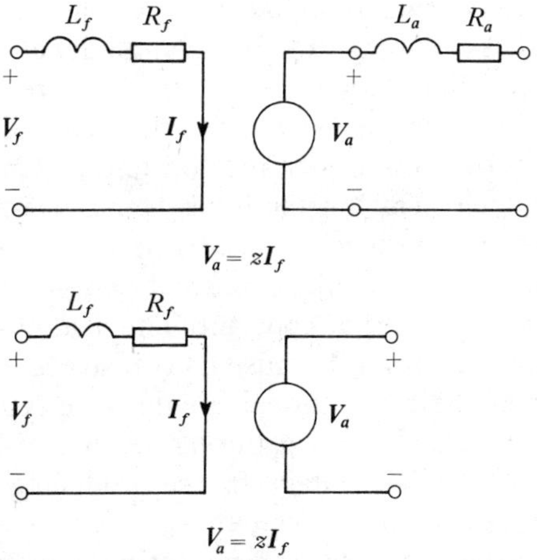

Fig. 12.19 Equivalent circuit of a simple d.c. generator with iron loss and saturation neglected: (*a*) complete; (*b*) simplified.

which the iron losses are neglected. If the armature resistance and inductance are regarded as part of the load to be connected across the output terminals, the circuit takes the form of fig. 12.19(*b*). The transfer function between the field voltage $\boldsymbol{V}_f$ and the armature e.m.f. $\boldsymbol{V}_a$ is the bracketed expression in the equation

$$\boldsymbol{V}_a = \left(\frac{z}{L_f \mathrm{s} + R_f}\right)\boldsymbol{V}_f. \tag{12.7.2}$$

Let the voltage be applied to the field as a step function $V_f u(t)$, so that $\boldsymbol{V}_f = V_f/\mathrm{s}$; then

$$\boldsymbol{V}_a = \frac{zV_f}{\mathrm{s}(L_f \mathrm{s} + R_f)}, \tag{12.7.3}$$

from which

$$v_a = V_f\left(\frac{z}{R_f}\right)(1 - \mathrm{e}^{-R_f t/L_f}). \tag{12.7.4}$$

The final value of the output voltage is therefore the input multiplied by (z/R_f); this may be called the *gain* (K). L_f/R_f is the time-constant of the field circuit (τ_f). In terms of K and τ_f, (12.7.2) becomes

$$\boldsymbol{V}_a = \left(\frac{K}{\tau_f \mathrm{s} + 1}\right)\boldsymbol{V}_f, \tag{12.7.5}$$

while in the more familiar form in which α_f is written for $1/\tau_f$,

$$\boldsymbol{V}_a = \left(\frac{K\alpha_f}{\mathrm{s} + \alpha_f}\right)\boldsymbol{V}_f. \tag{12.7.6}$$

The type of generator known as an *amplidyne* consists essentially of two stages of the form discussed, in cascade. If an amplidyne supplies the field of an ordinary d.c. generator we get three stages. The resulting system is typical of many problems comprising a number of cascade-connected elements, each with its own values of time-constant and gain. When there are more than two such elements, instability becomes possible.

Figure 12.20 shows how feedback might be used to control the output voltage of a d.c. generator (essentially a constant voltage source) in such a way that the output *current* is nearly constant. The field of the generator is supplied from an amplidyne (used because its two stages provide a higher gain than a generator of the conventional, single-stage type). The output current i_4 from the generator is passed through a current-sensing device, depicted as a current-controlled voltage source producing a voltage $z_4 i_4$; this is connected in series with the d.c. source V_0 with opposed polarity, and the difference between the two voltages constitutes the input to the amplidyne. In the steady state i_4 assumes the value corresponding with a constant input voltage $V_0 - v_4$. If the load resistance R_L is reduced so that i_4 increases, $V_0 - v_4$ is reduced; this tends to bring down the value of i_4 again, after a time-lag associated with the several time-constants. The inductance of the

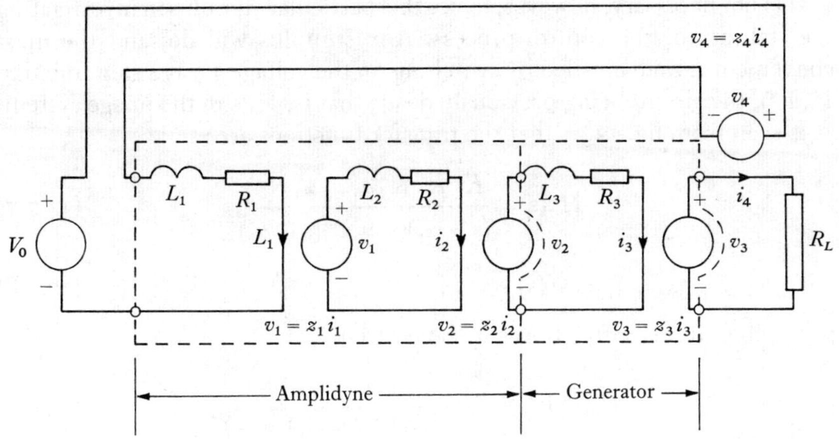

Fig. 12.20 Circuit for controlling the output voltage of a d.c. generator so that its output current may be nearly constant.

output circuit has been omitted because it is known that the time-constant of this circuit is negligible compared with the others.

Figure 12.21 shows the block diagram of the image system with the amplifier and feedback networks separated. The main energy sources are in the amplifier network, the controlled source in the feedback network being

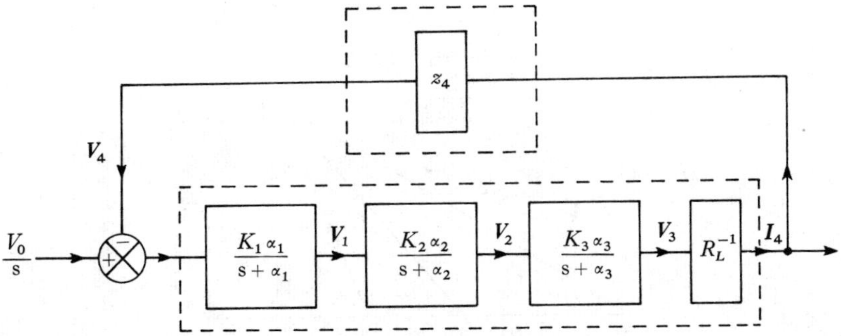

Fig. 12.21 Block diagram for the circuit of fig. 12.20 assuming that V_0 is switched on suddenly.

merely a buffer. The conditions (12.6.1,2) are satisfied, because both networks are unilateral and the feedback network has zero impedance on both the input and output sides. The transfer function of the amplifier network depends upon R_L, which must therefore be assumed to be virtually constant. It is true that we may wish to ascertain whether the system can hold i_4 nearly constant despite changes in R_L; but, if it is to be activated by short-circuiting part of R_L, it must be an infinitesimal part.

It is not necessary, however, to use this particular stimulus in investigating the stability of the control process. Any stimulus will do, and the most convenient is that afforded by switching on the voltage V_0 as a step function $V_0 u(t)$, the circuit being previously dead; that is, V_0/s in the image system. It is seen from fig. 12.21 that the transfer functions are

$$\boldsymbol{H}_0(\mathrm{s}) = \frac{K_1 K_2 K_3 \alpha_1 \alpha_2 \alpha_3 R_L^{-1}}{(\mathrm{s}+\alpha_1)(\mathrm{s}+\alpha_2)(\mathrm{s}+\alpha_3)}, \tag{12.7.7}$$

$$\boldsymbol{F}(\mathrm{s}) = z_4, \tag{12.7.8}$$

so the loop transfer function may be written

$$\boldsymbol{H}_0(\mathrm{s})\,\boldsymbol{F}(\mathrm{s}) = \frac{K\alpha_1 \alpha_2 \alpha_3}{(\mathrm{s}+\alpha_1)(\mathrm{s}+\alpha_2)(\mathrm{s}+\alpha_3)}, \tag{12.7.9}$$

where K (the overall gain) is given by

$$K = K_1 K_2 K_3 R_L^{-1} z_4. \tag{12.7.10}$$

The poles of $\boldsymbol{H}_0(\mathrm{s})\,\boldsymbol{F}(\mathrm{s})$ are located on the real axis at $\mathrm{s} = -\alpha_1, -\alpha_2, -\alpha_3$, and the Nyquist diagram may be plotted by the method indicated in fig. 12.22. The complex number $\mathrm{j}\omega + \alpha_1$ is represented by $r_1 \mathrm{e}^{\mathrm{j}\theta_1}$, where r_1, θ_1 are as shown; thus

$$\boldsymbol{H}_0(\mathrm{j}\omega)\,\boldsymbol{F}(\mathrm{j}\omega) = \frac{K\alpha_1 \alpha_2 \alpha_3}{r_1 r_2 r_3}\,\mathrm{e}^{-\mathrm{j}(\theta_1+\theta_2+\theta_3)}. \tag{12.7.11}$$

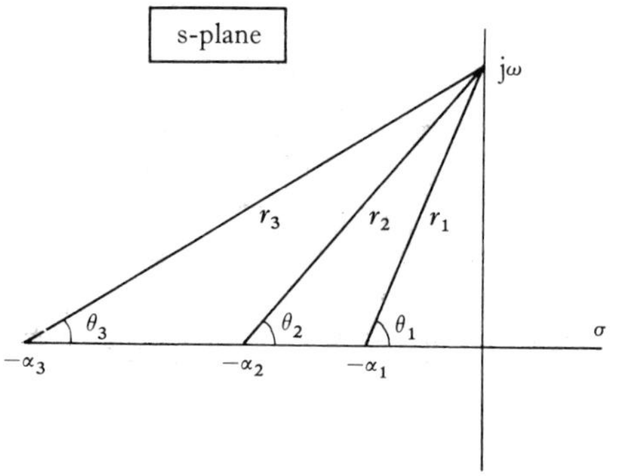

Fig. 12.22 Construction of the Nyquist locus for

$$\boldsymbol{H}_0(\mathrm{s})\boldsymbol{F}(\mathrm{s}) = \frac{K\alpha_1\alpha_2\alpha_3}{(\mathrm{s}+\alpha_1)(\mathrm{s}+\alpha_2)(\mathrm{s}+\alpha_3)}.$$

As ω increases from 0 to ∞, r_1, r_2 and r_3 increase from the initial values α_1, α_2, α_3, while θ_1, θ_2, θ_3 all increase from 0 to $\pi/2$; thus the locus starts at $s = K$ and moves initially downwards, spiralling towards the origin in such

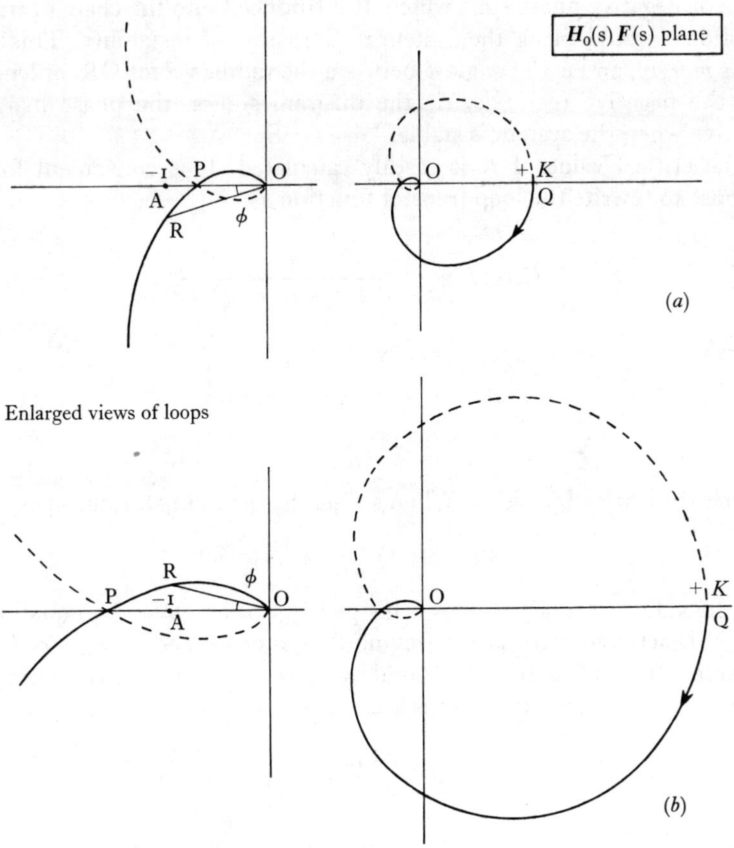

Fig. 12.23 Nyquist diagrams for

$$H_0(s)F(s) = \frac{K\alpha_1\alpha_2\alpha_3}{(s+\alpha_1)(s+\alpha_2)(s+\alpha_3)}.$$

(*a*) Stable case. (*b*) Unstable case.

a way that the radius vector turns through $3\pi/2$ (fig. 12.23). The negative frequencies give the same locus reflected in the real axis.

It is at once apparent that instability occurs if K is made too large. This is important, because the object of the circuit, namely to keep i_4 nearly constant, is most effectively attained by making K as large as possible. A great virtue of the method is that it shows not only whether the system is stable but, if so, by what margin. The ratio of the gain which brings the system to the verge of instability to the actual gain is called the *gain margin*; this is greater than 1 for a stable system, and is equal to the length ratio OA/OP (see fig. 12.23). A second margin which may be important is the angle of negative phase shift which, if introduced into the chain of transfer functions, would bring the system to the verge of instability. This is the *phase margin*, and is the angle ϕ between the radius vector OR, of length 1, and the negative real axis. As the diagram shows, the phase margin is positive when the system is stable.

The critical value of K is readily calculated. It is convenient for this purpose to rewrite the loop transfer function as

$$\boldsymbol{H}_0(\mathrm{s})\,\boldsymbol{F}(\mathrm{s}) = \frac{Kc_3}{\mathrm{s}^3 + c_1\,\mathrm{s}^2 + c_2\,\mathrm{s} + c_3}, \tag{12.7.12}$$

where

$$\left.\begin{aligned} c_1 &= \alpha_1 + \alpha_2 + \alpha_3, \\ c_2 &= \alpha_2\,\alpha_3 + \alpha_3\,\alpha_1 + \alpha_1\,\alpha_2, \\ c_3 &= \alpha_1\,\alpha_2\,\alpha_3, \end{aligned}\right\} \tag{12.7.13}$$

so that the c are all positive. When $\mathrm{s} = \mathrm{j}\omega$ the denominator becomes

$$(-c_1\,\omega^2 + c_3) + \mathrm{j}(-\omega^3 + c_2\,\omega),$$

and the values of ω which make this purely real are given by equating the second bracket to zero; $\omega = 0$ (giving the point Q) and $\omega = \pm\sqrt{c_2}$ (giving the point P). The length OP is found by writing $\mathrm{s} = \pm\mathrm{j}\sqrt{c_2}$ in (12.7.12), and the condition for stability is thus seen to be

$$\left|\frac{Kc_3}{-c_1\,c_2 + c_3}\right| < 1,$$

or (since $c_1\,c_2 > c_3$)

$$\frac{Kc_3}{c_1\,c_2 - c_3} < 1,$$

or

$$K < \frac{c_1\,c_2}{c_3} - 1. \tag{12.7.14}$$

If the designer requires freedom to make K as large as possible, he will seek to make $(c_1\,c_2/c_3)$ large. It so happens that, in the system which forms the basis of this example, a modification which is equivalent to reducing the time-constant of circuit loop 2 is feasible. If α_2 is treated as an independent variable, (12.7.13,14) enable the stability criterion to be rewritten in the form

$$K < A\alpha_2 + \frac{B}{\alpha_2} + C, \qquad (12.7.15)$$

where A, B, C are functions of α_1, α_3 only and are supposedly unalterable. It is evident that the object of having a large value of K combined with a stable system is served by making α_2 either sufficiently large or sufficiently small – that is, by making the time-constant of the adjustable circuit loop either as small or as large as possible. Thus the Nyquist method can be made to yield information on how a network should be modified in order that it may become more stable.

12.8 The range of validity of the feedback concept

The general conditions which render the feedback concept useful were set down at (12.6.1,2). In considering their meaning as applied to practical networks, it is convenient to think in terms of a sinusoidal constant-frequency input; the discussion can be generalized if necessary to cover a spectrum. The detailed consideration of one of the four cases will enable conclusions to be drawn about all four.

Figure 12.24(*a*) shows a non-unilateral voltage–current transfer network with its associated source. $\boldsymbol{Z}_1$ may comprise, as well as an impedance in the

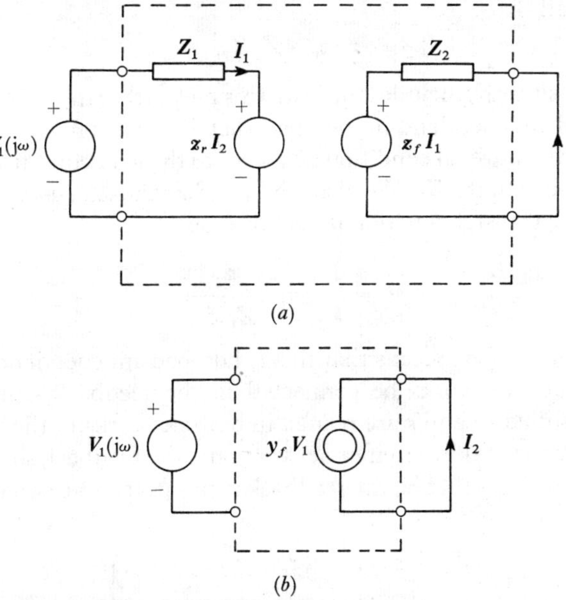

Fig. 12.24 (*a*) Non-ideal voltage–current transfer network. (*b*) Ideal counterpart of (*a*).

network, the impedance of an imperfect source; $\boldsymbol{Z}_2$ similarly contains both the network impedance and the external load. The general term *transmittance* will be used for the transfer functions of the controlled voltage sources. These can without difficulty be treated as complex numbers; when this is done, fig. 12.24(*a*) becomes (as shown in §8.4) a possible equivalent circuit for any linear twoport network, active or passive. The ratio of the amplitudes of the transmittances, z_f/z_r, is a dimensionless number which will be called the *transmittance ratio*. It is infinite for an ideal unilateral network, large for a typical amplifier network, and equal to unity for a passive network – the last statement being a consequence of the basic reciprocity relations (see table 8.5 in §8.11).

In an ideal voltage–current transfer network (fig. 12.24(*b*)) the impedances $\boldsymbol{Z}_1$, $\boldsymbol{Z}_2$ are infinite. The passage to the ideal occurs thus:

$$\boldsymbol{Z}_1 \to \infty; \qquad \boldsymbol{Z}_2 \to \infty; \qquad \boldsymbol{z}_r \to 0; \qquad \frac{\boldsymbol{z}_f}{\boldsymbol{Z}_1 \boldsymbol{Z}_2} \to \boldsymbol{y}_f. \tag{12.8.1}$$

The equations for the network of fig. 12.24(*a*) are

$$\left.\begin{aligned} \boldsymbol{Z}_1 \boldsymbol{I}_1 + \boldsymbol{z}_r \boldsymbol{I}_2 &= \boldsymbol{V}_1, \\ \boldsymbol{z}_f \boldsymbol{I}_1 + \boldsymbol{Z}_2 \boldsymbol{I}_2 &= 0. \end{aligned}\right\} \tag{12.8.2}$$

Eliminating $\boldsymbol{I}_2$ gives $\boldsymbol{V}_1 = \boldsymbol{Z}_i \boldsymbol{I}_1$, where $\boldsymbol{Z}_i$ (the input impedance) is

$$\boldsymbol{Z}_i = \boldsymbol{Z}_1 - \frac{\boldsymbol{z}_f \boldsymbol{z}_r}{\boldsymbol{Z}_2}. \tag{12.8.3}$$

If $\boldsymbol{z}_r$ is so small that

$$z_f z_r \ll Z_1 Z_2, \tag{12.8.4}$$

$\boldsymbol{Z}_i \approx \boldsymbol{Z}_1$; and since $\boldsymbol{Z}_2$ includes the load, this means that the input impedance is almost load-independent, in the sense that $\boldsymbol{Z}_i$ is virtually unchanged by a change from the no-load condition ($\boldsymbol{Z}_2 = \infty$) to the load condition associated with a particular finite $\boldsymbol{Z}_2$. We shall call (12.8.4) the *load-independent input condition*. The transfer function is found to be

$$\boldsymbol{y}_f = \frac{\boldsymbol{I}_2}{\boldsymbol{V}_1} = -\frac{\boldsymbol{z}_f}{\boldsymbol{Z}_i \boldsymbol{Z}_2}, \tag{12.8.5}$$

and the presence of $\boldsymbol{Z}_2$ means that this is not load-independent.

Let two such networks be connected in the feedback connexion (fig. 12.25). The suffixes *f* and *r* are related in both networks to the forward and reverse directions of the amplifier network, and the feedback source opposes the primary source – that is, the feedback is negative. The equations of this system are

$$\left.\begin{aligned} (\boldsymbol{Z}_1 + \boldsymbol{Z}_1')\boldsymbol{I}_1 + (\boldsymbol{z}_r + \boldsymbol{z}_r')\boldsymbol{I}_2 &= \boldsymbol{V}_1, \\ (\boldsymbol{z}_f + \boldsymbol{z}_f')\boldsymbol{I}_1 + (\boldsymbol{Z}_2 + \boldsymbol{Z}_2')\boldsymbol{I}_2 &= 0. \end{aligned}\right\} \tag{12.8.6}$$

Hence, eliminating I_1, we obtain the transfer function

$$y_f = \frac{I_2}{V_1} = \frac{-(z_f + z'_f)}{(Z_1 + Z'_1)(Z_2 + Z'_2) - (z_f + z'_f)(z_r + z'_r)}. \tag{12.8.7}$$

Dividing numerator and denominator by $(Z_1 + Z'_1)(Z_2 + Z'_2)$, we obtain

$$y_f = \frac{h_0}{1 + h_0 f}, \tag{12.8.8}$$

where

$$h_0 = \frac{-(z_f + z'_f)}{(Z_1 + Z'_1)(Z_2 + Z'_2)}, \tag{12.8.9}$$

and

$$f = (z_r + z'_r). \tag{12.8.10}$$

Equation (12.8.8) shows that even when there is some undesired forward transmission through the feedback network and some undesired backward transmission through the amplifier network, the transfer function can be

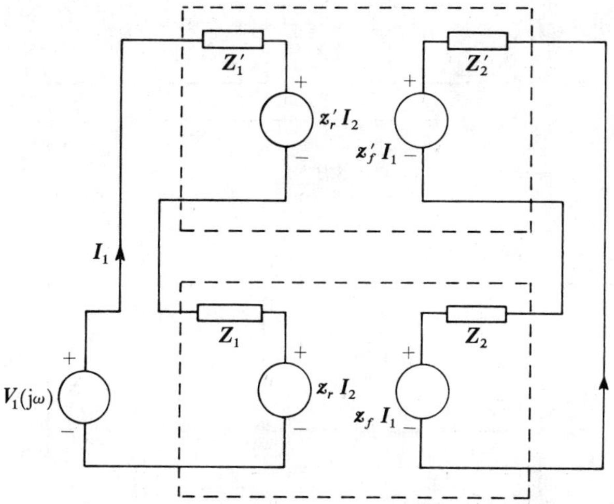

Fig. 12.25 Non-ideal voltage–current transfer network including negative feedback.

put in the typical feedback form. However, the usefulness of the equation is much reduced in these circumstances, for h_0, f are not identical with H_0, F, the separate transfer functions of the two networks. Both h_0 and f are associated with both networks; for example, z_r (a property of the amplifier network alone) appears in (12.8.10), which ideally should relate only to the feedback network. Fortunately, the desired association of f with the feedback network and $\boldsymbol{h}_0$ with the amplifier network is possible in many practical situations. There are two requirements; but if, in addition, a third requirement is met, it becomes possible not only to associate $\boldsymbol{f}$ and $\boldsymbol{h}_0$ predominantly with their respective networks but also to evaluate these quantities directly from the circuit.

The first condition is that the feedback should occur mainly through the feedback network. This is expressed by the equation

$$z'_r \gg z_r, \tag{12.8.11}$$

and it then follows that (12.8.10) becomes

$$\boldsymbol{f} \approx \boldsymbol{z}'_r = \boldsymbol{F} \text{ (say)}, \tag{12.8.12}$$

a property of the feedback network alone; namely, the open-circuit voltage produced at the left hand terminals of the feedback network per unit current injected at the right, as shown in fig. 12.26(*a*).

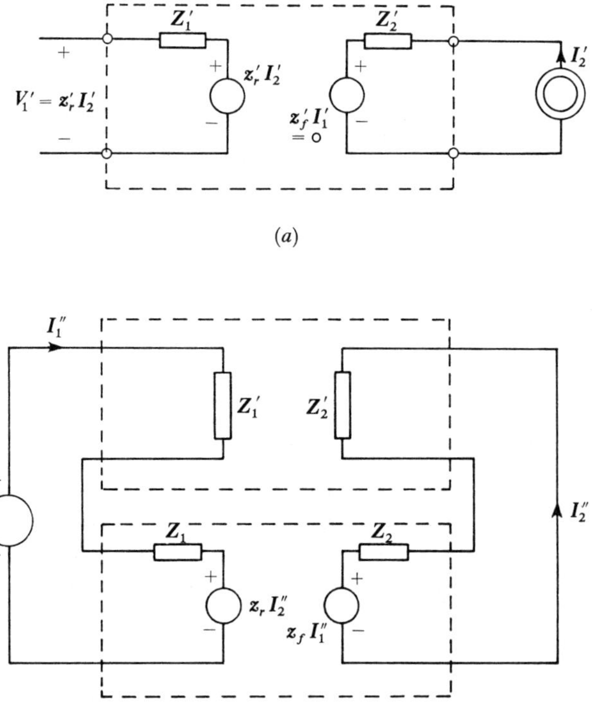

Fig. 12.26 Isolation of the feedback network (*a*) and modified amplifier network (*b*), for computation of $\boldsymbol{F}$ and $\boldsymbol{H}_0$: (*a*) Feedback network alone: $\boldsymbol{F} = \boldsymbol{V}'_1/\boldsymbol{I}'_2$; (*b*) Modified amplifier network alone: $\boldsymbol{H} = \boldsymbol{I}''_2/\boldsymbol{V}_1$.

The second condition is that the forward transmission should be predominantly due to the amplifier network; that is,

$$z_f \gg z'_f. \tag{12.8.13}$$

Equation (12.8.9) then becomes

$$\boldsymbol{h}_0 \approx \frac{-\boldsymbol{z}_f}{(\boldsymbol{Z}_1+\boldsymbol{Z}'_1)(\boldsymbol{Z}_2+\boldsymbol{Z}'_2)} = \boldsymbol{H}_0 \text{ (say)}. \tag{12.8.14}$$

This expression still contains the terms $\boldsymbol{Z}'_1$ and $\boldsymbol{Z}'_2$ which refer to the feedback network, but it may conveniently be regarded as the equation of a modified amplifier network in which the original impedances $\boldsymbol{Z}_1$ and $\boldsymbol{Z}_2$ have been increased by the loading caused at each port by the feedback network. It still remains, however, to equate $\boldsymbol{H}_0$ with some specific property of the modified amplifier network which can be computed directly from the circuit. It is shown below that $\boldsymbol{H}_0$ is the transfer function of the modified amplifier network, provided that one further condition is satisfied.

The modified amplifier network is shown in fig. 12.26(*b*) and its transfer function $(\boldsymbol{I}''_2/\boldsymbol{V}_1)$ is

$$\frac{\boldsymbol{I}''_2}{\boldsymbol{V}_1} = \frac{-\boldsymbol{z}_f}{(\boldsymbol{Z}_1+\boldsymbol{Z}'_1)(\boldsymbol{Z}_2+\boldsymbol{Z}'_2)-\boldsymbol{z}_f\boldsymbol{z}_r}. \tag{12.8.15}$$

This is nearly equal to $\boldsymbol{H}_0$, if

$$z_f z_r \ll |(\boldsymbol{Z}_1+\boldsymbol{Z}'_1)(\boldsymbol{Z}_2+\boldsymbol{Z}'_2)|. \tag{12.8.16}$$

Comparison of this inequality with (12.8.4) shows that it is the condition for the input impedance of the amplifier network to be load-independent, when the impedances associated with the feedback network have been added to the impedances $\boldsymbol{Z}_1$, $\boldsymbol{Z}_2$, associated with the amplifier network and its source and load. Though load-independence of the input impedance is not a self-evident requirement for the feedback approach to be valid, the fact that the condition (12.8.16) carries this meaning enables it to be expressed, like the other conditions, in general terms as follows.

The (non-unilateral) feedback network is to be replaced by an equivalent circuit of the appropriate canonical form given in fig. 12.27, and its impedances or admittances are to be transferred to the amplifier network. The conditions which make possible the feedback approach are then as follows:

The forward transmittance of the amplifier network greatly exceeds that of the feedback network. (12.8.17)

The reverse transmittance of the feedback network greatly exceeds that of the amplifier network. (12.8.18)

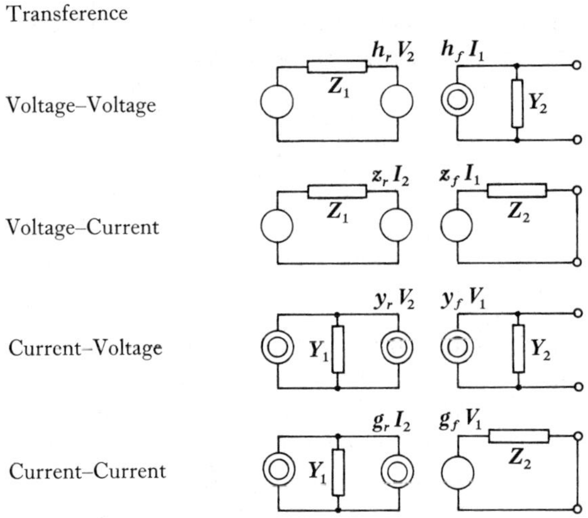

Fig. 12.27 Canonical forms for non-unilateral networks.

When an amplifier network is connected to a feedback network of the same canonical form, the appropriate parameters in each case and the feedback validity conditions are as tabulated. Primed letters denote quantities in the feedback network.

Transference	*Connexions*	*Parameters*	*Feedback conditions* (12.8.17)	(12.8.18)	(12.8.19)†
Voltage–voltage	Series–parallel	$[h] = \begin{bmatrix} Z_1 & h_r \\ h_f & Y_2 \end{bmatrix}$	$h_f \gg h'_f$	$h_r \ll h'_r$	$h_f h_r \ll Z_1 Y_2$
Voltage–current	Series–series	$[z] = \begin{bmatrix} Z_1 & z_r \\ z_f & Z_2 \end{bmatrix}$	$z_f \gg z'_f$	$z_r \ll z'_r$	$z_f z_r \ll Z_1 Z_2$
Current–voltage	Parallel–parallel	$[y] = \begin{bmatrix} Y_1 & y_r \\ y_f & Y_2 \end{bmatrix}$	$y_f \gg y'_f$	$y_r \ll y'_r$	$y_f y_r \ll Y_1 Y_2$
Current–current	Parallel–series	$[g] = \begin{bmatrix} Y_1 & g_r \\ g_f & Z_2 \end{bmatrix}$	$g_f \gg g'_f$	$g_r \ll g'_r$	$g_f g_r \ll Y_1 Z_2$

† In this column the quantities Y and Z are the total admittances or impedances of both networks when connected together, including the source and load.

The parameters of the amplifier network, as modified by the incorporation of the impedances or admittances of the feedback network, satisfy the load-independent input condition. (12.8.19)

The various forms of this last condition are included in the diagram. It is to be noted that the conditions (12.8.17–19) do not require the amplifier network to be put into the canonical form. The transmittances can be defined in terms of measurements made on open- or short-circuit, and the load-independent input condition amounts to a comparison, for the modified amplifier network, between the input impedance or admittance on open-circuit and on short-circuit.

A detailed consideration of the process of neglecting the small quantities in equations (12.8.8–10) would have shown that it was being tacitly assumed that $|(1+\boldsymbol{H}_0\boldsymbol{F})|$ is not itself small. This assumption may break down when positive feedback is used. The feedback approach may still be employed to

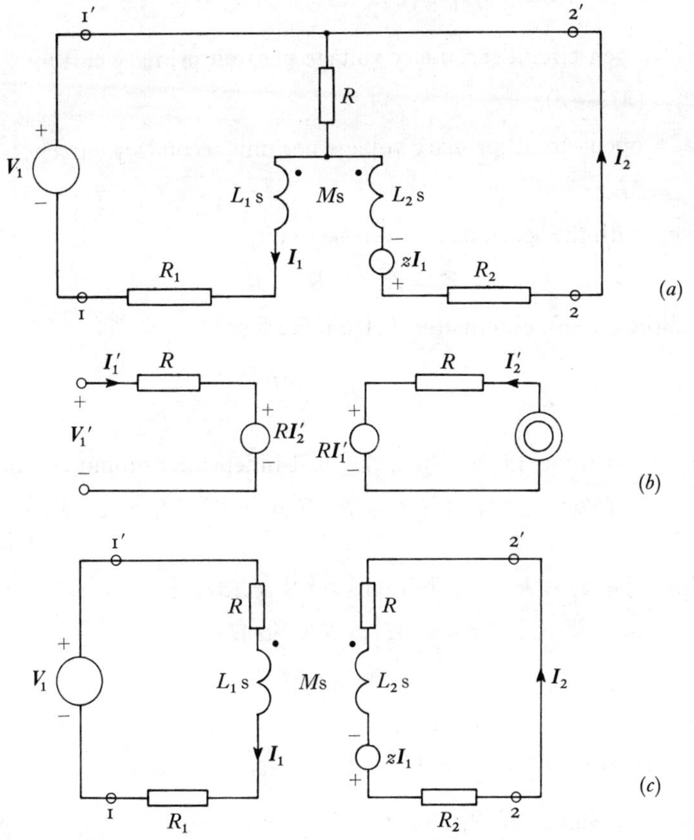

Fig. 12.28 (*a*) Voltage–current transfer network employing resistor *R* as feedback element. (*b*) Feedback network of (*a*) in canonical form. (*c*) Modified amplifier network derived from (*a*).

discover the threshold of instability, but may not exactly predict the behaviour of the actual circuit within the unstable zone.

As an example consider the circuit shown in fig. 12.28(*a*), where negative feedback is obtained by passing the output current through a resistor *R* which also forms part of the input circuit. The amplifier network is rendered non-unilateral by the presence of mutual inductance. The canonical form of the feedback network is shown in fig. 12.28(*b*), and the modified amplifier network, loaded with the impedances of the feedback network, in fig.

12.28(*c*). It will be seen that fig. 12.28(*b*) and (*c*) correspond with fig. 12.26(*a*) and (*b*) respectively.

For the modified amplifier network, the equations are

$$\left.\begin{aligned}(L_1 s + R_1 + R)\boldsymbol{I}_1 + Ms\boldsymbol{I}_2 = \boldsymbol{V}_1,\\ (Ms - z)\boldsymbol{I}_1 + (L_2 s + R_2 + R)\boldsymbol{I}_2 = 0.\end{aligned}\right\} \qquad (12.8.20)$$

Thus $\boldsymbol{z}_f$ = open-circuit secondary voltage per unit primary current

$= (Ms + z)$,

and $\boldsymbol{z}_r$ = open-circuit primary voltage per unit secondary current

$= Ms$.

Moreover, with the secondary on open-circuit,

$$\boldsymbol{Z}_i = L_1 s + R_1 + R,$$

and on short-circuit, eliminating $\boldsymbol{I}_2$ from (12.8.20),

$$\boldsymbol{Z}_i = L_1 s + R_1 + R - \frac{Ms(Ms - z)}{L_2 s + R_2 + R}.$$

At a particular frequency ω, then, the load-independent input condition is

$$|j\omega M(z - j\omega M)| \ll |(R_1 + R + j\omega L_1)(R_2 + R + j\omega L_2)|. \qquad (12.8.21)$$

Also, since $\boldsymbol{z}'_r = \boldsymbol{z}'_f = R$, the conditions (12.8.17, 18) are

$$|z - j\omega M| \gg R \gg |j\omega M|,$$

or

$$z \gg R \gg \omega M. \qquad (12.8.22)$$

With these, (12.8.21) is reduced to

$$(\omega M z)^2 \ll [(R_1 + R)^2 + (\omega L_1)^2]\,[(R_2 + R)^2 + (\omega L_2)^2]. \qquad (12.8.23)$$

For any given values of R_1, R_2, L_1, L_2, it is possible to assign ranges of values for z, R and M, such that (12.8.23) is satisfied for all frequencies. At any given frequency, values of z, R and M within these ranges may be chosen so that the inequalities (12.8.22) are also satisfied; but no fixed values of R and M can make $R \gg \omega M$ for all values of ω. At any given frequency, therefore, the conditions for the validity of the feedback representation may be satisfied; but this representation cannot be valid with a step function input, or for any other input which contains frequencies up to infinity.

In this simple example it would have been just as easy to write down the equations for the original circuit of fig. 12.28(*a*); but with more complicated amplifier networks the feedback approach offers a real analytical advantage, as well as giving a visual picture of the mode of functioning of the complete network.

12.9 The stability criterion of Routh and Hurwitz

A system is stable when, and only when, the poles of its transfer function have negative real parts. Thus the information as to whether it is stable is contained in the coefficients of the polynomial in s which forms the denominator of the transfer function. The stability criterion now to be discussed enables stability conditions to be deduced from the coefficients of the polynomial, without finding its roots.

Let the transfer function be written

$$\boldsymbol{H}(\mathrm{s}) = \frac{\mathrm{P(s)}}{\mathrm{Q(s)}}, \tag{12.9.1}$$

where P(s), Q(s) are polynomials without a common factor, and

$$\mathrm{Q(s)} = a_0\,\mathrm{s}^n + a_1\,\mathrm{s}^{n-1} + \ldots + a_{n-1}\,\mathrm{s} + a_n. \tag{12.9.2}$$

If the roots of the equation Q(s) = 0 do in fact have negative real parts, Q(s) is a product of factors of this form:

$$\begin{aligned}\mathrm{Q(s)} &= a_0(\mathrm{s}+\alpha_1)(\mathrm{s}+\alpha_2)\ldots(\mathrm{s}+\beta_1+\mathrm{j}\gamma_1)(\mathrm{s}+\beta_1-\mathrm{j}\gamma_1)\ldots,\\ &= a_0(\mathrm{s}+\alpha_1)(\mathrm{s}+\alpha_2)\ldots([\mathrm{s}+\beta_1]^2+\gamma_1^2)\ldots,\end{aligned}$$

where the quantities α, β, γ are all positive. The form of this product shows that every term from s^0 to s^n must be present and must have the same sign. Without loss of generality, it may be assumed that a_0 is positive; therefore

A necessary condition for the stability of a system is that the denominator of its transfer function, written with its first term positive, shall have all its terms positive and no term missing. (12.9.3)

This condition is not sufficient, however, when n is greater than 2. For example, the problem discussed in §12.7 would have given the overall transfer function

$$\boldsymbol{H}(\mathrm{s}) = \frac{\boldsymbol{H}_0(\mathrm{s})}{1 + \{Kc_3/(\mathrm{s}^3 + c_1\,\mathrm{s}^2 + c_2\,\mathrm{s} + c_3)\}},$$

so

$$\mathrm{Q(s)} = \mathrm{s}^3 + c_1\,\mathrm{s}^2 + c_2\,\mathrm{s} + (K+1)\,c_3. \tag{12.9.4}$$

Comparing this with (12.9.2), we write

$$c_1 = a_1/a_0, \qquad c_2 = a_2/a_0, \qquad (K+1)\,c_3 = a_3/a_0;$$

the stability condition (12.7.12), which may be rewritten

$$c_1 c_2 - (K+1)\,c_3 > 0, \tag{12.9.5}$$

then becomes

$$a_1 a_2 - a_0 a_3 > 0. \tag{12.9.6}$$

This condition can replace $a_2 > 0$, since when the other a are positive it can only be satisfied if $a_2 > 0$. The necessary and sufficient conditions for all the roots of the cubic $a_0 s^3 + a_1 s^2 + a_2 s + a_3 = 0$ to have negative real parts would thus appear to be

$$[a_0 > 0], \qquad a_1 > 0, \qquad a_1 a_2 - a_0 a_3 > 0, \qquad a_3 > 0. \qquad (12.9.7)$$

Although the approach has not been quite general (because $s^3 + c_1 s^2 + c_2 s + c_3$ was a product of real factors, which restricts Q(s) as defined by (12.9.4)), this conclusion is of general validity.

The necessary and sufficient conditions for all the roots of a polynomial equation of any degree to have negative real parts were discovered in 1877 by E. J. Routh, the Canadian-born Cambridge mathematician who was Senior Wrangler in the year (1854) in which Clerk Maxwell was Second Wrangler.† They were independently derived by the German A. Hurwitz in 1895; despite the difference in date his name is usually coupled with Routh's, and indeed polynomials possessing the stated property are commonly called *Hurwitz* polynomials! Routh's discussion consists essentially of an application of the theorem numbered (12.3.6) to the function Q(s) (which has no poles except at infinity) in relation to the stability discrimination contour. The result is given in the form of a rule for discovering how many roots have positive real parts; if we merely desire the condition that there shall be no such roots, the rule may be stated in the following simplified form:

1. Having adjusted the signs if necessary so as to make a_0 positive, arrange alternate coefficients in two rows, thus;

$$a_0, a_2, a_4, a_6, \ldots$$
$$a_1, a_3, a_5, a_7, \ldots$$

2. Form a third row by cross-multiplication, thus:

$$(a_1 a_2 - a_0 a_3), (a_1 a_4 - a_0 a_5), (a_1 a_6 - a_0 a_7), \ldots$$

3. Form a fourth row by applying the same cross-multiplication procedure to the second and third rows:

$$(a_1 a_2 - a_0 a_3) a_3 - a_1(a_1 a_4 - a_0 a_5), (a_1 a_2 - a_0 a_3) a_5 - a_1(a_1 a_6 - a_0 a_7), \ldots$$

4. The rows diminish in length by one term for each two rows. The process is continued until the two final rows have terms in the first column only; the resulting assemblage of numbers has been called a *Routh–Hurwitz array*. Routh's stability criterion may then be stated as follows:

The necessary and sufficient conditions for the stability of a system, in which the denominator of the transfer function is Q(s), are that all the

† See E. J. Routh, *Dynamics of a system of rigid bodies*, part 2, pp. 164–76, Macmillan (1884).

terms in the first column of the Routh–Hurwitz array derived from Q(s) shall be positive. (12.9.8)

The vanishing of a term before the full array is complete implies instability.

The criterion evidently lends itself to the investigation of numerical cases. For example, if $Q(s) = s^4 + 7s^3 + 17s^2 + 17s + 6$, the Routh–Hurwitz array is:

$$\begin{array}{ccc} 1 & 17 & 6 \\ 7 & 17 & \\ 102 & 42 & \\ 1440 & & \\ 60480 & & \end{array}$$

The system is therefore stable. The arithmetic may be simplified by cancelling a common factor from any row:

$$\begin{array}{ccc} 1 & 17 & 6 \\ 7 & 17 & \\ 102 \div 6 = 17 & 42 \div 6 = 7 & \\ 240 \div 240 = 1 & & \\ 7 & & \end{array}$$

Again, if $Q(s) = s^4 + 2s^3 + 7s^2 + 16s + 5$, the array is

$$\begin{array}{ccc} 1 & 7 & 5 \\ 2 & 16 & \\ -2 & 10 & \\ -52 & & \\ -520 & & \end{array}$$

The system is therefore unstable. It is unnecessary to proceed with the calculation after the appearance of the first negative sign.

As an algebraic example we may consider a negative feedback system with four time-constants; on the analogy of equation (12.9.4), this gives

$$Q(s) = s^4 + c_1 s^3 + c_2 s^2 + c_3 s + (K+1) c_4, \tag{12.9.9}$$

where K and the c are positive. The Routh–Hurwitz array is

$$\begin{array}{ccc} 1 & c_2 & (K+1)c_4 \\ c_1 & c_3 & \\ c_1 c_2 - c_3 & (K+1) c_1 c_4 & \\ c_1 c_2 c_3 - c_3^2 - (K+1) c_1^2 c_4 & & \\ (K+1) c_1 c_4 \times \text{the preceding.} & & \end{array}$$

Stability therefore requires $c_1 c_2 - c_3 > 0$ (a condition always satisfied when $s^4 + c_1 s^3 + c_2 s^2 + c_3 s + c_4$ is a product of real linear factors

$$(s + \alpha_1)(s + \alpha_2)(s + \alpha_3)(s + \alpha_4)),$$

and also

$$K > \left(\frac{c_1 c_2 c_3 - c_3^2}{c_1^2 c_4}\right) - 1. \tag{12.9.10}$$

12.10 Hurwitz polynomials

In the definition of stability in §12.4, a system was regarded as unstable when the application of a stimulus which was not sustained brought about a response which was sustained, even if that response did not increase. The marginal case would arise in a non-dissipative network without controlled energy sources, supposing that such a network could be constructed. If we view the dividing-line as separating, not stable networks from unstable, but passive from active, we are obliged to treat the marginal case as belonging to the passive or stable side, and to ask what difference this makes to the theory expounded so far.

If the response of a circuit to a spike function stimulus contains terms $A + B\cos\gamma t + C\sin\gamma t$, the transfer function, expressed in partial fractions, contains the terms

$$\frac{A}{s} + \frac{Bs + C\gamma}{s^2 + \gamma^2}.$$

The factors s, $(s^2 + \gamma^2)$, denoting poles on the imaginary axis at $s = 0$ and $\pm j\gamma$, therefore appear in the denominator of the transfer function. Factors such as s^2 or $(s^2 + \gamma^2)^2$ would indicate terms of the form t, $t\cos\gamma t$ or $t\sin\gamma t$; this would imply a steady increase in the amount of energy circulating from one part of the circuit to another, which cannot occur in a passive network. Thus the transfer function of a passive network may have poles to the left of the imaginary axis in the s-plane, and also poles of the first order or *simple poles* on the imaginary axis; but no others. For a network with a finite number of elements, the transfer function is expressible as a ratio of polynomials P(s)/Q(s), and the statement just made is equivalent to stating that Q(s) may have zeros of any order to the left of the imaginary axis or simple zeros upon it. It is convenient to use the term *Hurwitz polynomial* (already mentioned) for a polynomial possessing these properties, which are illustrated in fig. 12.29.

The general form of a Hurwitz polynomial is

$$Q(s) = s(s^2 + \gamma_1^2)(s^2 + \gamma_2^2) \ldots Q_1(s), \tag{12.10.1}$$

where $Q_1(s)$ contains the zeros to the left of the imaginary axis, and therefore (according to the theorem (12.9.3)) is a polynomial with every term positive and no term missing. Any or all of the added factors may be absent. From (12.10.1) can be deduced the following properties.

1. All the coefficients in a Hurwitz polynomial are positive.

2. It is only possible for the following categories of terms to be absent:
   (*a*) The constant term.
   (*b*) All odd powers of s.
   (*c*) All even powers of s. (12.10.2)

If a Hurwitz polynomial contains factors of the form s or $(s \pm j\gamma)$, the same quantities will be factors of the two polynomials formed by taking the even terms of the polynomial only, or the odd terms only. This also can be seen from (12.10.1).

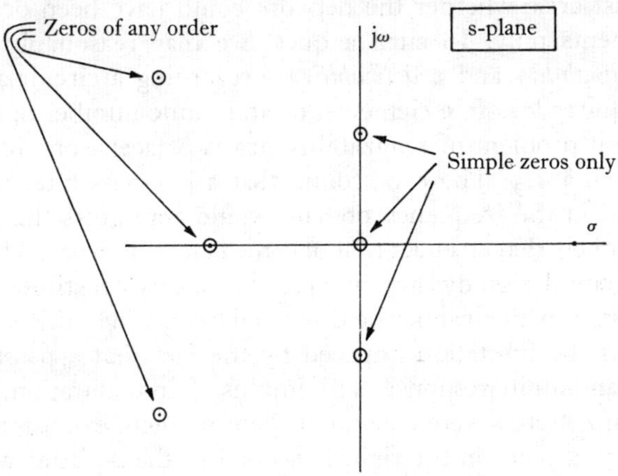

Fig. 12.29 Pattern of zeros for a typical Hurwitz polynomial.

The effect of factors such as $(s^2+\gamma^2)$ is to introduce zero terms into the Routh–Hurwitz array. Factors such as $(s^2+\gamma^2)^2$ would have the same effect, and we cannot here go into the intricate process of distinguishing the two cases; it is discussed by Routh in the book mentioned above. We content ourselves with the following statement:

If the Routh–Hurwitz array is constructed for a polynomial which satisfies the conditions (12.10.2), and all the terms of the first column are positive, the polynomial is a Hurwitz polynomial. If negative terms appear, it is not a Hurwitz polynomial. If zero appears, it may or may not be a Hurwitz polynomial. (12.10.3)

For example, the polynomial $s^5+2s^4+2s^3+4s^2+s+2$ has zero in the third row of the Routh–Hurwitz array; but it is not a Hurwitz polynomial in the sense defined, since its factors are $(s+j)^2(s-j)^2(s+2)$. This is readily apparent by separating the even and odd terms:

$$(s^5+2s^3+s)+(2s^4+4s^2+2),$$

because each bracket can be seen to contain the factor $(s^2+1)^2$.

12.11 Realizability

The complex plane plays a fundamental part, not only in the analysis of circuits, but in the inverse problem of synthesis – namely of designing circuits which shall have prescribed characteristics. An example of a synthesis procedure was given in §7.14, where the design of a maximally flat low-pass filter was discussed; but the network included controlled sources used as buffers, to enable elements to be added without interaction. Since the controlled sources were not required as energy supplies, the question must arise whether the network could have been designed with passive elements only. In such a quest we may reasonably distinguish between *realizability* and *practicability* – regarding a circuit as realizable even if it requires loss-free elements, or an infinite number of elements.

The general problem of realizability has two facets, one of which was encountered in §11.5. There we found that a low-pass filter which had a sudden cutoff in the frequency domain would transgress the principle of causality – namely that an effect cannot come before its cause. The limitation placed on circuit design by the principle of causality constitutes a large and difficult question which cannot be discussed here. When this is eliminated, there remains the limitation imposed by the fact that a passive network cannot have an infinite response to a stimulus of finite duration. This places on the design a more severe constraint than the mere exclusion of transfer functions having poles in the right hand half of the s-plane; we shall now seek to discover, in the simplest case, what that constraint is.

12.12 Positive-real functions

The problem which will be studied is that of determining the conditions under which a oneport network, having a prescribed impedance $\boldsymbol{Z}(\mathrm{s})$ or admittance $\boldsymbol{Y}(\mathrm{s})$, is realizable. (To cover the case of networks having other ports which are ignored, these quantities are often referred to as the *driving-point* impedance or admittance.)

If a network having impedance $\boldsymbol{Z}(\mathrm{s})$ is supplied with alternating current at frequency ω, we have

$$\boldsymbol{V}(\mathrm{j}\omega) = \boldsymbol{Z}(\mathrm{j}\omega)\boldsymbol{I}(\mathrm{j}\omega). \qquad (12.12.1)$$

$\boldsymbol{I}(\mathrm{j}\omega)$ may be taken as the reference quantity denoted by the real positive number I, and $\boldsymbol{Z}(\mathrm{j}\omega)$ may be written as real and imaginary parts $Z_R + \mathrm{j}Z_I$. Therefore the mean power input is given, as in (4.11.2), by

$$P = \tfrac{1}{2}\,\mathrm{Re}\,\boldsymbol{V}\boldsymbol{I}^* = \tfrac{1}{2}Z_R I^2, \qquad (12.12.2)$$

if $\boldsymbol{V}$, $\boldsymbol{I}$ are peak values. It is at once apparent that, for a passive network, Z_R can never be negative. The same applies to the real part, Y_R, of an admittance $\boldsymbol{Y}(\mathrm{j}\omega)$. These statements can be expressed in geometrical terms:

The impedance and admittance loci of a passive network cannot contain any points to the left of the imaginary axis. (12.12.3)

Figure 12.30 shows the pole–zero pattern in the s-plane, and the associated impedance locus in the $\boldsymbol{Z}(s)$-plane, for a particular network. The

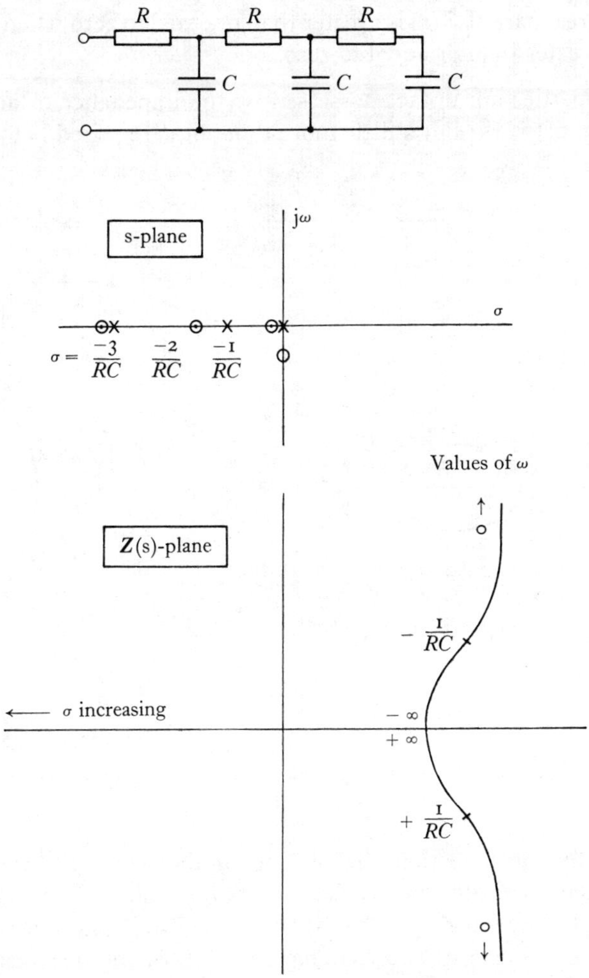

Fig. 12.30 Pole–zero pattern and impedance locus for the network shown.

latter diagram is a conformal map of the former, in which the complete map covers the $\boldsymbol{Z}(s)$-plane three times; the pole-and-zero-free region to the right of the imaginary axis in the s-plane is mapped to the right of the impedance locus. The real axis in the s-plane is mapped as the real axis in the $\boldsymbol{Z}(s)$-plane, since $\boldsymbol{Z}(s)$ is a ratio of two polynomials with real coefficients; the zeros and poles are mapped at the origin and at infinity respectively. The impedance locus is the map of the imaginary axis in the s-plane; this has been labelled with values of ω.

It would appear therefore, that $\boldsymbol{Z}(s)$ possesses two properties:
1. $\boldsymbol{Z}(s)$ is real when s is real.
2. The real part of $\boldsymbol{Z}(s)$ is greater than or equal to zero when the real part of s is greater than or equal to zero. (12.12.4)

These properties are in fact possessed by the impedance or admittance of any passive network; but a little more thought is required before accepting

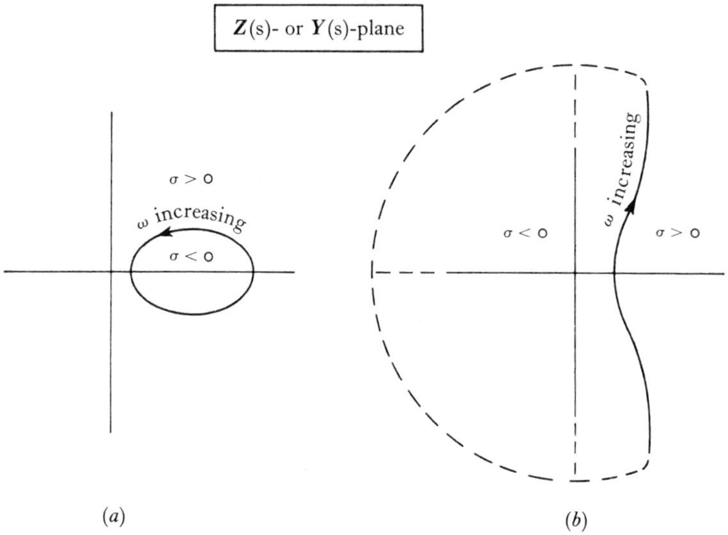

Fig. 12.31 Inadmissible types of impedance or admittance locus.

unreservedly the idea that the placing of the locus in accordance with (12.12.3) automatically means that points for which $\sigma > 0$ must lie in the right hand half of the $\boldsymbol{Z}(s)$ or $\boldsymbol{Y}(s)$-plane. The two possibilities which might vitiate this statement are shown in fig. 12.31(*a*), where the locus is described in such a way that points for which $\sigma > 0$ are mapped outside a closed region; and in fig. 12.31(*b*), where the locus is closed at infinity by an arc on the left hand side of the imaginary axis. The former type is ruled out by the fact that the point at infinity lies to the right of a path so described, which would mean that there were poles to the right of the imaginary axis in the s-plane – and we know that there are none. A passage of the locus through infinity, as in fig. 12.28(*b*), occurs whenever there are poles on the imaginary axis, and it becomes necessary to investigate briefly the origin and nature of such poles.

Consider an impedance function which has poles at $s = 0$, $s = \pm j\gamma$, so that it can be written

$$\boldsymbol{Z}(s) = \frac{P(s)}{s(s - j\gamma)(s + j\gamma)\, Q_1(s)}, \tag{12.12.5}$$

where P(s), $Q_1(s)$ are polynomials which do not have s or $(s \pm j\gamma)$ as factors. A partial fraction expansion of $\boldsymbol{Z}(s)$ yields the alternative form

$$\boldsymbol{Z}(s) = \frac{A}{s} + \frac{B}{s - j\gamma} + \frac{B^*}{s + j\gamma} + \frac{P_1(s)}{Q_1(s)}, \tag{12.12.6}$$

where $P_1(s)$ is another polynomial, and B* is the conjugate of B. The added terms control the behaviour of $\boldsymbol{Z}(s)$ in the immediate neighbourhood of the respective poles. Very near to the origin, for example, $\boldsymbol{Z}(s) \approx A/s$. A corresponding approximation derived from (12.12.5) is $\boldsymbol{Z}(s) \approx P(0)/\gamma^2 Q_1(0)s$, so that we must have

$$\left.\begin{aligned} A &= \frac{P(0)}{\gamma^2 Q_1(0)}, \\ \text{and similarly} \qquad B &= \frac{P(j\gamma)}{-2\gamma^2 Q_1(j\gamma)}. \end{aligned}\right\} \tag{12.12.7}$$

A, B are what remains of the function near the pole when the factor containing the pole is removed, and are called the *residues* of the function at the poles in question. A is evidently real. B looks like a complex number $B_R + jB_I$, but can be shown to be real as follows. By combining the two terms from the conjugate poles, we obtain the real expansion formula

$$\boldsymbol{Z}(s) = \frac{A}{s} + \frac{2(B_R s - B_I \gamma)}{s^2 + \gamma^2} + \frac{P_1(s)}{Q_1(s)}. \tag{12.12.8}$$

At the real frequency ω, this becomes

$$\boldsymbol{Z}(j\omega) = \frac{A}{j\omega} + \frac{2(B_R j\omega - B_I \gamma)}{\gamma^2 - \omega^2} + \frac{P_1(j\omega)}{Q_1(j\omega)}. \tag{12.12.9}$$

If $B_I \neq 0$, the term $-2B_I\gamma/(\gamma^2 - \omega^2)$ can contribute a negative real part, at frequencies either below or above γ; and this can be made as large as we please by making ω near enough to γ. Thus the assumption $B_I \neq 0$ conflicts with the fundamental requirement that $\boldsymbol{Z}(j\omega)$ must always have a positive real part; therefore $B_I = 0$, and B is real.

We can now show that both A and B must be positive. Let a step current $i(t) = I_1 u(t)$, or $\boldsymbol{I}(s) = I_1/s$, be forced through the circuit. The voltage

$\boldsymbol{Z}(s)\boldsymbol{I}(s)$ starts with the term AI_1/s^2, which is the image of $AI_1 t$. Increasing indefinitely with t, this must ultimately preponderate over all the other terms; but it only represents a positive input of energy if A is positive. Again, let a sinusoidal current $i(t) = I_1 \sin\gamma t u(t)$, or $\boldsymbol{I}(s) = I_1\gamma/(s^2+\gamma^2)$, be forced through. The voltage now contains the term $2BI_1\gamma s/(s^2+\gamma^2)^2$, which is the image of $BI_1 t\sin\gamma t$. This becomes the preponderant term, and only represents a positive input of energy if B is positive. We have therefore proved that

> The impedance $\boldsymbol{Z}(s)$ or admittance $\boldsymbol{Y}(s)$ of a passive network can have only simple poles on the imaginary axis in the s-plane, and the residues at those poles are real and positive. (12.12.10)

Physically the separation of the terms A/s and $2Bs/(s^2+\gamma^2)$ from $\boldsymbol{Z}(s)$ may be regarded as equivalent to separating $\boldsymbol{Z}(s)$ into three series-connected components as shown in fig. 12.32. This is not the same as proving that the

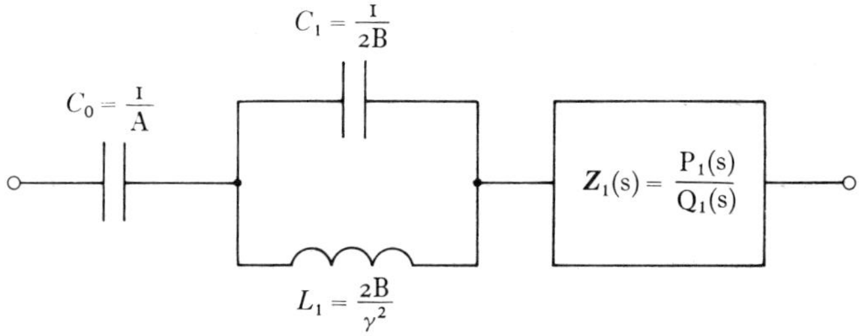

Fig. 12.32 Physical interpretation of an impedance function with poles on the imaginary axis.

circuit *must* look like this; if it were so, A and B would evidently be positive. What we have proved is that no ingenuity in the connexions of a passive network (which may include ideal transformers connected in either polarity) can cause it to simulate negative inductance or capacitance in the situations illustrated.

We can now return to considering the behaviour of the impedance or admittance locus at infinity. Take the impedance as expressed by (12.12.6), and let the pole at $s = j\gamma$ be circumvented, as in fig. 12.16, by a small semicircle to the right. On this,

$$s - j\gamma = r_0 e^{j\theta},$$

where θ changes from $-\pi/2$ to $+\pi/2$. The contribution to $\boldsymbol{Z}(s)$ is

$$\frac{B}{s-j\gamma} = \frac{B}{r_0} e^{-j\theta},$$

and this is the preponderant term since r_0 is very small. With B real and positive, the change of θ from $-\pi/2$ to $+\pi/2$ takes $\boldsymbol{Z}(s)$ round a large semi-circle of radius B/r_0 on the right hand half of the $\boldsymbol{Z}(s)$-plane. The locus shown in fig. 12.31(*b*) is therefore impossible.

By a process which may be called a closing of loopholes, we have now proved that the impedance or admittance of any passive network possesses the two properties numbered (12.12.4). Functions possessing these properties are called *positive-real functions* (the adjective is best hyphenated, since it describes a new property distinct from those described by the words 'positive' and 'real'); the abbreviation 'p.r. functions' is common. The definition is repeated here for convenience:

A function w(s) of a complex variable s is said to be positive-real if:
1. w(s) is real when s is real.
2. Re w(s) $\geqslant 0$ when Re s $\geqslant 0$. (12.12.11)

In establishing that the impedance or admittance of a passive network possesses this property, we have established the following conditions as equivalent; they are in fact both necessary and sufficient:

The necessary and sufficient conditions for a function w(s) to be positive-real are
1. w(s) is regular in the right hand half of the s-plane.
2. Any poles of w(s) on the imaginary axis of the s-plane are simple poles with positive, real residues.
3. Re w(jω) $\geqslant 0$ for all real values of ω. (12.12.12)

There is no necessity for these definitions to be confined to rational algebraic functions; but we are not envisaging any but rational algebraic functions here.

The converse proposition can also be proved; that, given any rational, algebraic, positive-real function of s, a network can be found having that function as its impedance or admittance. There will be no attempt to prove this here. Both proposition and converse were established by Otto Brune at the Massachusetts Institute of Technology, in what must surely be one of the most pregnant Ph.D investigations ever undertaken.† He unlocked the gate of the huge field of network synthesis. To synthesize a oneport passive network having a prescribed impedance or admittance, it is first necessary to establish that the proposed function is positive-real; after this, the function is split up, according to systematic rules, into portions whose physical embodiment is ascertainable. Fig. 12.33 illustrates the beginning of such a process; but we cannot follow this track very far, though an example will be given in the next section. In view of the fundamental importance of

† O. Brune, 'Synthesis of a finite 2-terminal network whose driving point impedance is prescribed', *Journal of Mathematics and Physics*, **10**, 191–236 (1931).

positive-real functions and their close relation with other topics discussed in this chapter, it has seemed right to introduce them; to conclude this section a few of their properties will be developed.

If $\boldsymbol{Z}(s)$ is the impedance of a network, $[\boldsymbol{Z}(s)]^{-1}$ is its admittance. Both are positive-real; therefore

If w(s) is a positive-real function, so is $[w(s)]^{-1}$. (12.12.13)

Impedances in series are added; therefore

If $w_1(s)$ and $w_2(s)$ are positive-real functions, so is $w_1(s)+w_2(s)$. (12.12.14)

In order that $\boldsymbol{Z}(s) \equiv P(s)/Q(s)$ and $\boldsymbol{Y}(s) \equiv Q(s)/P(s)$ may both satisfy conditions (1) and (2) of (12.12.12), it is necessary for P(s) and Q(s) to be both Hurwitz polynomials. Thus

If w(s), a ratio of two polynomials, is positive-real, both numerator and denominator must be Hurwitz polynomials. (12.12.15)

This condition is necessary but not sufficient; it guarantees that there shall only be simple poles on the imaginary axis, but not that the residue at those poles shall be real and positive.

If the lowest powers of s in P(s) and Q(s) differ by two or more, either P(s)/Q(s) or Q(s)/P(s) has a pole at the origin of order higher than the first. This is ruled out by condition (2) of (12.12.12). Similarly, if the highest powers of s differ by two or more, one of the ratios has a pole of order two or more at infinity. Therefore

If P(s)/Q(s) is positive-real, the lowest and the highest powers of s in P(s) and Q(s) respectively can differ at most by one. (12.12.16)

The foregoing will now be applied to determine whether a number of plausible impedance or admittance functions are realizable.

12.13 Some examples of the assessment of realizability

The criteria numbered (12.12.12) are like three sieves for catching unrealizable functions, each finer than its predecessor. Only functions which get through all three are realizable.

The requirement that w(s) shall be regular in the right hand half of the s-plane is met if the denominator of w(s) is a Hurwitz polynomial; we have shown that a later condition requires the same to be true of the numerator. The first test, therefore, is to ascertain whether P(s) and Q(s) are Hurwitz polynomials; if they satisfy (12.10.2) as preliminary conditions, they can be further tested by factorization or by the use of Routh's criterion.

To rule out the possibility of multiple poles on the imaginary axis, and to ascertain whether the residues at the simple poles are real and positive, it is necessary to identify what factors of Q(s) are of the form s or $s \pm j\gamma$. This will be done by separating the even and odd terms in Q(s),

$$Q(s) = Q_E(s) + Q_O(s), \tag{12.13.1}$$

and looking for common factors of form s or $s^2 + \gamma^2$ in $Q_E(s)$ and $Q_O(s)$. The function P(s)/Q(s) will already have been scrutinized to see that the highest and lowest powers of s in P(s) and Q(s) do not differ by more than 1.

As the final stage in proving that w(s) is positive-real, it is necessary to establish that the locus $w(j\omega)$ does not go to the left of the imaginary axis. This may be accomplished by direct plotting, which will involve the separation of the even and odd terms in P(s) and Q(s), since these contribute real and imaginary parts respectively when $s = j\omega$.

The following examples illustrate these points. A real constant has been omitted from each function, leaving w(s) as a ratio of polynomials in which the first coefficients are unity.

1. $$w(s) = \frac{s^2 + 5s + 6}{s^3 + 5s + 6}.$$

The denominator has one even term missing and one present. It therefore cannot be a Hurwitz polynomial, and w(s) is unrealizable. (The real factors are $(s+1)(s^2 - s + 6)$; the second term indicates conjugate poles to the right of the imaginary axis.)

2. $$w(s) = \frac{s^4 + 5s^2 + 4}{s^4 + 5s^2 + 6}.$$

The polynomials can be factorized into

$$\frac{(s^2+1)(s^2+4)}{(s^2+2)(s^2+3)},$$

which shows that they satisfy the Hurwitz condition of having only simple zeros on the imaginary axis and none to the right of it. The residue at $s = j\sqrt{2}$ is given by

$$\frac{(-2+1)(-2+4)}{(j2\sqrt{2})(-2+3)} = \frac{j}{\sqrt{2}}.$$

The imaginary residue shows that, although w(s) is a ratio of two Hurwitz polynomials, it is unrealizable. Quadratic factors of this form arise from series or parallel *LC* combinations, without resistance; if the function were realizable it would mean that resistance could be simulated by a circuit containing inductance and capacitance only, which is impossible.

3. $$w(s) = \frac{(s^2+1)(s^2+4)}{s(s^2+2)(s^2+3)}$$

is a form which is not open to the objection just mentioned. The polynomials are Hurwitz polynomials, and the residues at $s = 0$, $j\sqrt{2}$, $-j\sqrt{2}$, $j\sqrt{3}$, $-j\sqrt{3}$ are $\frac{2}{3}$, $\frac{1}{2}$, $\frac{1}{2}$, $-\frac{1}{3}$, $-\frac{1}{3}$. The presence of negative residues shows that w(s) is unrealizable.

4. The change of sign between the residues at $j\sqrt{2}$ and $j\sqrt{3}$ in the last example is due to the fact that the latter brings in an extra negative factor $(-3+2)$ in the denominator. In the somewhat similar form

$$w(s) = \frac{(s^2+1)(s^2+3)}{s(s^2+2)(s^2+4)},$$

each extra negative factor appearing in the denominator is balanced by one in the numerator; written in factors, the residues at $s = 0, j\sqrt{2}, j2$ are

$$\frac{(1)(3)}{(2)(4)} = \frac{3}{8}; \qquad \frac{(-1)(1)}{(j\sqrt{2})(j2\sqrt{2})(2)} = \frac{1}{8}; \qquad \frac{(-3)(-1)}{(j2)(-2)(j4)} = \frac{3}{16}.$$

The function thus passes the 'real positive residues' test. It can indeed be shown that in any loss-free network the poles and zeros (which necessarily lie on the imaginary axis) occur alternately, as they do here. $w(j\omega)$ is purely imaginary for all values of ω, so the condition $\operatorname{Re} w(j\omega) \geqslant 0$ is satisfied. Thus the function is realizable.

If we use a multiplying constant $(1/C)$ and write

$$\boldsymbol{Z}(s) = \frac{(s^2+1)(s^2+3)}{Cs(s^2+2)(s^2+4)}, \tag{12.13.2}$$

this may be expressed in partial fractions

$$\boldsymbol{Z}(s) = \frac{3}{8}\frac{1}{Cs} + \frac{1}{8C}\left(\frac{1}{s-j\sqrt{2}} + \frac{1}{s+j\sqrt{2}}\right) + \frac{3}{16C}\left(\frac{1}{s-j2} + \frac{1}{s+j2}\right),$$

$$= \frac{3}{8Cs} + \frac{1}{4C}\frac{s}{s^2+2} + \frac{3}{8C}\frac{s}{s^2+4}. \tag{12.13.3}$$

The terms are recognizable as the impedances of a capacitor and of two parallel *LC* circuits, and fig. 12.33 is a circuit which realizes the impedance (12.13.2).

5. As a final, more general example, consider

$$w(s) = \frac{s^2 + a_1 s + a_2}{s^2 + b_1 s + b_2}, \tag{12.13.4}$$

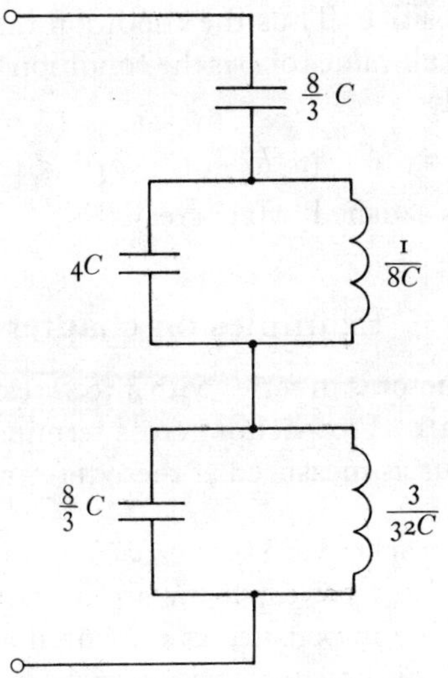

Fig. 12.33 Synthesis of network with impedance

$$\boldsymbol{Z}(s) = \frac{(s^2+1)(s^2+3)}{Cs(s^2+2)(s^2+4)}.$$

where all the coefficients are positive. This at once means that the polynomials are Hurwitz polynomials (the conditions for a quadratic were investigated in §12.4). Moreover they have no zeros on the imaginary axis. w(s) is therefore realizable provided that the real part of w(jω) never goes negative; and

$$\mathrm{w}(j\omega) = \frac{(a_2-\omega^2)+ja_1\omega}{(b_2-\omega^2)+jb_1\omega},$$

$$= \frac{[(a_2-\omega^2)+ja_1\omega][(b_2-\omega^2)-jb_1\omega]}{(b_2-\omega^2)^2+b_1^2\omega^2}.$$

The denominator being essentially positive, it is only necessary to examine the sign of the real part of the numerator, namely

$$(a_2-\omega^2)(b_2-\omega^2)+a_1b_1\omega^2,$$

or

$$\omega^4+(a_1b_1-a_2-b_2)\omega^2+a_2b_2. \tag{12.13.5}$$

This is a quadratic in ω^2. If its zeros are real, its form is

$$f(\omega^2) = (\omega^2-\alpha)(\omega^2-\beta),$$

and $f(\omega^2)$ is negative when ω^2 lies between α and β. If its zeros are complex, its form is

$$f(\omega^2) = (\omega^2-\alpha)^2+\beta^2,$$

which is always positive. Thus the condition that the expression (12.13.5) may be positive for all values of ω is the condition that its zeros are coincident or complex, namely

$$(a_1 b_1 - a_2 - b_2)^2 \leqslant 4a_2 b_2. \qquad (12.13.6)$$

If this condition is satisfied, w(s) is realizable.

Examples on chapter 12

12.1 A 1 mH inductor is in series with a 50 Ω resistor, and the combination is in parallel with a 1 μF capacitor across terminals AB. Derive and sketch the admittance locus as measured at these terminals.

12.2 A 1 H inductor and a 1 μF capacitor are in parallel across EF; a 1000 Ω resistor and a 1 μF capacitor are in parallel across FG. Find the poles and zeros of the impedance as measured across EG, and obtain the impedance locus. Show how the position of the locus relatively to the origin, and its behaviour at infinity, are related to the positions of the zeros and poles in the s-plane.

12.3 Show that the conformal transformation $w = (s - 1)/(s + 1)$ maps the right hand half of the s-plane on the interior of a circle of radius 1. What are the maps of the small circular contours $(\sigma - 0.5)^2 + \omega^2 = (0.01)^2$, $(\sigma - 2)^2 + \omega^2 = (0.01)^2$? Show that any straight line in the s-plane is mapped as a circular arc in the w-plane.

12.4 A system has the following forward transfer locus:

ω	10	20	30	40
$\boldsymbol{H}_0(j\omega)$	$-1.76\ -j0.4$	$-1.22\ +j0.12$	$-0.71\ +j0.28$	$-0.27\ +j0.22$

Show that with unity feedback it will be unstable. If the loop transfer function is modified by the addition of a phase advance network having a transfer function $(1 + 0.012\ s)$, determine the new loop transfer locus and estimate the gain margin of the system.

12.5 Show that a system having the loop transfer function

$$10\pi^2/(-\omega^2 + j\pi\omega)$$

is stable, and determine its phase margin.

12.6 Determine the conditions for the stability of a system in which the loop transfer function is given by

$$\boldsymbol{H}_0(j\omega)\,\boldsymbol{F}(j\omega) = \frac{K(1 + T_2 j\omega)}{j\omega(1 + T_1 j\omega)^2}.$$

Show that increasing T_2 changes the system from a conditionally stable to an inherently stable condition; find the value of T_2 at which this occurs, and sketch Nyquist diagrams for both conditions.

12.7 Determine the chief features of the loop transfer locus for a system having the loop transfer function

$$\boldsymbol{H}_0(s)\,\boldsymbol{F}(s) = \frac{1}{s(s^2+0.2s+1)},$$

and hence ascertain whether the system will be stable when the feedback loop is closed.

12.8 An active network has a forward transfer function

$$\boldsymbol{H}_0(s) = \frac{1}{(0.6s+1)(s+1)(3s^2+s+1)}$$

when the feedback loop is open; there is negative feedback of a fraction K of the output. Determine the leading features of the loop transfer locus, and hence find the value of K which brings the network to the verge of instability when the feedback loop is closed.

12.9 A field-effect transistor may be assumed to act like a three-terminal voltage-controlled current source with terminals G (gate), D (drain) and S (source). The controlling voltage $\boldsymbol{V}_{gs}$ is applied between G and S (G conventionally positive); this port is an open circuit. Connected between D and S is the current source $y_f \boldsymbol{V}_{gs}$ (directed towards S), with a resistance r_d in parallel.

Such a transistor is used in an amplifier circuit employing voltage-ratio feedback. Between D and S are connected two sets of resistors:

(i) The load resistor R_L, across which appears the output voltage $\boldsymbol{V}_o$ (D conventionally positive).

(ii) R_1, R_2 in series, with R_1 adjacent to D. The input voltage $\boldsymbol{V}_s$ is connected between G and the junction of R_1, R_2 (G conventionally positive).

Redraw this circuit with the amplifier and feedback networks separated with series–parallel connexions, and show that all the conditions of validity for the feedback approach are satisfied. Hence derive a formula for the voltage amplification $\boldsymbol{V}_o/\boldsymbol{V}_s$.

12.10 A system employing unity feedback has the forward transfer function

$$\boldsymbol{H}_0(s) = \frac{40}{(0.1s+1)(0.2s+1)(s+1)}.$$

By means of Routh's stability criterion, determine whether or not the system will be stable.

12.11 Ascertain whether the polynomial $s^5 + 8s^4 + s^3 + 12s^2 + 18s + 6$ is a Hurwitz polynomial. If it is not, determine the number of zeros in the right hand half of the s-plane.

12.12 The transfer function of a closed-loop system has the denominator

$$\frac{T_1 T^3}{16} s^4 + \frac{T_1 T^2}{4} s^3 + T_1 T s^2 + (T_1 + T) s + 1.$$

Determine the value of the time-constant T_1 in terms of the time-constant T in order that the system may be conditionally stable.

12.13 Which of the following functions can be seen to be non-positive-real by simple inspection? For the cases where the function is not positive-real, state why it is not.

$$(a)\ \frac{s + 0.2}{(s+1)^2}; \qquad (b)\ \frac{(s+1)(s-1)}{(s-1+j1)(s-1-j1)}; \qquad (c)\ -s;$$

$$(d)\ \frac{5+j4}{s+3+j4} + \frac{5-j4}{s+3-j4} + \frac{j6}{s+j3} + \frac{-j6}{s-j3}.$$

12.14 Are the following functions realizable as the impedances of passive networks?

$$(a)\ \frac{s^3 + 4s^2 + 4s + 3}{s^4 + 2s^3 + 4s^2 + 8s + 16}; \qquad (b)\ \frac{s^3 + 4s^2 + 4s + 3}{s^4 + 4s^3 + 8s^2 + 16s + 16}.$$

12.15 Two reactance functions $Z_1(s)$ and $Z_2(s)$ have zeros at $s = 0, \pm j2, \pm j4$ only. At these frequencies $dZ_1/ds = 1, 2, 3$ respectively, and $dZ_2/ds = 1, 4, 3$ respectively. If $Y_1 = 1/Z_1$ and $Y_2 = 1/Z_2$, show that the admittance $Y_1 - Y_2$ is realizable and find a network which realizes it.

12.16 Realize $Z(s) = \dfrac{24s^3 + 10s}{(2s^2+1)(3s^2+1)}$ in one way.

13 Distributed circuits

13.1 Periodic and distributed circuits

If a number of identical twoport networks are connected so that the output of the network numbered n is the input of that numbered $(n+1)$, the structure of the resulting network may be described as *periodic*. Fig. 13.1 shows two such networks; in each the series impedance of an element is $\boldsymbol{Z}$

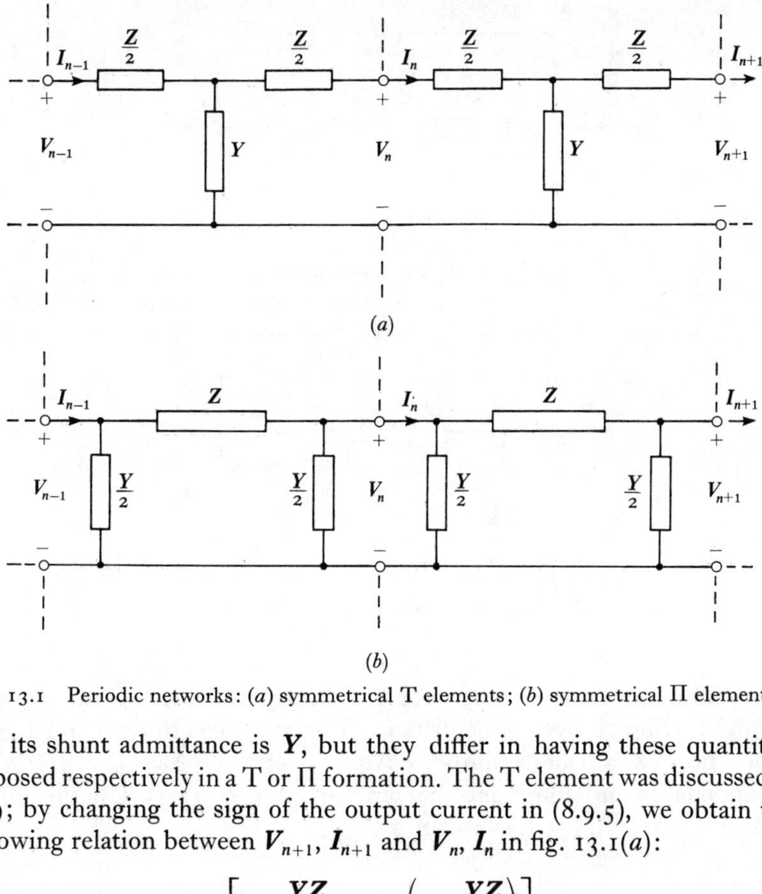

Fig. 13.1 Periodic networks: (*a*) symmetrical T elements; (*b*) symmetrical Π elements.

and its shunt admittance is $\boldsymbol{Y}$, but they differ in having these quantities disposed respectively in a T or Π formation. The T element was discussed in §8.9; by changing the sign of the output current in (8.9.5), we obtain the following relation between $\boldsymbol{V}_{n+1}$, $\boldsymbol{I}_{n+1}$ and $\boldsymbol{V}_n$, $\boldsymbol{I}_n$ in fig. 13.1(*a*):

$$\begin{bmatrix} \boldsymbol{V}_{n+1} \\ \boldsymbol{I}_{n+1} \end{bmatrix} = \begin{bmatrix} 1+\dfrac{\boldsymbol{YZ}}{2} & -\boldsymbol{Z}\left(1+\dfrac{\boldsymbol{YZ}}{4}\right) \\ -\boldsymbol{Y} & 1+\dfrac{\boldsymbol{YZ}}{2} \end{bmatrix} \begin{bmatrix} \boldsymbol{V}_n \\ \boldsymbol{I}_n \end{bmatrix}. \qquad (13.1.1)$$

In a similar manner, for the Π element of fig. 13.1(*b*):

$$\begin{bmatrix} V_{n+1} \\ I_{n+1} \end{bmatrix} = \begin{bmatrix} 1+\dfrac{YZ}{2} & -Z \\ -Y\left(1+\dfrac{YZ}{4}\right) & 1+\dfrac{YZ}{2} \end{bmatrix} \begin{bmatrix} V_n \\ I_n \end{bmatrix}. \tag{13.1.2}$$

Where the number of elements, N, is large, the performance of these circuits approximates to that of circuits in which the total series impedance

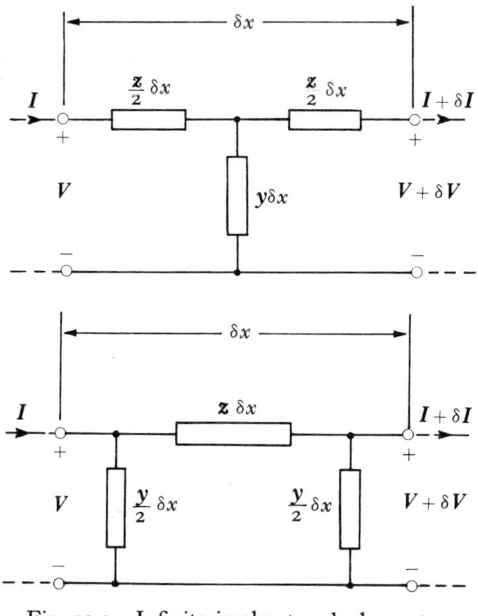

Fig. 13.2 Infinitesimal network elements.

NZ or shunt admittance NY, instead of being in N discrete units, is uniformly distributed over some distance d. A small length δx then contains impedance $NZ\delta x/d$ and admittance $NY\delta x/d$, or $z\delta x$, $y\delta x$, where z, y stand for impedance and admittance per unit length (fig. 13.2). For the infinitesimal T element, (13.1.1) becomes

$$\begin{bmatrix} V+\delta V \\ I+\delta I \end{bmatrix} = \begin{bmatrix} 1+\dfrac{yz}{2}\delta x^2 & -z\,\delta x\left(1+\dfrac{yz}{4}\delta x^2\right) \\ -y\,\delta x & 1+\dfrac{yz}{2}\delta x^2 \end{bmatrix} \begin{bmatrix} V \\ I \end{bmatrix},$$

or
$$\begin{bmatrix}\delta V\\ \delta I\end{bmatrix}=\begin{bmatrix}\dfrac{yz}{2}\delta x^2 & -z\,\delta x\left(1+\dfrac{yz}{4}\delta x^2\right)\\ -y\,\delta x & \dfrac{yz}{2}\delta x^2\end{bmatrix}\begin{bmatrix}V\\ I\end{bmatrix}.$$

Neglecting δx^2, one obtains

$$\begin{bmatrix}\delta V\\ \delta I\end{bmatrix}=\begin{bmatrix}0 & -z\,\delta x\\ -y\,\delta x & 0\end{bmatrix}\begin{bmatrix}V\\ I\end{bmatrix}, \tag{13.1.3}$$

and it is evident that the infinitesimal Π element leads to the same equation. The distribution of impedance and admittance within the infinitesimal element is thus immaterial, and the equations are much simplified; but the approximation has obscured the small difference which actually subsists between the two periodic networks.

Networks having distributed parameters also exist in their own right, the prime example being the transmission line (whether for power transmission or for telecommunication). The two basic forms of line are a pair of concentric cylinders (coaxial line) and a pair of parallel wires; the latter will be taken as an example, the wires being assumed to have a cross-section of radius a and to be spaced with a distance b between their centres, where $b \gg a$. It is proved in books on electromagnetism that two such conductors have a capacitance per unit length given by

$$c=\frac{\pi\epsilon_0}{\ln(b/a)}, \tag{13.1.4}$$

where ϵ_0 is the primary electric constant. Also if the conductors carry equal and opposite currents and thus form a circuit, it is found to have inductance equal to its length multiplied by

$$l=\frac{\mu_0}{\pi}\ln\frac{b}{a}, \tag{13.1.5}$$

where μ_0 is the primary magnetic constant.† The line is assumed to be in a vacuum, the supporting insulators – and also the end effects – being neglected. The point to notice is that the capacitance and inductance may be assumed to be distributed so that an element δx has capacitance $c\,\delta x$ and inductance $l\,\delta x$.

Physically the capacitance $c\,\delta x$ is associated with the electric flux passing between corresponding elements in the two wires (fig. 13.3). The question

† See, for example, G. W. Carter, *The electromagnetic field in its engineering aspects*, 2nd edition, pp. 42 and 160, Longmans (1967).

arises whether account must not also be taken of capacitance between different elements δx, $\delta x'$ in the same wire. In the electrostatic calculation leading to (13.1.4) these, being parts of the same conductor, are assumed to be at the same potential; thus the capacitance between δx and $\delta x'$ is, so to speak, short-circuited. The electric lines of force lie everywhere in planes perpendicular to the wires. Under dynamic conditions the two elements are not in general at the same potential, and it is not necessarily true that the lines of force are transverse to the wires; but it can be seen that if the lines of force were transverse, there could be no electrical association between δx and $\delta x'$ and therefore no effect from capacitance between them. A similar argument applies to the magnetic field; if the lines of force are transverse, forward mutual inductance (a difficult concept to define) need not be considered, because it can have no effect.

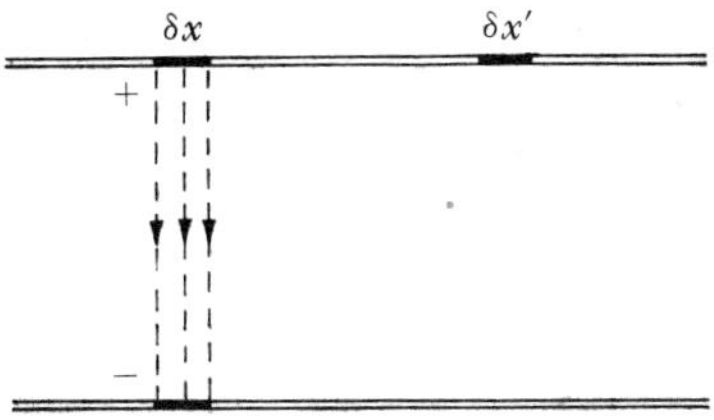

Fig. 13.3 Capacitance of parallel-wire line.

It is shown in books on guided electromagnetic wave propagation that, although the electric or magnetic field need not be transverse under dynamic conditions, it is possible for them both to be so.† This mode of operation is called *the TEM* (transverse electric and magnetic) *mode*; it is of great importance as being the only mode which can operate at all frequencies. Provided the conditions are such that the line operates in the TEM mode, it is a distributed circuit with series inductance and shunt capacitance. In practice the conductors will have resistance, distributed so that a length δx contains a resistance (for the two conductors) of $r\,\delta x$. It is customary also to take account of a possible leakage of current between the conductors over the supporting insulators, by assuming this to be also distributed so that, in parallel with the elementary capacitance, there is a conductance $g\,\delta x$. Thus the transmission line is represented by a sequence of the elements shown in fig. 13.4, in which the resistance and inductance associated with the whole loop are placed in one line. There is no need to divide $\boldsymbol{z}$ or $\boldsymbol{y}$ to make a T or Π element, because,

† See, for example, E. C. Jordan, *Electromagnetic waves and radiating systems*, Constable (1953). Chapter 7 (especially §7.05) deals with modes of propagation in a simple case; p. 211 extends the discussion to transmission lines.

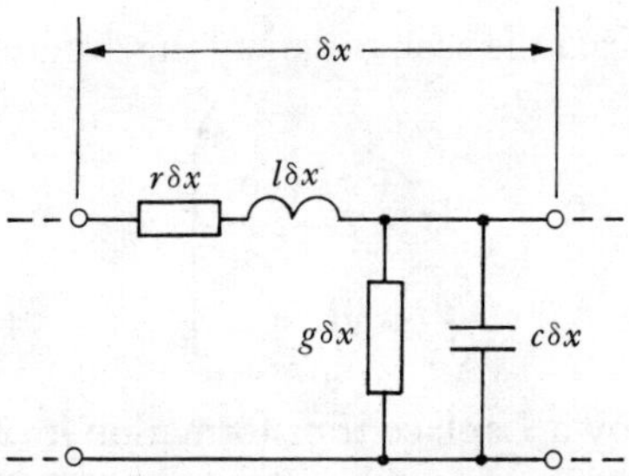

Fig. 13.4 Transmission line element.

to the first order in δx, the element shown is equivalent to both. At a given angular frequency ω,

$$\left.\begin{aligned} \boldsymbol{z} &= r + \mathrm{j}\omega l, \\ \boldsymbol{y} &= g + \mathrm{j}\omega c. \end{aligned}\right\} \tag{13.1.6}$$

From this point, ideal distributed circuits will be studied without further reference to their relationship with practical assemblages of wires or tubes; but the discussion just given is necessary in order that the limitations of the concept may be recognized. Thus the direct distributed-circuit representation of a transmission line is limited to the description of the TEM mode. It may be added that other modes of operation can be described by other distributed circuits which are less directly associated with the physical structure of the line, and are in effect models or analogues representing its performance.

13.2 Steady state a.c. operation of a distributed circuit of transmission line type

The discussion in §13.1 applies both to a.c. conditions (with phasor quantities) and to transient conditions (with Laplace transform quantities). In this and succeeding sections, the a.c. problem is investigated for the circuit of fig. 13.5; this is primarily regarded as a transmission line in which $\boldsymbol{z}$, $\boldsymbol{y}$ have the values given by (13.1.6), but is not necessarily so.

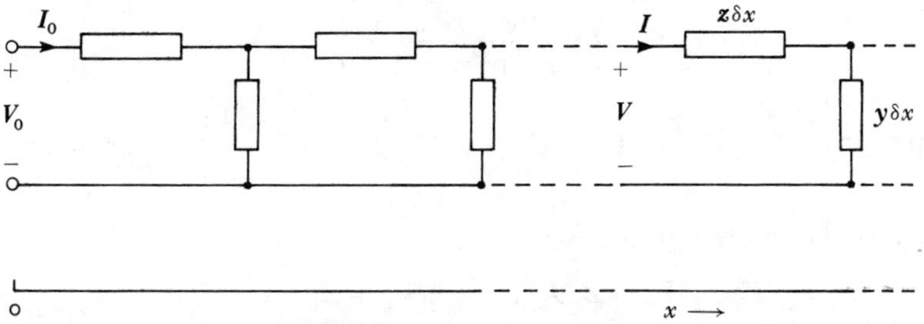

Fig. 13.5 Distributed circuit of transmission line type.

From (13.1.3) the circuit is seen to satisfy the differential equations

$$\left.\begin{aligned} \frac{d\boldsymbol{V}}{dx}+\boldsymbol{zI}=0, \\ \frac{d\boldsymbol{I}}{dx}+\boldsymbol{yV}=0. \end{aligned}\right\} \tag{13.2.1}$$

These will be solved by a Laplace transformation in x, in which the independent variable in the image world is denoted by p and transformed quantities are distinguished by bars; thus, corresponding with (5.4.5),

$$\mathscr{L}_x \boldsymbol{W}(x) = \int_0^\infty \boldsymbol{W}(x)\, e^{-px}\, dx = \overline{\boldsymbol{W}}(p). \tag{13.2.2}$$

It is to be noted that this treatment does not permit the use of negative values of x; but they are easily avoided. The image equations derived from (13.2.1) are then

$$\left.\begin{aligned} (p\overline{\boldsymbol{V}}-\boldsymbol{V}_0)+\boldsymbol{z}\overline{\boldsymbol{I}}=0, \\ (p\overline{\boldsymbol{I}}-\boldsymbol{I}_0)+\boldsymbol{y}\overline{\boldsymbol{V}}=0, \end{aligned}\right\} \tag{13.2.3}$$

from which

$$\overline{\boldsymbol{V}}=\frac{p\boldsymbol{V}_0-\boldsymbol{zI}_0}{p^2-\boldsymbol{yz}},$$

and

$$\overline{\boldsymbol{I}}=\frac{p\boldsymbol{I}_0-\boldsymbol{yV}_0}{p^2-\boldsymbol{yz}}. \tag{13.2.4}$$

Let

$$\sqrt{(\boldsymbol{yz})}=\gamma=\alpha+j\beta, \tag{13.2.5}$$

the sign being so chosen that α is positive; also let

$$\sqrt{\left(\frac{\boldsymbol{z}}{\boldsymbol{y}}\right)}=\boldsymbol{Z}_0, \quad \sqrt{\left(\frac{\boldsymbol{y}}{\boldsymbol{z}}\right)}=\boldsymbol{Y}_0 \tag{13.2.6}$$

again with the real part positive. Equations (13.2.4) may then be written

$$\left.\begin{aligned} \overline{\boldsymbol{V}}=\frac{p\boldsymbol{V}_0-\gamma\boldsymbol{Z}_0\boldsymbol{I}_0}{p^2-\gamma^2}, \\ \overline{\boldsymbol{I}}=\frac{p\boldsymbol{I}_0-\gamma\boldsymbol{Y}_0\boldsymbol{V}_0}{p^2-\gamma^2}. \end{aligned}\right\} \tag{13.2.7}$$

Moreover $\mathscr{L}_x^{-1}\left(\dfrac{p}{p^2-\gamma^2}\right)=\tfrac{1}{2}(e^{\gamma x}+e^{-\gamma x}),\ \mathscr{L}_x^{-1}\left(\dfrac{\gamma}{p^2-\gamma^2}\right)=\tfrac{1}{2}(e^{\gamma x}-e^{-\gamma x});$

therefore the solution in the x-world is

$$\left.\begin{aligned} \boldsymbol{V}=\tfrac{1}{2}(\boldsymbol{V}_0-\boldsymbol{Z}_0\boldsymbol{I}_0)\,e^{\gamma x}+\tfrac{1}{2}(\boldsymbol{V}_0+\boldsymbol{Z}_0\boldsymbol{I}_0)\,e^{-\gamma x}, \\ \boldsymbol{I}=\tfrac{1}{2}(\boldsymbol{I}_0-\boldsymbol{Y}_0\boldsymbol{V}_0)\,e^{\gamma x}+\tfrac{1}{2}(\boldsymbol{I}_0+\boldsymbol{Y}_0\boldsymbol{V}_0)\,e^{-\gamma x}. \end{aligned}\right\} \tag{13.2.8}$$

The significance of this is most clearly seen by considering a line which extends from $x = 0$ to $x = +\infty$. As α is positive, $|e^{\gamma x}|$ tends to infinity with x. Since an infinite response is physically impossible, $\boldsymbol{V}_0$ and $\boldsymbol{I}_0$ must be related by

$$\boldsymbol{V}_0 - \boldsymbol{Z}_0 \boldsymbol{I}_0 = 0, \qquad (13.2.9)$$

which may also be written $\boldsymbol{I}_0 - \boldsymbol{Y}_0 \boldsymbol{V}_0 = 0$. $\boldsymbol{V}$, $\boldsymbol{I}$ then reduce to a single exponential term:

$$\boldsymbol{V} = \boldsymbol{V}_0 e^{-\gamma x}; \qquad \boldsymbol{I} = \boldsymbol{I}_0 e^{-\gamma x} = \boldsymbol{Y}_0 \boldsymbol{V}_0 e^{-\gamma x}. \qquad (13.2.10)$$

The term $e^{-\gamma x}$ is $e^{-\alpha x} e^{-j\beta x}$. As x increases, both voltage and current decrease in amplitude on account of the factor $e^{-\alpha x}$, and simultaneously undergo a change of phase on account of the factor $e^{-j\beta x}$. When β is positive the change is in the lagging sense, and the voltage or current phasor varies

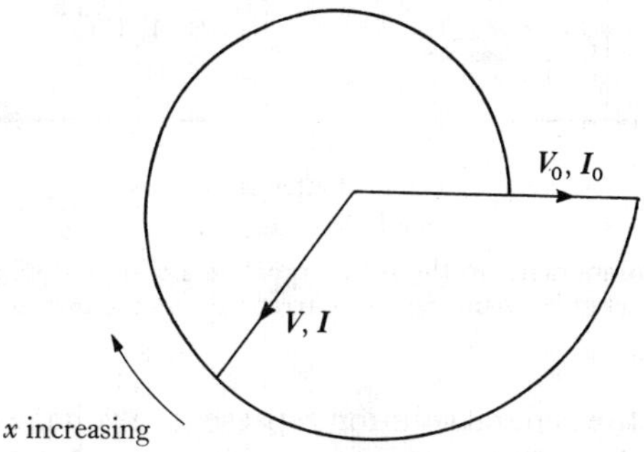

Fig. 13.6 Variation of voltage and current phasors in a line of infinite length.

with increasing x as shown in fig. 13.6. The ratio of voltage to current is $\boldsymbol{Z}_0$ at every point.

The complex quantity γ is commonly called the *propagation constant*; but *propagation coefficient* is preferable, since γ varies with ω. Its real part α is the *attenuation coefficient*, and the imaginary part β is the *phase coefficient*. When a quantity is reduced by a factor e^{-1}, it is said to suffer an attenuation of one *neper*; thus α is measured in nepers per unit length. Writing $\alpha + j\beta$ for γ in (13.2.10) and taking moduli, we obtain $V = V_0 e^{-\alpha x}$, or

$$\text{Attenuation in nepers} = \alpha x = \ln\left(\frac{V_0}{V}\right). \qquad (13.2.11)$$

In practice the neper is seldom employed, the *decibel* (mentioned in chapter 7) being preferred. The attenuation in bels is the logarithm to base 10 of the power ratio; thus

$$\text{Attenuation in decibels} = 10 \log_{10}\left(\frac{V_0^2}{V^2}\right) = 20 \log_{10}\left(\frac{V_0}{V}\right). \qquad (13.2.12)$$

Since $\log_{10}(V_0/V) = 0.4343 \ln(V_0/V)$, 1 neper = 8.68 db. The phase coefficient β is measured in radians per unit length.

The quantity $\boldsymbol{Z}_0$ has the dimensions of impedance, and is called the *characteristic impedance*, or sometimes, in relation to power lines, the *surge impedance*. In general it is complex, and is a function of ω. It is important to notice that $\boldsymbol{Z}_0$ denotes a property associated equally with every element of the line, not an 'impedance' in the normal lumped-circuit sense. $\boldsymbol{Y}_0$ will be the *characteristic admittance*.

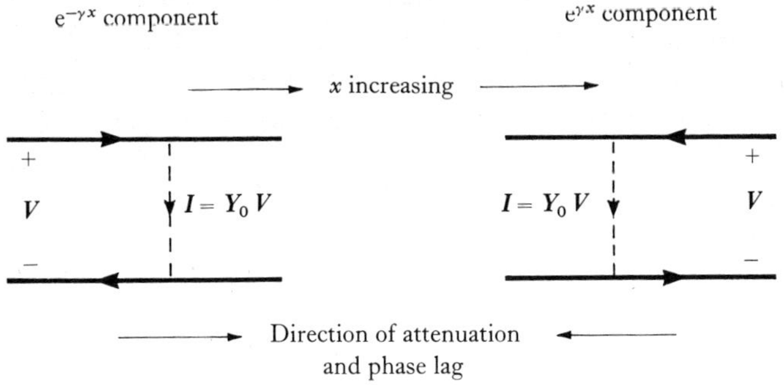

Fig. 13.7 The two components in the solution of the a.c. line problem. The circuits for current are completed by displacement current across the interconductor space.

The $e^{\gamma x}$ term in the general solution represents a voltage or current which is attenuated and changed in phase as x diminishes. In this component,

$$\frac{\boldsymbol{V}}{\boldsymbol{I}} = \frac{(\boldsymbol{V}_0 - \boldsymbol{Z}_0 \boldsymbol{I}_0)\,e^{\gamma x}}{(\boldsymbol{I}_0 - \boldsymbol{Y}_0 \boldsymbol{V}_0)\,e^{\gamma x}} = -\boldsymbol{Z}_0;$$

thus with a positive voltage is associated a negative current. The two components in the general solution are represented diagrammatically in fig. 13.7.

13.3 Distributed circuits without losses

The significance of the solution may be made clearer by reference to lossless circuits. The lossless transmission line has the series resistance r and shunt conductance g equal to zero, so that

$$\boldsymbol{z} = j\omega l, \qquad \boldsymbol{y} = j\omega c. \tag{13.3.1}$$

The line is illustrated in fig. 13.8(a). From equation (13.2.5),

$$\alpha + j\beta = \sqrt{(-\omega^2 lc)} = j\omega\sqrt{(lc)},$$

$$\alpha = 0, \qquad \beta = \omega\sqrt{(lc)}. \tag{13.3.2}$$

The component $e^{-\gamma x}$ occurring in the infinite line thus becomes $e^{-j\beta x}$, representing phase lag but no attenuation with increasing x. In the time

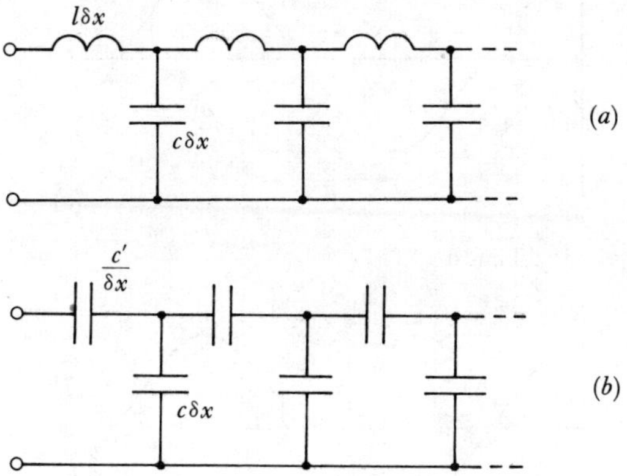

Fig. 13.8 Lossless distributed circuits: (*a*) transmission line; (*b*) attenuation network.

domain, the application of a voltage $V_0 \cos \omega t$ at $x = 0$ gives rise at the point x to

$$v = V_0 \cos(\omega t - \beta x) = V_0 \cos \omega \left(t - \frac{x}{u}\right), \tag{13.3.3}$$

where
$$u = \frac{\omega}{\beta} = (lc)^{-1/2}. \tag{13.3.4}$$

If v attains a certain value at $x = 0$ at a given instant t_0, it attains the same value at $x = x_1$ at the time t_1 given by

$$t_1 - \frac{x_1}{u} = t_0;$$

that is, (x_1/u) seconds later. This means that the voltage (and current) constitutes a *travelling wave*, u being the velocity of travel. For a lossless line, as (13.3.4) shows, u is independent of ω; thus a non-sinusoidal variation of voltage, equivalent to a spectrum of sinusoidal components, is transmitted without distortion or attenuation.

Equation (13.3.3) may be diagrammatically represented as in fig. 13.9, in two ways:

(*a*) as a function of t, for various values of x,

(*b*) as a function of x, for various values of t.

As a spatial picture which varies with time, the latter is allied to visual perception; the wave pattern is 'seen' to move from left to right with velocity u. One cycle in fig. 13.9(*b*) occupies a distance called the *wavelength*, λ; this

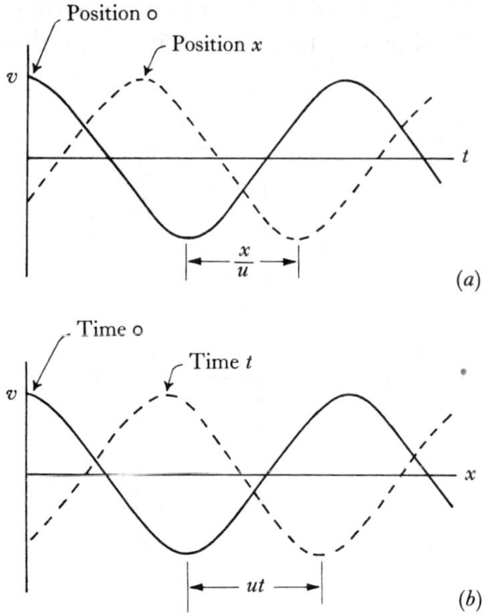

Fig. 13.9 Representation of a travelling voltage wave: (a) as a function of time; (b) as a function of position.

is the distance travelled by the wave in one time-cycle of the alternating voltage, so

$$\lambda = u\left(\frac{2\pi}{\omega}\right) = \frac{2\pi}{\beta},$$

or

$$\beta = \frac{2\pi}{\lambda}. \qquad (13.3.5)$$

The characteristic impedance, from (13.2.6), is

$$\boldsymbol{Z}_0 = \sqrt{\left(\frac{l}{c}\right)}. \qquad (13.3.6)$$

It is therefore resistive and frequency-independent. The symbols Z_0, Y_0 will be used for $\boldsymbol{Z}_0$, $\boldsymbol{Y}_0$ when these are necessarily real.

For the particular vacuum-insulated line whose constants are given by (13.1.4,5)

$$u = (\epsilon_0 \mu_0)^{-1/2} = 3 \times 10^8 \text{ m/s nearly.} \qquad (13.3.7)$$

This is the velocity of light in free space. All forms of ideal lines in a vacuum transmit waves with this velocity; in a medium of permittivity κ_e the velocity is reduced by a factor $\kappa_e^{-1/2}$. Practical air-insulated lines have u a little less than 3×10^8 m/s, partly owing to the capacitance of the supporting insulators, partly because of losses (see 13.4.6 below), and in slight measure because the permittivity of air is 1.0006.

The characteristic impedance is a function of the form and dimensions of the line. Typically it is a few hundred ohms for an air-insulated parallel-wire line, and a few tens of ohms for a coaxial conductor system, over the range of geometrical formations usually found.

There is another type of solution which can arise with lossless linelike circuits; it may be exemplified by considering the network of fig. 13.8(*b*). In this,

$$\boldsymbol{z} = \frac{-\mathrm{j}}{\omega c'}, \qquad \boldsymbol{y} = \mathrm{j}\omega c. \tag{13.3.8}$$

Therefore

$$\alpha + \mathrm{j}\beta = \sqrt{\left(\frac{c}{c'}\right)},$$

or

$$\alpha = \sqrt{\left(\frac{c}{c'}\right)}, \qquad \beta = 0. \tag{13.3.9}$$

(Note that the dimensions of c' are those of capacitance × length, whereas the dimensions of c are capacitance ÷ length). This network therefore brings about attenuation without change of phase, and may be described as an attenuation network. The characteristic admittance, evidently capacitive, is

$$\boldsymbol{Y}_0 = \mathrm{j}\omega\sqrt{(cc')}. \tag{13.3.10}$$

Reference to the first introduction of the travelling wave velocity in (13.3.4) shows that, when β is zero, u is infinite. Propagation through the present network is therefore instantaneous. Physically, however, the propagation of signals at a velocity greater than 3×10^8 m/s is impossible; the network of fig. 13.8(*b*) could not in practice be realized without the presence of inductance, which would make the velocity finite.

More complex networks may change from the transmission to the attenuation mode or vice versa as the frequency is varied. Thus it may be seen by inspection that the network of fig. 13.10(*a*) approximates to 13.8(*a*) as ω tends to zero and to 13.8(*b*) as ω tends to infinity; transmission takes place at the lower frequencies, and the network therefore constitutes a *low-pass filter*. Similarly, the network of fig. 13.10(*b*) is a *high-pass filter*. Considering, for example, the former, we have

$$\boldsymbol{z} = \frac{\mathrm{j}\omega l}{1 - \omega^2 lc'}, \qquad \boldsymbol{y} = \mathrm{j}\omega c. \tag{13.3.11}$$

There is evidently a critical frequency ω_c given by

$$\omega_c = (lc')^{-1/2}, \tag{13.3.12}$$

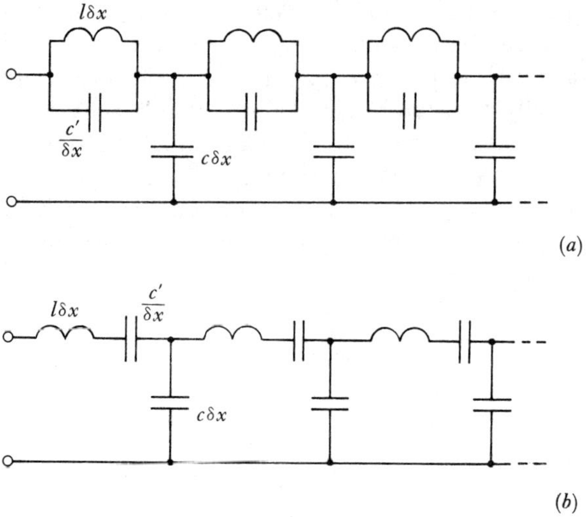

Fig. 13.10 (*a*) Low-pass filter network. (*b*) High-pass filter network.

so that we may write

$$\boldsymbol{z} = \frac{\mathrm{j}\omega l}{1-(\omega/\omega_c)^2}. \tag{13.3.13}$$

Thus

$$(\alpha + \mathrm{j}\beta)^2 = \frac{-\omega^2 lc}{1-(\omega/\omega_c)^2}, \tag{13.3.14}$$

and there are two cases to be considered. If $\omega < \omega_c$,

$$\alpha = 0, \qquad \beta = \omega\sqrt{\left(\frac{lc}{1-(\omega/\omega_c)^2}\right)}. \tag{13.3.15}$$

In this mode, travelling waves are transmitted without attenuation, but with a velocity which tends to zero as ω approaches ω_c. When $\omega > \omega_c$,

$$\alpha = \omega\sqrt{\left(\frac{lc}{(\omega/\omega_c)^2-1}\right)}, \qquad \beta = 0. \tag{13.3.16}$$

The signal is now disseminated instantaneously through the network, but attenuated; the attenuation coefficient is greatest when ω just exceeds ω_c, diminishing at high frequencies towards $\omega_c\sqrt{(lc)}$, or $\sqrt{(c/c')}$, as in (13.3.9).

The characteristic impedance, resistive for $\omega < \omega_c$, tends to infinity as ω_c is approached. When $\omega > \omega_c$ it becomes a capacitive reactance which is large when ω just exceeds ω_c and diminishes towards the value $-\mathrm{j}/\omega\sqrt{(cc')}$, in conformity with (13.3.10).

It is not necessary to give the corresponding analysis for the high-pass filter; and it must be emphasized that the subject of filter networks constitutes a large specialized field. The concept of the two modes of operation of a linelike network is merely the key which unlocks the door.

13.4 Transmission lines with losses – two special cases

For a dissipative transmission line, from (13.1.6) and (13.2.5,6),

$$\gamma = \sqrt{[(r + j\omega l)(g + j\omega c)]}, \qquad (13.4.1)$$

and

$$\boldsymbol{Z}_0 = \sqrt{\left(\frac{r + j\omega l}{g + j\omega c}\right)}. \qquad (13.4.2)$$

Two cases deserve special mention.

(a) The distortionless line

If the constants satisfy

$$\frac{r}{g} = \frac{l}{c} = k^2 \text{ (say)}, \qquad (13.4.3)$$

then

$$r + j\omega l = k^2(g + j\omega c).$$

Thus

$$\gamma = k(g + j\omega c) = \sqrt{(rg)} + j\omega\sqrt{(lc)}, \qquad (13.4.4)$$

and

$$\boldsymbol{Z}_0 = k = \sqrt{\left(\frac{l}{c}\right)}. \qquad (13.4.5)$$

Comparison of these equations with (13.3.2,6) shows that the only difference is in the attenuation coefficient, then zero, now $\sqrt{(rg)}$. Since β contains the factor ω, all frequency components are transmitted with the same velocity; since α is independent of ω, all are attenuated equally. Therefore a non-sinusoidal variation of voltage (and current) is transmitted without distortion, though attenuated.

With normally used materials, r/g exceeds l/c. To make a line distortionless l is increased deliberately to the value given by (13.4.3); it can be shown that this has another desirable result, in that for given values of r, g and c this value of l makes the attenuation a minimum. In cables the increase of l is brought about by a wrapping of ferromagnetic wire or tape; in air-insulated lines the effective inductance is increased by inserting coils at intervals. This procedure, called *loading*, was suggested by Heaviside. With tongue in cheek he proposed the name *heavification*; but unlike *inductance*, *impedance*, and other Heaviside words, this has not gained acceptance.

(b) The low-loss line

If at a particular frequency $r/\omega l$ and $g/\omega c$ are small, approximations may be obtained for γ and $\boldsymbol{Z}_0$, which throw light on the ways in which practical lines depart from the ideal. To the second order in $r/\omega l$,

$$(r+\mathrm{j}\omega l)^{1/2} = (\mathrm{j}\omega l)^{1/2}\left[1+\frac{1}{2}\frac{r}{\mathrm{j}\omega l}-\frac{1}{8}\left(\frac{r}{\mathrm{j}\omega l}\right)^2\right],$$

$$= (\mathrm{j}\omega l)^{1/2}\left[\left(1+\frac{1}{8}\frac{r^2}{\omega^2 l^2}\right)-\frac{\mathrm{j}}{2}\frac{r}{\omega l}\right].$$

$(g+\mathrm{j}\omega c)^{1/2}$ takes a similar form. Thus

$$\gamma = \mathrm{j}\omega(lc)^{1/2}\left[\left(1+\frac{1}{8}\frac{r^2}{\omega^2 l^2}\right)-\frac{\mathrm{j}}{2}\frac{r}{\omega l}\right]\left[\left(1+\frac{1}{8}\frac{g^2}{\omega^2 c^2}\right)-\frac{\mathrm{j}}{2}\frac{g}{\omega c}\right],$$

$$= \tfrac{1}{2}\surd(lc)\left(\frac{r}{l}+\frac{g}{c}\right)+\mathrm{j}\omega\surd(lc)\left[1+\frac{1}{8\omega^2}\left(\frac{r}{l}-\frac{g}{c}\right)^2\right]. \qquad (13.4.6)$$

The first order effect of r and g is therefore to introduce attenuation without altering the transmission velocity; the line is still effectively distortionless. The second order effect is to slow down the transmission velocity, especially at the lower frequencies.

The first order effect on $\boldsymbol{Z}_0$ is seen from

$$\boldsymbol{Z}_0 = \left(\frac{r+\mathrm{j}\omega l}{g+\mathrm{j}\omega c}\right)^{1/2},$$

$$= \left(\frac{l}{c}\right)^{1/2}\left(1-\frac{\mathrm{j}}{2}\frac{r}{\omega l}\right)\left(1+\frac{\mathrm{j}}{2}\frac{g}{\omega c}\right),$$

$$= \left(\frac{l}{c}\right)^{1/2}\left[1-\frac{\mathrm{j}}{2\omega}\left(\frac{r}{l}-\frac{g}{c}\right)\right]. \qquad (13.4.7)$$

Except in the distortionless case $r/l = g/c$, $\boldsymbol{Z}_0$ is no longer either resistive or frequency-independent. Again the effect is most marked at the lower frequencies; this is self-evident, since the 'smallness' of r and g is measured by comparing them with ωl and ωc respectively.

13.5 Distributed circuits of finite length

In §13.2 the general equations for the voltage and current in a linelike distributed circuit were obtained in terms of exponentials $e^{\pm\gamma x}$. If, in (13.2.8), we replace these by hyperbolic functions $\cosh\gamma x$, $\sinh\gamma x$, we obtain

$$\left.\begin{aligned} \boldsymbol{V} &= \boldsymbol{V}_0\cosh\gamma x - \boldsymbol{Z}_0\boldsymbol{I}_0\sinh\gamma x,\\ \boldsymbol{I} &= -\boldsymbol{Y}_0\boldsymbol{V}_0\sinh\gamma x + \boldsymbol{I}_0\cosh\gamma x.\end{aligned}\right\} \qquad (13.5.1)$$

It has been taken for granted that $\boldsymbol{V}_0$, $\boldsymbol{I}_0$ denote the voltage and current of a source located at $x = 0$; but this is not necessary, and with a line of finite length d it is convenient to regard $x = 0$ as the location of the load, and $x = d$ as that of the source (fig. 13.11). Quantities at the source end will be

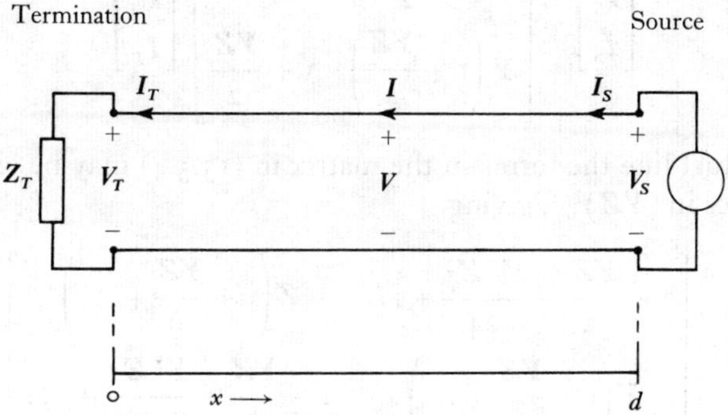

Fig. 13.11 Conventions for a finite line.

distinguished by the subscript S; at the load end, by T (termination). When this convention is used, it is customary still to regard a current flowing towards the termination as positive; the effect of this is to change the signs of the currents in equations (13.5.1), so that in matrix notation they take the form

$$\begin{bmatrix} \boldsymbol{V} \\ \boldsymbol{I} \end{bmatrix} = \begin{bmatrix} \cosh \gamma x & \boldsymbol{Z}_0 \sinh \gamma x \\ \boldsymbol{Y}_0 \sinh \gamma x & \cosh \gamma x \end{bmatrix} \begin{bmatrix} \boldsymbol{V}_T \\ \boldsymbol{I}_T \end{bmatrix}. \tag{13.5.2}$$

In particular, when $x = d$,

$$\begin{bmatrix} \boldsymbol{V}_S \\ \boldsymbol{I}_S \end{bmatrix} = \begin{bmatrix} \cosh \gamma d & \boldsymbol{Z}_0 \sinh \gamma d \\ \boldsymbol{Y}_0 \sinh \gamma d & \cosh \gamma d \end{bmatrix} \begin{bmatrix} \boldsymbol{V}_T \\ \boldsymbol{I}_T \end{bmatrix}. \tag{13.5.3}$$

Since $\gamma = (\boldsymbol{yz})^{1/2}$, $\gamma d = (\boldsymbol{y}d)^{1/2}(\boldsymbol{z}d)^{1/2}$, where $\boldsymbol{y}d$, $\boldsymbol{z}d$ are the total admittance and impedance of the line. Denoting these by $\boldsymbol{Y}$, $\boldsymbol{Z}$, we can write

$$\begin{bmatrix} \boldsymbol{V}_S \\ \boldsymbol{I}_S \end{bmatrix} = \begin{bmatrix} \cosh (\boldsymbol{YZ})^{1/2} & \boldsymbol{Z}_0 \sinh (\boldsymbol{YZ})^{1/2} \\ \boldsymbol{Y}_0 \sinh (\boldsymbol{YZ})^{1/2} & \cosh (\boldsymbol{YZ})^{1/2} \end{bmatrix} \begin{bmatrix} \boldsymbol{V}_T \\ \boldsymbol{I}_T \end{bmatrix}. \tag{13.5.4}$$

From (13.2.6),

$$\boldsymbol{Z}_0 = \sqrt{\left(\frac{\boldsymbol{z}}{\boldsymbol{y}}\right)} = \sqrt{\left(\frac{\boldsymbol{Z}}{\boldsymbol{Y}}\right)}. \tag{13.5.5}$$

Under certain conditions it is possible to approximate the behaviour of a finite line by treating $\boldsymbol{Y}$ and $\boldsymbol{Z}$ as though lumped in a T formation like one of the elements in fig. 13.1(*a*), or in a Π formation like one of the elements in fig. 13.1(*b*). This is spoken of as nominal T or Π representation, and was briefly mentioned in §3.12. To examine the validity of this it will suffice to consider the nominal Π network, for which, in a relation analogous to (13.1.2),

$$\begin{bmatrix} V_S \\ I_S \end{bmatrix} = \begin{bmatrix} 1+\dfrac{YZ}{2} & Z \\ Y\left(1+\dfrac{YZ}{4}\right) & 1+\dfrac{YZ}{2} \end{bmatrix} \begin{bmatrix} V_T \\ I_T \end{bmatrix}. \tag{13.5.6}$$

For the actual line the terms in the matrix in (13.5.4) may be expanded as power series in $(YZ)^{1/2}$, giving

$$\begin{bmatrix} V_S \\ I_S \end{bmatrix} = \begin{bmatrix} 1+\dfrac{YZ}{2}+\dfrac{Y^2Z^2}{24}+\dots & Z\left(1+\dfrac{YZ}{6}+\dots\right) \\ Y\left(1+\dfrac{YZ}{6}+\dots\right) & 1+\dfrac{YZ}{2}+\dfrac{Y^2Z^2}{24}+\dots \end{bmatrix} \begin{bmatrix} V_T \\ I_T \end{bmatrix}. \tag{13.5.7}$$

Comparison between (13.5.6) and (13.5.7) shows a similarity when $|YZ|$ is small, the maximum discrepancy occurring where $1+(YZ/6)+\dots$ in the line is replaced by 1 in the Π model. If this represents a 1 per cent error,

$$|YZ| = |\gamma^2 d^2| = 6/100,$$

or $|\gamma d| = 0.245$. For a lossfree line $|\gamma| = 2\pi/\lambda$, so the condition for an error less than 1 per cent is

$$\frac{d}{\lambda} < \frac{0.245}{2\pi} = 0.039.$$

On this criterion, therefore, the approximation is valid when the length of the line is less than 3.9 per cent of the wavelength – say 230 km for a 50 Hz overhead power line, or 4.7 km for an open telephone line operating up to 2500 Hz.

Whether the line transmits power or communication signals, the ratio of voltage to current at the load end is determined by the prescribed terminating impedance Z_T. The problem posed by the line may take several closely allied forms, such as: Given a voltage V_T at the termination, what are the voltage and current at the source? or, Given a constant voltage V_S at the source, how is V_T affected by the disconnexion of Z_T? or, What is the relation between the input impedance $Z_S = V_S/I_S$ and the terminating impedance Z_T? The formal solutions are easily obtained by using (13.5.3), but their physical significance is disguised by the difficulty of visualizing the nature of $\cosh\gamma d$ and $\sinh\gamma d$ when γ is complex; it is much more easily grasped when the line is lossless, and this case will be fully developed in the next section.

When the data are numerical, $\cosh\gamma d$ and $\sinh\gamma d$ are to be obtained from tables of real circular and hyperbolic functions by means of the identities

$$\left.\begin{aligned}\cosh(\alpha+\mathrm{j}\beta)\,d &\equiv \cosh\alpha d\cos\beta d+\mathrm{j}\sinh\alpha d\sin\beta d,\\ \sinh(\alpha+\mathrm{j}\beta)\,d &\equiv \sinh\alpha d\cos\beta d+\mathrm{j}\cosh\alpha d\sin\beta d.\end{aligned}\right\} \qquad (13.5.8)$$

A single example will suffice. It will relate to a power line, because at power frequencies r and $l\omega$ are usually comparable. The assumption of losslessness is more likely to be valid at the higher frequencies used in communication.

A single-phase line supplies power at 100 kV, 50 Hz, to a load of 5000 kVA and power factor 0.8 lagging. The total resistance of the line is 48 Ω, inductance 0.48 H, conductance negligible, capacitance 0.75 μF. It is required to find the voltage and current at the source.

From the data given,

$$\boldsymbol{Y}=\mathrm{j}235.6\ 10^{-6}=235.6\ 10^{-6}\angle 90^\circ,$$

$$\boldsymbol{Z}=48+\mathrm{j}150.8=158.2\angle 72.3^\circ.$$

Hence $$\boldsymbol{YZ}=\gamma^2d^2=3.728\ 10^{-2}\angle 162.3^\circ,$$

and so $$\gamma d=0.1931\angle 81.2^\circ=0.0296+\mathrm{j}0.1908,$$

since α is by definition positive (see (13.2.5)).

Thus $$\begin{aligned}\cosh\gamma d&=\cosh 0.0296\cos 0.1908+\mathrm{j}\sinh 0.0296\sin 0.1908,\\ &=1.0004\times 0.982+\mathrm{j}0.0296\times 0.1891,\\ &=0.982+\mathrm{j}0.0056,\\ &=0.982\angle 0.3^\circ.\end{aligned}$$

Similarly $\sinh\gamma d=0.1914\angle 81.3^\circ$.

Also $$\boldsymbol{Z}_0=\frac{\gamma d}{\boldsymbol{Y}}=\frac{0.1931\angle 81.2^\circ}{235.6\ 10^{-6}\angle 90^\circ}=818\angle -8.8^\circ,$$

so that $$\boldsymbol{Y}_0=1.223\ 10^{-3}\angle 8.8^\circ.$$

From (13.5.3), therefore,

$$\begin{bmatrix}\boldsymbol{V}_S\\ \boldsymbol{I}_S\end{bmatrix}=\begin{bmatrix}0.982\angle 0.3^\circ & 156.5\angle 72.5^\circ\\ 234.2\ 10^{-6}\angle 90.1^\circ & 0.982\angle 0.3^\circ\end{bmatrix}\begin{bmatrix}\boldsymbol{V}_T\\ \boldsymbol{I}_T\end{bmatrix}.$$

With the given load,

$$\boldsymbol{V}_T\ \text{(reference phasor)}=10^5\angle 0^\circ$$

and $$\boldsymbol{I}_T=50\angle -36.9^\circ.$$

Therefore

$$\begin{aligned} V_S &= 98.2\ 10^3 \angle 0.3^\circ + 7.8\ 10^3 \angle 35.6^\circ \\ &= \{(98.2 + j0.5) + (6.3 + j4.5)\}\ 10^3 \\ &= \{104.5 + j5.0\}\ 10^3, \quad \text{or} \quad 104.6 \angle 2.8^\circ \text{ kV}. \end{aligned}$$

And

$$\begin{aligned} I_S &= 23.4 \angle 90.1^\circ + 49.1 \angle -36.6^\circ, \\ &= (0 + j23.4) + (39.4 - j29.3), \\ &= 39.4 - j5.9, \quad \text{or} \quad 39.8 \angle -8.5^\circ \text{ A}. \end{aligned}$$

The nominal Π approximation gives $V_S = 104.7 \angle 2.8^\circ$ kV, $I_S = 39.8 \angle -8.6^\circ$ A. A line with the given values would be about 180 km long.

A question which might well have been asked is to determine the *regulation* of the line, namely the rise in voltage at the termination when the load is disconnected. If the open circuit voltage at the termination is V_{T0}, (13.5.3) gives

$$V_S = V_T \cosh \gamma d + Z_0 I_T \sinh \gamma d = V_{T0} \cosh \gamma d,$$

since V_S must be assumed constant; from which

$$\frac{V_{T0}}{V_T} = 1 + \frac{Z_0}{Z_T} \tanh \gamma d. \tag{13.5.9}$$

For a communication line, a more likely question would have been to determine the input impedance; this is given by

$$Z_S = \frac{V_S}{I_S} = \frac{V_T \cosh \gamma d + Z_0 I_T \sinh \gamma d}{Y_0 V_T \sinh \gamma d + I_T \cosh \gamma d},$$

$$= \frac{(Z_T/Z_0) + \tanh \gamma d}{1 + (Z_T/Z_0) \tanh \gamma d} \cdot Z_0. \tag{13.5.10}$$

The expressions (13.5.9,10) are functions of the same two quantities: $\tanh \gamma d$ and (Z_T/Z_0). The latter, which is the ratio of Z_T to an impedance inherent in the line, is called the *normalized* terminating impedance. If ζ_T is a complex number such that

$$\frac{Z_T}{Z_0} = \tanh \zeta_T, \tag{13.5.11}$$

(13.5.10) gives the input impedance, in normalized form, as

$$\frac{Z_S}{Z_0} = \tanh (\zeta_T + \gamma d). \tag{13.5.12}$$

Tanh ζ is therefore a function of great importance in the solution of a.c. line problems – not only those mentioned, but many others.

13.6 Graphical representation of functions of a complex variable. The Smith chart

The numerical evaluation of functions of a complex variable from tables of functions of real variables, as performed above, is laborious; but diagrams of the type already used in §12.1 to illustrate conformal transformations will provide a quick and convenient graphical method of attaining the same result. The function

$$\mathrm{w} = \tanh \zeta, \tag{13.6.1}$$

where $\mathrm{w} = u + jv$, $\zeta = \xi + j\eta$, has just been shown to be of particular importance.

In fig. 13.12 this transformation is illustrated by a plot in the plane of w, in which the co-ordinate lines for constant values of u, v and of ξ, η are shown

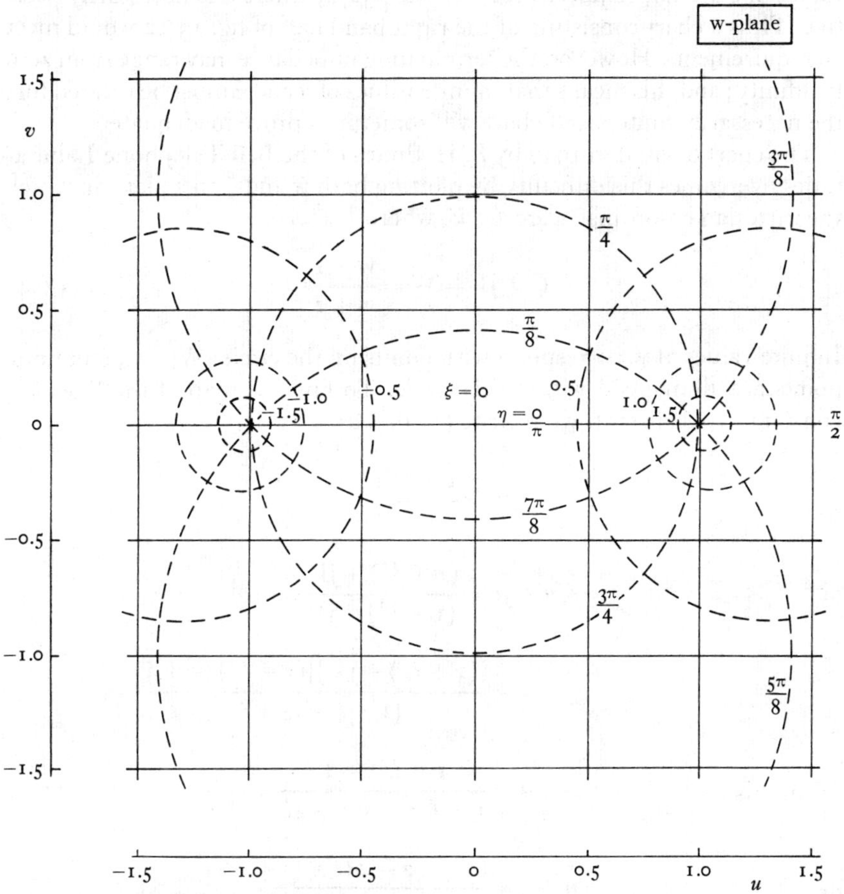

Fig. 13.12 Cartesian chart for the function w = tanh ζ.

as two superimposed meshworks, the latter being dotted. It may be verified that the equations of the circles of constant ξ or η are

$$u^2 + v^2 - 2(\coth 2\xi)\, u + 1 = 0, \tag{13.6.2}$$

and

$$u^2 + v^2 + 2(\cot 2\eta)\, v + 1 = 0. \tag{13.6.3}$$

These are orthogonal systems of coaxial circles with limiting points at $(\pm 1, 0)$. For clarity the numerical scales of u and v are removed to the edge of the diagram. To evaluate $\tanh\zeta$, it is merely necessary to identify the point having the co-ordinates (ξ, η) as an intersection of two circles, and read off its (u, v) co-ordinates on the edge scales.

The diagram contains a map of the ζ-plane over the range $-\infty \leqslant \xi \leqslant \infty$, $0 \leqslant \eta \leqslant \pi$. If η lies outside these limits, an appropriate multiple of π must be added or subtracted. In transmission line or other allied problems, negative values of ξ are not required; for $\gamma d = (\alpha + j\beta)d$, where α is necessarily positive. Thus a chart consisting of the right hand half of fig. 13.12 would meet all requirements. However, the terminating impedance may range from zero to infinity; and this means that infinite values of u and v must be catered for; the necessarily finite-sized chart will sometimes prove inadequate.

The chart devised in 1939 by P. H. Smith of the Bell Telephone Laboratories overcomes this difficulty by plotting both w and ζ on a plane in which the cartesian co-ordinates are U, V, where

$$U + jV = \mathrm{W} = \frac{\mathrm{w} - 1}{\mathrm{w} + 1}. \tag{13.6.4}$$

Infinite values of w correspond with points on the circle $|\mathrm{W}| = 1$, and finite points are all mapped within this circle. To find the map of the lines $u =$ constant, $v =$ constant, we rewrite (13.6.4) as

$$\mathrm{w} = \frac{1 + \mathrm{W}}{1 - \mathrm{W}},$$

or

$$u + jv = \frac{(1 + U) + jV}{(1 - U) - jV},$$

$$= \frac{[(1 + U) + jV][(1 - U) + jV]}{(1 - U)^2 + V^2}.$$

From this

$$u = \frac{1 - U^2 - V^2}{1 + U^2 + V^2 - 2U},$$

or

$$u + 1 = \frac{2 - 2U}{1 + U^2 + V^2 - 2U}.$$

Therefore the map of the line $u =$ constant is

$$U^2 + V^2 - 2U + 1 = \frac{2 - 2U}{u + 1},$$

or
$$U^2 + V^2 - \frac{2u}{u+1} . U + \frac{u-1}{u+1} = 0. \qquad (13.6.5)$$

This is a circle having its centre at $(u/(u+1), 0)$. Whatever the value of u, the point $(1, 0)$ lies on the circle. In a similar manner, $v =$ constant gives the system of circles

$$U^2 + V^2 - 2U - \frac{2}{v} . V + 1 = 0. \qquad (13.6.6)$$

The centre of the circle v is at $(1, 1/v)$, and again it necessarily passes through $(1, 0)$. The circles defined by (13.6.5) are orthogonal coaxial systems in which the limiting points coincide at $(1, 0)$.

The map of the lines $\xi =$ constant, $\eta =$ constant is found by writing $\tanh \zeta$ for w in (13.6.4), so that

$$\mathrm{W} = \frac{\tanh \zeta - 1}{\tanh \zeta + 1} = -\mathrm{e}^{-2\zeta}.$$

From this
$$U + \mathrm{j}V = -\mathrm{e}^{-2\xi}(\cos 2\eta - \mathrm{j} \sin 2\eta),$$
$$= \mathrm{e}^{-2\xi}[\cos(\pi - 2\eta) + \mathrm{j} \sin(\pi - 2\eta)]. \qquad (13.6.7)$$

Thus $(\mathrm{e}^{-2\xi}, \pi - 2\eta)$ are polar co-ordinates in the W-plane. $\xi =$ constant is a circle with centre at the origin and radius between zero and unity; $\eta =$ constant is a radial straight line; all such lines are covered by giving η values from 0 to π.

The chart is plotted in outline in fig. 13.13(*a*), in which full and dotted lines are maps of the full and dotted lines in fig. 13.12. The lines $u =$ constant, $v =$ constant are labelled on the chart itself, while external scales are provided for ξ, η. Commercially published Smith charts such as fig. 13.13(*b*) embody a number of scales for special calculations, among which the reader may have difficulty in identifying those which are fundamental; ξ itself may not be scaled, but a uniform radial scale extending from 0 at the centre to 1 at the periphery is a scale of $\mathrm{e}^{-2\xi}$, the radial polar co-ordinate. η is usually scaled as $(\eta/2\pi)$ from 0 to 0.5. The dotted co-ordinate lines are not printed, since these co-ordinates are easily derived from the scales with straight-edge and dividers.

As an example of the use of the Smith chart for the direct reading of $\tanh \gamma d$ we will evaluate (13.5.9), namely

$$\frac{\boldsymbol{V}_{T0}}{\boldsymbol{V}_T} = 1 + \frac{\boldsymbol{Z}_0}{\boldsymbol{Z}_T} \tanh \gamma d,$$

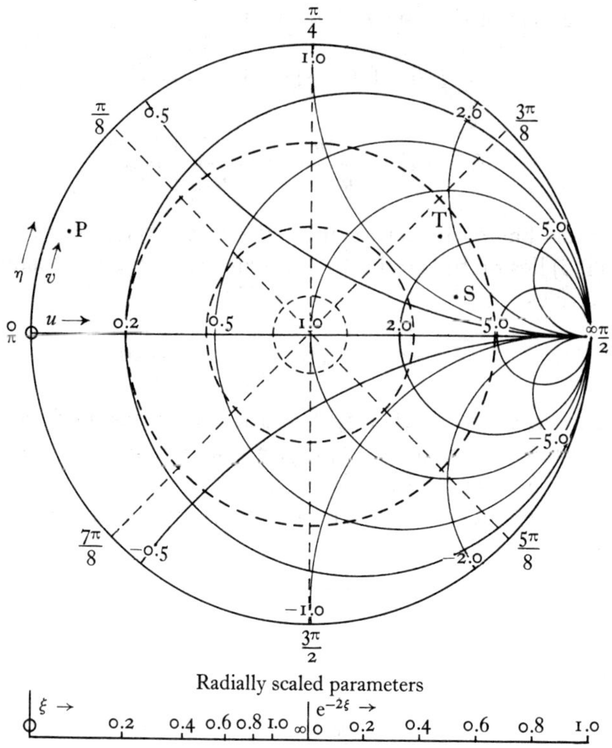

Fig. 13.13 (*a*) Outline Smith chart: (A plot of $w = \tanh \zeta$ on the plane of $W = (w - 1)/(w + 1)$);

where, in the problem discussed, $\boldsymbol{Z}_0 = 818 \angle -8.8°$, $\boldsymbol{Z}_T = 2000 \angle 36.9°$, and $\gamma d = 0.0296 + j0.1908$. The point P, at which $\xi = 0.0296$, $\eta = 0.1908$, (or $e^{-2\xi} = 0.943$, $\eta/2\pi = 0.0304$), is located by means of the radial and angular scales; $\tanh \gamma d$ is then read off from the (u, v) co-ordinates of P. On a full-sized chart it is easy to obtain the accuracy denoted by

$$\tanh \gamma d = u + jv = 0.03 + j0.19$$

$$= 0.19_2 \angle 81.0°,$$

from which
$$\frac{V_{T0}}{V_T} = 1 + \frac{818 \times 0.192}{2000} \angle (81.0° - 8.8° - 36.9°)$$

$$= 1 + 0.0785 \angle 35.3°$$

$$= 1.064 + j0.045, \quad \text{or} \quad 1.065 \angle 2.4°.$$

The object of the equation was to determine V_{T0}/V_T; it is now seen that this ratio is 1.065, so that the voltage at the termination rises by 6.5 per cent when

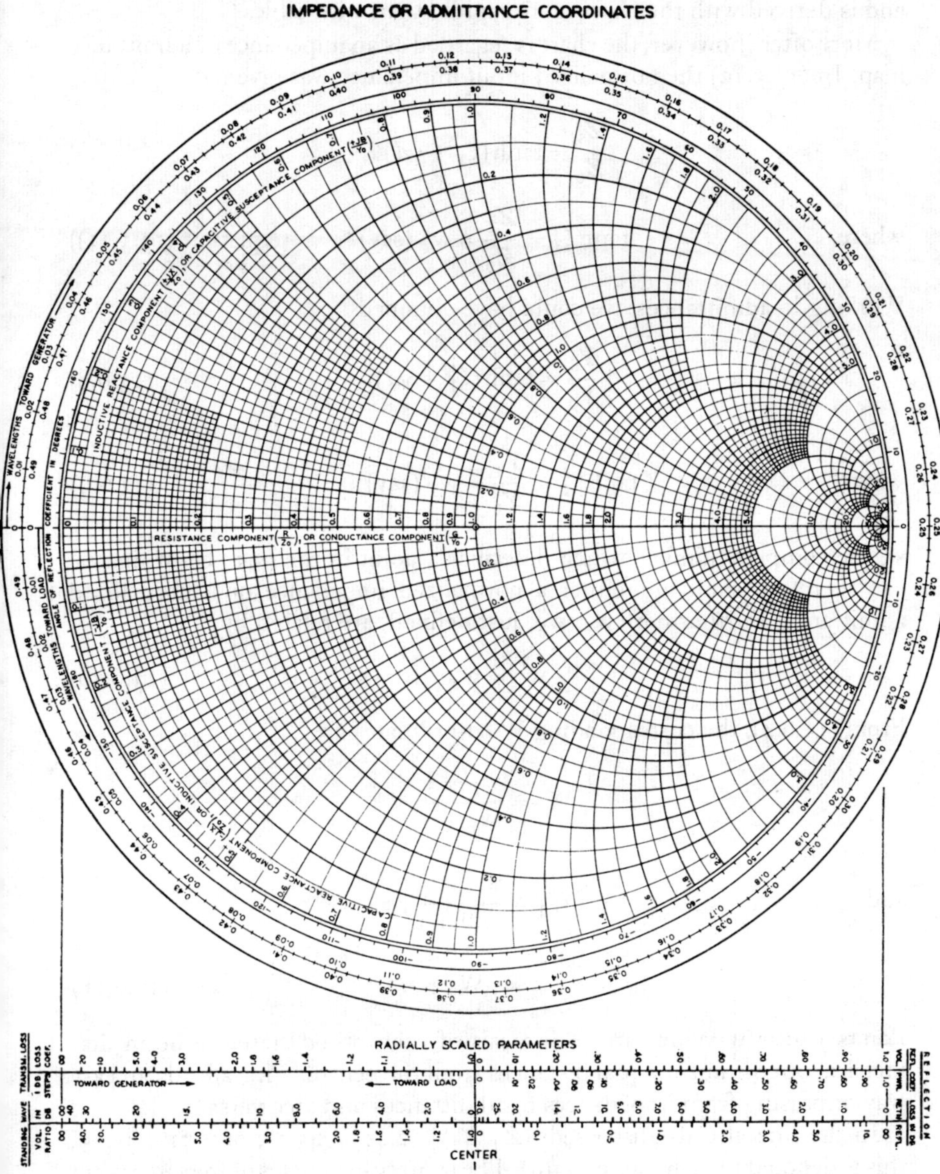

Fig. 13.13 (*b*) fully scaled Smith chart. See *Electronics* **17**, No. 1 (Jan. 1944), pp. 130–3 and 318–25.

the load is disconnected. This value is correct to the significant figures given, and is derived with the chart more rapidly than with tables.

More often, however, the chart is regarded as an impedance or admittance map. In (13.5.12) the normalized input impedance was given as

$$\frac{\boldsymbol{Z}_S}{\boldsymbol{Z}_0} = \tanh(\zeta_T + \gamma d)$$

where $$\tanh \zeta_T = \frac{\boldsymbol{Z}_T}{\boldsymbol{Z}_0} = \mathrm{w}_T \text{ (say).} \qquad \text{(equation (13.5.11))}$$

Working in admittances, we could have obtained

$$\frac{\boldsymbol{Y}_S}{\boldsymbol{Y}_0} = \tanh(\zeta_T' + \gamma d) \tag{13.6.8}$$

where $$\tanh \zeta_T' = \frac{\boldsymbol{Y}_T}{\boldsymbol{Y}_0} = \mathrm{w}_T' \text{ (say);} \tag{13.6.9}$$

w, the complex number which describes a point in the (u, v) co-ordinates, can therefore stand for either a normalized impedance or a normalized admittance. Furthermore w_T, w_T', just defined, are related by

$$\mathrm{w}_T' = \mathrm{w}_T^{-1}. \tag{13.6.10}$$

From (13.6.4), the corresponding W co-ordinates are

$$\mathrm{W}_T = \frac{\mathrm{w}_T - 1}{\mathrm{w}_T + 1},$$

and $$\mathrm{W}_T' = \frac{\mathrm{w}_T^{-1} - 1}{\mathrm{w}_T^{-1} + 1},$$

$$= -\mathrm{W}_T. \tag{13.6.11}$$

Points whose w-values are reciprocals of each other therefore lie in diametrical opposition at equal distances from the centre. This affords an easy way of passing from impedances to admittances and vice versa.

In the line already discussed, $(\boldsymbol{Z}_T/\boldsymbol{Z}_0) = 2.445\angle 45.7^\circ$, or $1.71 + j1.75$; this is denoted by T in fig. 13.13(*a*). The (ξ, η) co-ordinates of the same point give

$$\zeta_T = 0.26_8 + j1.26_5;$$

also $$\gamma d = 0.030 + j0.191,$$

so $$\zeta_T + \gamma d = 0.29_8 + j1.45_6,$$

defining a point S, whose (u, v) co-ordinates give the normalized value of the input impedance:

$$\frac{\boldsymbol{Z}_S}{\boldsymbol{Z}_0} = 3.0_2 + \mathrm{j}1.0_9.$$

The relation between the points S and T is to be noted. Since both components of γd are positive, the (ξ, η) co-ordinates of S are greater respectively than those of T. This means that S lies nearer the centre than T, and the movement from T to S is a clockwise rotation about the centre. If the position of S were calculated for lines of different lengths, the points so obtained would lie on a spiral (fig. 13.14). For a very long line the spiral comes close

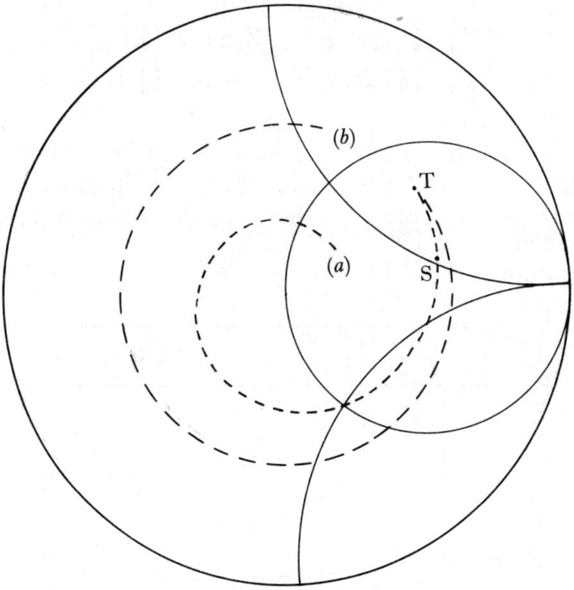

Fig. 13.14 Smith chart plot of the input impedances of lines of various length with the same terminating impedence (denoted by point T): (*a*) line with losses; (*b*) lossless line.

to the centre, giving an input impedance approximating to $\boldsymbol{Z}_0$ whatever the terminating impedance. Other linelike distributed circuits may give other loci, for example a spiral of oppositely directed involution, or (for the attenuation network of fig. 13.8(*b*)) a radial straight line. All are alike in moving from the termination point T nearer to the centre, except for certain lossless circuits; the lossless transmission line is of especial importance, and its locus is a circle through T, with centre at the origin, described clockwise. The use of this locus will be exemplified in the next section.

13.7 The lossless transmission line of finite length

In practical transmission lines attenuation cannot be absent; but at high frequencies the attenuation per wavelength may be negligible, and this gives practical as well as analytical importance to the discussion of the lossless line.

As was shown in §13.3, such a line has

$$\gamma = \mathrm{j}\beta = \mathrm{j}\omega\sqrt{(lc)} = \mathrm{j}\frac{2\pi}{\lambda}. \qquad \text{(equations (13.3.2,5))}$$

Note that β is positive. The equations of the finite line, namely (13.5.2), therefore take the form

$$\begin{bmatrix} \boldsymbol{V} \\ \boldsymbol{I} \end{bmatrix} = \begin{bmatrix} \cos\beta x & \mathrm{j}Z_0 \sin\beta x \\ \mathrm{j}Y_0 \sin\beta x & \cos\beta x \end{bmatrix} \begin{bmatrix} \boldsymbol{V}_T \\ \boldsymbol{I}_T \end{bmatrix}. \qquad (13.7.1)$$

Z_0, Y_0 are real quantities (13.3.6). This solution in terms of cosines and sines describes stationary envelopes within which the alternations of voltage or current occur (fig. 13.15); it is therefore described as a *standing-wave*

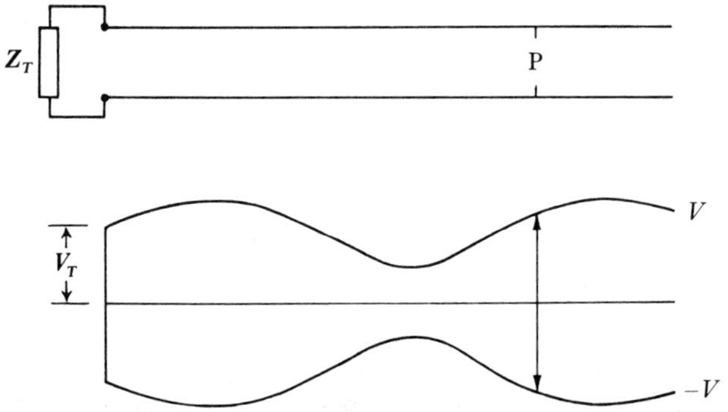

Fig. 13.15 Standing wave of voltage on a transmission line. The alternation of voltage at P takes place between the positive and negative maxima shown.

solution. This term is in contradistinction to the *travelling waves* denoted by $e^{\pm \mathrm{j}\beta x}$. The superposition of two oppositely-travelling waves sets up a standing wave, and there are thus two alternative ways of describing the same phenomena. Before investigating the pattern of standing waves in general terms we will examine three important special cases.

(a) The open-circuited line

When $\boldsymbol{Z}_T = \infty$, $\boldsymbol{I}_T = 0$, and (13.7.1) gives

$$\boldsymbol{V} = \boldsymbol{V}_T \cos\beta x, \qquad \boldsymbol{I} = \mathrm{j}Y_0 \boldsymbol{V}_T \sin\beta x. \tag{13.7.2}$$

The standing-wave patterns for voltage and current are illustrated in fig. 13.16(*a*) by plotting V and I to a base of βx or $2\pi x/\lambda$, the so-called *electrical length* between a point on the line and its termination. It must be understood that the voltage or current alternations take place, as in fig. 13.15, within an envelope consisting of the curve shown and its mirror image (not shown).

The input impedance of the line of length d is given by

$$\boldsymbol{Z}_S = \frac{\boldsymbol{V}_S}{\boldsymbol{I}_S} = -\mathrm{j}Z_0 \cot\beta d. \tag{13.7.3}$$

In fig. 13.16(*b*) this is illustrated by plotting X_S ($=\mathrm{j}^{-1}\boldsymbol{Z}_S$) as a function of the electrical length. The line simulates a pure reactance, which may take any value either inductive or capacitive.

(b) The short-circuited line

When $\boldsymbol{Z}_T = 0$, so that $\boldsymbol{V}_T = 0$, the corresponding equations are seen to be

$$\boldsymbol{V} = \mathrm{j}Z_0 \boldsymbol{I}_T \sin\beta x, \qquad \boldsymbol{I} = \boldsymbol{I}_T \cos\beta x, \tag{13.7.4}$$

and

$$\boldsymbol{Z}_S = \mathrm{j}Z_0 \tan\beta d. \tag{13.7.5}$$

These are shown in fig. 13.17. Again X_S takes all values.

It is clear that the open- and short-circuited lines constitute essentially the same problem; conditions occurring at x in the open-circuited line are also found at $x + (\lambda/4)$ in the short-circuited line. In the Smith chart for impedance values, both are represented by the circle which forms the outer periphery, on which the open- and short-circuit points lie respectively at the right and left hand ends of the horizontal diameter.

Of particular interest is the line for which $\beta d = \pi/2$, or $d = \lambda/4$ (the *quarter-wavelength line*). When terminated by an open-circuit, this has an input impedance of zero and so looks like a short-circuit; when short-circuited it looks like an open-circuit. On the Smith chart it is represented by a semicircular arc on the periphery; the extremities of this are diametrically opposed, and therefore represent normalized impedance values which are reciprocals of each other. It follows that when a quarter-wavelength line is terminated with $\boldsymbol{Z}_T$, the input impedance $\boldsymbol{Z}_S$ is such that

$$\frac{Z_0}{\boldsymbol{Z}_S} = \frac{\boldsymbol{Z}_T}{Z_0},$$

or

$$\boldsymbol{Z}_S \boldsymbol{Z}_T = Z_0^2. \tag{13.7.6}$$

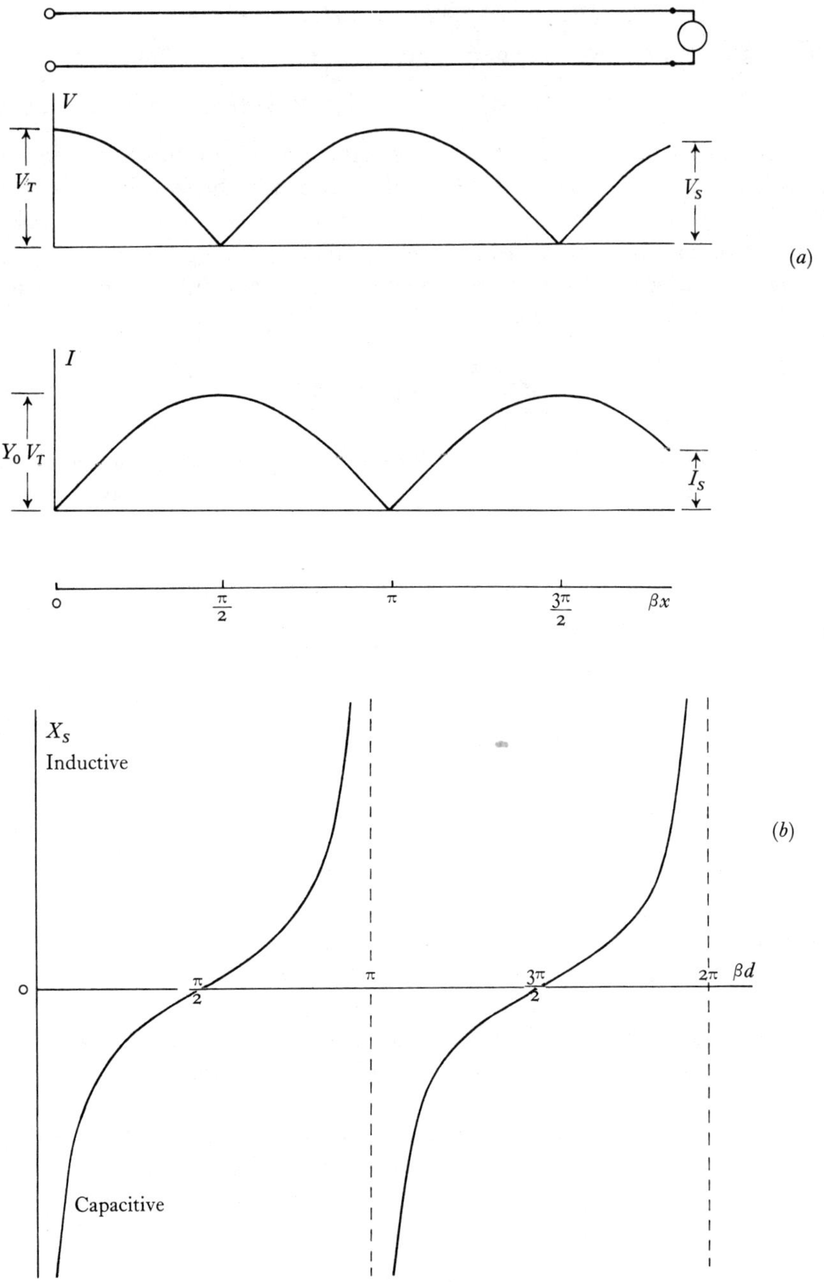

Fig. 13.16 Open-circuited line: (*a*) standing-wave patterns; (*b*) input reactance as a function of length.

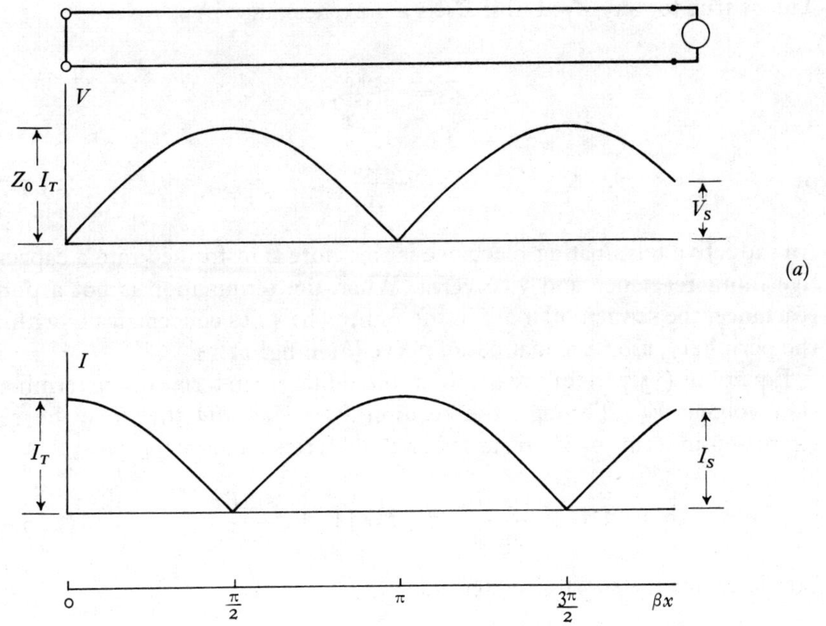

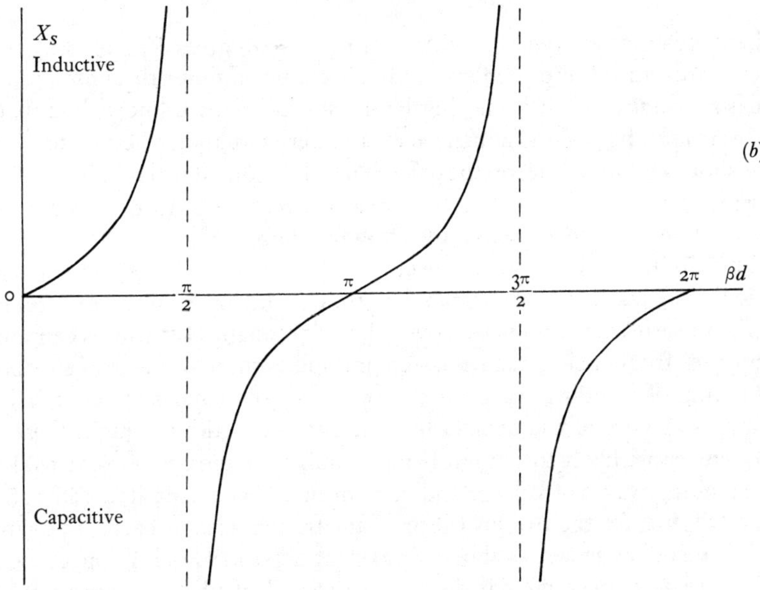

Fig. 13.17 Short-circuited line: (*a*) standing-wave patterns; (*b*) input reactance as a function of length.

This is true for any $\boldsymbol{Z}_T$; but if $\boldsymbol{Z}_T$ is a pure reactance $\mathrm{j}X_T$,

$$\boldsymbol{Z}_S = -\mathrm{j}\frac{Z_0^2}{X_T},$$

or

$$X_S = -\frac{Z_0^2}{X_T}. \qquad (13.7.7)$$

An inductive terminating reactance is therefore transformed into a capacitive input reactance and vice versa. When the termination is not a pure reactance, the semicircular arc in the Smith chart lies concentrically within the periphery, after the manner of curve (*b*) in fig. 13.14.

Equations (13.7.2) for the open-circuited line are in terms of the termination voltage $\boldsymbol{V}_T$. Through the relation $\boldsymbol{V}_S = \boldsymbol{V}_T \cos\beta d$ they may be re-expressed in terms of $\boldsymbol{V}_S$, regarded as the reference phasor:

$$\boldsymbol{V} = V_S \frac{\cos\beta x}{\cos\beta d}, \qquad \boldsymbol{I} = \mathrm{j}Y_0 V_S \frac{\sin\beta x}{\cos\beta d}. \qquad (13.7.8)$$

Similarly, for the short-circuited line

$$\boldsymbol{V} = V_S \frac{\sin\beta x}{\sin\beta d}, \qquad \boldsymbol{I} = -\mathrm{j}Y_0 V_S \frac{\cos\beta x}{\sin\beta d}. \qquad (13.7.9)$$

When βd is $\pi/2$ or any odd multiple of it, (13.7.8) gives V, I tending to infinity; thus an infinitesimal stimulus V_S creates a finite response, and the line is *resonant*. For the same lengths the short-circuited line is described as *anti-resonant*, because V and I are everywhere less than or equal to V_S, I_S. The short-circuited line, on the other hand, is resonant when βd is a multiple of π; if open-circuited it then becomes anti-resonant. All these statements assume that the line is excited by a voltage source.

This discussion assumes a constant frequency and a length which can be varied. If we take a line of constant length, whether open- or short-circuited, there is a series of frequencies at which it is resonant. A stimulus containing a range of frequencies, such as a step or a spike, therefore excites a response consisting of standing waves at the resonant frequencies; theoretically an infinite sequence of frequencies occurs, but in practice the higher resonant frequencies are likely to be increasingly inhibited by losses. Fig. 13.18 shows the standing waves of voltage and current in an open-circuited and a short-circuited line, for the two lowest resonant frequencies of each. The curves are drawn below as well as above the axis, because in a lossless line a negative ordinate has a meaning – it denotes a reversal of phase. A reader who is familiar with acoustic waves in organ-pipes will recognize that they are analogous with transmission lines; if voltage in the line is taken as the analogue of pressure in the pipe, the short-circuited line is the analogue of an

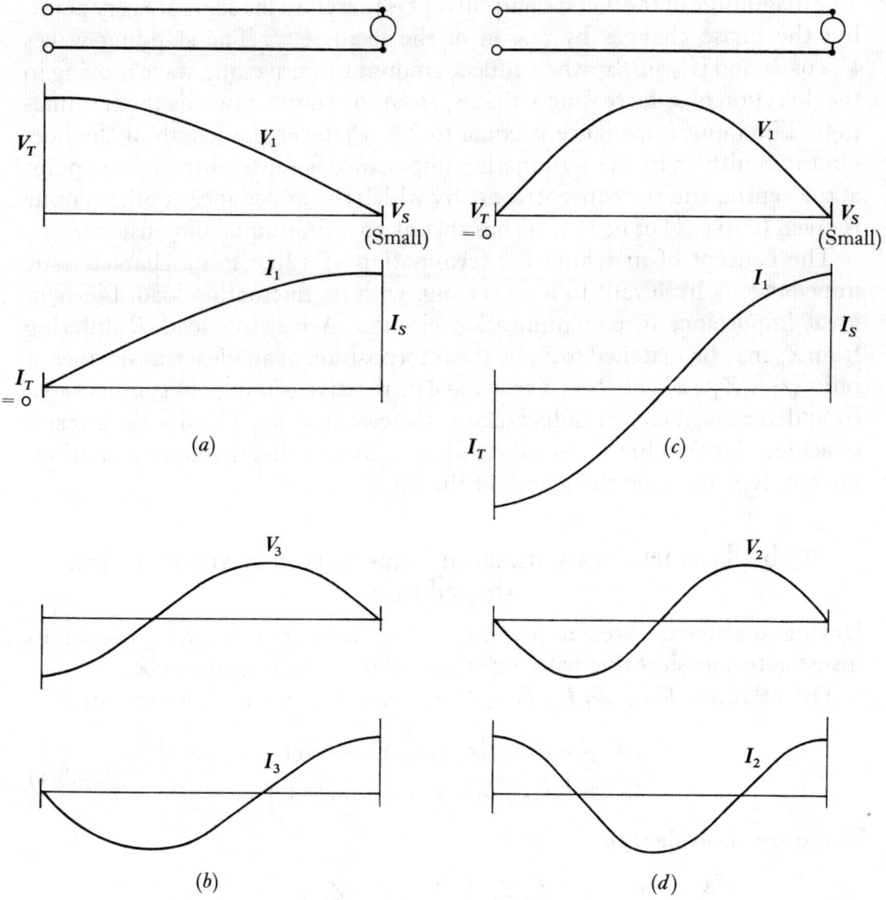

Fig. 13.18 Standing waves due to resonance on a line: (*a*) open-circuited line, fundamental; (*b*) open-circuited line, third harmonic; (*c*) short-circuited line, fundamental; (*d*) short-circuited line, second harmonic.

open pipe (having zero pressure rise at the open end), and the open-circuited line of a closed pipe. The short-circuited line resonates to its fundamental frequency and all integral multiples of it; the open-circuited line, to the fundamental and odd harmonics only.

(c) The matched line

A line which is terminated by an impedance equal to its characteristic impedance is said to be *matched*; for a lossless line the terminating impedance must be a resistance Z_0. In this case (13.7.1) reduces to

$$\left.\begin{aligned} \boldsymbol{V} = \boldsymbol{V}_T(\cos\beta x + \mathrm{j}\sin\beta x) = \boldsymbol{V}_T\,\mathrm{e}^{\mathrm{j}\beta x}, \\ \boldsymbol{I} = \boldsymbol{I}_T\,\mathrm{e}^{\mathrm{j}\beta x}. \end{aligned}\right\} \qquad (13.7.10)$$

The magnitude of the voltage and current is therefore the same at every point, but the phase changes by reason of the factor $e^{j\beta x}$. The standing waves $\boldsymbol{V}_T \cos\beta x$ and $j\boldsymbol{V}_T \sin\beta x$, when added, amount to a travelling wave moving in the direction of x decreasing – that is, from the source towards the termination. The input impedance is equal to Z_0, whatever the length of the line. On the Smith chart the terminating impedance is represented by the point at the centre, and the concentric arc by which the impedance at other points is given (curve (*b*) of fig. 13.14) has shrunk to infinitesimal dimensions.

The concept of matching the termination of a line to its characteristic impedance is irrelevant to a power line with its fluctuating load, but is of great importance in communication circuits. A resistive load R differing from Z_0 may be matched to Z_0 by the interposition of an ideal transformer of ratio $\sqrt{(R/Z_0)}$; a load $R+jX$ may have its reactive component compensated by added capacitance or inductance as the case may be. The desirable result so achieved is the elimination of standing waves, so that the input conditions are not dependent on the length of the line.

13.8 Lossless transmission lines with any terminating impedance

Having discussed three important special terminations, we proceed to investigate a lossless line with a general terminating impedance $\boldsymbol{Z}_T$.

The relations $\boldsymbol{V}_T = \boldsymbol{Z}_T \boldsymbol{I}_T$, $\boldsymbol{I}_T = \boldsymbol{Y}_T \boldsymbol{V}_T$ enable (13.7.1) to be rewritten

$$\left.\begin{aligned} \boldsymbol{V} &= \boldsymbol{V}_T(\cos\beta x + jZ_0 \boldsymbol{Y}_T \sin\beta x), \\ \boldsymbol{I} &= \boldsymbol{I}_T(\cos\beta x + jY_0 \boldsymbol{Z}_T \sin\beta x). \end{aligned}\right\} \qquad (13.8.1)$$

These are equivalent to

$$\left.\begin{aligned} \boldsymbol{V} &= \tfrac{1}{2}\boldsymbol{V}_T[(1+Z_0\boldsymbol{Y}_T)e^{j\beta x} + (1-Z_0\boldsymbol{Y}_T)e^{-j\beta x}], \\ \boldsymbol{I} &= \tfrac{1}{2}\boldsymbol{I}_T[(1+Y_0\boldsymbol{Z}_T)e^{j\beta x} + (1-Y_0\boldsymbol{Z}_T)e^{-j\beta x}]. \end{aligned}\right\} \qquad (13.8.2)$$

The ratio of the coefficients of $e^{-j\beta x}$ and $e^{j\beta x}$ is

$$\left.\begin{aligned} \frac{1-Z_0\boldsymbol{Y}_T}{1+Z_0\boldsymbol{Y}_T} &= \frac{\boldsymbol{Z}_T - Z_0}{\boldsymbol{Z}_T + Z_0} = \boldsymbol{\Gamma}\ \text{(say)}, \\ \text{and} \quad \frac{1-Y_0\boldsymbol{Z}_T}{1+Y_0\boldsymbol{Z}_T} &= \frac{\boldsymbol{Y}_T - Y_0}{\boldsymbol{Y}_T + Y_0} = -\boldsymbol{\Gamma}. \end{aligned}\right\} \qquad (13.8.3)$$

Using this, we may rewrite (13.8.2) as

$$\left.\begin{aligned} \boldsymbol{V} &= \boldsymbol{V}_1(e^{j\beta x} + \boldsymbol{\Gamma} e^{-j\beta x}), \\ \boldsymbol{I} &= \boldsymbol{I}_1(e^{j\beta x} - \boldsymbol{\Gamma} e^{-j\beta x}), \end{aligned}\right\} \qquad (13.8.4)$$

where $\quad \boldsymbol{V}_1 = \tfrac{1}{2}\boldsymbol{V}_T(1+Z_0\boldsymbol{Y}_T), \qquad \boldsymbol{I}_1 = \tfrac{1}{2}\boldsymbol{I}_T(1+Y_0\boldsymbol{Z}_T), \qquad (13.8.5)$

so that, as may be verified,

$$\boldsymbol{V}_1 = Z_0 \boldsymbol{I}_1. \tag{13.8.6}$$

The complex number Γ denotes the magnitude ratio and phase difference between a travelling wave leaving the termination and one arriving there; it is therefore called the *reflexion coefficient* of the termination. Writing

$$\Gamma = |\Gamma| \, e^{2j\psi} \tag{13.8.7}$$

– the reason for the index $2j\psi$ will appear – we obtain

$$\boldsymbol{V} = \boldsymbol{V}_1 \, e^{j\beta x}[1 + |\Gamma| \, e^{-2j(\beta x - \psi)}],$$
$$= \boldsymbol{V}_1 \, e^{j\beta x}\{[1 + |\Gamma| \cos 2(\beta x - \psi)] - j[|\Gamma| \sin 2(\beta x - \psi)]\}.$$

Taking moduli,

$$\left.\begin{aligned} V &= V_1 \sqrt{[1 + 2|\Gamma| \cos 2(\beta x - \psi) + |\Gamma|^2]}, \\ \text{and similarly} \quad I &= I_1 \sqrt{[1 - 2|\Gamma| \cos 2(\beta x - \psi) + |\Gamma|^2]}. \end{aligned}\right\} \tag{13.8.8}$$

In fig. 13.19, V/V_1 and I/I_1 are plotted as functions of electrical lengths βx measured from the termination for a particular value of $|\Gamma|$, which necessarily lies between 0 and 1. The ordinates of each curve vary between the limits

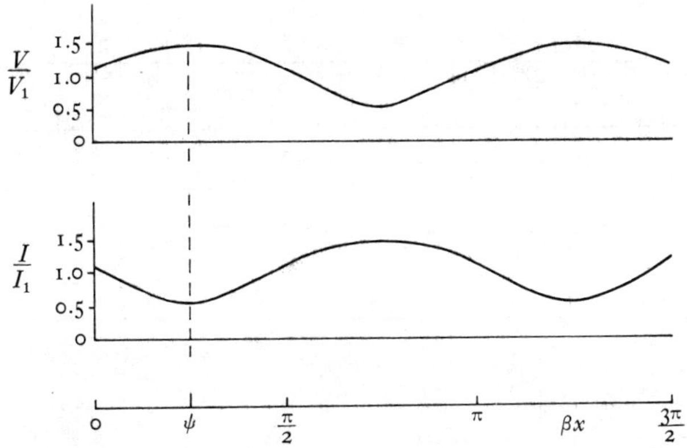

Fig. 13.19 Standing-wave patterns for a line with termination such that $|\Gamma| = 0.5$.

$|1 \pm |\Gamma||$, and the angle ψ, assumed to lie between 0 and π, measures the electrical distance between the termination and the first voltage maximum, or the first current minimum. The curves are not sine waves; the minima are more sharply curved than the maxima, so that locating the standing waves experimentally is best done by observation of the minima. This feature is clearly seen in the curves for the open-circuited line (fig. 13.16) or short-circuited line (fig. 13.17), which should be compared with fig. 13.19; they represent respectively the special cases in which $\Gamma = 1$ and -1.

The ratio between the maximum and minimum amplitude of the standing waves of voltage or current is

$$S = \frac{1 + |\Gamma|}{1 - |\Gamma|}. \tag{13.8.9}$$

S is known as the *standing-wave ratio*; it varies between unity for a matched line, and infinity for a line which is either open- or short-circuited. The term *voltage standing-wave ratio* (abbreviated VSWR) is also used for S to emphasize its significance as a ratio of voltages or currents, as against S^2 which measures the same ratio on a power basis. Observation of standing waves is made by means of a probe, which measures the variations of electric field and therefore of voltage. Without measuring the absolute value of the electric field strength, S is determined as a ratio of a maximum to a minimum reading, and $|\Gamma|$ is then deduced from the inverse of (13.8.9):

$$|\Gamma| = \frac{S - 1}{S + 1}. \tag{13.8.10}$$

Electrical lengths are determined by assigning the value π to the observed distance between successive minima in the standing-wave pattern (fig. 13.20). The phase angle 2ψ of the reflexion coefficient may be measured by

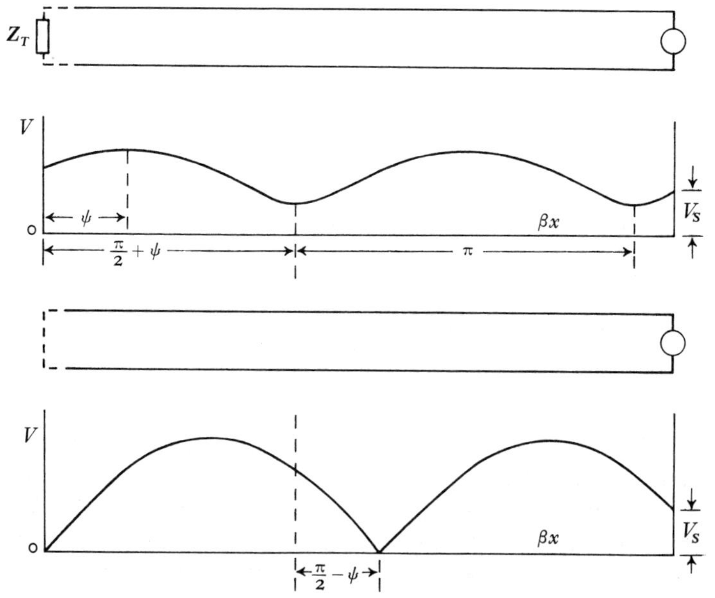

Fig. 13.20 Measurements on a standing-wave pattern.

observing the shift in the standing-wave pattern when the termination is replaced by a short circuit; as the diagram indicates, this is $(\pi/2)-\psi$. Such a procedure obviates the need for making measurements close to the termination, where the basic conditions mentioned in §13.1 (that the electric and magnetic fields shall be transverse) may not be satisfied. When Γ has been determined in both magnitude and phase, the normalized value of $\boldsymbol{Z}_T$ follows if Z_0 is known; and from (13.7.3) and (13.7.5), Z_0 is seen to be the geometric mean of the input impedances measured with open-circuited and short-circuited terminations. In this way the value and effective location of the terminating impedance are both deduced from measurements made on the standing-wave pattern.

The Smith chart is again useful here. If the normalized terminating impedance is w_T, (13.8.3) may be written

$$\Gamma=\frac{w_T-1}{w_T+1},$$

and (13.6.4) then shows that

$$\Gamma=W_T. \tag{13.8.11}$$

Thus if T is the point denoting the terminating impedance (fig. 13.21), $|\Gamma|$ and 2ψ are the polar co-ordinates of T with respect to the centre. In (13.6.7)

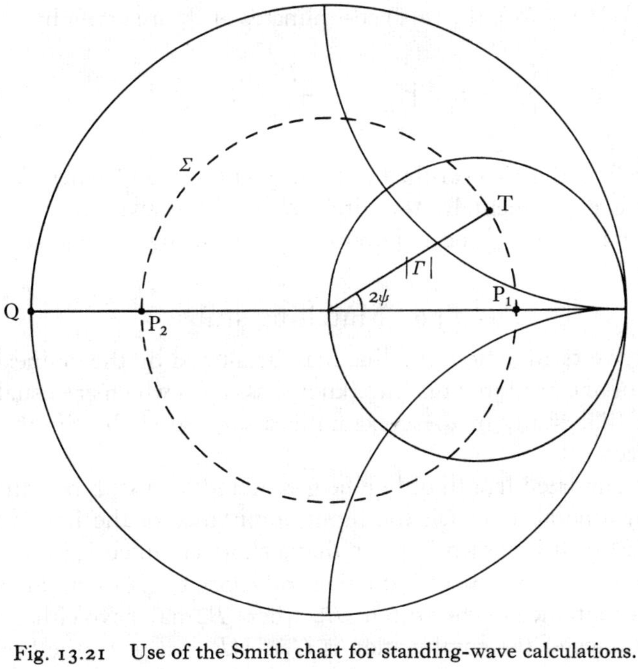

Fig. 13.21 Use of the Smith chart for standing-wave calculations.

it was seen that the polar co-ordinates for W were $(e^{-2\xi}, \pi - 2\eta)$; if, then, the point T has co-ordinates (ξ_T, η_T) in the ζ co-ordinate system,

$$|\Gamma| = e^{-2\xi_T}, \tag{13.8.12}$$

and

$$\psi = \frac{\pi}{2} - \eta. \tag{13.8.13}$$

The ratio of voltage to current at any point of a line so terminated is given by the (u, v) co-ordinates of the corresponding point on the dotted circle Σ through T. The magnitude of the ratio is a maximum at P_1, which is the 'voltage maximum, current minimum' point, and a minimum at P_2, the 'voltage minimum, current maximum'. The arc TP_1 subtends an angle 2ψ, corresponding with the electrical length ψ between the termination and the voltage maximum; for angles in the Smith chart are double the electrical lengths – a complete rotation of 2π corresponds with a movement of half a wavelength.

The standing-wave ratio is immediately derived from the diagram as

$$S = \frac{1+|\Gamma|}{1-|\Gamma|} = \frac{QP_1}{QP_2}, \tag{13.8.14}$$

where Q is the point shown. It can also be read off on the scales; for, as $w = (1+W)/(1-W)$, the (u, v) co-ordinates of P_1 are given by

$$u_1 + jv_1 = \frac{1+|\Gamma|}{1-|\Gamma|} = S + j0. \tag{13.8.15}$$

S is therefore the u-co-ordinate of the point P_1; and when S has been determined experimentally the circle Σ can be drawn, and the point T (which determines $\boldsymbol{Z}_T$) located upon it by measuring the angle ψ.

13.9 Matching stubs

The effective termination of a line may be altered by the connexion, near the load, of one or more branches (known as *stubs*) which are usually short-circuited. The study of these exemplifies the use of the Smith chart for admittances.

Let a terminated length of line be given, and let a stub be connected in parallel at a point at which the input admittance of the line is $G_1 + jB_1$ (fig. 13.22(*a*)). It is shown in §13.7 that a short-circuited lossless line takes all values of reactance, both inductive and capacitive, at lengths up to $\lambda/2$; thus the admittance of the stub is jB_s, where B_s may have either sign, and the admittance of the combination is $G_1 + j(B_1 + B_s)$. In the Smith chart for admittances (fig. 13.22(*b*)), the input admittance of the terminated line is denoted by P_1, at which

$$w = u_1 + jv_1 = \frac{G_1 + jB_1}{Y_0},$$

and that of the combination by P'_1, at which

$$w = u_1 + j(v_1 + v_s) = \frac{G_1 + j(B_1 + B_s)}{Y_0},$$

so that P'_1 and P_1 lie on the same circle $u = u_1$.

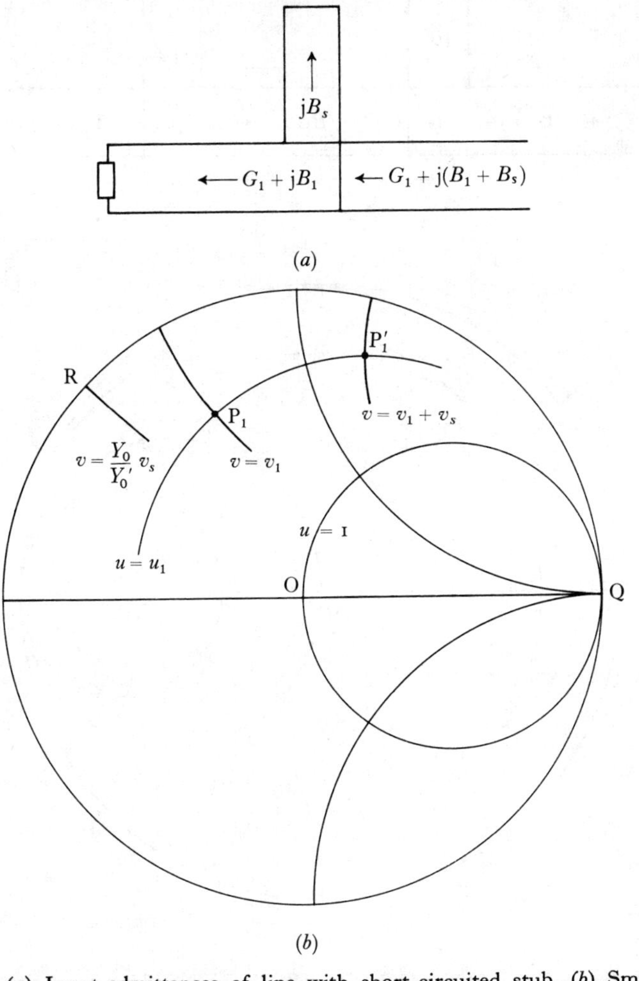

Fig 13.22 (*a*) Input admittances of line with short-circuited stub. (*b*) Smith chart calculation on (*a*).

It is not necessary for the stub to have the same characteristics as the main line. To see this, we will consider how the admittance change from P_1 to P'_1 might be effected by the use of a stub of characteristic admittance Y'_0. v_s is given; therefore jB_s (the actual admittance of the stub) must equal jY_0v_s. Normalized with respect to Y'_0 this is $j(Y_0/Y'_0)v_s$. The electrical length of a stub possessing this normalized admittance is given by the arc QR, measured clockwise round the peripheral circle from the short-circuit point Q to the point R at which $v = Y_0v_s/Y'_0$.

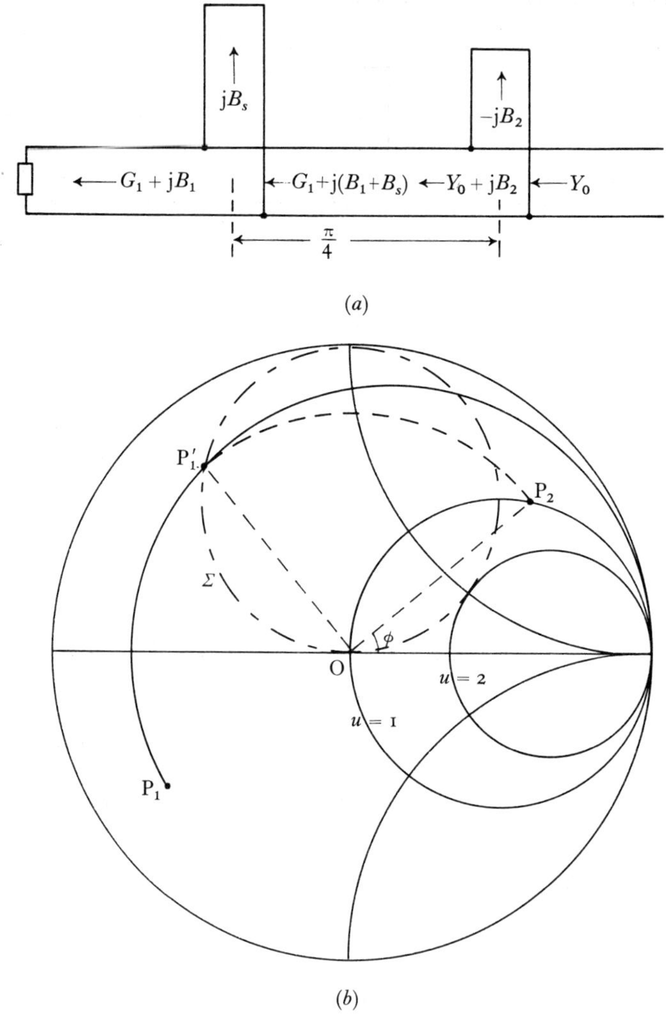

Fig. 13.23 (*a*) Input admittances of line matched by means of two short-circuited stubs. (*b*) Smith chart calculation on (*a*).

The central point O corresponds with the matched condition. It is evidently possible for P'_1 to coincide with O only if the starting-point P_1 happens to lie on the circle $u = 1$; thus a single stub in a fixed location is of little use as a matching device. Two separated stubs are of much more use; the practical condition is where the stubs are fixed in position but adjustable as to length, and the electrical distance between them will be assumed for the purpose of illustration to be $\pi/4$ (fig. 13.23(*a*)). In the Smith chart (fig. 13.23(*b*)), P_1 denotes the normalized admittance $(G_1 + jB_1)/Y_0$, and the first stub, of admittance jB_s, changes this to the value denoted by P'_1. From here an arc P'_1P_2, subtending 90 degrees on a circle concentric with O, accounts for the effect of the line of length $\pi/4$ between the stubs; and P_2 lies on the circle $u = 1$, so that a second stub of suitable length cancels the imaginary component of the admittance and causes the line, from this point onwards, to be matched.

Will this process work for all starting-points P_1? To answer this question we must carry out the process backwards, starting from the circle $u = 1$ and drawing quadrantal arcs from every point so as to find all the possible positions for P'_1. If ϕ is the angle shown, the polar co-ordinates of P_2 are $(\cos\phi, \phi)$, and of P'_1, $(\cos\phi, \phi + \pi/2)$; therefore the locus of P'_1 is

$$r = \sin\theta, \tag{13.9.1}$$

which is the circle Σ shown. It will be found that Σ touches the circle $u = 2$; thus the condition for double-stub matching to be possible, with stubs $\pi/4$ apart, is that the starting-point P_1 must lie outside the circle $u = 2$, or in other words $u_1 \leqslant 2$. This restricts the range of terminations which can be matched. It can, however, be shown that all terminations can be matched if three fixed stubs are employed.

13.10 Transient phenomena in transmission lines

Part of the importance of the steady state a.c. theory of transmission lines is that it applies to other linelike networks; the sole requisite is that the network can be represented as a sequence of elements having series impedance $\mathbf{z}\delta x$ and shunt admittance $\mathbf{y}\delta x$. Thus, at a fixed frequency, both the networks of fig. 13.10 are exactly equivalent to a lossless transmission line regarded in a general sense; where they differ is in behaving differently as the frequency is varied. When the circuits are excited by a non-repetitive stimulus such as a step or a spike, a spectrum of frequencies is present; the analysis must therefore be restricted to one form of circuit at a time, and in this section the transmission line proper is considered.

The passage from the steady-state a.c. case to the transient problem might be made via the Fourier transform; but, as was pointed out in chapter 11, the Laplace transform gives a solution which is simpler analytically – here it must be a double transform, replacing the variables t and x by image

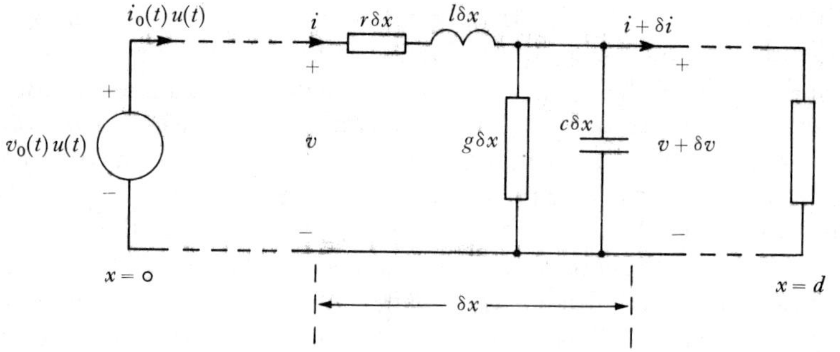

Fig. 13.24 Transmission line excited by a general voltage source.

variables s and p. The line will be considered to have a source $v_0(t)$, $i_0(t)$, located at $x=0$ and switched on at $t=0$, so that it is described by $v_0(t)u(t)$, $i_0(t)u(t)$ (fig. 13.24). The equations for the element δx are

$$\delta v+(r\,\delta x)\,i+(l\,\delta x)\frac{\partial i}{\partial t}=0,$$

and
$$\delta i+(g\,\delta x)(v+\delta v)+(c\,\delta x)\frac{\partial(v+\delta v)}{\partial t}=0;$$

when second order small quantities are disregarded, these lead to the differential equations

$$\left.\begin{aligned}\frac{\partial v}{\partial x}+ri+l\frac{\partial i}{\partial t}&=0,\\ \frac{\partial i}{\partial x}+gv+c\frac{\partial v}{\partial t}&=0.\end{aligned}\right\}\qquad(13.10.1)$$

When either i or v is eliminated, it is found that the other satisfies the equation

$$\frac{\partial^2 v}{\partial x^2}=lc\frac{\partial^2 v}{\partial t^2}+(lg+rc)\frac{\partial v}{\partial t}+rgv.\qquad(13.10.2)$$

In the lossless line $r=g=0$, and the equation reduces to

$$\frac{\partial^2 v}{\partial x^2}=lc\frac{\partial^2 v}{\partial t^2}.\qquad(13.10.3)$$

Another important idealized limiting case is the *non-inductive and leakage-free cable*, in which $l=g=0$; in this,

$$\frac{\partial^2 v}{\partial x^2}=rc\frac{\partial v}{\partial t}.\qquad(13.10.4)$$

Equations (13.10.3,4) exemplify two of the great equations of mathematical physics, the *wave propagation equation* and the *diffusion equation* respectively. They will be discussed separately, and will be found to lead to two contrasting ways of transmitting signals.

For a line in which v and i are zero at every point when $t = 0$, the Laplace transform with respect to t is obtained by simply replacing v, i by $\boldsymbol{V}$, $\boldsymbol{I}$ and $\partial/\partial t$ by s ((5.5.3), with $w_0 = 0$). Thus (13.10.1) is transformed into

$$\left.\begin{aligned} \frac{\partial \boldsymbol{V}}{\partial x} + (r + \mathrm{s}l)\boldsymbol{I} = 0, \\ \frac{\partial \boldsymbol{I}}{\partial x} + (g + \mathrm{s}c)\boldsymbol{V} = 0. \end{aligned}\right\} \tag{13.10.5}$$

These are exactly of the form of (13.2.1) but with

$$\left.\begin{aligned} \boldsymbol{z} = \boldsymbol{z}(\mathrm{s}) = r + \mathrm{s}l, \\ \boldsymbol{y} = \boldsymbol{y}(\mathrm{s}) = g + \mathrm{s}c, \end{aligned}\right\} \tag{13.10.6}$$

– that is, with jω replaced by s. The analysis therefore continues as before, leading to

$$\left.\begin{aligned} \boldsymbol{V} = \tfrac{1}{2}(\boldsymbol{V}_0 - \boldsymbol{Z}_0 \boldsymbol{I}_0)\,\mathrm{e}^{\gamma x} + \tfrac{1}{2}(\boldsymbol{V}_0 + \boldsymbol{Z}_0 \boldsymbol{I}_0)\,\mathrm{e}^{-\gamma x}, \\ \boldsymbol{I} = \tfrac{1}{2}(\boldsymbol{I}_0 - \boldsymbol{Y}_0 \boldsymbol{V}_0)\,\mathrm{e}^{\gamma x} + \tfrac{1}{2}(\boldsymbol{I}_0 + \boldsymbol{Y}_0 \boldsymbol{V}_0)\,\mathrm{e}^{-\gamma x}. \end{aligned}\right\} \tag{13.10.7}$$

Here $$\boldsymbol{V}_0 = \mathscr{L}_t v_0(t)\,u(t), \qquad \boldsymbol{I}_0 = \mathscr{L}_t i_0(t)\,u(t), \tag{13.10.8}$$

also $$\gamma = \sqrt{(\boldsymbol{y}\boldsymbol{z})} = \sqrt{[(r + \mathrm{s}l)(g + \mathrm{s}c)]}, \tag{13.10.9}$$

and $$\boldsymbol{Z}_0 = \sqrt{\left(\frac{\boldsymbol{z}}{\boldsymbol{y}}\right)} = \sqrt{\left(\frac{r + \mathrm{s}l}{g + \mathrm{s}c}\right)}. \tag{13.10.10}$$

But whereas these solutions can be immediately interpreted when $\mathrm{s} = \mathrm{j}\omega$, here the major difficulty is still to come – that of getting back from the s-world to the t-world.

With $\mathrm{s} = \sigma + \mathrm{j}\omega$, γ and $\boldsymbol{Z}_0$ are complex numbers; we may write

$$\gamma = \alpha(\sigma, \omega) + \mathrm{j}\beta(\sigma, \omega),$$

and, as in the steady-state a.c. theory, α is taken as positive. Thus the $\mathrm{e}^{\gamma x}$ terms cannot appear when the line is infinitely long; they would represent a wave reflected back from a termination, and there is no termination. In this case

$$\boldsymbol{V}_0 = \boldsymbol{Z}_0 \boldsymbol{I}_0, \tag{13.10.11}$$

and $$\boldsymbol{V} = \boldsymbol{V}_0\,\mathrm{e}^{-\gamma x}, \qquad \boldsymbol{I} = \boldsymbol{I}_0\,\mathrm{e}^{-\gamma x}. \tag{13.10.12}$$

The interpretation of these transform equations will be considered through special cases.

(a) The lossless line

When $r = g = 0$, $\gamma = s\sqrt{(lc)}$, or

$$\left.\begin{aligned}\gamma &= s/u,\\ \text{where}\quad u &= (lc)^{-1/2}\end{aligned}\right\} \tag{13.10.13}$$

Thus
$$V = V_0\, e^{-sx/u}. \tag{13.10.14}$$

This expression for V has the form of the right hand side of equation (10.8.1), which was

$$\mathscr{L} f(t-T)\, u(t-T) = e^{-Ts} F(s),$$

or
$$\mathscr{L}^{-1} e^{-Ts} F(s) = f(t-T)\, u(t-T). \tag{13.10.15}$$

Therefore

$$\mathscr{L}_t^{-1} V_0\, e^{-sx/u} = v = v_0\left(t - \frac{x}{u}\right) u\left(t - \frac{x}{u}\right). \tag{13.10.16}$$

The voltage variation $v_0(t)u(t)$, impressed at $x = 0$ by the source, is thus reproduced at x with a time-delay of x/u (fig. 13.25(*a*)). Equation (13.10.16) may be compared with (13.3.3), derived for a sinusoidal stimulus; the absence of distortion of a non-sinusoidal waveform is deducible from the fact that the velocity of propagation of a sinusoidal wave is independent of its time-frequency. The current wave is propagated similarly; since $Z_0 = \sqrt{(l/c)}$, a constant independent of s, the shapes of the voltage and current waves are identical.

(b) The distortionless line

When $r/g = l/c$, we may write

$$\frac{r}{l} = \frac{g}{c} = \alpha. \tag{13.10.17}$$

In this case
$$\gamma = (s+\alpha)\sqrt{(lc)} = \frac{s+\alpha}{u}, \tag{13.10.18}$$

and
$$V = V_0\, e^{-(s+\alpha)x/u}. \tag{13.10.19}$$

This differs from (13.10.14) only by the factor $e^{-\alpha x/u}$, which does not contain s and therefore behaves like a constant in the operation $\mathscr{L}_t^{-1}$. Hence

$$v = \mathscr{L}_t^{-1} V_0\, e^{-(s+\alpha)x/u} = e^{-\alpha x/u}\, \mathscr{L}_t^{-1} V_0\, e^{-sx/u},$$

or
$$v = e^{-\alpha x/u}\, v_0\left(t - \frac{x}{u}\right) u\left(t - \frac{x}{u}\right). \tag{13.10.20}$$

The impressed voltage variation is thus transmitted without change of shape, but with its amplitude diminished on account of the factor $e^{-\alpha x/u}$ (fig. 13.25(*b*)). Again $Z_0 = \sqrt{(l/c)}$, and the current wave has the same shape and the same attenuation as the voltage wave.

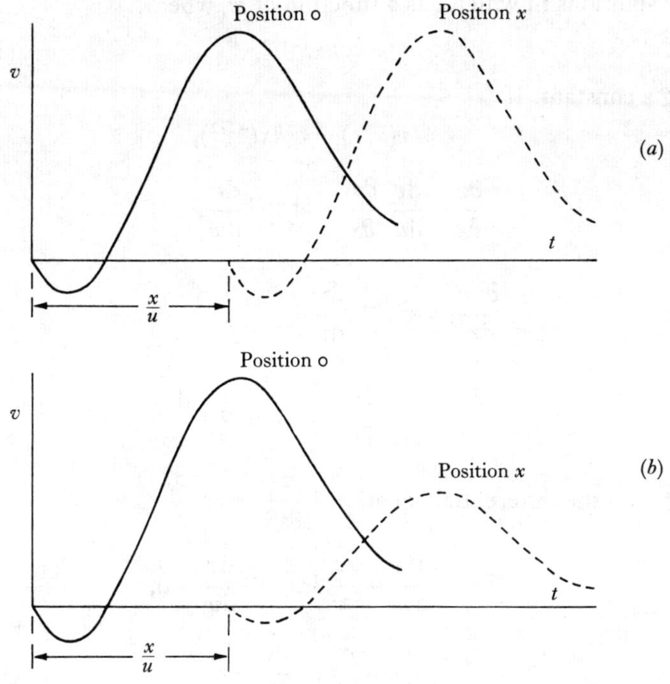

Fig. 13.25 Representation of a travelling voltage wave as a function of time: (*a*) lossless line; (*b*) distortionless line with losses.

(*c*) *The non-inductive and leakage free cable*

The ideal case $l = g = 0$ is inherently unrealizable, because circuits cannot be made without inductance. It is important, however, as a limiting case, and leads to a different mode of signal propagation. In treating resistance rather than inductance as the controlling term of the series impedance, we are effectively assuming $r \gg l\omega$; thus the departure from the ideal will occur in respect of the higher-frequency components of the signal. The word 'cable' is appropriate because the inductance of a line is lowest when the space for magnetic flux is confined.

When $l = g = 0$,

$$\gamma = \sqrt{(rcs)}, \tag{13.10.21}$$

and so

$$V = V_0 e^{-x\sqrt{(rcs)}}. \tag{13.10.22}$$

This is quite unlike any Laplace transform hitherto discussed. To interpret it we go back to the differential equation

$$\frac{\partial^2 v}{\partial x^2} = rc\frac{\partial v}{\partial t}, \qquad \text{(equation (13.10.4))}$$

and try solutions in which v is a function of w, where

$$w = kxt^{-1/2}, \tag{13.10.23}$$

k being a constant. If

$$v = v(x, t) - v(kxt^{-1/2}),$$

then
$$\frac{\partial v}{\partial x} = \frac{dv}{dw}\cdot\frac{\partial w}{\partial x} = kt^{-1/2}\frac{dv}{dw},$$

$$\frac{\partial^2 v}{\partial x^2} = k^2 t^{-1}\cdot\frac{d^2 v}{dw^2},$$

and
$$\frac{\partial v}{\partial t} = \frac{dv}{dw}\cdot\frac{\partial w}{\partial t} = -\tfrac{1}{2}kxt^{-3/2}\frac{dv}{dw}.$$

If v satisfies the differential equation $\dfrac{\partial^2 v}{\partial x^2} = rc\dfrac{\partial v}{\partial t}$,

$$k^2 t^{-1}\cdot\frac{d^2 v}{dw^2} + \frac{rc}{2}\cdot kxt^{-3/2}\frac{dv}{dw} = 0,$$

or
$$\frac{d^2 v}{dw^2} + \frac{rc}{2k^2}\cdot w\cdot\frac{dv}{dw} = 0. \tag{13.10.24}$$

The disappearance of x and t explicitly, and their replacement by w, shows that some kind of solution in which v is of the postulated form can be found. The constant k is at our disposal, and will be chosen to make $(rc/2k^2) = 2$, so that

$$w = \frac{x}{2}\sqrt{\left(\frac{rc}{t}\right)}. \tag{13.10.25}$$

Equation (13.10.24) is then

$$\frac{d^2 v}{dw^2} + 2w\cdot\frac{dv}{dw} = 0,$$

and is solved by using the integrating factor $e^{\int 2w\,dw}$, or e^{w^2}. We obtain

$$\frac{d}{dw}\left(e^{w^2}\cdot\frac{dv}{dw}\right) = 0,$$

or $$e^{w^2}\cdot\frac{dv}{dw} = \text{constant} = C_1' \text{ say.}$$

Then $$\frac{dv}{dw} = C_1' e^{-w^2},$$

and $$v = C_0 + C_1' \int e^{-w^2}\, dw, \tag{13.10.26}$$

where C_0 is another constant.

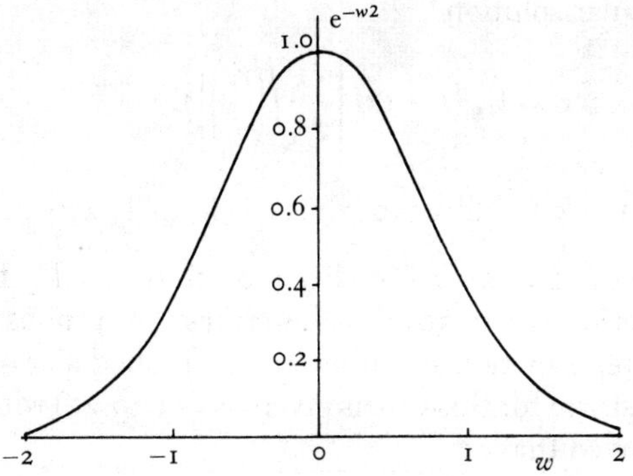

(*a*)

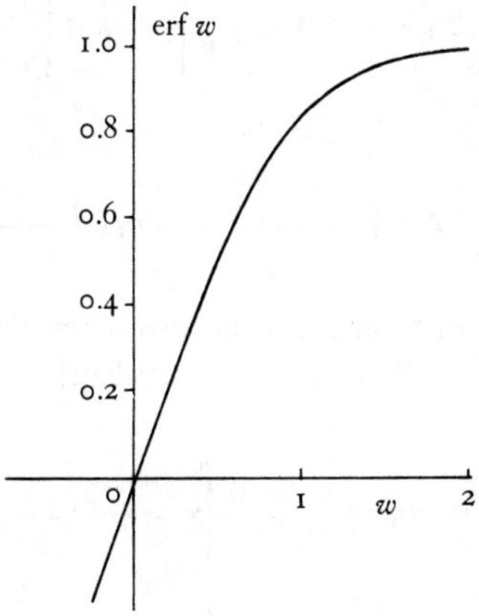

(*b*)

Fig. 13.26 (*a*) Gaussian distribution function e^{-w^2}. (*b*) Error function erf w.

e^{-w^2} is the *Gaussian distribution function* occurring in the theory of statistics (fig. 13.26(*a*)). It can be shown that

$$\int_0^\infty e^{-w^2}\, dw = \tfrac{1}{2}\sqrt{\pi}, \tag{13.10.27}$$

and the *error function* erf w (fig. 13.26(*b*)) is defined by the definite integral

$$\operatorname{erf} w = \frac{2}{\sqrt{\pi}} \int_0^w e^{-w^2}\, dw, \tag{13.10.28}$$

so that erf $0 = 0$, erf $\infty = 1$. The solution (13.10.26) may therefore be written

$$v = C_0 + C_1 \operatorname{erf} w, \tag{13.10.29}$$

where $C_1 = C_1'\sqrt{\pi}/2$; for the replacement of the indefinite by the definite integral amounts at most to an adjustment in C_0. v, so defined, varies between C_0 at $w = 0$ and $C_0 + C_1$ at $w = \infty$; putting $C_0 = V_0$, $C_1 = -V_0$, we obtain the particular solution

$$v = V_0\left\{1 - \operatorname{erf}\left[\frac{x}{2}\sqrt{\left(\frac{rc}{t}\right)}\right]\right\}. \tag{13.10.30}$$

When $t = 0$, $\frac{x}{2}\sqrt{\left(\frac{rc}{t}\right)} = \infty$ for all $x > 0$; thus v is initially zero at every point. When $x = 0$ the bracket is zero for all $t > 0$; thus $v = V_0$ for all $t > 0$, or $v = V_0 u(t)$. Equation (13.10.30) thus describes the response of the line, initially dead, to a step function stimulus $V_0 u(t)$ applied at $x = 0$.

The Laplace transform for this case is given by (13.10.22) with $\boldsymbol{V}_0 = V_0/\mathrm{s}$. It has thus been proved that

$$\mathscr{L}_t^{-1}\frac{V_0}{\mathrm{s}}\, e^{-x\sqrt{(rc\mathrm{s})}} = V_0\left\{1 - \operatorname{erf}\left[\frac{x}{2}\sqrt{\left(\frac{rc}{t}\right)}\right]\right\},$$

or in general terms

$$\mathscr{L}^{-1}\left(\frac{e^{-a\sqrt{\mathrm{s}}}}{\mathrm{s}}\right) = 1 - \operatorname{erf}\left(\frac{a}{2\sqrt{t}}\right). \tag{13.10.31}$$

The function on the right hand side is zero when $t = 0$; therefore differentiation in the t-world corresponds with multiplication by s in the s-world. It follows that

$$\mathscr{L}^{-1} e^{-a\sqrt{\mathrm{s}}} = \frac{d}{dt}\left[1 - \operatorname{erf}\left(\frac{a}{2\sqrt{t}}\right)\right],$$

$$= \frac{a}{2\sqrt{\pi}}\, t^{-3/2}\, e^{-a^2/4t} \tag{13.10.32}$$

Both these transforms are included in the list in the appendix (nos. 42 and 40 respectively).

Equation (13.10.32) enables the response of the cable to a spike function stimulus to be found – possibly the most vivid way of picturing its behaviour. With such a stimulus,

$$\boldsymbol{V}_0 = \text{constant} = \Phi_0 \text{ (say)}, \tag{13.10.33}$$

so equation (13.10.22) becomes $V = \Phi_0 \mathrm{e}^{-x\sqrt{(rcs)}}$. Taking the inverse transform back into the t-world,

$$v = \frac{\Phi_0\sqrt{(rc)}}{2\sqrt{\pi}} x t^{-3/2} \mathrm{e}^{-rcx^2/4t}. \tag{13.10.34}$$

The nature of this solution is most clearly seen if the independent variables x, t are replaced by x, ξ, where

$$\xi = \frac{rcx^2}{4t}. \tag{13.10.35}$$

It is found that

$$v = \frac{4}{\sqrt{\pi}} \frac{\Phi_0}{rcx^2} \xi^{3/2} \mathrm{e}^{-\xi}. \tag{13.10.36}$$

The product $\xi^{3/2}\mathrm{e}^{-\xi}$ is zero at $\xi = 0$ and ∞ (see p. 411, footnote); for any given x this means $t = \infty$ and 0, respectively. v is a maximum at given x when $\partial v/\partial \xi = 0$, which gives $\xi = 3/2$, or

$$t = \tfrac{1}{6} rcx^2. \tag{13.10.37}$$

The value of the maximum is

$$\frac{4}{\sqrt{\pi}} \frac{\Phi_0}{rcx^2} \left(\frac{3}{2}\right)^{3/2} \mathrm{e}^{-3/2}.$$

or

$$v_{\max}(x) = 0.925 \left(\frac{\Phi_0}{rcx^2}\right). \tag{13.10.38}$$

This response, shown in fig. 13.27(*a*), is evidently very different from the travelling waves hitherto discussed. The maximum amplitude falls off as x^{-2}; the time taken to reach that maximum is proportional to x^2, so that the 'wave' progresses more and more slowly. The width of the pulse between half-amplitude points also increases as x^2.

Equation (13.10.37) may be written

$$t = \tfrac{1}{6}(rx)(cx),$$
$$= \tfrac{1}{6} \text{ (Resistance up to } x\text{) (Capacitance up to } x\text{)}.$$

This is the basis of Lord Kelvin's once well-known 'KR law' for telegraphy, discovered by him in 1854 when the practicability of an Atlantic telegraph cable was being considered. The law stated that the speed of signalling (in a pulsed code) would be inversely proportional to the product of the total capacitance 'K' and total resistance 'R'. In a cable in which applied short pulses of voltage are spread out in the manner shown, it is evident that the speed of signalling will be slow; nevertheless, a speed of 20 to 50 five-letter words per minute was obtained in the early Atlantic cables, in which the time $t = rcd^2/6$ was in the range 0.35 to 0.7 s.

The response to a step function stimulus, as given by (13.10.30), is the time-integral of the response to a spike and is shown in fig. 13.27(*b*). The

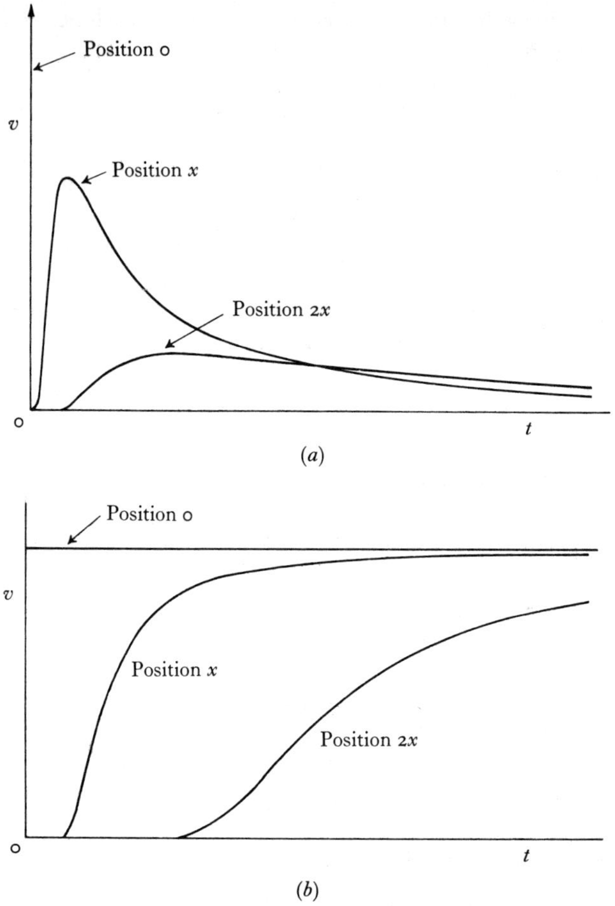

Fig. 13.27 Response of a non-inductive leakage-free cable: (*a*) to a spike stimulus; (*b*) to a step stimulus.

same transmission of a quantity by diffusion is found in a number of other physical systems; for example, in the response of the temperature at points within a lagged metal rod to a temperature change applied at the end, and in the penetration of magnetic flux into a conducting material. The current waveforms can be derived from the voltage equations through the relation

$$i = -\frac{1}{r}\frac{\partial v}{\partial x}. \qquad \text{(equation (13.10.1))}$$

The three cases discussed are the principal ones. The general problem of the lossy line, which may be said to combine the propagation and diffusion modes, is also amenable to Laplace transform analysis but will not be discussed here.

13.11 Transient phenomena in terminated lines

The general formula for the characteristic impedance of a line is

$$\boldsymbol{Z}_0 = \sqrt{\left(\frac{r+\mathrm{s}l}{g+\mathrm{s}c}\right)}. \qquad \text{(equation (13.10.10))}$$

This reduces to a constant in the special cases of the lossless and the distortionless line, but is otherwise an irrational function of s. No lumped network has such a function as its impedance; therefore only the two special lines mentioned can be matched with a lumped terminating impedance. The discussion which follows is confined to the lossless line.

As with the a.c. problem, it is convenient to take the terminating impedance $\boldsymbol{Z}_T = \boldsymbol{Z}_T(\mathrm{s})$ as located at $x = 0$; the energy source is assumed to be a voltage source $\boldsymbol{V}_S(\mathrm{s})$ at $x = d$. In (13.10.7) as applied to this problem, $\boldsymbol{V}_0 = \boldsymbol{V}_T$ and $\boldsymbol{I}_0 = \boldsymbol{Y}_T \boldsymbol{V}_T$; also $\gamma = \mathrm{s}/u$; we thus obtain equations closely analogous to (13.8.2):

$$\left.\begin{aligned} \boldsymbol{V} &= \tfrac{1}{2}\boldsymbol{V}_T[(1+Z_0\boldsymbol{Y}_T)\,\mathrm{e}^{\mathrm{s}x/u} + (1-Z_0\boldsymbol{Y}_T)\,\mathrm{e}^{-\mathrm{s}x/u}], \\ \boldsymbol{I} &= \tfrac{1}{2}\boldsymbol{I}_T[(1+Y_0\boldsymbol{Z}_T)\,\mathrm{e}^{\mathrm{s}x/u} + (1-Y_0\boldsymbol{Z}_T)\,\mathrm{e}^{-\mathrm{s}x/u}]. \end{aligned}\right\} \qquad (13.11.1)$$

As in §13.8, we define the reflexion coefficient Γ as

$$\Gamma = \Gamma(\mathrm{s}) = \frac{\boldsymbol{Z}_T - Z_0}{\boldsymbol{Z}_T + Z_0} = -\frac{\boldsymbol{Y}_T - Y_0}{\boldsymbol{Y}_T + Y_0}, \qquad (13.11.2)$$

so that (13.11.1) becomes

$$\left.\begin{aligned} \boldsymbol{V} &= \boldsymbol{V}_1(\mathrm{e}^{\mathrm{s}x/u} + \Gamma\mathrm{e}^{-\mathrm{s}x/u}), \\ \boldsymbol{I} &= \boldsymbol{I}_1(\mathrm{e}^{\mathrm{s}x/u} - \Gamma\mathrm{e}^{-\mathrm{s}x/u}), \end{aligned}\right\} \qquad (13.11.3)$$

where $$\boldsymbol{V}_1 = \tfrac{1}{2}\boldsymbol{V}_T(1+Z_0\boldsymbol{Y}_T), \qquad \boldsymbol{I}_1 = \tfrac{1}{2}\boldsymbol{I}_T(1+Y_0\boldsymbol{Z}_T), \qquad (13.11.4)$$

so that $$\boldsymbol{V}_1 = Z_0\boldsymbol{I}_1. \qquad (13.11.5)$$

At $x = d$, $\boldsymbol{V} = \boldsymbol{V}_S$; therefore

$$\boldsymbol{V}_S = \boldsymbol{V}_1(\mathrm{e}^{\mathrm{s}d/u} + \Gamma\mathrm{e}^{-\mathrm{s}d/u}),$$

and in terms of $\boldsymbol{V}_S$,

$$\left.\begin{aligned}\boldsymbol{V} &= \boldsymbol{V}_S\left(\frac{e^{sx/u}+\Gamma e^{-sx/u}}{e^{sd/u}+\Gamma e^{-sd/u}}\right),\\ \boldsymbol{I} &= Y_0\boldsymbol{V}_S\left(\frac{e^{sx/u}-\Gamma e^{-sx/u}}{e^{sd/u}+\Gamma e^{-sd/u}}\right).\end{aligned}\right\} \quad (13.11.6)$$

In particular, at the termination,

$$\left.\begin{aligned}\boldsymbol{V}_T &= \frac{\boldsymbol{V}_S(1+\Gamma)}{e^{sd/u}+\Gamma e^{-sd/u}},\\ \boldsymbol{I}_T &= \frac{Y_0\boldsymbol{V}_S(1-\Gamma)}{e^{sd/u}+\Gamma e^{-sd/u}}.\end{aligned}\right\} \quad (13.11.7)$$

A step function stimulus $v_S(t) = V_S u(t)$, $\boldsymbol{V}_S = V_S/s$, will now be assumed, and several special problems examined in detail.

(a) Resistive termination

When $\boldsymbol{Z}_T = R_T$, Γ does not contain s, but is a real number lying between ± 1. This means that a wave reflected from the termination will have the same time-shape as the incident wave. Equation (13.11.7) for $\boldsymbol{V}_T$ may be written

$$\boldsymbol{V}_T = \frac{V_S(1+\Gamma)}{s}e^{-sT}(1+\Gamma e^{-2sT})^{-1}, \quad (13.11.8)$$

where T is written for d/u, the time taken by a wave to travel along the line. Expansion of the last bracket by the binomial theorem gives

$$\boldsymbol{V}_T = \frac{V_S(1+\Gamma)}{s}[e^{-sT}-\Gamma e^{-3sT}+\Gamma^2 e^{-5sT}-\ldots]. \quad (13.11.9)$$

The corresponding function of t is a sequence of delayed step functions:

$$v_T = V_S(1+\Gamma)[u(t-T)-\Gamma u(t-3T)+\Gamma^2 u(t-5T)-\ldots]. \quad (13.11.10)$$

This is illustrated in fig. 13.28 for a negative and a positive Γ. Each added time-delay $2T$ corresponds with the time taken for a new reflected wave to travel from the termination to the source and back.

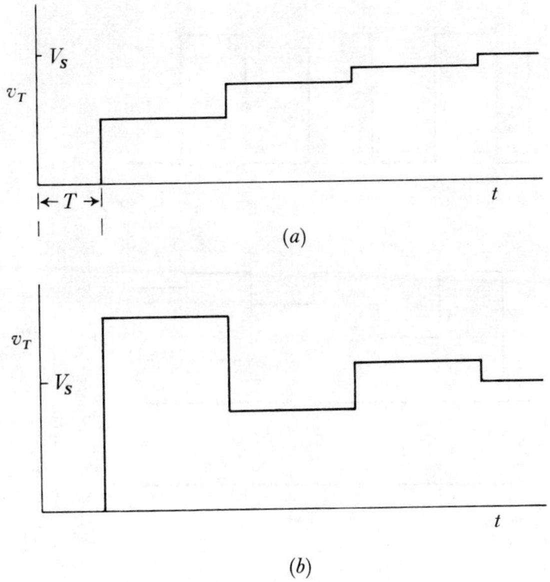

Fig. 13.28 Voltage at a resistive termination due to the application of a step function voltage: (*a*) $R_T = \frac{1}{3}Z_0$, $\Gamma = -\frac{1}{2}$; (*b*) $R_T = 3Z_0$, $\Gamma = +\frac{1}{2}$.

The voltage at the mid-point of the line is obtained by writing $T/2$ for x/u in the first equation of (13.11.6):

$$\begin{aligned} \boldsymbol{V}_M &= \frac{V_S}{\mathrm{s}} \mathrm{e}^{-\mathrm{s}T/2}(1 + \Gamma \mathrm{e}^{-\mathrm{s}T})(1 + \Gamma \mathrm{e}^{-2\mathrm{s}T})^{-1} \\ &= \frac{V_S}{\mathrm{s}} [\mathrm{e}^{-\mathrm{s}T/2} + \Gamma \mathrm{e}^{-3\mathrm{s}T/2} - \Gamma \mathrm{e}^{-5\mathrm{s}T/2} - \Gamma^2 \mathrm{e}^{-7\mathrm{s}T/2} + \dots], \end{aligned} \tag{13.11.11}$$

and

$$v_M = V_S[u(t-T/2) + \Gamma u(t-3T/2) - \Gamma u(t-5T/2) - \Gamma^2 u(t-7T/2)\dots]. \tag{13.11.12}$$

This is illustrated in fig. 13.29 for terminations from short-circuit through to open-circuit. All the curves are identical for the first $3T/2$, because the mid-point does not 'know' what the termination is until the first wave has had time to go there and bring back a message.

Figure 13.29(*e*) illustrates the statement of §13.7 that an open-circuited line excited by a voltage source is resonant at frequencies for which βd is an odd multiple of $\pi/2$ – that is, at the fundamental frequency $1/4T$ or any odd harmonic of it. Figure 13.29(*a*) illustrates the fact that the fundamental frequency of a short-circuited line excited by a voltage source is $1/2T$,

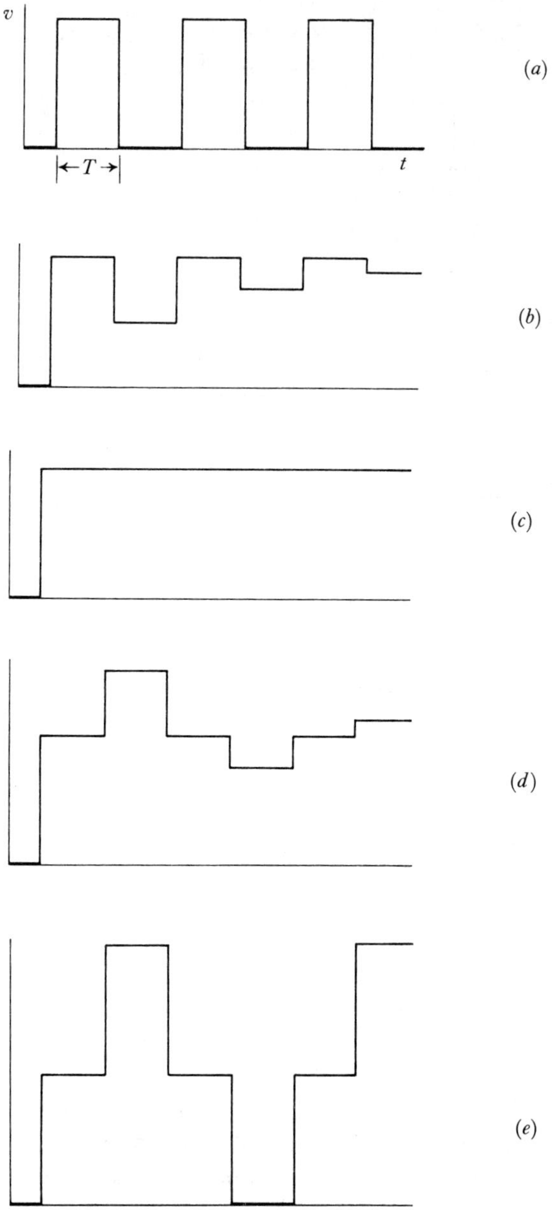

Fig. 13.29 Voltage at the mid-point of a resistively terminated line due to the application of a step function voltage: (a) Short-circuit, $\Gamma = -1$; (b) $\Gamma = -\frac{1}{2}$; (c) Matched termination, $\Gamma = 0$; (d) $\Gamma = +\frac{1}{2}$; (e) Open-circuit, $\Gamma = +1$.

twice the frequency of the same line when open-circuited; the odd harmonics of this frequency are also visible, but the even harmonics of voltage, present elsewhere in the line, have a node at the mid-point. The solution in terms of a series of harmonics, when possible, may be obtained from (13.11.6) in a manner which is exemplified by the open-circuit case ($\Gamma = 1$). In this,

$$\boldsymbol{V} = \frac{V_S}{\mathrm{s}}\left(\frac{\mathrm{e}^{\mathrm{s}x/u} + \mathrm{e}^{-\mathrm{s}x/u}}{\mathrm{e}^{\mathrm{s}d/u} + \mathrm{e}^{-\mathrm{s}d/u}}\right),$$

$$= \frac{V_S \cosh(\mathrm{s}x/u)}{\mathrm{s}\cosh(\mathrm{s}d/u)}. \qquad \text{(see (12.3.2))}$$

Also $$\cosh \mathrm{s}T \equiv \left[1 + \left(\frac{2\mathrm{s}T}{\pi}\right)^2\right]\left[1 + \left(\frac{2\mathrm{s}T}{3\pi}\right)^2\right] \dots \text{ ad infinitum.}$$

(see (12.3.3))

This factorization of the denominator is used to form the basis of a partial fraction expansion; to make this possible, an explicit formula for the partial fraction coefficients will first be found.

If $$\frac{\mathrm{P(s)}}{\mathrm{Q(s)}} = \frac{\mathrm{P(s)}}{\mathrm{K(s-s}_a)(\mathrm{s-s}_b)\dots(\mathrm{s-s}_g)}$$

$$\equiv \frac{\mathrm{A}}{\mathrm{s-s}_a} + \frac{\mathrm{B}}{\mathrm{s-s}_b} + \dots + \frac{\mathrm{G}}{\mathrm{s-s}_g},$$

where $\mathrm{s}_a, \mathrm{s}_b, \dots, \mathrm{s}_g$ are all different, the usual procedure gives

$$\frac{\mathrm{P(s)}}{\mathrm{K}} \equiv \mathrm{A(s-s}_b)\dots(\mathrm{s-s}_g) + \mathrm{B(s-s}_a)(\mathrm{s-s}_c)\dots(\mathrm{s-s}_g) + \dots$$

whence, writing $\mathrm{s = s}_a$,

$$\mathrm{A} = \frac{\mathrm{P(s}_a)}{\mathrm{K(s}_a - \mathrm{s}_b)\dots(\mathrm{s}_a - \mathrm{s}_g)},$$

and so on for B, C, ..., G. Now

$$\ln \mathrm{Q(s)} = \ln \mathrm{K} + \ln(\mathrm{s-s}_a) + \ln(\mathrm{s-s}_b) + \dots + \ln(\mathrm{s-s}_g),$$

so, differentiating w.r.t. s, $$\frac{\mathrm{Q'(s)}}{\mathrm{Q(s)}} = \frac{1}{\mathrm{s-s}_a} + \frac{1}{\mathrm{s-s}_b} + \dots + \frac{1}{\mathrm{s-s}_g},$$

or $$\mathrm{Q'(s)} = \mathrm{K}[(\mathrm{s-s}_b)\dots(\mathrm{s-s}_g) + (\mathrm{s-s}_a)(\mathrm{s-s}_c)\dots(\mathrm{s-s}_g) + \dots]$$

and $$\mathrm{Q'(s}_a) = \mathrm{K(s}_a - \mathrm{s}_b)\dots(\mathrm{s}_a - \mathrm{s}_g),$$

the other terms vanishing. Thus the coefficient A may be written $P(s_a)/Q'(s_a)$, and the expansion is expressible by the formula

$$\frac{P(s)}{Q(s)} = \sum_j \frac{P(s_j)}{(s-s_j)\,Q'(s_j)}. \qquad (13.11.13)$$

In problems with a small number of factors this formula is hardly worth using, as it does not cover multiple poles; but it comes into its own when, as here, the poles are of the first order but their number is infinite. Its applicability to the extended case will be assumed.

In the problem now in question, $P(s) = V_S \cosh(sx/u)$, or $V_S \cosh(sTx/d)$; $Q(s) = s\cosh sT$, which has zeros at $s = 0, \pm j\pi/2T, \pm 3j\pi/2T$, etc. Therefore, using (13.11.13), with $Q'(s) = \cosh sT + sT\sinh sT$,

$$\frac{V_S\cosh(sTx/d)}{s\cosh sT} = V_S\left\{\frac{1}{s} + \frac{\cosh(j\pi x/2d)}{(s-j\pi/2T)(j\pi/2)\sinh(j\pi/2)} - \frac{\cosh(-j\pi x/2d)}{(s+j\pi/2T)(j\pi/2)\sinh(-j\pi/2)} + \frac{\cosh(3j\pi x/2d)}{(s-3j\pi/2T)(3j\pi/2)\sinh(3j\pi/2)} - \dots\right\},$$

$$= V_S\left\{\frac{1}{s} - \frac{4}{\pi}\left[\left(\cos\frac{\pi x}{2d}\right)\left(\frac{s}{s^2+(\pi/2T)^2}\right) - \frac{1}{3}\left(\cos\frac{3\pi x}{2d}\right)\left(\frac{s}{s^2+(3\pi/2T)^2}\right) + \dots\right]\right\}. \qquad (13.11.14)$$

From this,

$$v = V_S u(t)\left\{1 - \frac{4}{\pi}\sum_{n=1}^{\infty}\frac{(-1)^n}{(2n+1)}\cos\frac{(2n+1)\pi x}{2d}\cos\frac{(2n+1)\pi t}{2T}\right\}. \qquad (13.11.15)$$

This does not compare with the travelling-wave solution as a means of calculating waveforms; many terms would have to be added to get anywhere near the waveform of fig. 13.29(*e*); but it does link the present theory with what has gone before, and shows what harmonics are present and of what magnitude.

(b) Capacitive termination

When $\boldsymbol{Z}_T = 1/C_T s$,

$$\Gamma = \frac{(1/C_T s) - Z_0}{(1/C_T s) + Z_0},$$

$$= -\left(\frac{s-\alpha}{s+\alpha}\right), \tag{13.11.16}$$

where $$\alpha = (Z_0 C_T)^{-1}. \tag{13.11.17}$$

Equations (13.11.8,9) are applicable to this case also, with Γ now a function of s. Equation (13.11.9) may be written

$$\boldsymbol{V}_T = V_S[\boldsymbol{F}_1(s)\,e^{-sT} + \boldsymbol{F}_3(s)\,e^{-3sT} + \boldsymbol{F}_5(s)\,e^{-5sT} + \ldots], \tag{13.11.18}$$

where

$$\left.\begin{aligned} \boldsymbol{F}_1(s) &= \frac{1+\Gamma}{s} = \frac{2\alpha}{s(s+\alpha)}, \\ \boldsymbol{F}_3(s) &= -\Gamma \boldsymbol{F}_1(s) = \frac{2\alpha(s-\alpha)}{s(s+\alpha)^2}, \\ \boldsymbol{F}_5(s) &= -\Gamma \boldsymbol{F}_3(s) = \frac{2\alpha(s-\alpha)^2}{s(s+\alpha)^3}, \quad \text{etc.} \end{aligned}\right\} \tag{13.11.19}$$

The expansions of $\boldsymbol{F}_1$, $\boldsymbol{F}_3$, etc. are found to be

$$\left.\begin{aligned} \boldsymbol{F}_1(s) &= \frac{2}{s} - \frac{2}{s+\alpha}, \\ \boldsymbol{F}_3(s) &= \frac{-2}{s} + \frac{2}{s+\alpha} + \frac{4\alpha}{(s+\alpha)^2}, \\ \boldsymbol{F}_5(s) &= \frac{2}{s} - \frac{2}{s+\alpha} - \frac{8\alpha^2}{(s+\alpha)^3}, \\ \boldsymbol{F}_7(s) &= \frac{-2}{s} + \frac{2}{s+\alpha} + \frac{4\alpha}{(s+\alpha)^2} - \frac{8\alpha^2}{(s+\alpha)^3} \\ &\quad + \frac{16\alpha^3}{(s+\alpha)^4}, \quad \text{etc.} \end{aligned}\right\} \tag{13.11.20}$$

The corresponding functions of t are

$$\left.\begin{aligned} f_1(t) &= 2u(t)[1 - e^{-\alpha t}], \\ f_3(t) &= 2u(t)[-1 + (1 + 2\alpha t)e^{-\alpha t}], \\ f_5(t) &= 2u(t)[1 - (1 + 2\alpha^2 t^2)e^{-\alpha t}], \\ f_7(t) &= 2u(t)[-1 + (1 + 2\alpha t - 2\alpha^2 t^2 + \tfrac{4}{3}\alpha^3 t^3)e^{-\alpha t}], \text{ etc.} \end{aligned}\right\} \quad (13.11.21)$$

In terms of these functions, the solution is

$$v_T = V_S[f_1(t-T) + f_3(t-3T) + f_5(t-5T) + \ldots]. \quad (13.11.22)$$

Fig. 13.30 illustrates this solution for a particular ratio of T and $Z_0 C_T$, the two time-constants associated with the system. The voltage oscillating

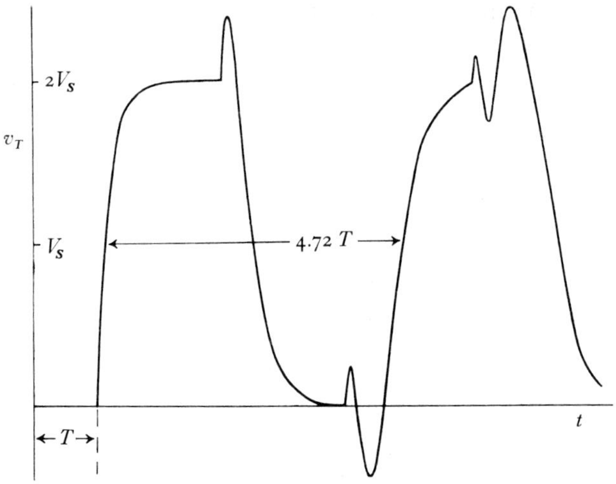

Fig. 13.30 Voltage at a capacitive termination due to the application of a step function voltage $V_S u(t)$ ($\alpha T = 5$).

about the applied value V_S is to be expected in this lossless system, but its magnitude is not confined between the limits 0 and $2V_S$, as would be the case if the line were replaced by an inductance.

The frequencies present may be found by writing $\boldsymbol{Y}_T = j\omega C_T$ in (13.8.1):

$$\boldsymbol{V} = \boldsymbol{V}_T(\cos\beta x - \omega Z_0 C_T \sin\beta x). \quad (13.11.23)$$

The resonant frequencies are those at which a finite $\boldsymbol{V}_T$ is compatible with an infinitesimal $\boldsymbol{V}_S$; namely, those for which

$$\cos\beta d - \omega Z_0 C_T \sin\beta d = 0,$$

or

$$\cot\omega T - \omega Z_0 C_T = 0. \quad (13.11.24)$$

The functions $\cot \omega T$ and $\omega Z_0 C_T$, or $\omega T/\alpha T$, are plotted as functions of ωT in fig. 13.31; the intersections P_1, P_2, ... give the natural frequencies, which do not form a harmonic series. In the absence of C_T the frequencies

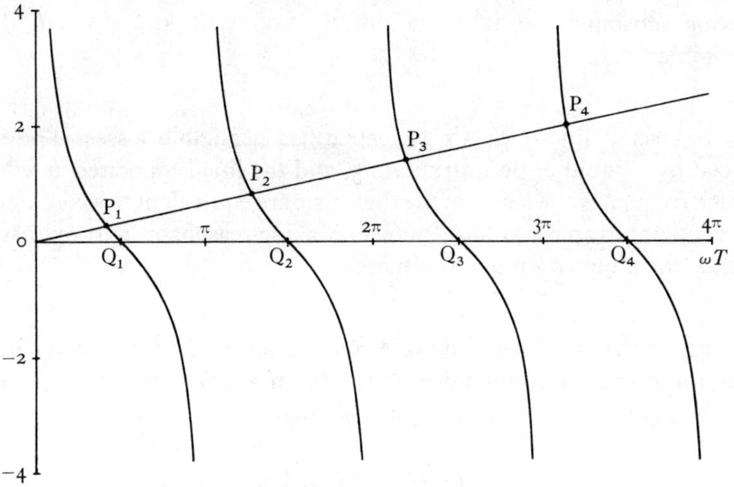

Fig. 13.31 Determination of the frequencies present in fig. 13.30.

would have been those denoted by Q_1, Q_2, ... The terminal capacitance therefore lowers the frequencies; in the case shown ($\alpha T = 5$) the fundamental period is increased from $4T$ to $4.72T$ – an effect discernible in fig. 13.30.

From the examples given it will be evident that the solution of the problem of transient voltages and currents in lines and linelike networks reduces to a problem of interpreting functions of s. From the work of Heaviside onward, indeed, the need to analyse distributed networks has been a notable incentive to the development of the Laplace transform calculus. In this book the subject must be left here; for it is no part of our aim to treat any topic exhaustively. Rather we have tried to unlock as many gates as possible, in the hope that the reader, attracted by glimpses of new territory, may explore it for himself.

Examples on chapter 13

13.1 A lossless line, when short-circuited at the termination, is found to have an inductive reactance of 1000 Ω when viewed from the source. When open-circuited it appears as a capacitive reactance of 360 Ω. Determine Z_0, and also the length of the line in wavelengths.

13.2 A lossless quarter-wavelength line is open at one end and short-circuited at the other. Connexion is made to the two conductors at a distance

x from the closed end, and the impedance is measured. Prove that, provided x is not zero, the value obtained is independent of x.

13.3 At a frequency of 796 Hz, compute $\boldsymbol{Z}_0$, γ, u and λ for a line having the following constants: $l = 2230$ μH/km, $c = 0.0055$ μF/km, $r = 6.46$ Ω/km, g negligible.

13.4 A coaxial line with air dielectric has negligible losses. The air is replaced by a liquid of permittivity 2.5, and the line is operated at 1 MHz, at which frequency the losses in the dielectric are equivalent to a conductance 1 per cent of the capacitive susceptance. Making reasonable approximations, calculate the attenuation per kilometre.

13.5 Show that a lossless line of length d, terminated by a capacitor C, has an input impedance identical with that of a short-circuited line of the same characteristic impedance Z_0 but of length

$$d - \frac{1}{\beta}\cot^{-1}(\omega C Z_0).$$

13.6 At a certain frequency, a transmission line has $\boldsymbol{Z}_0 = 553\angle 60^\circ$ Ω and $\gamma = 0.005 + j0.03$ neper/km. The line is 60 km long. A sinusoidal e.m.f. of 100 V at the specified frequency is applied to one end. Find:

(*a*) the voltage at the termination, if open-circuited;

(*b*) the current at the termination, if short-circuited.

13.7 A lossless line, having $Z_0 = 50$ Ω, is terminated by a resistance of 100 Ω. Find the shortest length of this line (in wavelengths) which may be used to produce at the input terminals an impedance having a maximum negative reactance component, and obtain the value of this impedance.

13.8 A lossless air-insulated line has $Z_0 = 600$ Ω and is operated at a frequency of 1592 Hz. Its length is 20 km, and it is terminated with a capacitor of 0.5 μF. Find the input impedance as given by a nominal T representation, and compare the result with that obtained from the Smith chart.

13.9 A lossless line of length d has capacitances C_1, C_2, connected across its input and output ends respectively. For a given frequency ω and given value of C_1, determine the value of C_2 which gives rise to an infinite input impedance.

13.10 A lossless line of length d has an inductive reactance $\mathrm{j}X$ connected between its conductors at its mid-point. Show that the condition for the overall ***ABCD*** matrix to be of the form

$$\begin{bmatrix} 0 & -\mathrm{j}\boldsymbol{K} \\ -\mathrm{j}\boldsymbol{K}^{-1} & 0 \end{bmatrix}$$

is $\tan\beta d = -2Y_0X$; and find the resulting value of $\boldsymbol{K}$.

13.11 A lossless line terminated by a load $\boldsymbol{Z}_L$ has a standing-wave ratio of 3. The line constants are $l = 2\times10^{-6}$ H/m, $c = 6.25\times10^{-12}$ F/m, and the excitation frequency is 300 MHz. If the distance between the load and the first maximum is 32 cm, determine $\boldsymbol{Z}_L$.

13.12 A slotted coaxial line with $Z_0 = 50\,\Omega$ is used to measure an unknown impedance $\boldsymbol{Z}_L$. The line is first short-circuited and the minimum voltage positions are noted. The short circuit is then replaced by the impedance $\boldsymbol{Z}_L$; the voltage minima are observed to have been shifted towards the load by a distance 0.125λ, and the VSWR is 2. Find the value of $\boldsymbol{Z}_L$.

13.13 Standing-wave measurements on a lossless line give a VSWR of 2 and a voltage minimum 0.25λ from the load. Determine the normalized load impedance; also the stub length and point of connexion for a single short-circuited shunt stub, of normalized characteristic impedance $\sqrt{2}$, which matches the load to the line.

13.14 The standing-wave ratio on a $50\,\Omega$ line is measured to be 2.6, and a voltage minimum is found at 0.17λ from the load.

(*a*) Find the load impedance.

(*b*) Design a single-stub tuner to match the line, with freedom both as to the location and length of the stub.

(*c*) Design a double-stub tuner to match the line, given that the stubs are 0.3λ and 0.55λ from the load.

All lines are lossless and have the same characteristic impedance.

13.15 Two long transmission lines A and B are connected in series, but between them is a cable C 1.5 km long. The lines have $l = 1.6\times10^{-6}$ H/m, $c = 10$ pF/m; for the cable the corresponding figures are $l' = 5\times10^{-7}$ H/m, $c' = 89$ pF/m. A rectangular voltage wave of magnitude 10 kV and long duration travels along A towards the cable. Find the magnitude of the second voltage step occurring at the junction of the cable and the line B. What will be the voltage at the junction of line A and the cable 21 μs after the initial voltage reaches this point?

13.16 An overhead line 60 km long, having $Z_0 = 800\ \Omega$, is connected to a fork, both branches of the fork being cable for which $Z_0' = 200\ \Omega$. The velocity of travelling waves is 3×10^8 m/s on the line and 1.5×10^8 m/s on the cable. One branch of the fork, 15 km long, is open-circuited; the other, 7.5 km long, is terminated by a resistance of 100 Ω. A step function voltage $V_0 u(t)$ is applied to the overhead line at the end remote from the fork. Trace out the travelling wave system, and so find the voltage at the fork for values of t from zero to 450 μs.

13.17 Into a long overhead line, having $Z_0 = 500\ \Omega$, is inserted a surge modifying circuit. This consists of two coils of 50 μH inserted in series into each conductor, together with a capacitor of 5000 pF connected between the junctions of the coils in one conductor and the other. The mutual inductances between the coils are negligible. A rectangular voltage wave of amplitude V_1 approaches from the left, arriving at the surge modifier at $t = 0$. Calculate the subsequent voltage at the right hand terminals of the surge modifier.

13.18 Derive a formula for the input current of a resistive inductanceless cable, infinitely long, to which a step voltage V_0 is applied from time $t = 0$. Hence show that, when the signal has penetrated so far that the voltage at a certain point x is 0.5 V_0, the input current is about 0.54 I, where I is the ultimate value of the input current to a cable short-circuited at the point x and subjected to a voltage V_0 indefinitely sustained.

13.19 An infinitely long cable has constants $r = 0.5\ \Omega$/km, $c = 0.2\ \mu$F/km, $l = g = 0$. A step function voltage signal is applied to one end. Find the time at which the voltage 1000 km from the source attains half the applied value. Compare this time with that obtained by an approximation in which the first 1000 km are represented by a nominal T element, the infinite remainder being represented by a second similar element (open-circuited).

Appendix 1 Tables of Laplace transforms

The Laplace transformation,

$$W(s) = \int_0^\infty w(t)\, e^{-st}\, dt \qquad \text{(equation (5.4.5))}$$

implies the assumption that $w(t) = 0$ for $t < 0$. It is therefore taken for granted that every function of t is multiplied by $u(t)$.

Not all the formulae have been explicitly derived in the text; but all can be derived by the methods given. From the hundreds available these are selected for their likely occurrence in circuit problems. Many others, not listed, can be easily derived by applying one of the general theorems in list A.1 to a transform given in list A.2.

A.1 General theorems

	Original	*Transform*	*Notes*
1.	$w(t)$	$\boldsymbol{W}(\mathrm{s})$	
2.	$\dfrac{\mathrm{d}w}{\mathrm{d}t}$	$\mathrm{s}\boldsymbol{W}(\mathrm{s}) - w_0$	Eqn (5.5.3). $w = w_0$ when $t = 0$
3.	$\dfrac{\mathrm{d}^2 w}{\mathrm{d}t^2}$	$\mathrm{s}^2\boldsymbol{W}(\mathrm{s}) - \mathrm{s}w_0 - w_0'$	$\dfrac{\mathrm{d}w}{\mathrm{d}t} = w_0'$ when $t = 0$
4.	$\dfrac{\mathrm{d}^n w}{\mathrm{d}t^n}$	$\mathrm{s}^n\boldsymbol{W}(\mathrm{s}) - \mathrm{s}^{n-1} w_0 - \mathrm{s}^{n-2} w_0' - \ldots - w_0^{(n-1)}$	$\dfrac{\mathrm{d}^{n-1} w}{\mathrm{d}t^{n-1}} = w_0^{(n-1)}$ when $t = 0$
5.	$\int_0^t w(\tau)\,\mathrm{d}\tau$	$\dfrac{\boldsymbol{W}(\mathrm{s})}{\mathrm{s}}$	Eqn (5.5.4)
6.	$\mathrm{e}^{-\alpha t}\int_0^t \mathrm{e}^{\alpha\tau} w(\tau)\,\mathrm{d}\tau$	$\dfrac{\boldsymbol{W}(\mathrm{s})}{\mathrm{s}+\alpha}$	Derived by solving the differential equation $\mathrm{d}y/\mathrm{d}t + \alpha y = w(t)$, using the integrating factor $\mathrm{e}^{\alpha t}$
7.	$\mathrm{e}^{-\alpha t} w(t)$	$\boldsymbol{W}(\mathrm{s}+\alpha)$	Eqn (10.7.4)
8.	$\dfrac{\partial}{\partial\alpha} w(t,\alpha)$	$\dfrac{\partial}{\partial\alpha}\boldsymbol{W}(\mathrm{s},\alpha)$	Eqn (10.7.1)
9.	$tw(t)$	$-\dfrac{\mathrm{d}\boldsymbol{W}}{\mathrm{d}\mathrm{s}}$	Derived by differentiating the transform eqn (5.4.5) w.r.t. s (applies to 9 and 10)
10.	$t^n w(t)$	$(-1)^n\dfrac{\mathrm{d}^n\boldsymbol{W}}{\mathrm{d}\mathrm{s}^n}$	

11.	$t^{-1} w(t)$	$\int_s^{\infty} \boldsymbol{W}(s)\, ds$	If the integral exists. Derived by integrating eqn (5.4.5) w.r.t. s
12.	$w(t-T)$	$e^{-Ts}\boldsymbol{W}(s)$	Provided $T > 0$ – eqn (10.8.1)
13.	$\int_0^t w_1(\tau)\, w_2(t-\tau)\, d\tau$	$\boldsymbol{W}_1(s)\boldsymbol{W}_2(s)$	The convolution theorem – eqn (10.9.3)

Though not transforms, these general theorems are conveniently included:

14. $\lim_{t\to\infty} w(t) = \lim_{s\to 0} s\boldsymbol{W}(s)$ if the limits exist – eqn (10.7.6)

15. $w(0) = \lim_{s\to\infty} s\boldsymbol{W}(s)$ if the limit exists – eqn (10.7.7)

16. $\dfrac{P(s)}{Q(s)} = \sum_j \dfrac{P(s_j)}{(s-s_j)Q'(s_j)}$ – the general formula for expansion in partial fractions, applicable when the zeros s_j of Q(s) are all simple – eqn (13.11.13)

A.2 Specific transform-pairs

	Original	*Transform*	*Notes*
17.	$\delta(t)$	1	
18.	$u(t)$	$\dfrac{1}{s}$	Eqn (5.6.1)
19.	t	$\dfrac{1}{s^2}$	The factor $u(t)$ is taken for granted in this and all succeeding originals, except nos. 38–9 and 57–9

A.2 Specific transform-pairs—*continued*

	Original	Transform	Notes
20.	$\dfrac{t^{n-1}}{(n-1)!}$	$\dfrac{1}{s^n}$	Eqn (5.6.2)
21.	$\dfrac{1}{\sqrt{(\pi t)}}$	$\dfrac{1}{\sqrt{s}}$	A generalization of no. 20, with $(-\frac{1}{2})! = \sqrt{\pi}$. Higher half-powers are obtained by using the identity $(n+1)! \equiv (n+1) . n!$, starting at $n = -\frac{1}{2}$
22.	$e^{-\alpha t}$	$\dfrac{1}{s+\alpha}$	Eqn (5.6.4)
23.	$\dfrac{e^{-\alpha t} - e^{-\beta t}}{\beta - \alpha}$	$\dfrac{1}{(s+\alpha)(s+\beta)}$	
24.	$\cosh \alpha t$	$\dfrac{s}{s^2-\alpha^2}$	
25.	$\sinh \alpha t$	$\dfrac{\alpha}{s^2-\alpha^2}$	
26.	$t\, e^{-\alpha t}$	$\dfrac{1}{(s+\alpha)^2}$	Eqn (5.7.16)
27.	$\dfrac{t^{n-1}\, e^{-\alpha t}}{(n-1)!}$	$\dfrac{1}{(s+\alpha)^n}$	
28.	$\cos \beta t$	$\dfrac{s}{s^2+\beta^2}$	Eqn (5.8.2)

29.	$\sin \beta t$	$\dfrac{\beta}{s^2 + \beta^2}$	Eqn (5.8.3)
30.	$\dfrac{t}{2\beta} \sin \beta t$	$\dfrac{s}{(s^2 + \beta^2)^2}$	Derived from no. 28 via no. 8
31.	$\dfrac{\sin \beta t - \beta t \cos \beta t}{2\beta^3}$	$\dfrac{1}{(s^2 + \beta^2)^2}$	Derived from no. 29 via no. 8
32.	$e^{-\alpha t} \cos \beta t$	$\dfrac{s + \alpha}{(s + \alpha)^2 + \beta^2}$	Eqn (5.7.12)
33.	$e^{-\alpha t} \sin \beta t$	$\dfrac{\beta}{(s + \alpha)^2 + \beta^2}$	Eqn (5.7.13)
34.	$e^{-\alpha t} \left(\cos \beta t - \dfrac{\alpha}{\beta} \sin \beta t \right)$	$\dfrac{s}{(s + \alpha)^2 + \beta^2}$	
35.	$P_n(2 e^{-\alpha t} - 1)$	$\dfrac{(s - \alpha)(s - 2\alpha) \ldots (s - n\alpha)}{s(s + \alpha)(s + 2\alpha) \ldots (s + n\alpha)}$	n is a positive integer; P_n is a Legendre polynomial
36.	$P_n(\cos \beta t)$	$\dfrac{(s^2 + \beta^2)(s^2 + 3^2 \beta^2) \ldots (s^2 + (n-1)^2 \beta^2)}{s(s^2 + 2^2 \beta^2)(s^2 + 4^2 \beta^2) \ldots (s^2 + n^2 \beta^2)}$	n is an even positive integer
37.	$P_n(\cos \beta t)$	$\dfrac{s(s^2 + 2^2 \beta^2)(s^2 + 4^2 \beta^2) \ldots (s^2 + (n-1)^2 \beta^2)}{(s^2 + \beta^2)(s^2 + 3^2 \beta^2) \ldots (s^2 + n^2 \beta^2)}$	n is an odd positive integer
38.	$\delta(t - T)$	e^{-Ts}	Derived from no. 17 via no. 12
39.	$u(t - T)$	$\dfrac{e^{-Ts}}{s}$	Derived from no. 18 via no. 12
40.	$\dfrac{a}{2\sqrt{\pi}} t^{-3/2} e^{-a^2/4t}$	$e^{-a\sqrt{s}}$	Eqn (13.10.32)

A.2 Specific transform-pairs—*continued*

	Original	Transform	Notes
41.	$\frac{1}{\sqrt{\pi}} t^{-1/2} e^{-a^2/4t}$	$\frac{e^{-a\sqrt{s}}}{\sqrt{s}}$	Derived from no. 40 via no. 9
42.	$1 - \operatorname{erf} \frac{a}{2\sqrt{t}}$	$\frac{e^{-a\sqrt{s}}}{s}$	Eqn (13.10.31). erf x is the error function
43.	$\frac{1}{\sqrt{\pi t}} - \sqrt{\alpha}\,\{e^{\alpha t}(1 - \operatorname{erf}\sqrt{(\alpha t)})\}$	$\frac{1}{\sqrt{s} + \sqrt{\alpha}}$	
44.	$e^{\alpha t}(1 - \operatorname{erf}\sqrt{(\alpha t)})$	$\frac{1}{\sqrt{s}(\sqrt{s} + \sqrt{\alpha})}$	
45.	$\frac{1}{\sqrt{\alpha}}\{1 - e^{\alpha t}(1 - \operatorname{erf}\sqrt{(\alpha t)})\}$	$\frac{1}{s(\sqrt{s} + \sqrt{\alpha})}$	
46.	$\delta(t) - \sqrt{(\alpha/t)}\, J_1(2\sqrt{(\alpha t)})$	$e^{-\alpha/s}$	$J_n(x)$ is the Bessel function of the first kind of order n
47.	$J_0(2\sqrt{(\alpha t)})$	$\frac{e^{-\alpha/s}}{s}$	
48.	$J_0(\alpha t)$	$\frac{1}{\sqrt{(s^2 + \alpha^2)}}$	
49.	$I_0(\alpha t)$	$\frac{1}{\sqrt{(s^2 - \alpha^2)}}$	$I_n(x)$ is the Bessel function with imaginary argument, defined by $I_n(x) = j^{-n} J_n(jx)$
50.	$e^{-\alpha t} I_0(\alpha t)$	$\frac{1}{\sqrt{(s^2 + 2\alpha s)}}$	Derived from no. 49 via no. 7

51.	$J_n(\alpha t)$	$\dfrac{1}{\sqrt{(s^2+\alpha^2)}}\left\{\dfrac{\sqrt{(s^2+\alpha^2)}-s}{\alpha}\right\}^n$	$n \geqslant 0$
52.	$I_n(\alpha t)$	$\dfrac{1}{\sqrt{(s^2-\alpha^2)}}\left\{\dfrac{s-\sqrt{(s^2-\alpha^2)}}{\alpha}\right\}^n$	$n \geqslant 0$
53.	$e^{-\alpha t} I_n(\alpha t)$	$\dfrac{1}{\sqrt{(s^2+2\alpha s)}}\left\{\dfrac{\sqrt{(s+2\alpha)}-\sqrt{s}}{\sqrt{(2\alpha)}}\right\}^{2n}$	$n \geqslant 0$. Derived from no. 52 via no. 7
54.	$\dfrac{n}{t} J_n(\alpha t)$	$\left\{\dfrac{\sqrt{(s^2+\alpha^2)}-s}{\alpha}\right\}^n$	$n > 0$
55.	$\dfrac{n}{t} I_n(\alpha t)$	$\left\{\dfrac{s-\sqrt{(s^2-\alpha^2)}}{\alpha}\right\}^n$	$n > 0$
56.	$\dfrac{n}{t} e^{-\alpha t} I_n(\alpha t)$	$\left\{\dfrac{\sqrt{(s+2\alpha)}-\sqrt{s}}{\sqrt{(2\alpha)}}\right\}^{2n}$	$n > 0$
57.	$J_0(\beta\sqrt{(t^2-T^2)})\,u(t-T)$	$\dfrac{e^{-T\sqrt{(s^2+\beta^2)}}}{\sqrt{(s^2+\beta^2)}}$	The factor $u(t-T)$ in nos. 57–9 denotes a time-lag in the response, as occurs in nos. 38–9
58.	$I_0(\beta\sqrt{(t^2-T^2)})\,u(t-T)$	$\dfrac{e^{-T\sqrt{(s^2-\beta^2)}}}{\sqrt{(s^2-\beta^2)}}$	
59.	$e^{-\alpha t} I_0[\beta\sqrt{(t^2-T^2)}]\,u(t-T)$	$\dfrac{e^{-T\sqrt{[(s+\alpha)^2-\beta^2]}}}{\sqrt{[(s+\alpha)^2-\beta^2]}}$	Derived from no. 58 via no. 7

Appendix 2 Prefixes denoting decimal multiples or sub-multiples

Factor	*Prefix*	*Symbol*
10^{12}	tera	T
10^{9}	giga	G
10^{6}	mega	M
10^{3}	kilo	k
10^{-3}	milli	m
10^{-6}	micro	μ
10^{-9}	nano	n
10^{-12}	pico	p
10^{-15}	femto	f
10^{-18}	atto	a

Solutions of examples

Chapter 1

1.1 (*a*) 6617 Ω; (*b*) 0.7614 Ω.

1.2 1 A; 20 V; 20 W; 5 Ω.

1.3 2 kΩ; 1.25 kΩ; 7.5 kΩ.

1.4 $I = \dfrac{2(R-6)}{38R+71}$; 6 Ω.

1.5 −0.846 A.

1.6 Fractions of slider length from negative end: 0.766, 0.410, 0.591.

1.7 +5.18% or −4.83%.

1.8 4 V.

1.9 0.456 A; 3.22 V.

1.10 −7.12 V.

1.11 (*a*) 102.4 V; (*b*) 95.2 V.

1.12 3.72 mA.

Chapter 2

2.1 1.5 A; 2.25 J.

2.2 100 V; 63.2 V; 100 V.

2.3 $v_c = \dfrac{V_s t}{C(R_s + r_0)}$ when $\mu = 1$

$$= \frac{V_s}{C(R_s + r_0)}\left(\frac{1 - e^{-\alpha t}}{\alpha}\right) \text{ when } \mu \neq 1, \text{ where}$$

$$\alpha = \frac{1-\mu}{C(R_s + r_0)}.$$

2.4 (*a*) (i) 4.70 H, (ii) 1.30 H;
(*b*) (i) 0.982 H, (ii) 0.273 H.

2.5 (*a*) 12.2 H; (*b*) 1.56 As^{-1} from E to F in the short-circuit.

2.6 1 V; 0.118 V.

2.8 (*a*) 4.52 Ω; (*b*) (i) 4.52 mH, (ii) 6.76 mH.

2.9 $\dfrac{C_1 V_1}{C_1 + C_2}$.

2.10 $\dfrac{(L_1 + M) I_1}{L_1 + L_2 + 2M}$.

Chapter 3

3.1 (*a*) 7 V; (*b*) 5 V; (*c*) 5 V if the ratio of the frequencies can be expressed as a ratio of two integers – otherwise the value is undefined.

3.2 $i = 5 + 1.77\sin(\omega t - 45°) + 0.32\sin(3\omega t - 72°)$; 53.5 V; 5.16 A.

3.3 375 W; 0.250 lagging; 205 μF.

3.4 3.68 Ω; 2.77 Ω; 415 μF; 41.7 A.

3.5 1.0001 mH; 62.838 kΩ.

3.7 1040 V; 61.4 V.

3.8 204 V.

3.9 72.4 mH; 71.2 Ω.

3.11 (*a*) 505 A, 0.827 lagging; (*b*) 362 μF; (*c*) 0.862 leading.

3.12 153.9 kV; 202 A; 0.743 lagging.

Chapter 4

4.1 (*a*) 4.13 $\angle 9°40'$ Ω; (*b*) 0.434 V_s, leading $\boldsymbol{V}_s$ by 164°38′; 0.986 lagging.

4.3 $$\frac{\boldsymbol{V}_2}{\boldsymbol{V}_1} = \frac{1}{(1 - \omega^2 R_1 R_2 C_1 C_2) + j\omega(R_1[C_1 + C_2] + R_2 C_2)}.$$
See fig. S.1.

4.4 (*a*) 0.868 V_s, lagging $\boldsymbol{V}_s$ by 49°24′; (*b*) V_s, lagging $\boldsymbol{V}_s$ by 90°.

4.5 58.5 $\angle 18°49'$ mA.

4.6 0.944 $\angle 173°52'$ A; 0.962 $\angle 182°56'$ A.

4.7 3.32 $\angle -91°24'$ A; 2.17 $\angle 17°28'$ A.

4.8 $R_1 = R_2 = \sqrt{(L/C)}$.

4.9 14.4 $\angle -32°$ Ω.

4.10 2.39 A.

4.11 32.87 kV (line), lagging the e.m.f.s by 2°15′; 10.39 $\angle -44°21'$ A, 7.49 $\angle -7°1'$ A.

4.12 $P = (ad + be) - \omega C(ah - bf)$;
$Q = (-ae + bd) - \omega C(af + bh)$.

Chapter 5

5.1 (*a*) $\frac{3}{2}e^{-t/2} + \frac{5}{2}e^{-5t/2}$; (*b*) $3\cos 2t + 2\cos 3t$;
(*c*) $e^{-2t}(\cos 3t + 2\sin 3t)$; (*d*) $2e^{-t} + 3e^{-2t}\cos 2t$;
(*e*) $2e^{-t} + t\,e^{-2t}$.

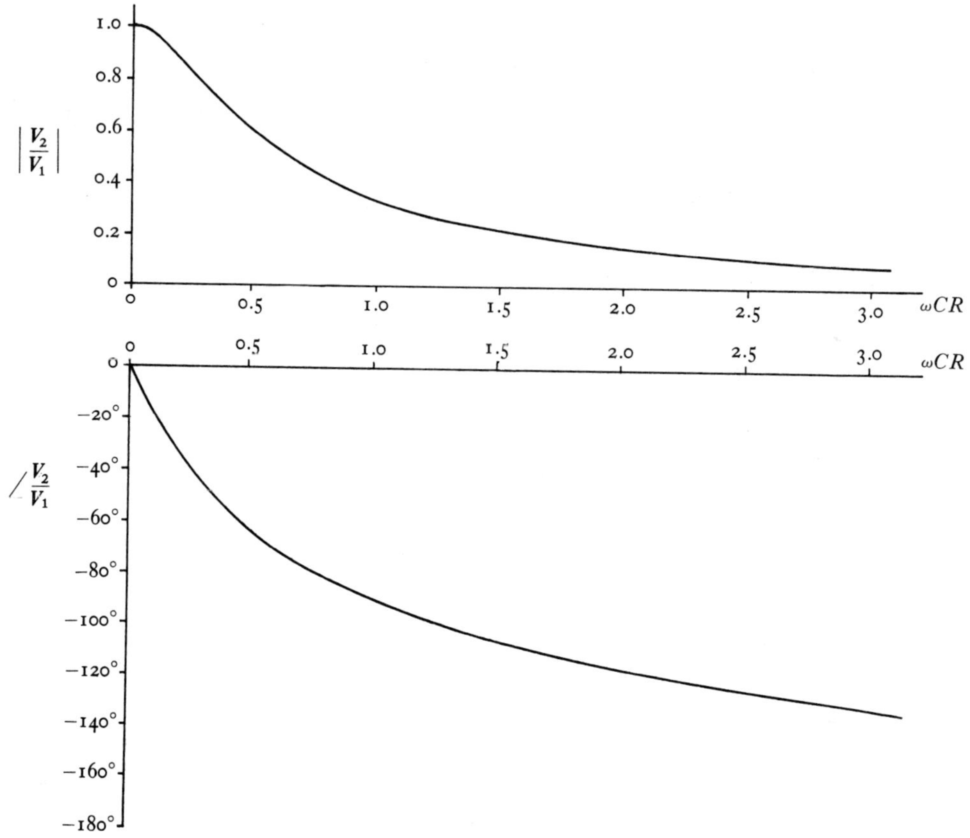

Fig. S.1

5.2 $\frac{V_1}{R}(1-e^{-\alpha t})$ in the inductor; $\frac{V_1}{2R}e^{-\alpha t}$ in the parallel resistor;

$\frac{V_1}{R}(1-\frac{1}{2}e^{-\alpha t})$ in the series resistor; where $\alpha = R/2L$.

5.3 $V_1(1-1.171\,e^{-0.38t}+0.171\,e^{-2.62t})$.

5.4 $R_1 = R_2 = \sqrt{\left(\frac{L}{C}\right)}$.

5.5 $1.204\,(e^{-73.9t}-e^{-406.1t})$ A; 5.13 ms.

5.6 $\frac{V_1}{R}\left\{\cos\omega t - e^{-\alpha t}\left(\cos\beta t - \frac{\alpha}{\beta}\sin\beta t\right)\right\}$, where $\alpha = (R/2L)$, $\beta^2 = \omega^2 - \alpha^2$.

5.7 $25 - 10(\cos 10t + 2\sin 10t) - 5e^{-10t}(3\cos 10t - \sin 10t)$ C.

5.8 $\frac{s-\alpha}{2(s+\alpha)}$, where $\alpha = 1/RC$;

$$V_1 \frac{\alpha\omega}{\alpha^2+\omega^2}\left\{\cos\omega t - \left(\frac{\alpha^2-\omega^2}{2\alpha\omega}\right)\sin\omega t - e^{-\alpha t}\right\}.$$

5.9 $\frac{V_1}{R}\left(\frac{\alpha e^{-\beta t} - \beta e^{-\alpha t}}{\alpha - \beta}\right)$, where $\alpha + \beta = R/L$, $\alpha\beta = 1/LC$.

5.10 $22.2\{e^{-10\,000t} - \cos 314t + 31.8 \sin 314t\}$ V.

5.11 $v_1(t) = V_1 u(t)$, where $V_1 = \sqrt{\left(\frac{L_2}{L_1}\right)} R_1 i_{20}$;

$i_1(t) = \sqrt{\left(\frac{L_2}{L_1}\right)} i_{20} u(t)$. It is advisable to assume $L_1 L_2 \neq M^2$ at first, equating these quantities later. Alternatively, assuming $L_1 L_2 = M^2$ throughout, the model in fig. 2.11(*a*) can be used.

5.12 $\frac{V_1}{2L\omega}(\sin\omega t + \omega t \cos\omega t)$; 0.91 CV_1.

Chapter 6

6.1 450 W.

6.2 1.80 $\angle -72.6°$A; 1.81 $\angle -79.6°$ A.

6.3 57.2 $\angle -4°44'$ V in series with 489 $\angle -23°33'$ Ω; 0.117 $\angle 18°49'$ A.

6.4 0.99 $\angle 48.4°$ mA.

6.5 300 Ω; 261 $\angle -29.5°$ Ω (300 Ω resistance in parallel with 6 μF capacitance).

6.6 104.7 V; 2.26 A.

6.8 Circuit ABCD (fed at AC, detector at BD), with AB: L in series with R; CD: C in parallel with R; BC and DA: R only. The dual of two such bridges fed in parallel is two such bridges fed in series.

6.9 4.47; 102 W; 208 W.

6.10 L must be increased and C reduced by a factor of 50.

6.11 $\frac{(1-\omega^2 LC) + j\omega R_L C(2-\omega^2 LC)}{R_L(1-\omega^2 LC) + j\omega L}$; 40 mH, 0.0025 μF.

Chapter 7

7.1 $\frac{V_2}{V_1} = \frac{R_2}{R_1+R_2}\frac{1+j(\omega/\omega_1)}{1+j(\omega/\omega_a)}$, where $\omega_1 = \frac{1}{R_1 C_1}$, $\omega_a = \frac{R_1+R_2}{R_1 R_2(C_1+C_2)}$;

−0.83 db up to $\omega_1/10$, decreasing at 20 db per decade up to ω_1, constant thereafter; zero at $\omega_1/100$, decreasing at 45° per decade up to $\omega_1/10$, constant up to ω_1, increasing at 45° per decade up to 10 ω_1, zero thereafter; −17.8 db, −39°.

7.2 (*a*) 5.03 (1 − j) μS; (*b*) 5.03 μS; (*c*) 5.03 (1 + j) μS; see fig. S.2.

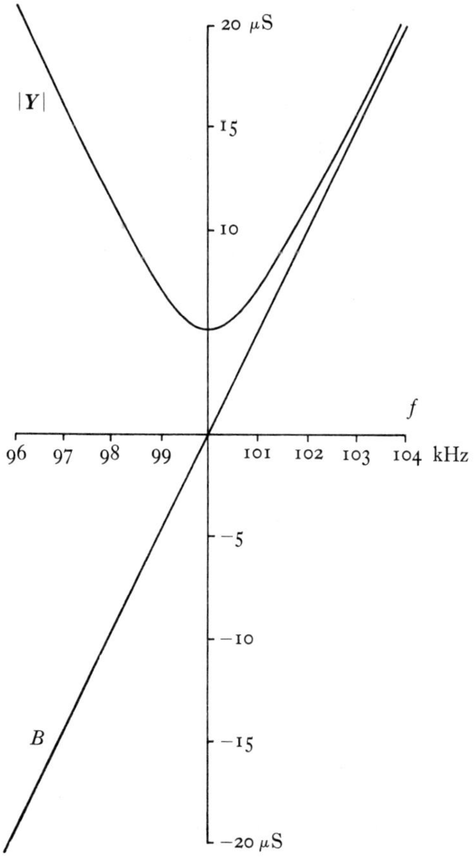

Fig. S.2

7.3 $\sqrt{\left(\frac{1}{LC}\left[\frac{L-R_1^2C}{L-R_2^2C}\right]\right)}; \frac{R}{L}$.

7.5 $\frac{1}{R}\sqrt{\left(\frac{L}{C}\right)}$; 0.354 mW.

7.6 5 pF; 1.21 MΩ.

7.8 871 pF, 8.7 pF; 1.7 kHz.

7.9 (*a*) Pole at $-1/(R_1+R_2)C_3$, zero at the origin; (*b*) 5.3 Hz; (*c*) zero at the origin, poles at $-(\alpha \pm j\beta)$, where $2\alpha = \dfrac{C_1G_2+C_2G_1+C_3(G_1+G_2)}{C_2C_3+C_3C_1+C_1C_2}$,

$\alpha^2+\beta^2 = \dfrac{G_1G_2}{C_2C_3+C_3C_1+C_1C_2}$; (*d*) 1.60 kHz, 5.2 Hz.

7.10 (*a*) It can be shown that the conditions stated are satisfied if

$\dfrac{R_1}{L} = \dfrac{G_2}{C} = \omega_m\sqrt{\left(\dfrac{R_1G_2}{1-R_1G_2}\right)}$; hence $L = 173$ mH, $C = 0.0433\,\mu$F. An elegant proof of the above is obtainable by the method of Lagrange's multipliers; the maximum of $|\boldsymbol{V}_2/\boldsymbol{V}_1|$, a function of two variables, is to be found subject to the condition that the frequency at the maximum must be ω_m. (*b*) 627 mH, 0.0199 μF; or 79.7 mH, 0.157 μF.

7.11 20 mH, 0.01 μF.

7.12 $L_1 = 21.0\ \mu$H, $C_1 = 11.3$ pF, $L_2 = 12.1\ \mu$H, $C_2 = 22.5$ pF; or $L_1 = 24.2\ \mu$H, $C_1 = 11.3$ pF, $L_2 = 10.5\ \mu$H, $C_2 = 22.5$ pF.

Chapter 8

8.1 $$-\frac{X_C^2}{(X_C-X_L)^2}\cdot\frac{1}{R+j\left[X+\dfrac{2X_CX_L}{X_C-X_L}\right]}.$$

8.2 (*a*) $$\begin{bmatrix} 1-2\times10^{-6}\,\omega^2 & -j2\omega \\ j2\times10^{-6}\,\omega(1-10^{-6}\,\omega^2) & -(1-2\times10^{-6}\,\omega^2)\end{bmatrix}.$$

(*b*) $$\frac{\boldsymbol{a}_{12}-R_L\boldsymbol{a}_{11}}{\boldsymbol{a}_{22}-R_L\boldsymbol{a}_{21}}.$$

(*c*) Yes; no; no.

8.3 $$\begin{bmatrix} j(\omega L_1-1/\omega C) & j(\omega M-1/\omega C) \\ j(\omega M-1/\omega C) & j(\omega L_2-1/\omega C)\end{bmatrix};\ \omega = (MC)^{-1/2}.$$

8.4 1.22 kΩ, 21.8 kΩ.

8.5 $\boldsymbol{h}_i = 49.7$ kΩ, $\boldsymbol{h}_r = 57.4\ 10^{-3}$, $\boldsymbol{h}_f = 47.1$, $\boldsymbol{h}_o = 56.6\ \mu$S; 305 k$\Omega$; -0.927 mA.

8.6 $\boldsymbol{A} = 1+\boldsymbol{B}_1(2\boldsymbol{C}_1+jB_C\boldsymbol{A}_1)$, $\boldsymbol{B} = \boldsymbol{B}_1(2\boldsymbol{A}_1+jB_C\boldsymbol{B}_1)$, $\boldsymbol{C} = \boldsymbol{A}_1(2\boldsymbol{C}_1+jB_C\boldsymbol{A}_1)$; 406 MW; 134 MVAr (leading), 477 MW, 32 MVAr (lagging). Another solution is theoretically possible, but will not be attained if the sending-end voltage is varied unidirectionally from 275 to 300 kV.

8.7 $g_i = 5$ mS, $g_r = -0.7$, $g_f = -40$, $g_o = 10$ kΩ; −29.4 V.

8.9
$$\begin{bmatrix} R_1 + R_2 - \dfrac{j}{\omega C_4} & -R_2 & \dfrac{j}{\omega C_4} \\ -R_2 & R_2 + R_3 + R_5 + j\omega L_5 & -R_3 \\ \dfrac{j}{\omega C_4} & -R_3 & R_3 + \dfrac{R_6}{1 + j\omega R_6 C_6} - \dfrac{j}{\omega C_4} \end{bmatrix}.$$

The dual is obtained by replacing R by G, L by C and C by L; for the network see fig. S.3.

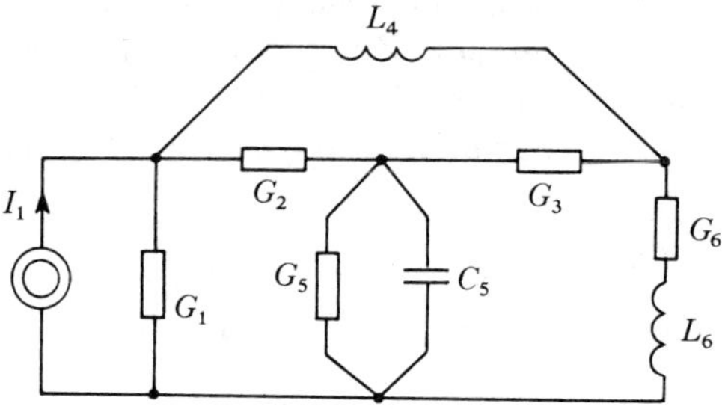

Fig. S.3

8.10 $\boldsymbol{z}_{11} = -\boldsymbol{Y}_{11}/|\boldsymbol{y}|$, $\boldsymbol{z}_{1P} = -\boldsymbol{Y}_{1P}/|\boldsymbol{y}| = \boldsymbol{z}_{P1}$, $\boldsymbol{z}_{PP} = -\boldsymbol{Y}_{PP}/|\boldsymbol{y}|$, where $\boldsymbol{Y}_{PQ}$ is the cofactor of $\boldsymbol{y}_{PQ}$ in $|\boldsymbol{y}|$; $\boldsymbol{z}_{11} = (s^2 + 3s + 2)/F(s)$, $\boldsymbol{z}_{12} = (s + 2)/F(s)$ $= \boldsymbol{z}_{21}$, $\boldsymbol{z}_{22} = (s^2 + 5s + 5)/F(s)$, where $F(s) = s^3 + 6s^2 + 9s + 3$.

8.11 About 0.73 MHz.

Chapter 9

9.1 $f(\theta) = \dfrac{4}{\pi}(\cos\alpha\sin\theta + \tfrac{1}{3}\cos 3\alpha\sin 3\theta + \tfrac{1}{5}\cos 5\alpha\sin 5\theta + \cdots)$; $\alpha = \pi/6$.

9.2 $$\frac{1}{100} + \frac{2}{\pi}\sum_{n=1}^{\infty}\frac{1}{n}\sin\frac{n\pi}{100}\cos\frac{2n\pi t}{1000}.$$

The range of frequencies is arbitrary. Retention of terms up to $n = 3000$ includes harmonics of amplitude down to about 1% of the fundamental.

9.3 $$f(\theta) = \frac{1}{\pi} + \tfrac{1}{2}\cos\theta + \frac{2}{\pi}\left\{\frac{\cos 2\theta}{1\cdot 3} - \frac{\cos 4\theta}{3\cdot 5} + \frac{\cos 6\theta}{5\cdot 7}\cdots\right\}.$$

(i) 100π V; (ii) 10.3 H, assuming that the ripple voltage across R is dependent only upon L and C, and that only the fundamental component of the ripple is significant.

9.4 $f_1(\theta) = \frac{9}{\pi^2}\{\sin\theta + \frac{1}{25}\sin 5\theta - \frac{1}{49}\sin 7\theta - \frac{1}{121}\sin 11\theta + \cdots\};$

$$f_2(\theta) = \frac{6\sqrt{3}h}{\pi^2}\{\sin\theta - \frac{1}{25}\sin 5\theta + \frac{1}{49}\sin 7\theta - \frac{1}{121}\sin 11\theta + \cdots\};$$

$$h = \sqrt{3}/2.$$

9.5 $\frac{1}{\pi}\{\sin\theta - \frac{1}{2}\sin 2\theta + \frac{1}{3}\sin 3\theta - \frac{1}{4}\sin 4\theta + \cdots\};$

$$\tfrac{1}{2} - \frac{1}{\pi}\{\sin\theta + \tfrac{1}{2}\sin 2\theta + \tfrac{1}{3}\sin 3\theta + \cdots\}.$$

9.6 $\frac{\tau}{T}\left\{\frac{1}{2} + \sum_{n=1}^{\infty}\left(\frac{\sin(n\pi\tau/2T)}{n\pi\tau/2T}\right)^2 \cos\frac{2n\pi t}{T}\right\}.$

9.7 $f(x) = 2\alpha^2 d^2\left\{\frac{\sin(\pi x/d)}{\pi(\pi^2 + \alpha^2 d^2)} - \frac{\sin(2\pi x/d)}{2\pi(4\pi^2 + \alpha^2 d^2)} + \frac{\sin(3\pi x/d)}{3\pi(9\pi^2 + \alpha^2 d^2)}\cdots\right\}.$

9.8 $f(\theta) = \frac{12\delta}{\pi}\left\{1 + k\cos\theta + \sum_{n=1}^{\infty}\frac{\sin 12n\delta}{12n\delta}\,[k\cos(12n-1)\theta\right.$

$$\left. + 2\cos 12n\theta + k\cos(12n+1)\theta]\right\}.$$

Obtain the spectrum of the unmodulated pulse-train, and multiply by the modulating function $1 + k\cos\theta$.

Chapter 10

10.1 (a) $\delta(t) - e^{-t}(2\cos 2t - \sin 2t)$;
(b) $\delta(t) + (a-b)[2 + (a-b)t]e^{-bt}$;
(c) $t e^{-\alpha t}\sin\beta t$; (d) $t^2\cos\beta t$;
(e) A sequence of positive half-cycles of $\sin t$.

10.2 $\frac{\alpha^2 e^{-\beta t} - \beta^2 e^{-\alpha t}}{\alpha^2 - \beta^2} V_1$, where $\alpha + \beta = R/L$, $\alpha\beta = 1/LC$

10.3 $\frac{V_1}{R}\{\cos(\phi+\theta)\sin(\phi-\theta)e^{-(\alpha_1+\alpha_2)t} + \sin\theta\cos(\omega t - \theta)\}$, where

$\alpha_1 = 1/RC_1$, $\alpha_2 = 1/RC_2$, $\tan\theta = \omega/(\alpha_1+\alpha_2)$, $\tan\phi = \omega/\alpha_2$.

10.4 $\frac{RI_0(Ls + r)(RCs + 1)}{s[RLCs^2 + (L + RrC)s + (R + r)]};$

$$\frac{RrI_0}{R+r}\left\{1 + \frac{R}{r}e^{-\alpha t}\left(\cos\beta t + \frac{\alpha}{\beta}\sin\beta t\right)\right\},$$

where $2\alpha = \frac{r}{L} + \frac{1}{RC}$, $\alpha^2 + \beta^2 = \frac{R+r}{RLC}$.

10.5 $417 \sin(3.75 \times 10^4 t)$, assuming that the pulse can be treated as a spike.

10.6 $e^{-t} - e^{-t/2}\left[\cos\frac{t\sqrt{3}}{2} - \frac{1}{\sqrt{3}}\sin\frac{t\sqrt{3}}{2}\right]$.

10.7 $V_1[1-(1+\alpha t)e^{-\alpha t}]$.

10.8 $\lambda\left(\frac{L \pm M}{L}\right)[2L - (L \pm M)e^{-Rt/L}]$.

10.9 $\frac{s^2+1}{s^2+4s+1}$; $\omega = 1$.

10.10 $\frac{2I_1}{\omega_1 C}\sin\frac{\omega_1 T}{2}\cos\omega_1\left(t - \frac{T}{2}\right)$, where $t = 0$ at the beginning of the pulse; $\frac{I_1 T}{C}\cos\omega_1 t$.

10.11 $\frac{e^{-\alpha t} - 1 + \alpha t}{\alpha^2}$.

10.12 $\frac{1}{60}\sin\frac{3t}{20}$ MV; $280\sin\left(\frac{3t}{20} - 0.865\right) + 246e^{-t/5} - 33e^{-2t/3}$ V; where t is in μs in each case.

10.13 $v = 100 - 110e^{-0.2t}$ from $t = 0$ to 0.477;
$= 1.961(1 - e^{-10.2(t-0.477)})$ from $t = 0.477$ to 1;
$= 1.951e^{-10.2(t-1)}$ from $t = 1$ to ∞,
where t is in ms.

Chapter 11

11.1 $\frac{T}{2}\left(\frac{\sin \omega T/4}{\omega T/4}\right)^2$.

11.2 $\frac{2\pi T}{\pi^2 - \omega^2 T^2}\cos\frac{\omega T}{2}$.

11.3 $\frac{1}{\sqrt{(\alpha^2+\omega^2)}}e^{-j\theta}$, where $\theta = \tan^{-1}\frac{\omega}{\alpha}$; $\frac{2\alpha}{\alpha^2+\omega^2}$.

11.4 $\sqrt{(2\pi)}\,Te^{-\omega^2 T^2/2}$.

11.5 (i) $\frac{\sin \omega T/2}{\omega T/2}$; (ii) $\frac{j8}{\omega T}\left\{\frac{\sin \omega T/2}{\omega T/2} - \cos \omega T/2\right\}$;

(iii) $\dfrac{12}{\omega^2 T^2}\left\{\dfrac{\sin \omega T/2}{\omega T/2} - \cos \omega T/2\right\}$.

11.6 $\dfrac{6}{\omega^2 T}\left(\cos\dfrac{\omega T}{6} - \cos\dfrac{\omega T}{2}\right)$. Differentiate the waveform successively until a sequence of spikes is obtained.

11.8 The output is

$$A\left\{f(t-T) + \frac{a}{2} f\left(t - T + \frac{\pi}{\omega_c}\right) + \frac{a}{2} f\left(t - T - \frac{\pi}{\omega_c}\right)\right\}.$$

11.9 $\dfrac{2\omega_1}{\pi}\left(\dfrac{\sin \omega_1 t}{\omega_1 t}\right)\cos \omega_0 t$.

11.10 $A(\omega_1) = (1+\omega_1^6)^{-1/2}$; $\phi(\omega_1) = -\tan^{-1}\dfrac{2\omega_1 - \omega_1^3}{1 - 2\omega_1^2}$; where $\omega_1 = 10^{-6}\omega$.

$i = \dfrac{Q_1}{2\pi}\displaystyle\int_{-\infty}^{\infty} A(\omega_1)\, e^{j[\phi(\omega_1)+\omega_1 t_1]}\, d\omega_1$, where $t_1 = 10^6 t$, and $Q_1 = 10^6 Q$, Q being the strength of the spike.

$$i = Q_1\left\{e^{-t_1} - e^{-t_1/2}\left(\cos\frac{t_1\sqrt{3}}{2} - \frac{1}{\sqrt{3}}\sin\frac{t_1\sqrt{3}}{2}\right)\right\}.$$

11.11 $\dfrac{27/8}{\sigma^3 + 3\sigma^2 + (9/2)\sigma + (27/8)}$, where $\sigma = \dfrac{s}{4.10^4} + \dfrac{4.10^4}{s}$.

$\dfrac{1}{\pi}\displaystyle\int_{-\infty}^{\infty} \boldsymbol{H}(\omega)\frac{\sin(\omega/2000)}{\omega}\, e^{j\omega t}\, d\omega$, where $\boldsymbol{H}(\omega)$ is the function obtained by writing $s = j\omega$ in the expression above; assuming that the pulse lasts from $t = -(1/2)\, 10^{-3}$ to $+(1/2)\, 10^{-3}$ s.

Chapter 12

12.1 See fig. S.4.

12.2 Poles at $s = -10^3$ and $\pm j10^3$; zeros at infinity and at $s = -250 \pm j661$. See fig. S.5. The locus goes through the origin because the impedance function has a zero at infinity. The passage through infinity with $Z_R = 500\ \Omega$ is associated with the poles at $\pm j10^3$.

12.3 Small circles: centre $w = -\frac{1}{3}$, radius $(\frac{8}{9})\ 10^{-2}$;
centre $w = +\frac{1}{3}$, radius $(\frac{2}{9})\ 10^{-2}$.

12.4 Approximately 1.12.

12.5 The transfer locus is shown in fig. S.6, and the phase margin is 18°.

12.6 When $T_2 > \frac{1}{2}T_1$ the transfer locus is of the form shown in fig. S.7(*a*), giving stability for all values of K (inherent stability). When $T_2 < \frac{1}{2}T_1$ the form is that of fig. S.7(*b*), giving stability on condition that $K(\frac{1}{2}T_1 - T_2) < 1$.

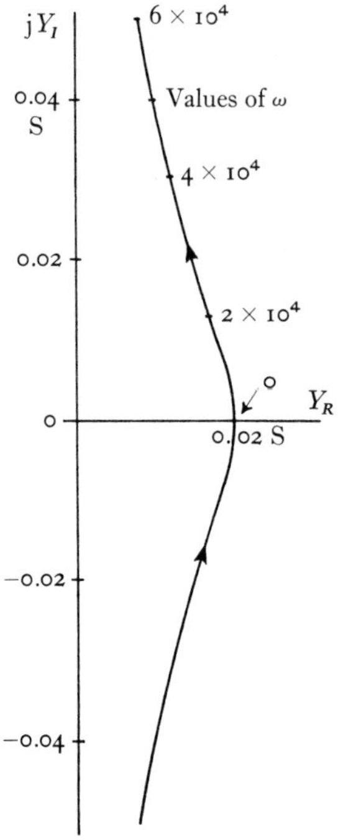

Fig. S.4

12.7 The transfer locus is shown in fig. S.8, and indicates instability.

12.8 The transfer locus is shown in fig. S.9; for stability K must be less than $\frac{88}{81}$, ≈ 1.08.

12.9 See fig. S.10. To show that condition (12.8.19) is satisfied it is necessary to imagine an impedance connected across GS and allowed to tend to infinity.

$$\text{Amplification} = \frac{-y_f R_L'}{1 + y_f R_L' R_2/(R_1 + R_2)},$$

where $$\frac{1}{R_L'} = \frac{1}{r_d} + \frac{1}{R_L} + \frac{1}{R_1 + R_2}.$$

12.10 Unstable.

12.11 Not Hurwitz; two zeros in the right-hand half of the s-plane, as can be shown by mapping the stability discrimination contour on the f(s)-plane and noting that it encircles the origin clockwise twice.

12.12 $T_1 > 0{\cdot}434\,T$.

12.13 (*b*) Because the function has poles in the positive half s-plane.
(*c*) Because the real part is evidently <0 when $\sigma > 0$.
(*d*) Because the residues at the poles $s = \pm j3$ are complex.

12.14 (*a*) No; the denominator is non-Hurwitz.
(*b*) Yes.

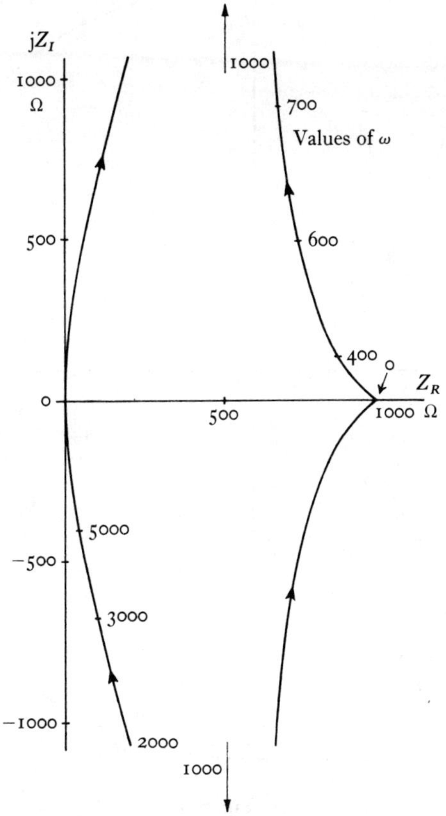

Fig. S.5

12.15 Inductance 2 H in series with capacitance 1/8 F.

12.16 (Inductance 4 H in parallel with capacitance 1/2 F) in series with (inductance 6 H in parallel with capacitance 1/2 F).

Chapter 13

13.1 600 Ω; 0.164 + any multiple of 0.5.

13.2 $Z_i = \infty$ for all non-zero x.

13.3 Assuming $2\pi \times 796 = 5000$: $\boldsymbol{Z}_0 = 684\ \angle -15°\ \Omega$; $\alpha = 42.4 \times 10^{-3}$ db/km, $\beta = 18.2 \times 10^{-3}$ rad/km; $u = 275 \times 10^3$ km/s; $\lambda = 345$ km.

13.4 1.44 db/km.

13.6 $263\ \angle -128°$ V; $0.177\ \angle -154°$ A.

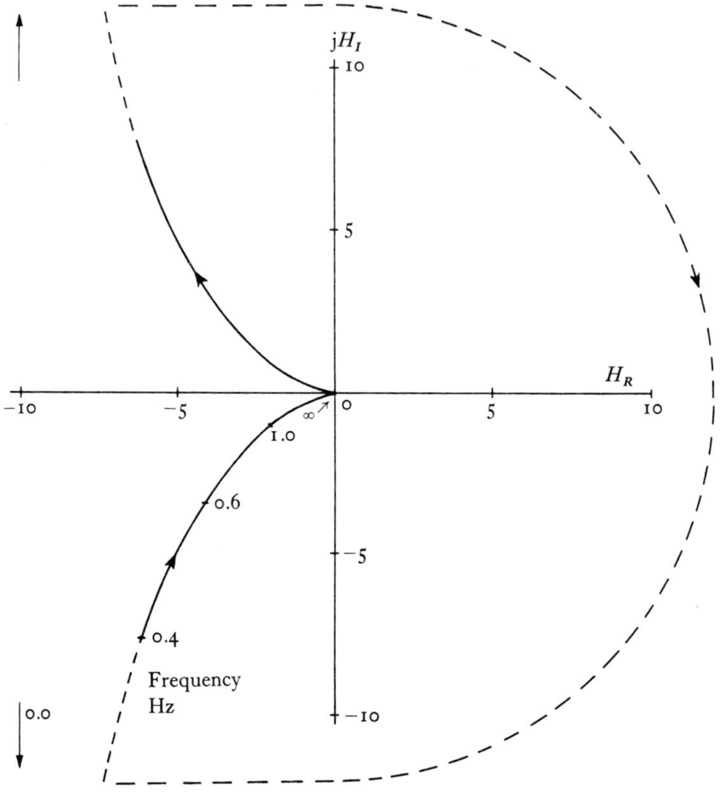

Fig. S.6

13.7 $0.0738\ \lambda$; $62.5 - j37.5\ \Omega$.

13.8 $j200\ \Omega$; $j216\ \Omega$.

13.9 $$\frac{C_1 + (Y_0/\omega)\tan\beta d}{\omega Z_0 C_1 \tan\beta d - 1}$$

13.10 $-Z_0 \tan\beta d/2$.

13.11 $394\ \angle -50°\ \Omega$.

13.12 $(40 + j30) = 50\ \angle 37°\ \Omega$.

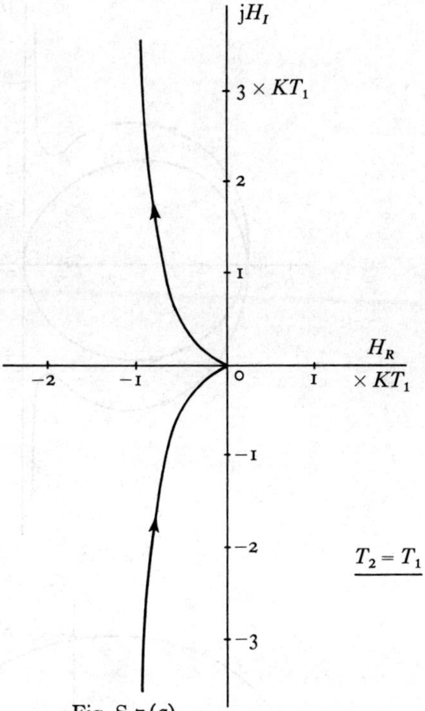

Fig. S.7 (*a*)

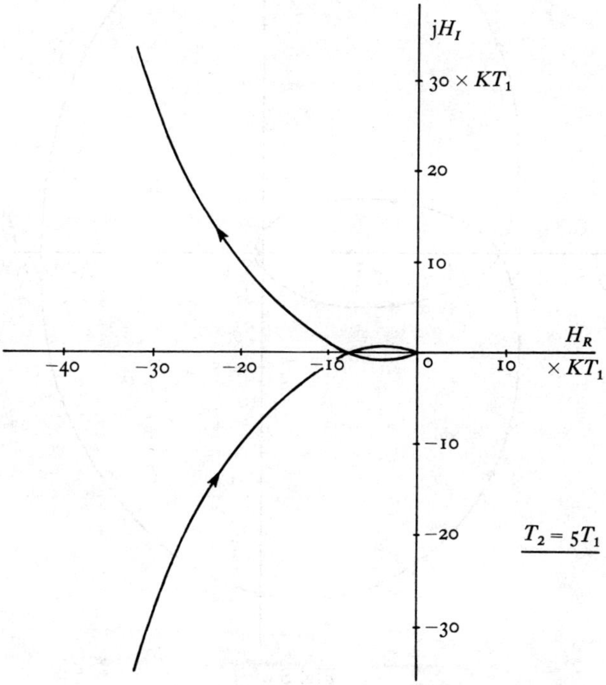

Fig. S.7(*b*)

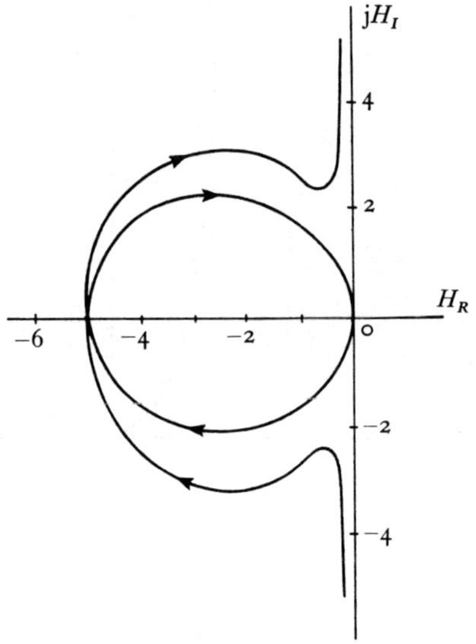

Fig. S.8

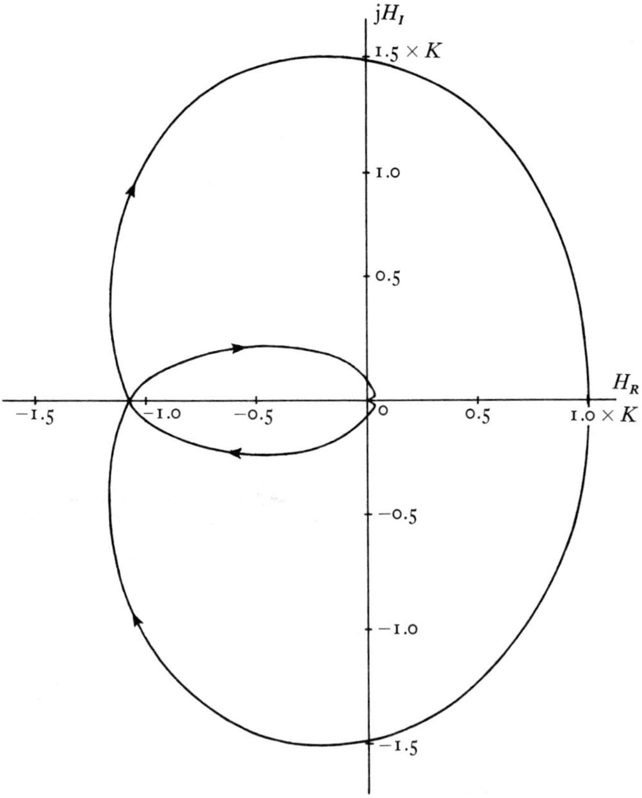

Fig. S.9

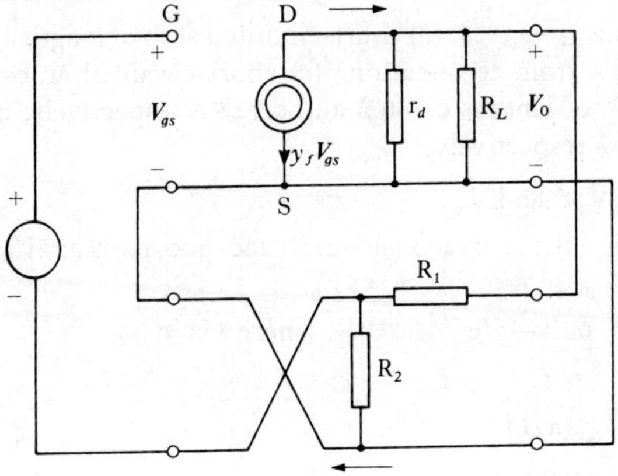

Fig. S.10

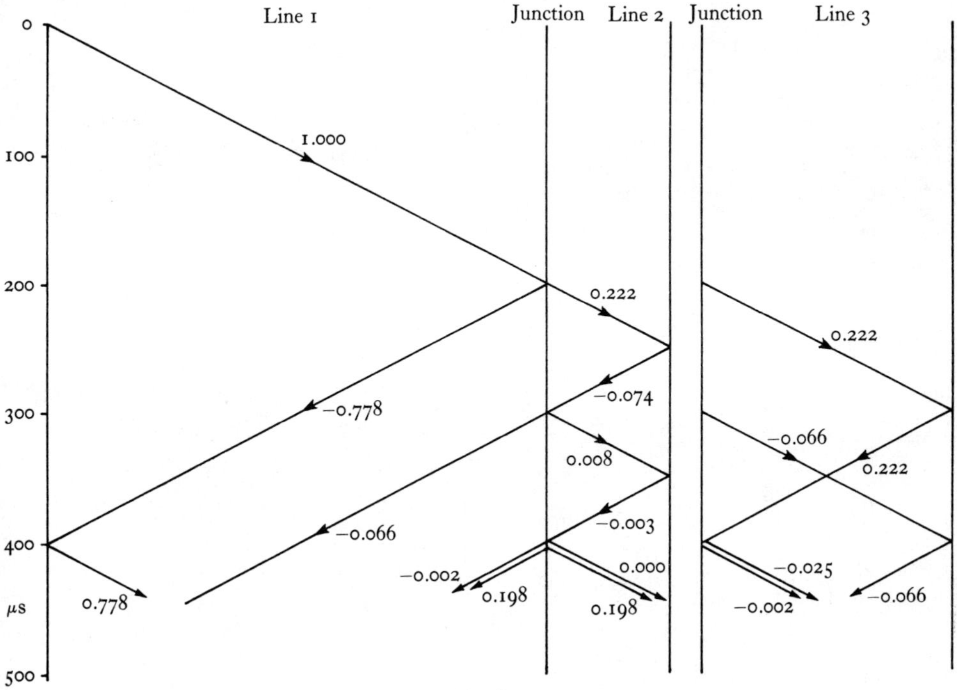

Fig. S.11

13.13 2; stub length 0.125 λ, distant 0.152 λ from termination.

13.14 (*a*) 76.2 $\angle -43^\circ\ \Omega$; (*b*) short-circuited stub of length 0.126 λ, distant 0.082 λ from termination; (*c*) short-circuited stubs at 0.3 λ and 0.55 λ, of lengths 0.286 λ and 0.148 λ respectively, or 0.389 λ and 0.352 λ respectively.

13.15 2.49 kV; 6.80 kV.

13.16 See fig. S.11; 0–200 μs zero, 200–300 μs $0.222V_0$, 300–400 μs $0.156V_0$, 400–500 μs $0.351V_0$.

13.17 $V_1\{1 - e^{-5t} - \frac{5}{3}(e^{-t} - e^{-4t})\}$, where t is in μs.

13.18 $i = V_0 \sqrt{\left(\dfrac{C}{\pi r t}\right)}$.

13.19 0.110 s; 0.093 s.

Index